GEOLOGY AND PRODUCTION OF HELIUM AND ASSOCIATED GASES

GEOLOGY AND PRODUCTION OF HELIUM AND ASSOCIATED GASES

STEVEN A. TEDESCO
Running Foxes Energy Inc., Centennial, CO, United States
Starfox Helium Inc., Hamburg, Germany

ELSEVIER

Elsevier
Radarweg 29, PO Box 211, 1000 AE Amsterdam, Netherlands
The Boulevard, Langford Lane, Kidlington, Oxford OX5 1GB, United Kingdom
50 Hampshire Street, 5th Floor, Cambridge, MA 02139, United States

ISBN: 978-0-323-90988-4

For information on all Elsevier publications visit our website at https://www.elsevier.com/books-and-journals

Publisher: Candice Janco
Acquisitions Editor: Amy Shapiro
Editorial Project Manager: Leticia M. Lima
Production Project Manager: Kumar Anbazhagan
Cover Designer: Miles Hitchen

Typeset by TNQ Technologies

Contents

5. Processing procedures to extract helium, nitrogen, and carbon dioxide

6. Reserves and future trends in helium

Acknowledgments

In putting together any textbook or project, there are a number of people that help in a variety of ways. I want to thank Robert S Hoar and Dr. Anita Thapalia for their critical review and suggestions in editing the six chapters of the book. I want to thank my Elsevier book team of Leticia Lima (my book editor), Kumar Anbazhagan, and Dinesh N for their patience and guidance through the process. I also want to thank Alton Brown, Tyler Wiseman, and Mike Poirier for information and comments provided early on in the process. Finally, thanks to Mario Barche and Achim Weber for their help in pursuing helium with me.

Introduction

If everyone is thinking alike, then somebody isn't thinking. **General George S. Patton.**

Welcome to the wonderful world of helium. If one thinks finding oil and gas is hard, helium is even more difficult. The above quote by General Patton is apt because the helium niche of the petroleum industry has suffered for a century thinking of exploration and exploitation as if it is found like petroleum accumulations. Helium deposits have almost always have been found by accident on occasion and not been found when specifically sought after. There is a plethora of data in the literature on using helium and other noble gas isotopes related to the age or source of groundwater or petroleum deposits. Unfortunately, to date, isotopes are not a practical pathfinder tool for finding new deposits. As will be discussed in Chapters 3, 4, and 6, there are some general characteristics for helium accumulations, but there are numerous exceptions or geologic oddities that leave one fascinated but unable to use them as models. There is the perceived correlation with nitrogen which is seemingly coincidental as helium cannot, except under extreme conditions, combine with other elements. But there is no doubt that reservoirs with high nitrogen or carbon dioxide content can indicate the possible presence of anomalous helium. Helium is never found without nitrogen, except in the rare cases it is found with carbon dioxide. Nitrogen is found in many areas without significant helium. Carbon dioxide is found in many places without any significant or any helium. The fact that at least helium can be found in economic quantities in the subsurface has to be considered a good thing. It is unfortunate that the other noble gases are not in a similar situation. Without helium, the modern-day society would either not exist or be under a significantly handicap in terms of functionality.

In 1988 I was working with surface geochemical methods in petroleum exploration and realized the lack of any coherent text on the subject. In 1994, Van Nostrand published my book on the subject. In writing this book on helium I had forgotten how difficult it was to write a textbook, bring together many sources, and cut through the weeds and trees trying to find the forest. About 5 years ago I started looking at helium first as a curiosity and then as a commodity. I found in my research that helium and the other noble gases, from an explorationist perception, were lacking in any coherent synthesis, let alone a professional book that brings all of the pieces together and provides some sort of coherent analysis or suggestions for the explorationist. There are other books, that are very good on helium that are part history, present economics, future trends, and generalities on generally super-sized fields. There are excellent specific texts on existing helium deposits in Arizona, New Mexico, Wyoming, Kansas, and Utah. What I tried to do in this book is bring together a number of diverse texts and articles into a more central synthesis covering numerous areas of helium productive fields with associated nitrogen and carbon dioxide. Collecting together information from the United States was relatively easy as published articles and books on helium productive fields for specific states are available. There is a fair amount of information from Canada, but then there is a significant lack of information from almost all other countries. China has become more open about its helium geology, whereas Russia has not. Poland, Australia, Qatar, and Algiers have enough data for analysis simply. A great source of information for helium is the American Association of Petroleum Geologists and related organizations such as the Rocky Mountain Association of Geologists and Kansas Geological Survey. An additional exceptionally strong source is Elsevier with its many journals that discuss a wide variety of helium investigations and datasets.

Many general books on helium have focused on the larger fields or areas that have always made news. The Texas Panhandle-Hugoton Field in the Central United States is a geological and pressure oddity, huge in areal size, and yielding over 400 BCF of helium gas. I occasionally get asked, do you think there is another Texas Panhandle-Hugoton Field? Only a small number of places in the United States and Canada could make that happen. This is contemplated against the millions of wells drilled in the United States and Canada. Unfortunately, it is even less likely as all of the potential locations that have rocks are too young, are tectonically active, or are in failed rift systems with limited drilling and no encouragement of any gas. Examples are the basins and ranges in the United States, the Midcontinent Rift System in the Central United States, the Rough Creek Fault System-Rome Trough in Kentucky, the Cincinnati Arch in Ohio, Kentucky, or Tennessee, other arches around the United States and Canada, or potentially really deep drilling target formations such as in the Pinedale area in Wyoming. Would it likely be found elsewhere such as China or Russia or Africa, etc.? Anything is possible if the right conditions exist, but my personal estimation is it is unlikely. The types of fields remaining to be found in the United States and Canada are probably small consisting of 1–10 wells. The other remaining super-sized fields or facilities, such as Big Piney-La Barge, North and South Pars, Amur, and Hassi R'Mel, are in many respects similar to the Panhandle-Hugoton Field being of low percentage of helium, world class, producing or processing large volumes of natural gas and/or carbon dioxide.

In putting together the book, I tried to organize it based on what an explorationist would look at in terms of needs in the tool box. Chapters 1 and 2 are designed to provide basic background. Detail chemistry can be found elsewhere. Noble gases chemistry is almost a nonword when it comes to discussing compound that they can form. They rarely combine with anything and act very independently of all other elements. It is fortunate that certain minerals such as amphibole can hold helium, but this ruins the calculation of how much helium is on the planet, probably a lot more than anyone thinks. Calculating the amount of helium and other noble gases present on planet is theoretical and any number out there is unverifiable like any good climate model. Carbon dioxide, uses and nitrogen are quite the opposite of the noble gases and despite generally being waste products they are so vital to life. Chapter 2 presents a short discussion of nitrogen and carbon dioxide, uses and sources in the subsurface.

Chapter 3 is the heart of the book in terms of bringing together numerous helium and carbon dioxide producing fields in various areas. Russia lacks data for individual fields and unfortunately the discussion is brief. Canada is going through a rebirth in the search for helium; unfortunately there are only a few examples that have had data available over the years. Australia, while possessing potential for helium seems to lack the exploration commitment for their industry to drill and take risk. This is somewhat surprising as the

Aussie mining and petroleum industry has an excellent reputation of being the tip of the spear in exploration. China has had more and more articles about their petroleum industry and helium and in addition they have shed their history of suspect science to become a major scientific and dependable source or contributor. Europe is in a quandary with only limited resources in Poland which are depleting and becoming more and more dependent upon Russia and the United States. The Middle East is fortunate that it has the largest long-term potential helium reserves locked up in North/South Pars field. Unfortunately, it is controlled by two very different countries who disagree on every level in terms of religion and political focus. Added to the development of this field is a triad of world powers and associated neighbors that have their own agendas that create volatile markets in helium. South America may have helium but presently it is likely in low volumes related to large-scale natural gas production. Africa is the great unknown. The East Africa Rift system potentially can provide a major source, but the risks are high as it has young pore water (see Chapter 3) which has a pretty definitive history of not containing economic helium.

Chapter 4 is focused on exploration methods which, for the most part, has limited choices available. The two pathfinder exploration tools, nitrogen and carbon dioxide, are so common that their ability in finding helium is essentially nonexistent. Seismic is great at finding structures or stratigraphic traps but it cannot specifically target helium. Using existing gas wells to analyze for helium is not necessarily a definitive exploration tool but it allows a good start. Aeromagnetics provides a potential regional exploration tool as helium accumulations can potentially show a probable relation to many basement structures, extensive faulting or intrusives in some areas. There are exceptions whereby helium accumulations are not always associated with the basement as defined by aeromagnetics. Petrophysics to date cannot measure helium in well logs. Log analysis can mislead and be misinterpreted just as seismic, subsurface geology, drill stem tests, etc. Production data, though is the ultimate absolute measurement of results, is not an exploration tool.

As an explorationist and producer when delving into a play, getting the product to market in some respects can be the most critical element. When talking about helium with most people in the petroleum business, they think of cryogenic plants as the way helium is extracted. Chapter 5 goes through the processes used to extract helium as well as nitrogen and carbon dioxide in a general sense. In reality for most deposits, it's the pressure swing adsorption process that varies from $2M to $6M (US) versus the cryogenic units that are in excess of $20M (US) that are the workhorses of the small deposits. The extraction process for helium is generally simple but controlled by the input gas composition, pressure, and temperature. The chapter is for understanding the process of removing helium, nitrogen, and carbon dioxide in general terms. In many respects for the producer, the helium extraction process cost is the major stumbling block to pursing the element. The what if, the great unknown, if the project does not work and how much will be lose, always casts doubt and prevents many projects from going forward. It is always safe to do nothing in the petroleum business and follow the other persons lead. Many projects do fail because a number of factors could not be considered because they were unknown or ignored. An alternative to buying your own personal helium extraction machine is to ship the product in a raw form to a processing plant. These processing plants can be very lucrative for the facility but not always for the operator in terms of revenue.

Chapter 6 discusses the economics of helium consumption and future trends. Data used are generally from Canada and the United States, but in bits and pieces or nonexistent from most other countries. Because of the pandemic, the consumption and growth of the helium industry along with the world economy slowed. With the world economy opening up again, the consumption will go back to normal, whatever that means. The Amur plant in Russia going online is considered by some as providing sufficient supply for years to come. However, as with many countries, there are no data to verify the extent of its sources. Case in point is the Saudi Arabian government for decades has had the same amount of oil reserves and has provided no data to verify one way or another.

Hopefully the reader will enjoy having a lot of field data in one place along with the concepts and ideas, some new, presented here. Understanding of helium exploration is an ongoing process. There is a lot of excitement with early stages of exploration and exploitation in Canada, parts of the United States, and East Africa Rift System. The growth of helium consumption will continue, and if fusion takes off, the need for it will grow by a thousand-fold. Then it is off to the moon to mine for helium!!!

CHAPTER

1

Helium

Abbreviations

^{3}Helium ^3He
^{4}Helium ^4He
Aluminum Al
Argon Ar
Billions of cubic feet BCF
Carbon dioxide CO_2
Centigrade C
Cubic meters m^3
East Africa Rift System EARS
Fahrenheit F
Feet ft
Helium He
Hydrogen H
Kelvin K
Kilometers Km
Krypton Kr
Lithium Li
Megapascal Pressure Unit Mps
Meters m
Methane CH_4
Methane plus other hydrocarbons CH_4+
Mile Mi
Millions of cubic feet of gas MMCF
Millions of cubic feet of Helium MMCFHe
Millions of years MY
Neon Ne
Nitrogen N_2
Oxygen O
Parts per billion ppb
Parts per million ppm
Parts per trillion ppt
Pounds per square inch psi
Silica Si
Thousands of cubic feet of gas MCF
Thousands of years Kp
Water H_2O
Xenon Xe

https://doi.org/10.1016/B978-0-323-90988-4.00002-5

1.1 The helium atom and isotopes

Helium, a noble gas, is the second most abundant element in the universe except for on Earth where it is rare, a contradiction that is yet to be explained. The name Helium (He) is from the Greek word Helios and the element was named by J.N. Lockyer in 1871. Helium (Crockett et al., 1973; Greenwood and Earnshaw, 1997; Rumble, 2020):

(1) Is a colorless, odorless, tasteless, nontoxic, inert, monatomic gas;
(2) Is the first in the noble gas group in the periodic table;
(3) Has melting point of −272.20 °C (−457.96 °F), a boiling point of −268.928 °C (−425.02 °F), density is 0.1786 g/L at standard temperature and pressure;
(4) Averages 5.2 parts per million (ppm) in the Earth's atmosphere. After Neon (Ne), Helium is the second least reactive noble gas and element (Lewars, 2008);
(5) Has the smallest size of any element; diffuses through solids at a rate three times that of air and around 65% greater than hydrogen (H) (Hampel, 1968);
(6) Has an unusually stable nucleus arranged into complete shells. The stability and low energy of the electron cloud state in helium accounts for the element's chemical inertness. This defines the inability of He to combine with other elements except at high pressures and temperatures;
(7) The energetic stability of the He nucleus accounts for the ease of the ^{4}He isotope's production in atomic reactions that involve either heavy-particle emission or fusion;
(8) Helium, because of its electron shell and lack of inability to combine with other elements, has the lowest boiling and melting points of all elements;
9) Conventional wisdom is that helium originated during the Big Bang of the Universe and is estimated to be 23% of the total volume; overall and 2% of the weight of all elements in the universe.
(10) Helium's index of refraction is closer to unity than that of any other gas (Stone and Stejskal, 2004).
(11) Helium heats up when allowed to freely expand and thus has a negative Joule-Thomson coefficient at normal ambient temperatures. Below He's Joule-Thomson inversion temperature, approximately 32−50 K at 1 atm, it cools upon free expansion (Hampel, 1968). Helium can be liquefied through expansion cooling once it is precooled beyond the Joule-Thomson inversion.
(12) Helium is the least water-soluble gas with a solubility of 0.70797 $x_2/10^{-5}$ (Weiss, 1971; Scharlin et al., 1998). Helium's inert nature makes it questionable when it is supposedly in solution whether it is actually in solution or just present. This actually is something to consider when discussed in later chapter He's presence in petroleum or groundwater systems. However, there are three compounds that have lower solubility than He. These compounds are (Scharlin et al., 1998):
(a) CF_4 (tetrafluoromethane) fraction solubility is 0.3802 $x_2/10^{-5}$;
(b) SF_6 (sulfur hexafluoride) fraction solubility is 0.4394 $x_2/10^{-5}$; and
(c) C_4F_8 (octafluorocyclobutane) fraction solubility is 0.2372 $x_2/10^{-5}$

Helium, unlike any other element, will remain in a liquid phase down to absolute zero at normal pressures as a direct effect of quantum mechanics. Zero-point energy of the He system is too high to allow freezing. Solid He only occurs at a temperature of 1−1.5 K (about −272°C or −457°F) at about 25 bar (2.5 MPa) of pressure (Department of Physics, 2021). The refractive indices of the two He phases, solid and liquid, are often hard to distinguish, since they are nearly the same. The solid has a sharp melting point and a crystalline structure, but it is highly compressible; applying pressure in a laboratory can decrease its volume by more than 30% (Lide, 2005). Helium has a bulk modulus of about 27 MPa [it is ~100 times more compressible than water (Grilly, 1973)]. Solid He has a density of 0.214 ± 0.006 g/cm^3 at 1.15 K and 66 atm; the projected density at 0 K and 25 bar (2.5 MPa) is 0.187 ± 0.009 g/cm^3 (Henshaw, 1958). At higher temperatures, He will solidify with sufficient pressure. At room temperature, this requires about 114,000 atm (Pinceaux et al., 1979).

Helium found on Earth has different properties than most extraterrestrial He found in plasma state. In the plasma state, He's electrons are not bound to its nucleus, resulting in very high electrical conductivity. The solar wind together with ionized He particles interact with the Earth's magnetosphere, giving rise to Birkeland currents of plasma charged particles. These particles cause atmospheric auroras which are highly influenced by magnetic and electric fields (Hampel, 1968).

Like all noble gases, He is sourced from the atmosphere, sedimentary rocks, the crust, and the mantle. Helium associated with present economic reserves are predominantly sedimentary or crust-dominated in source (Ballentine and Lollar, 2002; Gilfillan et al., 2008). It is estimated that every year 3000 M^3 (150,994 MCF) of He is generated on the Earth by normal decay processes.

The two stable isotopes of He are: (1) rare ^{3}He, "primordial helium," left over from the formation of the Earth in the mantle, and (2) prolific ^{4}He, "radiogenic helium," which is a product of the alpha decay of ^{235}Uranium (U), ^{238}Uranium (U), and ^{235}Thorium (Th) (Mamyrin and Tolstikhin, 1984; Oxburgh et al., 1986). Lithium (Li) also sheds He during natural decay, but the amount contributed to the total He present is considered negligible. Carbonaceous mudstones typically contain U and Th within their organic material resulting in their characteristic high response on density-neutron and gamma logs. The radioactive minerals within the carbonaceous mudstones are thought to be shedding He during natural decay. There are additional seven unstable He isotopes that can be produced, but they rapidly decay into other substances. The shortest-lived heavy He isotope is ^{5}He with a half-life 7.6×10^{-22} s; ^{6}He has a half-life of 0.8 s, emits a halo, and decays by emitting a beta particle. ^{7}Helium decays by emitting a beta particle as well as a gamma ray. ^{7}Helium and ^{8}He are created in certain types of nuclear reactions.

There is one ^{3}He atom to every one million ^{4}He atoms in the Earth's atmosphere (Emsley, 2001). The ^{3}He isotope is largely associated with active degassing during tectonic activity related to volcanoes or mid-oceanic ridges of the Earth (Craig et al., 1975; Craig and Lupton, 1976; O'Nions and Oxburgh, 1988; Torgersen, 1989). Most of the ^{3}He present on the Earth has been present since its formation, though some fall to Earth trapped in cosmic dust. Stable ^{3}He atoms are produced in fusion reactions from H, but it is a very small fraction compared to ^{4}He. The other terrestrial major source of ^{3}He occurs in the upper few meters of the crust, is thermal neutron captured by ^{6}Li in predominantly clay-rich areas (^{6}Li(n,α)^{3}H(β-) → ^{3}He) (Hiyagon and Kennedy, 1992; Ozima and Podosek, 2001; Ballentine and Burnard, 2002). Trace amounts are also produced by the beta decay of tritium from processes employed by the US and Russian military. The importance of ^{3}He/^{4}He isotopes ratio is used to investigate the origin of rocks and the composition of the Earth's mantle and will be discussed in later chapters. The ratios can vary by a factor of 10. The more ^{3}He present the more likely that the He present was derived from mantle rocks.

^{3}Helium is much more abundant in stars as a product of nuclear fusion. ^{3}Helium is found in large amounts on the lunar surface, 1.4 ppb and 15 ppb in sunlit areas, and may contain concentrations as much as 50 ppb in permanently shadowed regions, deposited from the solar wind (Dilmare, 2004; Slyuta et al., 2007; and Williams, 2007; Beike, 2011; Schmitt, 2013, 2014; Ambrose and Cutright, 2017). The Moon's surface contains ^{3}He at concentrations on the order of 10 ppb, which is much higher than the approximately 5 ppt found in the Earth's atmosphere. Wittenberg et al. (1986) proposed to explore the moon, mine ^{3}He from the lunar regolith, and use the ^{3}He for fusion (see Chapter 3).

1.2 History

The history of He has been covered extensively in many texts such as Nuttall et al. (2012); and Sears (2015), and a brief summary will be given here. Helium was first detected by Isaac Newton through the use of a spectrometer looking at the Sun. The spectrometer allowed sunlight to pass through a prism and split into several lines that are identified today as Fraunhofer lines. The lines or refractory wavelengths of the Sun's light help define its elemental composition (Nuttall et al., 2012; Sears, 2015). The lines are labeled A through K of which the D line represents He. During the 1800s the spectrometer gave way to the spectroscope that allowed greater definition of what elements the lines or wavelengths represented. J.N. Lockyer was the first to actually identify and given credit for the discovery of He. In 1896 Sir Williams Ramsey became the first to identify He from an earthly source when He. Sir Ramsey analyzed an unknown gas generated from dissolution of uraninite by sulfuric acid, (Sears, 2015).

Helium was first found to be a natural occurring element in the spring of 1903 by a company that was drilling just off main street for oil and gas in Dexter, Kansas (Fig. 1.1; Nuttall et al., 2012; Sears, 2015). The company hit a strong gas flow, estimated at 9 MMCFGPD, from

FIGURE 1.1 Helium plant in Dexter, Kansas reprinted with permission from the East Cowley Historical Society, Winfield, Kansas.

a sandstone reservoir at approximately 400 ft (120 m) which most likely came from the Layton sandstone. The Layton sandstone designation is based on mechanical logs submitted to the State of Kansas from later oil and gas wells that were drilled adjacent to the Dexter He discovery (Fig. 1.2). The original "Dexter" well is located in the northeast of the southeast quarter of Section 13, Township 33 South Range 6 East in Cowley County, Kansas, USA. Petroleum drilling in the area has produced approximately 1MM bbls of oil from the Lansing-Kansas City, Cherokee, Mississippian, and gas from the Admire and Shawnee group reservoirs as of December 2019. Actual gas production since the Dexter He discovery is unknown, as it occurred several decades before the State of Kansas began keeping oil, gas, and He records.

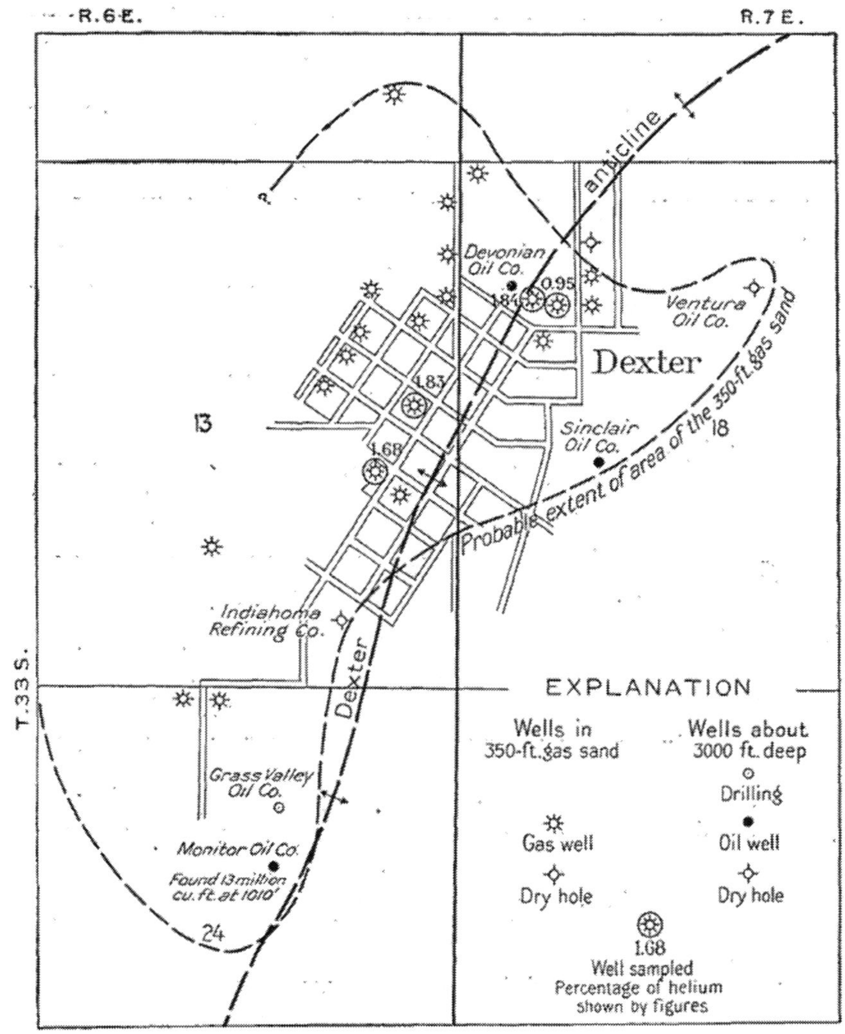

FIGURE 1.2 Location of the Dexter He well discovery, Dexter, Cowley County, Kansas, USA, in 1903 (Rogers, 1921).

The "Dexter gas" was unburnable containing only 15% CH_4+ plus 72% N_2+ and small amount of inert residue or gases. Of these remaining gases, 13% was later identified as containing He (Nuttall et al., 2012; Sears, 2015). The He present was a curiosity but it resulted in the sampling of 44 wells in Missouri and Kansas by Hamilton P. Cady and David F. McFarland in 1903. Cady and McFarland (1903) noted a strong relationship between He and N_2. The result of the survey also demonstrated to Cady and McFarland (1906, 1907a and 1907b) the great abundance of He available in the central USA.

Prior to 1914, He had little application beyond experimental uses. At the time, it had, no commercial value, even though it was present relatively, for the time, in "unlimited" quantities (Nuttall et al., 2012; Sears, 2015). World War I changed the demand for He with the publication by G. Austerweil (1914) of "Die angewandte Chemie in der Luftahrt" which in English means "Applied Chemistry in Aviation" (Sears, 2015). In chapter one of Austerweil's text was a section entitled: "Helium als ballonfull gas" or "Helium as balloon-filling gas" that mentions the advantages of using noninflammable He as a replacement for H in balloons. At the time of Austerweil's publication the only supply of He that Austerweil was aware of was limited to his laboratory in Germany. Austerweil had no knowledge of the He discovered in Dexter, Kansas, or in other areas in the United States (Sears, 2015).

Events driven by World War I began an arms race in He (Nuttall et al., 2012; Sears, 2015). A German Zeppelin was hit by bullets from Allied forces while on a combat mission but did not explode. It was assumed by several observers that the Germans had discovered an inflammable, light gas, such as He (Moore, 1926). The British government quickly moved forward in search of sources of He throughout the empire, but specifically in Canada. The first He extraction plant was built by the British in Hamilton, Ontario, Canada, and was capable of delivering 87% pure He with the remainder being N_2 (Nuttall et al., 2012; Sears, 2015). By the time the plant was fully operational, the natural gas production from fields in Ontario was in decline, and so the plant moved to Bow Island Gas Field, Calgary, Alberta, Canada. The He plant at Bow Island eventually, post-war, produced a total of 60,000 cubic feet of He that was 60%–90% pure (Nuttall et al., 2012; Sears, 2015).

The US entry into He extraction started when Clifford Winslow Seibel, a student at the University of Kansas, who reluctantly conducted research on rare gases from Dexter, Kansas, USA, for his then Professor Hamilton Cady (Rogers, 1921; Nuttall et al., 2012; Sears, 2015). From this work Dr. Richard B. Moore and Lathrop Parsons of the US Bureau of Mines (USBM) recognized the significance of Seibel's studies, and reported it to the head of the USBM War Investigations group, George A. Burnell. Mr. Burnell quickly identified the Petrolia Field, Wichita Falls, Texas, USA, as a potential candidate for He as it had 20%–30% N_2, and 1% He (Nuttall et al., 2012; Sears, 2015). In addition, the USBM's chief metallurgist, Frederick Gardner, was brought in to the group. This subsequently led to a letter sent to Major Charles de Forest Chandler of the Balloon Division of the Aviation Section of the Signals Corp, US Army (Sears, 2015). The US entry in to World War I had caused anxiety in the United States because of the German use of zeppelins and the USA had no similar weapon. The navy also became interested in He for its airships to be used for scouting, submarine detection, and rescue (Nuttall et al., 2012; Sears, 2015).

The difficulty for all participants in the war was how to produce He commercially. In the USA, Seibel's cost for extracting a cubic foot of He was $2,500 (1917 dollars), which translates into several millions of dollars of product to fill a single airship. A Fred H. Norton developed the first He extraction facility for a cost of $28,000 (1917 dollars) that could produce 5000 cubic feet of He a day (Nuttall et al., 2012; Sears, 2015). The USA and the British determined that a mixture of 90% H to 10% He did not ignite. US Navy and Army and the British Admiralty eventually commissioned three competing processes, the Joule-Thomson effect, the Claude Cycle, and the Norton process to provide He. Two He facilities were developed in the USA, one in Ft. Worth, Texas, and one at the Petrolia Field in Wichita Falls, Texas, USA. The Joule-Thomson achieved 70% pure He, and with reprocessing it was increased to a 92% purity (Manning, 1919). The Claude Cycle achieved a 70% purity of He and to achieve a 92% the He purity was sent to the Joule-Thomson plant for reprocessing (Manning, 1919). The Norton process only achieved a 20% He purity level at very high costs (Manning, 1919) and was scrapped in 1919. The Claude Cycle facility remained operational with the Joule-Thomson process; owned by Linde Air Products and was designated to be the primary extraction facility 20 days before the end of World War I. Another result of World War I was a nationwide sampling of numerous gas wells in the USA for He only (Nuttall et al., 2012; Sears, 2015).

The US Army determined that it would be in the country's best interest to continue production of He post World War I (Sears, 2015). In the 1920s and 1930s great strides were made in storing He underground, repurifying contaminants out of He products, and lowering costs of processing. The Helium Conservation Act of March 3, 1925, was established to place He production under government control and set prices. It was determined that 500 MMCFHe were lost each year through natural gas production (Congress, 1924). The USBM conducted a survey of natural gas wells in the United States from 1919 to 1933 which led to three He discoveries (Fig. 1.3): the Cliffside Field near Amarillo, Texas, the Helium Reserve Number One near Emery, Utah, and Helium Number Two in Grand County, Utah, USA (Seibel and Kennedy, 1934). The first United States Bureau of Mines (USBM) facility for He

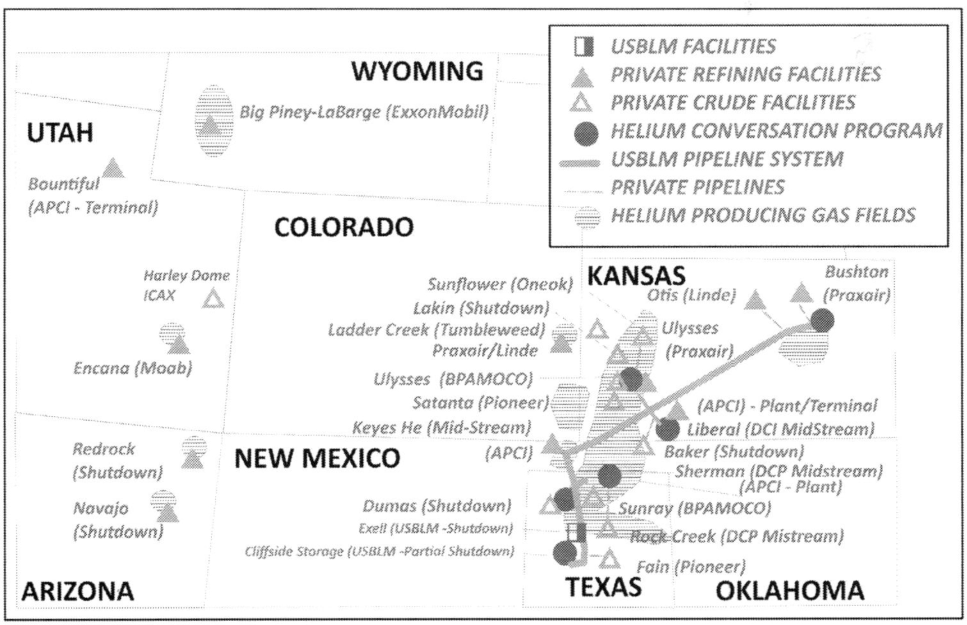

FIGURE 1.3 Location of He supply sources and major plants in the USA (modified from National Research Council, 2010).

processing was placed at Cliffside Field in Texas (Sears, 2015). In the 1930s the US military moved away from airships as airplanes were more reliable and versatile creating a surplus of He right before World War II (Nuttall et al., 2012; Sears, 2015).

The 1920s also saw private investment into He exploration, extraction, and processing at Model Dome, Colorado, and at Dexter, Kansas (Fig. 1.3) (Nuttall et al., 2012; Sears, 2015). These private facilities would eventually be bought by the USBM and dismantled in 1944. During this time the Petrolia Field near Wichita, Texas (Fig. 1.3) was declining in natural gas production and with the establishment of the Cliffside facility was timely in taking over production of He. The US Congress acted again in 1937 with respect to He with The Helium Act of 1937 to address some issues caused by the Helium Act of 1925 (US Congress, 1937).

With the advent of World War II, new technologies demanded larger amounts of He. Helium was necessary for diving using $He-N_2$ mixtures, welding, and naval and barrage balloons (Nuttall et al., 2012; Sears, 2015). To meet wartime demands a second plant was established on the southside of the Texas Panhandle Field by the USBM called Exell (Fig. 1.3). Also, a number of facilities were added at Rattlesnake Field near Shiprock, New Mexico, the Channing area in Moore, Potter, Oldham and Hartley counties, Texas, the Otis area, Rush County, Kansas and Cunningham area, Kingman County, Kansas (Fig. 1.3). The Manhattan Project also used a substantial amount of He without which there would have been no atomic bomb (Sears, 2015).

With the end of World War II, the demand for He dropped dramatically and all but the Exell Plant stopped processing and producing. Through USBM efforts, excess He was stored in the Cliffside Field. It was not until 1950 that the consumption of He increased, due to a rebounding economy and technological developments. This increase in demand would cause the Rattlesnake, Otis, and Amarillo plants to be restarted (Nuttall et al., 2012; Sears, 2015). The Rattlesnake Field plant was abandoned after a short period of production, 42 MMCFHe, due to encroaching water within the reservoir. The original estimate of He potential was 800 MMCFHe for the Rattlesnake Field (Royston and Wheeler, 1956). The Keyes Field in Cimarron County, Oklahoma (Fig. 1.3) had tremendous natural gas reserves but was inhibited from producing for several years due to a high N_2 content until a joint venture put in a facility to process N_2 out of the gas. In 1958 the USBM estimated that four BCFHe was being lost annually from natural gas production (Nuttall et al., 2012; Sears, 2015). When the Keyes Field was finally placed on line in 1959, it caused supply to exceed demand and allowed the US Bureau of Mines (USBM) to store He at Cliffside again (Nuttall et al., 2012; Sears, 2015).

The Helium Act of 1960 was put into place to allow purchases by the government of He from private sources (US Congress, 1960). The He would be processed at either the Exell or Keyes facilities (Nuttall et al., 2012; Sears, 2015). Congress also had the price for He increased from \$19 to \$35 per 1000 cubic foot (Sears, 2015). As part of this legislation a 450-mile-long pipeline was built that transported the He gas from several areas of the Panhandle Hugoton Field (PHF) (Fig. 1.3) to the Cliffside facility. The He was extracted at the Cliffside facility, processed and returned to the sellers via the same pipeline system (Nuttall et al., 2012; Sears, 2015). In 1962 Ker-McGee began producing He from the Pinto Dome Field, Apache County, Arizona (Fig. 1.3), whose product was eventually sold to end-users on the West Coast of the USA. Canada entered the market with He production from the Swift Current Field in Saskatchewan, Canada. The Helium Act of 1960 had the unintended consequences of setting a top price for He which allowed private companies to come in and produce it for less (Nuttall et al., 2012; Sears, 2015). Another decline in consumption of He occurred at the end of the Vietnam War, which caused a decline in He use, resulting in the government to shutting-in, selling-off, or abandoning several facilities. Domestic demand for He by 1970, prior to the end of the Vietnam War, was in significant decline but export sales were rapidly increasing. New plants had started up in Poland, France and Russia (Sears, 2015). The Helium Act of 1960 was terminated in 1971 as sufficient supply of He had been achieved. In 1976 the Dineh-bi-Keyah field in Apache County, Arizona (Fig. 1.3), a oil field, began producing He (Nuttall et al., 2012; Sears, 2015).

The Big Piney-La Barge Field, Sublette County, Wyoming, USA, discovered in 1962 (Fig. 1.3), has become the biggest present-day producer of He domestically. The Big Piney-La Barge Field gas contains 66% carbon dioxide (CO_2), 21% methane (CH_4), 5% hydrogen sulfide(H_2S), and the remaining 8% a combination of nitrogen (N_2) and He (Nuttall et al., 2012; Sears, 2015). The gas requires extensive processing to extract CO_2, N_2, and H_2S. At the time of the Big Piney-La Barge field discovery there were no uses for either the CO_2 or H_2S. Mobil Oil (now Exxon-Mobil) determined that the CO_2 could be used in secondary and tertiary enhanced oil recovery and began drilling additional wells in 1978 (Nuttall et al., 2012; Sears, 2015). Helium was not produced until 1986 after the government and Mobil came to agreement on pricing for said gas. The Shute Creek Processing Facility at the Big Piney-La Barge Field produces 1.5 MMCFHePD or 540 MMCFHePY.

Helium demand continued to drop or remain stable through the 1990s in the USA. During that time the Helium Privatization Act of 1996 was signed into law by Congress. The Act called for the privatization of He sales, shutting down of government run He facilities and selling off of the He reserve of 30 BCFHe stored at Cliffside facility by 2015 (Sears, 2015). According to the act, 600 MMCFHe of the He reserve was to remain in the US government's control. Helium sold from the reserve was sold at higher prices than He purchased from private companies suggesting that supply and demand was in relative parity.

In 2000 US private suppliers began to see a decline in He production and prices began exceeding the 1953 government established fixed price of $35 (Nuttall et al., 2012; Sears, 2015). Algeria and Qatar established new He processing plants that handled very low percentages of the gas, similar to PHF, that moved very large volumes of natural gas making He recovery economic. By 2013 Cliffside Facility was ready to close but the Helium Stewardship Act enacted by Congress prevented that and it remained opened until 2020 when it was shut down. The USBM had auctions in 2014 for He with a realized price of $161.32 per MCF reflecting the high demand and short supply. The Stewardship Act further has allowed the government to sell off He at competitive prices even with the knowledge that the reserves will run out.

Since the mid-2010s, several private companies in the United States and Canada have been exploring new sources of He with mixed results. Presently, southeast Asia has only one source of He, from a processing facility at Darwin, Australia. That plant will go off line in the next couple of years. The Algiers He field is also in significant decline. Tanzania has become a focus of He exploration along the southern portion of East African Rift System. Russia and Qatar are bringing on new sources of He that may temporarily cause supply exceeding He demand. However, He demand is outstripping supply due to the growth in industries such as computers, medical analysis technologies, semiconductors, private space companies, fusion, welding.

1.3 Uses

Helium has many uses in present day society. Those uses are growing exponentially as He characteristics become more and more critical tomany industrial applications. It is estimated that 25% of He consumed in the United States is by the military and that its use is growing with the development of high-tech weapons. Since the concentration of He in air is very minimal (5.2 ppm), industrial extraction from air is presently not economically viable. Helium is typically extracted from He-bearing natural gas, CO_2, or N_2 rich gases. It is also extracted from the vent gases at liquefied natural gas (LNG) processing plants. Five He producing fields and processing facilities in the world alone supply about 80% of the world's supply. Production outages or slowdowns from just one of these five fields of processing facilities would have a severe impact on the global He supply. The United States is still the major producer of He, but several new sources are developing elsewhere such as a major processing plant in Russia and additional processing facilities at the South Pars Field, Qatar. Helium is a global product that except for the recent pandemic from March 2020 to present is in tight supply and any disruptions create shortages and dislocation. Helium consumption worldwide is indicated in Fig. 1.4, and it is presently dominated by North America and Europe, with demand in China and rest of Asia growing.

1.3.1 General uses

Helium is used to detect leaks in high-vacuum and high-pressure equipment (Fig. 1.4 and Tables 1.1 to 1.3). Welding is one of the largest uses for gaseous He where it provides an inert shield to protect the weld zone from the atmosphere when working with aluminum (Al) and

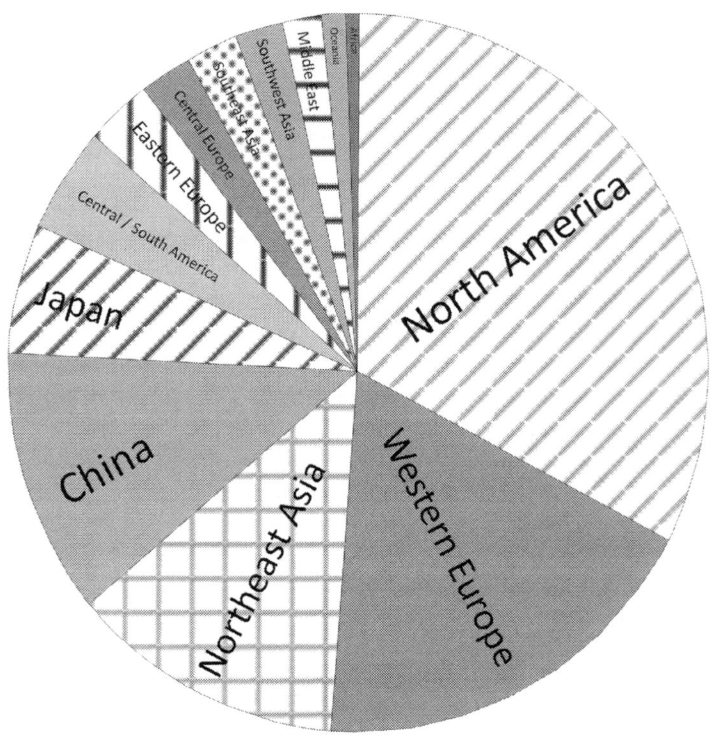

FIGURE 1.4 World Consumption of Helium in 2018 by region or country (modified from IHS Markit, 2019).

TABLE 1.1 Properties and applications of Helium.

Properties	Applications
Lowest boiling point, does not solidify at atmospheric pressure and 0 K	Liquid cooling of Tc superconductors; purging liquid H systems
Second lightest element after H	Lifting medium for balloons and airships
Smallest molecular size, with the greatest permeation rate of any substance	Leak detection; supersonic wind tunnels (high Reynolds number)
Chemically inert	Carrier gas, analytical uses, semiconductors
Very high specific heat and thermal conductivity	Faster cooling in vacuum furnaces; gaseous cooling—fiber optics; nuclear reactors
Radiologically inert (no radioactive isotope caused by neutrons)	Heat transfer medium in nuclear fission or fusion reactors
Highest ionization potential	Metal-arc and plasma-arc welding
Extremely low solubility in water	Deep-sea diving gases
Very high sonic velocity	Metal coating

Modified after Cai, Z., Clarke, R.H., Nuttall, W.J., 2012. Applications, prices and substitutions, In: Nuttall, W.J., Clarke, R.H., Glowacki, B.A. (Ed.). The Future of Helium as a Natural Resource, Chapter 9, Routledge, 376pp. pp. 134–156.

copper (Cu). The two major welding processes that use He are gas tungsten arc welding (GTAW) and gas metal arc welding (GMAW). Helium is also commonly used to pressurize and purge systems of unwanted gases. Helium is used in supersonic wind tunnels because it is thermally inert and calorically perfect in nature. Also has a speed of sound three times that of normal air (Emsley, 2001), and high value of heat to capacity ratio.

Helium is used to condense H and O_2 in the production of rocket fuel. It is also used to purge fuel and oxidizer from support vehicles and to cool H fuel in space vehicles. Helium can be added to O_2 tanks so that divers can breathe more easily. This is especially important for deep diving (over 450 ft, (136) m below sea level). This He mix reduces the risk of narcosis and high-pressure nervous syndrome.

1.3.2 Medical

Helium is critical in the medical industry (Berganza and Zhang, 2013) for multiple types of equipment and procedures (Fig. 1.4 and Tables 1.1 to 1.3). Helium is mixed with O_2 to help in the treatment of asthma, emphysema, and other respiratory problems by helping to reduce airways resistance to gas flow and respiratory muscle work load, until more definitive treatments can be implemented. Helium is used in surgery instead of CO_2 to inflate the abdomen of patients undergoing laparoscopic abdominal procedures due to its superiority in preventing respiratory acidosis in patients with comorbid conditions that would cause retention of CO_2 in tissue.

Helium is critical in magnetic resonance imagery (MRI) and used as cooling agent for superconducting magnets. MRIs use tremendous amount of energy, up to 11.75 T to scan the human body and provide an image. Tesla is the derived unit of the magnetic induction or magnetic flux density (shades of the movie *Back to the Future*) in the International System of Units.

TABLE 1.2 Gaseous Helium applications.

General industry	Specific uses
Medicine	"Heliox" medical therapies: He^4 asthma and neonatal distress treatment; ^3He (lung function MRI imaging)
Diving	"Heliox" breathing mixtures for deep-sea diving
Protective atmospheres	Detection gas for leak detection
	Optical fiber manufacturing
	Semiconductor manufacturing
	Rare metal smelting
	Heat treatment of metals or components
	Shielding gas for cutting and welding
	Working gas for laser applications
Heat transfer medium	Heat exchange in gas-cooled nuclear reactors
Transportation	Balloon and airship inflation
	Inert gas in space sensors
	Airport neutron detectors (He^3)
Laser technology	Resonator gas component
Metrology	Carrier and purging gas in gas chromatography
	Calibration and test gas in analysis
Military	Heat-seeking and guided missiles
	The US navy uses helium to erratic noisy signals
	Cool thermographic cameras and other equipment
	Purge and pressurize propulsion systems
	Weather monitoring and certain detection equipment

Modified after Cai, Z., Clarke, R.H., Nuttall, W.J., 2012. Applications, prices and substitutions, In: Nuttall, W.J., Clarke, R.H., Glowacki, B.A. (Ed.). The Future of Helium as a Natural Resource, Chapter 9, Routledge, 376pp. pp. 134–156.

Helium is used to freeze tissues with minimum issues in terms of reactions and low solubility to blood or other fluids. The cost of alternative methods is over 100 times more expensive. Coburn and Maze (2008) found that in treating patients with traumatic brain injury using He at elevated pressures resulted in a neuroprotective effect that limits tissue degeneration.

1.3.3 Electronics

Helium is critical in the electronics industry and its use is expanding (Fig. 1.4 and Tables 1.1 to 1.3). Helium is used in the following ways:

(a) Control atmospheres when the levels of different gases have to be kept constant.
(b) As a protective gas when growing silicon and germanium crystals and when producing Ti and Zr.

TABLE 1.3 Liquid helium applications.

General industry	Specific uses
Metrology	Nuclear magnetic resonance analysis (NMR)
	Superconducting magnetic field probes
	Cooling of sensors
Medicine	Magnetic resonance imaging (MRI)
	Magnetoencephalography
	Mixed with oxygen for respiratory treatment
Research and development	Nuclear fusion and particular physics
	Low temperature physics
	Superconducting energy storage
	Superconducting materials and sensors
	Ultralow temperature material research
	Materials testing
Technology and industrial applications	Strong field magnetic separator
	Magnetic melt stabilization
	Particle accelerators
	Infrared detectors
	Josephson junction devices

Modified after Cai, Z., Clarke, R.H., Nuttall, W.J., 2012. Applications, prices and substitutions, In: Nuttall, W.J., Clarke, R.H., Glowacki, B.A. (Ed.). The Future of Helium as a Natural Resource, Chapter 9, Routledge, 376pp. pp. 134–156.

(c) He-Ne lasers. These lasers can be used to read barcodes. Helium lasers have been mostly replaced by diode lasers.
(d) As a heat-transfer medium in some gas-cooled nuclear reactors.
(e) To age date rocks that contain U and Th.
(f) As a carrier gas in gas chromatography.
(g) In solar telescopes to reduce the distortion caused by temperature fluctuations in between the lenses.
(h) Helium is critical in building components for, and assembling MRIs, and in semiconductors, LCDs and in fiber optic industries, which are centered in China, India, South Korea, Taiwan, and the Middle East.

Many electronic devices use semiconductor chips of Si and Ge composition that provide significant consistency and efficiency for carrying out computing, storing data, and communication functions that require very stable conditions. The Si and Ge chips are grown in He-rich environments that provide a stable environment that no other element or gaseous mixtures can provide, minimizing flaws and defects. Helium has a high thermal conductivity that conducts heat very effectively during the growth process of these crystals. As the dimensions of the circuitry on the semi-conductors continues to shrink, the high quality necessary for precision crystal growth would be impossible to achieve without the control that a He atmosphere can provide. The electronics industry will use considerably more He to satisfy its cooling needs with the introduction of the 450-mm diameter (large diameter) Si or Ge wafer. Without SI and Ge chips, high-tech civilization would not function thus representing the continued need for He.

1.3.4 Welding and leak detection

Helium is the optimal tracer gas to find leaks because it is nontoxic, inert, noncondensable, nonflammable, and normally present in the atmosphere at the extremely low trace amount of 5.2 ppm. Because of its small atomic size, He will pass easily through leaks (Fig. 1.4 and Tables 1.1 to 1.3). The only molecule smaller than He is H which is not inert, very volatile, and therefore not usable for leak detection. A mass spectrometer leak detector (MSLD) is used to locate and measure the size of leaks into or out of a system or containing device. Helium is introduced in to the system or container material that is suspected of leaking. An He dectector monitors the presence of abnormal amounts of He by partial pressure and the amount of leakage are displayed on a meter. The two main methods to test material that may be leaking are using vacuum testing (outside-in) and pressure testing (inside-out). Vacuum testing takes the material to be tested, evacuates it with a separate pumping system for larger volumes, or within the detector itself for smaller volumes. Helium is applied to the suspected leak site(s) of the part using a spray probe with an adjustable flow. Pressure testing materials use He or an He and air combination, the potential leak sites are scanned using a sniffer probe connected to the inlet of the leak detector.

Industrial process tools that use vacuum systems or pressure systems must be tested or checked on a regular basis for leaks. Examples of vacuum systems are vacuum furnaces, vacuum coaters, electron microscopes, glove boxes, linear accelerators, electron beam and ion beam process equipment, semiconductor process equipment, and laser process equipment. Examples of pressurized systems are power plants, gas handling systems, bioreactors, liquid gas facilities, underground tanks, underground cables and pipes, submarines and space capsules.

1.3.5 Cryogenics

The largest use of He is in cryogenics. The reason is its low boiling point, weak interatomic reactions, extremely low molecular weight and low density that yield interesting properties when it is cooled below the critical temperature of 5.2 K (Woods and Crowley, 1973). Helium is used for the following:

- Cool superconducting magnets in MRI scanners which consume 20% of all He product;
- Steer charged particles in particle accelerators;
- Widely used in the retail industry in refrigeration systems.
- Protect structures of proteins under high-flux X-rays, it is more effective than N_2 systems;
- As a superfluid for quantum computing;
- In dilution refrigerators that have no moving parts.

Helium at ambient pressure and at absolute Kelvin does not form a solid and its vibrational energies are comparable to very weak interatomic binding interactions (Yang and Ellis, 2012). This prevents lattice formation. Helium's fluid characteristics consist of two phases commonly referred to as He I and He II. Helium I display thermodynamic and hydrodynamic properties of classical fluids, along with various quantum characteristics. At 2.17 K and below, which is called the lambda point (T_λ) He I will transition to He II and become a quantum superfluid with zero viscosity (Yang and Ellis, 2012).

Helium under extreme conditions such as when it is cooled beyond T_λ has the ability to form a new state of matter, known as a Bose-Einstein Condensate (BEC), in which the atoms literally lose all their energy. When He is below T_λ the liquid surface becomes smoother, indicating the transition from liquid to superfluid (Charlton et al., 2008). The atoms begin to aggregate creating a volume of equivalent density and energy without any energy to transfer (Penrose and Onsager, 1956). Based on observations, liquid He only exhibits superfluidity because it contains isolated islands of BECs, which have well-defined magnitude and phase (Haussman, 1994). Quantum energy associated with compressional waves, such as the vibration of a crystal lattice is called a Phonon (P). Helium's elementary excitation in a superfluid is called a Roton (R). The PR modes in the BEC's have the same energy resulting in zero-point He vibrational energies in preventing lattice formation (Bossy et al., 2008). Cryogenic He systems can move heat from a volume of relatively low temperature to a volume of relatively high temperature. This appears to contradict the second law of thermodynamics but experiments have confirmed that this phenomenon is related to the heat associated with the phase change between liquid and gaseous He (Urban et al., 2013). The use of He in cryogenics is rapidly evolving and expanding as its properties allow it to be used in ways other elements and compounds do not.

1.3.6 Balloons, blimps, and airships

Helium is best known for its use in balloons for parties, birthdays, special occasions, etc. Helium is also critical in airships and blimps which, among other uses fly over sporting events and advertise various consumer products. When mixed with argon, it's also used to inflate airbags. It has been suggested that He in balloons could be used to elevate solar panels above the ground to provide power. Google had a project called Loon where by it proposed using He-filled balloons to bring the internet to rural areas (Google, 2021) A number of issues for this type of He use is the durability of the balloons under high winds, cost per kilowatts of electricity generated, problems with other forms of air transportation, problems with visual pollution and probably many other issues not yet defined (Zephyrsolutions, 2015). Google in January of 2021 canceled the Loon project for unspecified reasons (Spacenews, 2021).

Airships, blimps (Zeppelins), or dirigible balloons are lighter-than-air aircrafts that use a lifting gas that is less dense than the surrounding air to keep them airborne. The three main types of airships are rigid, semirigid, and nonrigid. Nonrigid airships are commonly call blimps. Blimps rely on internal pressure to maintain their shape. Semi-rigid airships also rely on internal pressure to maintain their shape, but the structure is also supported by a metal keel at the bottom of the structure. Rigid airships, also called zeppelins, have a structural framework that keeps the shape and the lifting gas is contained in one or more cells within the airship. The first steam-powered airship flew in September of 1852 approximately 51 years before the Wright brothers' first flight. Balloons or blimps were used in the Civil War for observation by the US military. Airships were used extensively in World War I by Germany but used H for their lifting medium. The decline of airships was accelerated in the 1920s and 1930s after a number of crashes occurred. Most notably, the German airship the Hindenburg (which was filled with H) burst into flames while landing in New Jersey in 1937, killing 35 people on board. In World War II these same types of aircraft (filled with He or a He-H mix) were used to prevent bombers from flying low over cisties and populated area, surveillance and anti-submarine warfare with a mixture of H and He gases. Prop driven and then jet driven planes and eventually helicopters technology advancements post World War II made blimp and airship technology obsolete. In latter part of the 20th century blimps have been used in a very limited capacity for advertising, freight transportation, tourism, camera platforms for sporting events, geological surveys, aerial observation, interdiction platforms, TV coverage, tourism, and some research purposes because of the ability to hover for long periods of time. Airships have now generally disappeared at the end of the century because of tremendous costs to construct, maintain and fly them. Airships require up to $100,000(US) of He per trip. With recent He shortages, price increases and the rise of drones makes most airships likely obsolete. At present about 128 people are licensed to fly airships of which 17 people are paid to fly them according to the Federal Aviation Administration. There are only 25 blimps in the USA that are still in existence of which only about half of them are still used for advertising purposes. Another competing technology that has had a significant impact is drones, which can hover at extremely low costs relative to He filled airships. The use of He filled balloons at children's parties have remained strong to present day.

1.3.7 Earthquake detection

Helium has been used for the detection of earthquakes, generally after the fact either through analysis of groundwater or the soil section (Guerra and Lombardi, 2001). Movement along fault systems releases stress, which opens new pathways that allow rapid release of trapped He, H, CO_2, and N_2 (Wakita et al., 1980; Doering and Friedman, 1983; Sano et al., 1998, 2016; Deshpande and Gupta, 2013).

1.3.8 Fusion

^{3}Helium is critical for fusion power generation of electricity and heat (Kulcinski, 1989). Discussed here is a general overview of the fusion process as it is out of the scope of this book. The fusion process combines two light nuclei to form a heavier one which generates heat which in turn is converted to electricity. The fusion process requires a confined environment that needs sufficient time, temperature, and pressure to create a plasma that creates the process called the *Lawson criterion* (fusion) to occur. In stars the common fuel to create fusion

TABLE 1.4 Comparison of different types of fusion reactions.

Key technological features of fusion power plants				
	Deuteron–tritium (generates a neutron and a He molecule)	**Deuteron –deuteron**	**Deuteron–^{3}He**	**^{3}He–^{3}He**
Physics	Easiest (20 keV)	Next to easiest	Harder (60 keV)	Hardest (200 keV)
Radioactivity (vs. fission)	Same (1 day) 0.1% (100 years)	Same (1 day) 0.1% (100 years)	3% (1 day) 0.003% (100 years)	None (1 day) None (100 days)
First wall life (Matls development)	3–4 FPY's (Extensive)		Full lifetime (small)	Full lifetime (off-the-shelf)
Worse accident scenario	No offsitefatalities	No offsitefatalities	No Evacuationrequired	Inherentlysafe
Electrical efficiency (vs. fission)	= Same	Less	1.5 to 2 times higher	2 times higher
Projected cost of electricity (vs. fission)	= 1.5 times	Several times	= 1 to 1.5 times	< Fission

is through the element H which is under intense gravity resulting in extremely higher pressures, long confinement times, and very temperatures temperatures. The present proposed fuels for fusion facilities on earth are isotopes of H, deuterium and tritium, that actually require lower temperatures and pressures, and confinement time to reach the Lawson Criterion (Table 1.4). The preferred fuel for fusion is ^{3}He because it produces optimal results but unfortunately is rare on Earth. Only on the moon is sufficient quantities ^{3}He as a source for fusion reactors on Earth (Bradshaw and Hamacher, 2013). The amount of ^{3}He needed to run fusion reactions is greater than all He produced every year for just one facility. The temperature to heat the fuel in fusion reactors, whether H or He, require temperatures of tens of millions of degrees, which presents a major challenge in providing a successful container that won't break down.

Nuclear fusion as a source of energy is expected to have many advantages over fission. One of fusion's appeals is the low radioactivity in operation, little high-level nuclear waste, and increased safety. Present fusion processes require a sustainable combination of temperature, pressure, and time that has proven to be difficult to produce in a practical and economical manner. Since the 1940s there has been research in fusion but to date, no facility has been able to generate more electricity than was used to drive the fusion reaction. The longest period of time that a fusion reaction has been sustained was in mid-2021 by China's EAST facility for 102 s at 49.999 million degrees °C (Sciencealert, 2021). Another issue is managing the neutrons that are released during the reaction time which degrade many common materials over time used within the reaction chamber.

1.3.8.1 Medical hazards and industrial issues with helium

Helium in the atmosphere is nontoxic, plays no biological role, and is found in trace amounts in human blood. When inhaling He from a balloon causes an abrupt change in vocal sounds which is due to He having nearly three times the speed of sound of air (Emsley, 2001). The natural resonance frequency of a gas-filled cavity is proportional to the speed of sound in the gas so that when He is inhaled, a corresponding increase occurs in the resonant frequencies of the vocal tract, i.e., amplifier of vocal sound. After inhaling He gas, the muscles that control the voice box still move in the same way as when the voice box is filled with air; therefore, the fundamental frequency or pitch produces a direct vibration of the vocal folds that does not change. The high-frequency-preferred amplification does cause a change in timbre of the amplified sound, resulting in a reedy, ducklike vocal quality. The opposite effect, lowering resonant frequencies, can be obtained by inhaling a dense gas such as F_6S or Xe.

Death by He is extremely rare and is related to extreme misuse. Helium is a simple asphyxiant and so displaces O_2 needed for normal respiration (Emsley, 2001). Helium inhaled directly from pressurized cylinders or even balloon filling valves under high flow rate and pressure inhaled into the lungs can result in barotrauma which fatally ruptures tissue (Grassberger, and Krauskpf, 2007). One of the side effects of inhaling excessive amounts of He may temporarily turn a person blue (Balloonartists, 2020).

The first recorded death in the media from He asphyxiation was of a 15-year-old girl who inhaled He at a friend's birthday party in 1998 (Wikipedia, 2021). A man died of barotrauma in North Carolina in 2002. Inhaling He can be dangerous if done to excess and fatalities have been recorded in Vancouver in 2003 and two adults who suffocated in Florida (2006) (Wikipedia, 2021). A Californian youth in 2009 was found with a bag over his head, attached to a He tank, and in 2010 another teenager in Northern Ireland died of He asphyxiation. In 2012, a teenage girl in Eagle Point, Oregon, died of He barotrauma at a party and a girl from Michigan died of He from hypoxia later in the year (Wikipedia, 2021). In 2015, a 12-year-old after inhaling huge quantities of He as part of a game lost consciousness and fell into a coma as a result of air bubbles blocking the flow of blood to the brain (Wikipedia, 2021). Mixture of He and O_2 at high pressures at more than 20 atm or 2 MPa can lead to high-pressure nervous syndrome (Wikipedia, 2021). This can be alleviated by adding small amounts of N_2.

The safety issues for cryogenic He is that at extremely low temperatures can result in cold burns, and the liquid-to-gas expansion ratio can cause explosions if no pressure-relief devices are installed (Wikipedia, 2021). Even containers of He gas at 5 K to 10 K should be handled as if they contain liquid He due to the rapid and significant thermal expansion that occurs when He gas at less than 10 K is warmed to room temperature.

Helium concentrations in a localized atmosphere can cause issues with electronics. An IT worker at the Morris Hospital in Illinois discovered that He can cause problems for Apple iPhones and watches because several people complained that their IOS devices were not working properly and disabled. A recently installed MRI machine was leaking He and disabled iPhones and Apple watches after which a fresh MRI machine was installed (Potuck, 2018). Apple devices share a common family of MEMS resonator to reduce manufacturing costs. This type of clock likely leaks in He rich atmospheres, pushing the output frequency outside of the bounds that the main processors are designed to handle, rendering them nonfunctional. The devices may begin to function again when removed from the He rich atmosphere, but depending on the He concentration that leaked in, this could take anywhere from weeks to years. Apple mentions this issue in its user guide: *Exposing iPhone to environments having high concentrations of industrial chemicals, including near evaporating liquefied gasses such as helium, may damage or impair iPhone functionality.* Android systems for some reason are not effected.

1.4 Sources of helium

The source of He in numerous natural gas fields has been the subject of debate. Initially the source for all He was thought to be slow degassing from the mantle. The amount of He is relatively stable in the atmosphere and is at approximately 5.2 ppm and remains relatively constant suggesting that whatever amount of He that exits the Earth is quickly replaced. This suggests that the Earth is degassing at a constant rate except in areas where an excessive amount of He is being released from volcanoes, Mid-Oceanic Ridges, vents or geysers, and/or outcrops (Raitt et al., 1978). Some He in the atmosphere may come from cosmic sources attached as part of debris falling to Earth.

Once He has escaped into the atmosphere it has the ability to obtain escape velocity and leave the atmosphere for space. So He is assumed not to be building up through time in the atmosphere (Raitt et al., 1978). The complex function of photoionization related to the Earth's magnetic field and electric fields changes between summer and winter hemispheres and solar activity (Raitt et al., 1978). Nitrogen on the otherhand does not have the ability to achieve escape velocity and is retained in the Earth's atmosphere. Unlike He, N_2 is part of the life cycle, forms a multitude of compounds and is constantly being released from organic and inorganic sources and recycled.

During the Earth's formation ^{3}He, generated by the Sun, was trapped within the Earth's mantle and over time and has slowly leaked out. Concentrations of ^{4}He in the subsurface that are greater than the atmospheric 5.2 ppm are associated with groundwater, ancient brines, ancient pore water, natural gas fields, Carlin-type gold deposits, Mississippi Valley-type deposits, Witwatersrand gold deposits, other ore deposits, hydrothermal fluids, volcanic degassing, oil field brines, lakes, ice, sediments, coal measures, and lunar surfaces (Heaton, 1984; Torgersen and Clarke, 1985; Torgersen and Ivey, 1985; Creedy, 1988; Ballentine et al., 1991; Stute et al., 1992; Marty et al., 1993; Pinti and Marty, 1995; Stuart et al., 1995; Ballentine et al., 1996; Pettke et al., 1997; Winckler et al., 1997; Pinti and Marty, 1998; Burnard et al., 1999; Ballentine et al., 2002; Ballentine and Burnard, 2002; Ballentine and Sherwood-Lollar, 2002; Kipfer et al., 2002; Yang et al., 2003;; Poreda et al., 2004; Graupner et al., 2006; Williams, 2007; Gilfillan et al., 2008; Holzner et al., 2008; Ray et al., 2009; Zhou et al., 2012; Barry et al., 2013; Power et al., 2013; Winckler and Severinghaus, 2013; Caracausi and Paternoster, 2015; Lowenstern et al., 2015; Tomonaga et al., 2017; Danabalan, 2017; Halford, 2018; MHS, 2020).

(1) ^{3}He is derived from the mantle and is released during degassing events through time (Mamyrin and Tolstikhin, 1984; Hunt, 1996; Oxburgh et al., 1986). ^{3}He is thought to be scavenged from the upper mantle by partial melting or dehydration on a regional scale. The He is then emplaced in the crust or at the surface in basaltic bodies that are carrying U, Th, and K (O'Nions and Oxburgh, 1988). Fig. 1.5 is a diagram of the present concept of He moving from the mantle through the crust to the surface as presented by Wakita et al.

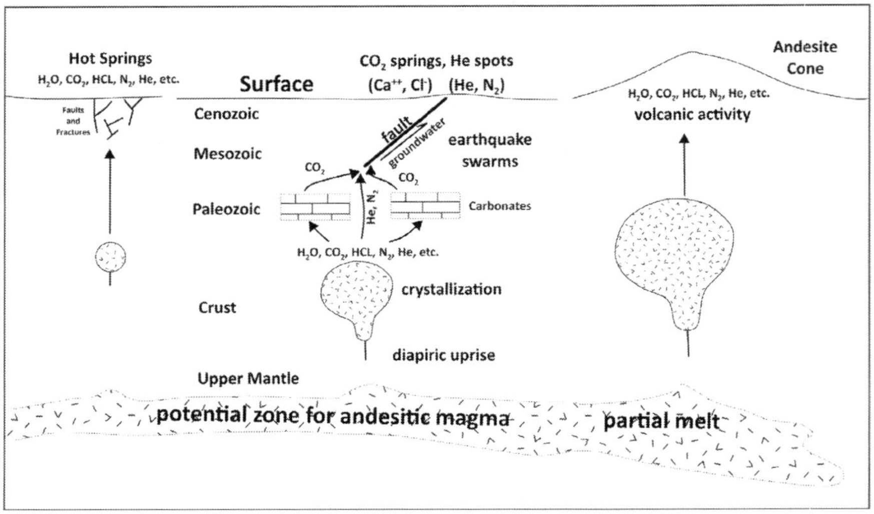

FIGURE 1.5 Source of He, CO_2, and N_2 from the upper mantle (modified from Wakita et al., 1978).

(1978) for areas in and around Japan but is applicable to any area. Anderson (1998) identified the He paradox which suggested that the lower mantle is devoid of ^4He and rich in ^3He. The lack of ^3He in hotspots such as Hawaii when compared to the Mid-Atlantic Oceanic Ridge does not reflect a lack of ^3He but rather the inability of the isotope to be degassed (Anderson, 1998). Hotspots high in He may be the result of differential partitioning of U and He into the magmas. The low ^4He concentrations in some areas indicated it is depleted in the crust or mantle (Wakita et al., 1978).

(2) ^4He is derived mainly from radiogenic decay of U and Th in crustal rocks in the Earth's asthenosphere and is called "magmatic" ^4He (Tolstikhin et al., 2016; Oxburgh et al., 1986; Jenden et al., 1988; Ballentine et al., 1991; Ozima, 1994; Hunt, 1996; Ballentine and Lollar, 2002). Fig. 1.6 is a generic model for He and noble gases migration (Ballentine et al., 2002; Halford, 2018; Pinti and Marty, 2000). Large concentrations of magmatic ^4He show a deep crustal input into the shallow reservoir and aquifer systems (Fig. 1.6). Seventy five percent of ^4He is produced in the upper 10 km of the crust due to preferential partitioning of the incompatible U and Th isotopes in the crust (Ballentine and Burnard, 2002). Transport of radiogenic ^4He out of basement or granitic rocks appears to be related to the presence of fracture and fault systems that can act as migration pathways for the movement through the impermeable granites in which it is generated. Crustal ^4He is usually found as trace amounts (<0.05 ppm) (Danabalan, 2017). Once in regional reservoirs or aquifers, brine and/or groundwater flow can move the helium to traps along its extent. Helium is not only transported via porous and permeable strata but along disconformities, unconformities, fractures, and fault systems (Danabalan, 2017; Halford, 2018).

(3) ^4He is derived from neutron capture reactions by hydrogen in Li-bearing sediments. There is no present evidence that Li produces significant amounts of He (Hiyagon and Kennedy, 1992; Mamyrin and Tolstikhin, 1984).

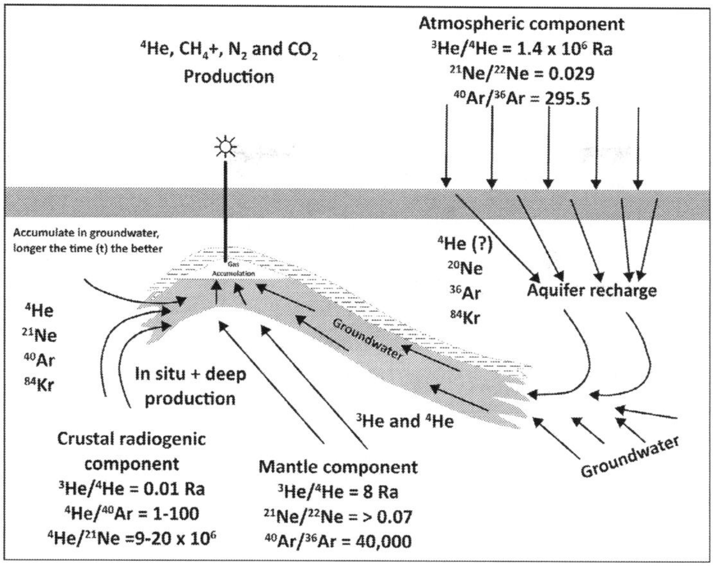

FIGURE 1.6 Diagram showing different isotopic components (atmosphere and air saturated water via aquifer recharge, crustal/radiogenic, and mantle) interacting with a petroleum system (modified from Ballentine et al., 2002; Halford, 2018; Pinti and Marty, 2000). Noble gases act as geochemical tracers in the subsurface. By understanding the primary components contributing noble gases into the system, the sources of various gases, migration pathways, and mixing of gases can be inferred.

(4) Helium trapped in natural gas reservoirs seems to be predominantly from crustal rocks. Contribution by decay from U and Th in carbonaceous mudstone or organic-rich clay shale seems to be very likely especially in Paleozoic rocks. However, the carbonaceous mudstones also generate hydrocarbons which dilute any significant volume of He that is generated, thus destroying these He deposits (Wang et al., 2020).

1.5 Helium isotopes

Helium isotopes are used as tracers for the identification of mantle reservoirs and mantle dynamics (Fig. 1.6) (Craig and Lupton, 1976; Lollar et al., 1993; Ballard and Lollar, 2002; Kurz and Jenkins, 1981; Stroncik et al., 2007; Kulongoski, and Hilton, 2011; Danabalan, 2017; Halford, 2018). He and noble gas ratios are very important for understanding various petroleum and groundwater systems. However, as an exploration tool in finding new deposits of He they have limited use. High $^3He/^4He$ ratios are thought to represent deep mantle material, whereas ratios around 8 RA (RA stands for the atmospheric $^3He/^4He$ ratio of 1.39×10^{-6}) are thought to be representative of the upper mantle (Table 1.5).

TABLE 1.5 Summary of $^3He/^4He$ ratios with corresponding source of the He.

$^3He/^4He$ ratio	Source
1.4×10^{-6}	Atmosphere
3.0×10^{-8}	Crustal rocks
6 to 8×10^{-6}	Volcanic arcs
8	Mantle plumes such as Hawaii, Yellowstone, and Iceland
5 to 6	Geothermal areas
0.1	Geothermal areas exclusively derived from crustal sources
15 to 60	Mid-Atlantic Ridges
3.1×10^{-4}	Deep Pacific sediments

Derived from Craig and Lupton (1976); Kurz and Jenkins (1981); Mamyrin and Tolstikhin (1984); Jenden (1989); Lollar et al. (1993); Ozima and Podosek (2001); Ballentine and Lollar (2002); Stroncik et al. (2007); Kulongoski and Hilton (2011); Danabalan (2017); Halford (2018).

Table 1.5 displays the ratios of $^3He/^4He$ that can help identify the source of the He. Uranium and Th decrease in percentage toward the crust-mantle boundary though natural decay, there are still potential magmatic sources of ^4He in the mantle. The lower mantle is assumed to be depleted of ^4He but rich in ^3He but the evidence for that can be considered circumstantial.

Based on He, Ne, and Ar fusion data from fresh, submarine volcanic glasses of several Mid-Atlantic Ridge off-axis seamounts (MORB), He isotopes can be susceptible to interferences during melt formation and evolution resulting in the decoupling of He from other elements, such as Ne or Pb (Stroncik et al., 2007). All He data evaluated was indistinguishable from that of the MORB range. In contrast, Ne isotopic compositions are much more primitive than MORB and in that context combined with He and Ar systematics show that the source region of these seamounts experienced a preferential loss of He compared to Ne and Ar. This He loss, combined with subsequent ^4He production, resulted in the decoupling of the He isotope from Ne and Pb (Stroncik et al., 2007). Thus, when considering He and Ne, only Ne has preserved the evidence that a primitive mantle component contributed to the formation of the investigated seamounts. As these seamounts are not fed from a mantle plume being derived from the deep mantle, the primitive Ne component resides within the upper mantle, implying that primitive noble gases are not necessarily only indicative of deep mantle material (Stroncik et al., 2007).

Barry et al. (2013) used $^3He/^4He$ to identify the source of He from cold "mazuku" CO_2 vents and hydrothermal gases and fluids from Rungwe Volcanic Province and associated East African Rift System (EARS), southern Tanzania. They concluded that the crustal ^4He was added during migration of mantle fluids to the surface. The longer the mantle fluids remained within the crust, the more ^4He was added. Several of the cold mazukus based on the data were isolated and no longer connected to the hydrothermal system (Barry et al., 2013; Danabalan, 2017). The hydrothermal system is consistently changing due to the activity of EARS constantly reopening or closing faults-fracture systems that feed the various He and CO_2 rich pathways and reservoirs. This concept of a hydrothermal plumbing system feeding fluids with associated He can be applied to other tectonic areas such as the Colorado Plateau and Big Piney-La Barge Fields, USA.

^{3}Helium is associated with worldwide magmatic activity and significant amounts of CO_2, both mantle and crustal derived (Poreda et al., 1988). Mantle CH_4 is not a required associated component but can be generated due to heat in the crust that thermally alters organic matter. Some magmas such as in the Barbados are strictly crustal driven with no faults or conduits to the mantle based on $^3He/^4He$ ratios (Poreda et al., 1988). Kulongoski et al. (2005) identified that the Morongo Basin in California which is divided into the Surprise Spring and Deadman basins, that have very different mantle ^3He and crustal ^4He concentrations. The subbasins are filled with Tertiary sediments that are underlain by Jurassic metamorphic and igneous basement rocks. Kulongoski et al. (2005) related the difference in $^3He/^4He$ ratios to increase faulting in the Deadman subbasin that was subject to episodic events. He concentrations varied over short distances and maybe related to advection as there is no underlying magma that has been identified. Kulongoski et al. (2005) determined that in the Tertiary strata the higher He and the resident times and concentrations were greater than the ^{14}C dating implies. The local tectonic activity is causing episodic movement of basinal fluids, He, and other noble gases into the Morongo Basin and subsequently completely controlling the groundwater system.

Another example is from Kotarba et al. (2020), who defined the sources of gases in the Miocene strata in the eastern part of the Polish Carpathian Foredeep as derived from Jurassic strata using isotopic data. Kotarba et al. (2020) determined that the He isotopes indicate He was from a deep source, the heavier noble gases, Ne, Ar, Kr, and Xe were atmospheric and in the local gas reservoirs due to groundwater recharge. Part of the CO_2 was generated from microbial processes. The N_2 and some CO_2 were generated by thermal transformation of organic matter and the CH_4 was from microbial processes. The source, migration, and accumulation history of the natural gas fields in this part of the Eastern Polish Carpathian Foredeep Basin based on isotopic analysis indicate a complex process that without noble gases analysis would not be clearly understood.

Isotopic analysis by Chen et al. (2019) report noble gas signatures of groundwater, hot springs, and bedrock samples from a major fault system that separates regional-scale blocks of accreted, continental materials in southern Taiwan. Despite the continental setting, the isotopic signatures argue for the presence of mantle-derived fluids, suggesting that the active fault system is deep-seated. This is consistent with deep, nonvolcanic tremors identified in the same area. They concluded that the mantle fluids are escaping along a crustal-scale fault system marked by clusters of nonvolcanic tremors directly beneath the southern Central Range. The evidence of these tremors and electrical conductivity anomalies along the strike of the fault recognized previously, correlated up dip with the surface trace of a major active fault (Chen et al., 2019).

Helium and other noble gases have been used successfully to study groundwater movement and history within aquifers and movement of petroleum (Andrews, 1985; Mazor and Bosch, 1987; Ballentine and Sherwood-Lollar, 2002; Kulongoski and Hilton, 2011). The inert nature of He and distinct nature of ^3He and ^4He and their ratio allow recognition of a number of factors that contribute to the total He budget of groundwater. He concentrations in groundwater often exceed the expected solubility values as a result of radiogenic production and the release from associated aquifer material in associated pore water (Kulongoski and

Hilton, 2011). By measuring the abundance of He in-situ in the groundwater, it is possible to measure the residence time (Brown, 2010; Brown, 2014; Hendry, et al., 2005). He that has migrated in from an extraneous source such as from the mantle, adjacent strata, volcanic activity, diffusion out of the aquifer, etc. can all affect the estimation of age of groundwater. The components of He include atmospheric in the form of air-equilibrated and air-entrained, terrestrial components in the aquifer, deep crustal and/or mantle He (Kulongoski and Hilton, 2011). These components can be used to define the chronology of the groundwater being sampled through the role of faults in degassing of the mantle and crust. It is necessary to address factors that may skew the age of the of estimation of the groundwater such as weathering, diffusion rates, introduction of mantle ^{3}He or ^{4}He. Using other isotopes to confirm estimation of groundwater age can help address this issue (Kulongoski and Hilton, 2011).

1.6 Helium transport

After He has been generated, it has been generally thought to move through the subsurface via (1) diffusion or a concentration gradient; (2) advection or pressure gradient, and (3) a combination (Halford, 2018). Diffusion is subject to variations to temperature, tortuosity, permeability, and the pore space fluid (Brown, 2010). Advection is dependent upon chemical changes related to metamorphism and hydrocarbon related to external sources such as magmatic fluids and/or aquifers (Ballentine and Burnard, 2002; Brown, 2010). Halford, 2018 proposed a combination of both in that the different mechanisms will occur at different times in He migration from source to the trap.

Helium has been shown to have a high rate of diffusion but in many areas the level of He in the rock section are greater than the rate of diffusion would allow (Bach et al., 1999). Pandey et al. (1974) studied the rate of diffusion in several cores specifically one from the Cliffside Field in the Texas Panhandle. The authors determined that across 11,000 acres over 50 years that 2.78×10^5 He would diffuse through the caprock and be lost. If extrapolated to 1 MY then approximately 8 BCF would have been lost (Pandey et al., 1974). The implication suggest there could not be an economic He left in the Texas Panhandle Field under pressured reservoir relative to the caprock's ability to retain the noble gas. Helium cannot diffuse through the caprock as it cannot exceed the capillary pressure of fracture, pores or uniformities and therefore it is retained.

Cherniak et al. (2015), studied the diffusion rates in carbonates and determined that calcite and magnesite have little capability to retain He whereas dolomite and aragonite have some ability under cooler conditions but their ability is impacted by grain size. Concentrations of He in the sedimentary rock in some areas, not only in economic He accumulations, is related to tectonic events that allow the release of higher volumes of He from the crust. Tardani et al., (2016) concluded in Southern Volcanic Zone in southern Chile related to magmas at depth that He, as well as C and N_2, movement is controlled by tectonic forces that release them along fault and fracture systems.

Deformation and metamorphism are important controls on noble gas movement. It is noted that the highest He content was found in finest grained, ultramylonite peridotites (Kurz et al. 2009). It was also found that crushing ultramylonite peridotites in a vacuum only releases a small portion of He and Ne for sampling and only by heating to 800 °C were most of the mantle-derived noble gases were released. Atmospheric He can be adsorbed into olivine and as determined by Protin et al. (2016), the process is irreversible at 900 °C suggesting that van der Waals forces are sufficient to bind it to the mineral. This suggests that atmospheric He has been adsorbed into amphibole and then can be recycled in to the crust via subduction zones (Poreda et al., 1988; Tolstikhin et al., 2016). Amphibole is a major carrier of He and other noble gases trapping them in its lattice. Apatite can hold significant amounts of He not through solubility but trapping in microvoids as defined by crushing and step-heating experiments (Zeitler et al., 2017). This suggest that there may be significantly more He and Ne present in the mantle than previously thought.

Helium found in pore H_2O rarely has a high enough concentration to form a gas phase in groundwater or buried water saturated sedimentary rocks (Brown, 2005). The greater the depth of the pore waters, the more likely there are higher concentrations of He. However, as shown in Chapter 3, there are numerous small, but shallow He rich deposits. As CH_4 moves through the aquifer it removes the He and when a trap is initially filled it is potentially He rich (Brown, 2005). But as more CH_4^+, N_2, and/or CO_2 enter the trap, the He will be diluted and/or may be forced into the seal rock.

Noble gas isotopes have been used successfully to define the source and migration histories of accumulated hydrocarbons, as they are not affected by chemical changes (Ballentine and Burnard, 2002; Holland and Gilfillan, 2013; Byrne et al., 2017, 2020). The noble gases are unaffected by biological activity, chemical alteration, or redox reactions but are affected by mixing, dissolution, phase partitioning, and diffusion. The data from noble gas analysis can be used to investigate and model oil, gas and water interactions, and timing and migrations of fluids and gases (Byrne et al., 2017, 2020). There use is limited in the He and hydrocarbon industry because analysis is expensive. Companies tend to be focused on what contributes to the bottom line. Until noble gases utility is established as a critical factor in exploration, exploitation, and production, essentially adding to the bottom line, they will remain an academic exercise.

An example of noble gases used in conjunction with a petroleum basin with CO_2, CH_4, and He was utilized by Hulston et al. (2001), in the Taranaki Basin in New Zealand. The Taranaki Basin is located between the northern and southern islands of New Zealand and is a prolific offshore petroleum producing province. The basin has relatively thin crust, adjacent andesite volcanoes, and is undergoing subduction in parts. Hulston et al. (2001) found that ^{3}He was near zero in the producing oil and gas fields in the eastern part of the basin but there was a significant contribution in the Maui and New Plymouth fields. The increase in ^{3}He was related to both increased thermal gradient, increased faulting, and thinning of the crust. The CO_2 was a combination of organic, marine, and mantle-derived sources whose overall compositional percentages depended where they were founded in the basin. Heavier hydrocarbons and lower CH_4 were found in areas of lower heat flow in the basin (Hulston et al., 2001). Isotope methodology can predict the composition of gases and fluids in a basin and help predict likely production results. An additional example was presented by Byrne et al. (2020) in the East Texas Basin, USA, identifying that mantle-derived ^{3}He and ^{40}Ar that were concentrated in shallower reservoirs sourced from the southern end of the basin. Byrne et al., (2020) has been proposed the mantle fluids were circulated throughout the basin by a diffusion-advection model.

1.7 Summary

Helium is a unique element that is critical in today's society. Helium unlike CH_4^+, CO_2, and N_2 is not reactive and cannot combine with other elements or compounds under any natural conditions. Helium's physical characteristics are unique and acts independently of other solids, fluids, and gases. The majority of He that is sourced on the planet is from U and Th decay in the crust. Helium in natural gas deposits based on isotopic work is generally all of crustal origin. Mantle He does contribute in small amounts to the overall He present in these various natural gas accumulations. Prior to release from the crust, He is thought to be primarily held in granitic bodies and then released during tectonic and magmatic activity. This is augmented by release from recycled metamorphic rocks that contain atmospheric and crustal-derived He. The amount of actual U, Th, and trapped He present in the crust is only an approximation and not based on actual observation. The literature shows the difficulty of estimating the amount of He in the crust. There are minerals such as apatite and amphibole that can contain, He thus high-grading areas of He which eventually can be released in large quantities that may be trapped in a shallower reservoir. The ability of minerals like amphibole to capture and hold He allows the recycling of atmospheric and crustal He through subduction zones and other tectonic processes that may explain the present estimated volumes of He on Earth.

Helium has a predictable stability and equilibrium in the atmosphere that suggests there is steady degassing of the Earth over the past 100 years of recorded measurement. Helium has a perceived association with N_2 and sometimes CO_2, which is probably coincidental rather than as a chemical or physical relationship. Carbon dioxide and N_2 can be generated and/or accumulated during similar processes, but the amount that is present with He in a natural gas accumulation is highly variable. Within the petroleum industry, as the He industry grows, and exploration, exploitation, and production increases; the amount of data will hopefully allow a greater understanding of its sources, migration, and accumulation.

References

Ambrose, W.A., Cutright, B.L., 2017. Living on the Moon: Lessons for Mars. AAPG, p. 24. Search and Discovery Article #70279.

Anderson, D.L., 1998. The helium paradox. Proc. Natl. Acad. Sci. U.S.A. 95, 4822–4827.

Andrews, J.N., 1985. The isotopic composition of radiogenic helium and its use to study groundwater movement in confined aquifers. Chem. Geol. 49 (1–3), 339–351.

Austerweil, G., 1914. Die angewandte Chemiein der Luftahart, Munchen and Berlin, pp. 8–9.

Bach, W., Naumann, D., Erzinger, J., 1999. A helium, argon, and nitrogen record of upper continental crust (KTB drill holes, Oberpfalz): Germany: implications for crustal degassing. Chem. Geol. 160, 81–101.

Ballentine, C.J., Burnard, P.G., 2002. Production, release, and transport of noble gases in the continental crust. Rev. Mineral. Geochem. 47, 481–538.

Ballentine, C.J., Sherwood-Lollar, B., 2002. Regional groundwater focusing of nitrogen and noble gases into the Hugoton Panhandle giant gas field. USA: Geochem. Cosmochim. Acta 66, 2483–2497.

Ballentine, C.J., O'Nions, R.K., Oxburgh, E.R., Horvath, F., Deak, J., 1991. Rare gas constraints on hydrocarbon accumulation, crustal degassing and groundwater flow in the Pannonian Basin. Earth Planet Sci. Lett. 105 (1–3), 229–246.

Ballentine, C.J., O'Nions, R.K., Coleman, M.L., 1996. A Magnus opus: helium, neon, and argon isotopes in a North Sea oilfield. Geochem. Cosmochim. Acta 60 (5), 831–849.

Ballentine, C.J., Burgess, R., Marty, B., 2002. Tracing fluid origin, transport and interaction in the crust. Rev. Mineral. Geochem. 47 (1), 539–614.

Balloonartists, 2020, balloonartists.com.au/helium-dangers.html.

Barry, P.H., Hilton, D.R., Fischer, T.P., de Moor, J.M., Mangasini, F., Ramirez, C., 2013. Helium and carbon isotope systematics of cold "mazuka" CO_2 vents and hydrothermal gases and fluids from Rungwe Volcanic Province, southern Tanzania. Chem. Geol. 141–156.

Beike, D., 2011. Lunar Helium-3 Value Chain: Investment and Funding. AAPG, p. 14. Search and Discovery Article #80154.

Berganza, C.J., Zhang, J.H., 2013. The role of helium gas in medicine. Med. Gas Res. 3 (18), 7.

Bossy, J., Pearce, J., Schober, H., Glyde, H., 2008. Phonon–Roton modes and localized Bose–Einstein Condensation in liquid helium under pressure in nanoporous media. Phys. Rev. Lett. 101 (2).

Bradshaw, A.M., Hamacher, T., 2013. Nuclear fusion and helium supply problem. Fusion Eng. Des. 88 (9–10), 2694–2697.

Brown, A.A., 2005. Origin of High Helium Concentrations in Dry Gas Water Fractionation, vol. 89. AAPG, Annual Convention, Calgary, AB, Canada.

Brown, A.A., 2010. Formation of High Helium Gases a Guide for Explorationist. AAPG. New Orleans Convention Poster, Search and Discovery Article 80115.

Brown, A.A., 2014. Compositional Gas Migrating through Water-satruated Rock, 41391. AAPG, Search and Discovery.

Burnard, P.G., Hu, R.Z., Turner, G., Bi, X.W., 1999. Mantle, crustal and atmospheric noble gases in Ailaoshan gold deposits, Yunnan Province, China. Geochim Cosmochim Acta 63 (10), 1595–1604.

Byrne, P.H.B., Lawson, M., Ballentine, C.J., 2017. Noble gases in conventional and unconventional petroleum systems. In: Lawson, M., Formolo, M.J., Eiler, J.M. (Eds.), Source to Seep: Geochemical Applications in Hydrocarbon Systems. Geological Society, London, Special Publications no. 468.

Byrne, P.H.B., Lawson, M., Ballentine, C.J., 2020. The use of noble gas isotopes to constrain subsurface fluid flow and hydrocarbon migration in the East Texas Basin. Geochem. Cosmochim. Acta 268, 186–208.

Cady, H.P., McFarland, 1903. Helium in Kansas natural gas. Trans. Kans. Acad. Sci. 20 (December).

Cady, H.P., McFarland, 1906. Helium in natural gas. Science, New Series 24. September 14, December.

Cady, H.P., McFarland, 1907a. The occurrence of Helium in natural gas and the composition of natural gas. J. Am. Chem. Soc. 29 (November).

Cady, H.P., McFarland, 1907b. Neon and Argon in natural gas. Trans. Kans. Acad. Sci. 21 (December), 64–65.

Cai, Z., Clarke, R.H., Nuttall, W.J., 2012. Applications, prices and substitutions. In: Nuttall, W.J., Clarke, R.H., Glowacki, B.A. (Eds.), The Future of Helium as a Natural Resource, Chapter 9. Routledge, pp. 134–156, 376pp.

Caracausi, A., Paternoster, M., 2015. Radiogenic helium degassing and rock fracturing: a case study of the southern Apennines active tectonic region. J. Geophys. Res.: Solid Earth 120 (4), 2200–2211.

Charlton, T.R., Dalgliesh, R.M., Kirichek, O., Langridge, S., Ganshin, A., McClintock, P.V.E., 2008. Neutron reflection from a Liquid helium surface. Low Temp. Phys. 34 (4), 316–319.

Chen, A., Shen, C., Byrne, T.B., Sano, Y., Takahata, N., Yang, T.F., Wang, Y., 2019. Mantle fluids associated with crustal-scale faulting in a continental subduction setting, Taiwan. Sci. Rep. 9, 10805. https://doi.org/10.1038/s41598-019-47070-2.

Cherniak, D.J., Amidon, W., Hobbs, D., Watson, E.B., 2015. Diffusion of helium in carbonates: effects of mineral structure and composition. Geochem. Cosmochim. Acta 165, 449–465.

Coburn, M., Maze, M., 2008. The neuroprotective effects of xenon and helium in an in vitro model of traumatic brain injury. Crit. Care Med. 36 (2), 588–595.

Congress, U.S., 1924. Conservation of Helium Gas: Hearings before Committee on Military Affairs House of Representatives, 68th Congress, 1st Session, H.R, p. 5722.

Congress, U.S., 1937. Public-No. 411-75th Congress, Chapter 895, 1st Session, S. 1567, Sect. Post World War I.

Congress, U.S., 1960. Helium Act Amendments of 1960. Pub. L. 86-777, Sept, 13th, 1960 74 Stat 918.

Craig, H., Lupton, J.E., 1976. Primordial neon, helium, and hydrogen in oceanic basalts. Earth Planet Sci. Lett. 31, 369–385.

Craig, H., Clarke, W.B., Beg, M.A., 1975. Excess 3He in deep water on the east pacific rise. Earth Planet Sci. Lett. 26 (2), 125–132.

Creedy, D.P., 1988. Geological controls on the formation and distribution of gas in British coal measure strata. Int. J. Coal Geol. 10 (1), 1–31.

Crockett, A.H., Smith, K.C., Bartlett, N., 1973. The Chemistry of Monatomic Gases, Pergamon, p. 208.

Danabalan, D., 2017. Helium Exploration Methodology for a Strategic Resource, PhD. Durham University, p. 295.

Department of Physics, 2021. Solid Helium. University of Alberta. http://www.phys.ualberta.ca/~therman/lowtemp/projects1.htm.

Deshpande, R.D., Gupta, S.K., 2013. Groundwater helium: an indicator of active tectonic regions along Narmada River, central India. Chem. Geol. 344, 42–49.

Dilmare, D., 2004. Mining the Moon (October 2004). Popular Mechanics (October).

Doering, W.P., Friedman, I., 1983. Survey of Helium in Natural Water Wells and Springs in Southwest Montana and Imperial Valley, California, Part V – Jan. 1 – Dec. 31, 1982. USGS. OFR 83-500, 32pp.

Emsley, J., 2001. Nature's Building Blocks. Oxford University Press, Oxford, pp. 175–179.

Gilfillan, S.M., Ballentine, C.J., Holland, G., Blagburn, D., Lollar, B.S., Stevens, S., Schoell, M., Cassidy, M., 2008. The noble gas geochemistry of natural CO_2 gas reservoirs from the Colorado Plateau and Rocky Mountain provinces, USA. Geochem. Cosmochim. Acta 72 (4), 1174–1198.

Google, 2021. Loon, Expanding Internet Connectivity with Stratospheric Balloons. https://x.company/projects/loon/.

Grassberger, M., Krauskpf, A., 2007. Suicidal asphyxiation with helium: report of three cases. Wien Klin Wochenschr v 119, 323–325.

Graupner, T., Niedermann, S., Kempe, U., Klemd, R., Bechtel, A., A., 2006. Origin of ore fluids in the Muruntau gold system: constraints from noble gas, carbon isotope and halogen data. Geochem. Cosmochim. Acta 70 (21), 5356–5370.

Greenwood, N.N., Earnshaw, A., 1997. Chemistry of the Elements. Butterworth-Heinemann, p. 1364.

Grilly, E.R., 1973. Pressure-volume-temperature relations in liquid and solid ^{4}He. J. Low Temp. Phys. 11 (1–2), 33–52.

Guerra, M., Lombardi, S., 2001. Soil-gas Method for Tracing Neotectonic Faults in Clay Basins: The Pisticci Field (Southern Italy).

Halford, H.D.T., 2018. Isotopic Analysis of Helium from Wells Located in the Four Corners Area, Southwestern US, CSM PHD Dissertation, p. 376.

Hampel, C.A., 1968. The Encyclopedia of the Chemical Elements. New York: Van Nostrand Reinhold, pp. 256–268.

Haussmann, R., 1994. Properties of a fermi liquid at the superfluid transition in the crossover region between BCS superconductivity and Bose–Einstein Condensation. Phys. Rev. B 49 (18), 12975–12983.

Heaton, T.H.E., 1984. Rates and sources of 4He accumulation in groundwater. Hydrol. Sci. J. 29 (1), 29–47.

Hendry, M.J., Kotzer, T.G., Solomon, D.K., 2005. Sources of radiogenic helium in a clay till aquitard and its use to evaluate the timing geologic events. Geochimica et Cosmochimica Acta 69 (2), 475–483.

Henshaw, D.B., 1958. Structure of solid helium by neutron diffraction. Phys. Rev. Lett. 109 (2), 328–330.

Hiyagon, H., Kennedy, B.M., 1992. Noble gases in CH4-rich gas fields, Alberta, Canada. Geochem. Cosmochim. Acta 56 (4), 1569–1589.

Holland, G., Gilfillan, S.M.V., 2013. Application of noble gases to the viability of CO_2 storage. In: Burnard, P. (Ed.), The Noble Gases as Geochemical Tracers, Advances in Isotopic Geochemistry. Springer-Verlag, pp. 177–223.

Holzner, C.P., McGinnis, D.F., Schubert, C.J., Kipfer, R., Imboden, D.M., 2008. Noble gas anomalies related to high-intensity methane gas seeps in the Black Sea. Earth Planet Sci. Lett. 265 (3), 396–409.

Hulston, J.R., Hilton, D.R., Kaplan, I.R., 2001. Helium and carbon isotope systematics of natural gases from Taranaki Basin, New Zealand. Appl. Geochem. 16, 419–436.

Hunt, J.M., 1996. Petroleum Geology and Geochemistry. W.H. Freeman press, p. 743.

IHS Markit, 2019, https://ihsmarkit.com.

Inhabitat, 2010, https://inhabitat.com/worlds-first-solar-powered-blimp-set-to-cross-the-english-channel/.

Jenden, P.D., Kaplan, I.R., 1989. Analysis of Gases in the Earth's Crust. Global Geochemistry Corp (Chapter 9), 44pp.

Jenden, P.D., Newell, K.D., Kaplan, L.R., Watney, W.L., 1988. Composition and stable-isotope geochemistry of natural gases from Kansas. Chem. Geol. 71, 117–147.

Kipfer, R., Aeschbach-Hertig, W., Peeters, F., Stute, M., 2002. Noble gases in lakes and ground waters. Rev. Mineral. Geochem. 47 (1), 615–700.

Kotarba, M.J., Sumino, H., Nagao, K., 2020. Origin of hydrocarbon and noble gases, carbon dioxide and molecular nitrogen in the Miocene strata of the eastern part of the Polish Carpathian Foredeep: isotopic and geologic approach. Appl. Geochem. 122, 21p.

Kulcinski, G.L., 1989. Importance of Helium-3 for the Future. NASA, pp. 150–171. Nas_techdoc_19000006527.

Kulongoski, J.T., Hilton, D.R., 2011. Chapter 15: applications of groundwater helium. In: Baskaran, M. (Ed.), Handbook of Environmental Isotope Geochemistry, Advances in Isotope Geochemistry. Springer-Verlag Berlin Heidelberg, pp. 285–304.

Kulongoski, J.T., Hilton, D.R., Izbicki, J.A., 2005. Source and movement of helium in the eastern Morongo groundwater basin: the influence of regional tectonics on crustal and mantle helium fluxes. Geochem. Cosmochim. Acta 69 (15), 3857–3872.

Kurz, M.D., Jenkins, W.J., 1981. The distribution of helium in oceanic basalt glasses. Earth Planet Sci. Lett. 53, 41–54.

Kurz, M.D., Warren, J.M., Curtice, J., 2009. Mantle deformation and noble gases: helium and neon in oceanic mylonites. Chem. Geol. 266, 10–18.

Lewars, E.G., 2008. Helium Compounds, Chapter 5, Modeling Marvels. Springer, Dordrecht, pp. 67–80.

Lide, D.R., 2005. CRC Handbook of Chemistry and Physics, 86th ed. CRC Press, Boca Raton (FL), p. 2544.

Lollar, B.S., Frappe, S.K., Fritz, P., Macko, S.A., Welhan, J.A., Blomvist, R., Lahermo, P.W., 1993. Evidence for bacterially generated hydrocarbon gas in Canadian Shield and Fennoscandian rocks. Geochem. Cosmochim. Acta 57, 5073–5085.

Lowenstern, J.B., Bergfeld, D., Evans, W.C., Hunt, A.G., 2015. Origins of geothermal gases at Yellowstone. J. Volcanol. Geoth. Res. 302, 87–101.

Mamyrin, B.A., Tolstikhin, I.N., 1984. Helium Isotopes in Nature. Elsevier, p. 289.

Manning, V.,H., 1919. Petroleum Investigations and Production of Helium. U.S. Bureau of Mines, Bulletin, 178C.

Marty, B., Torgersen, T., Meynier, V., O'Nions, R.K., Marsily, G., 1993. Helium isotope fluxes and groundwater ages in the Dogger Aquifer. Paris Basin, Water Res. Res. 29 (4), 1025–1035.

Mazor, E., Bosch, A., 1987. Noble gases in formational fluids from deep sedimentary basins: a review. Appl. Chem. 2, 621–627.

MHS, 2020. https://www.renergen.co.za/sproule-evaluation-of-certain-helium-prospective-resources-on-the-tetra4-virgina-gas-project-july-2020/.

Moore, R.B., 1926. Commercial production of radium, Mesothorium, and Helium. Ind. Eng. Chem. 18 (2), 203.

National Research Council, 2010. Selling the Nation's Helium Reserve. National Academies Press, Washington D.C., p. 140

Nuttall, W.J., Clarke, R.C., Glowacki, D.G., 2012. The Future of Helium as a Natural Resource. Routledge Publishing, p. 376.

Oxburgh, E.R., O'Nions, R.K., Hill, R.I., 1986. Helium isotopes in sedimentary basins. Nature 324, 632–635.

Ozima, M., 1994. Noble gas state in the mantle. Rev. Geophys. 32 (4), 405–426.

Ozima, M., Podosek, F.A., 2001. Noble Gas Geochemistry. Cambridge University Press, p. 302.

O'Nions, R.K., Oxburgh, E.R., 1988. Helium, volatile fluxes and the development of continental crust. Earth Planet Sci. Lett. 90 (3), 331–347.

Pandey, G.N., Tek, M.R., Katz, D.L., 1974. Diffusion of fluids through porous media with implications in petroleum geology. AAPG 58 (2), 291–303.

Penrose, O., Onsager, L., 1956. Bose-Einstein condensation and liquid helium. Phys. Rev. 104 (3), 576.

Pettke, T., Frei, R., Kramers, J.D., Villa, I.M., 1997. Isotope systematics in vein gold from Brusson, Val d'Ayas (NW Italy) (U+ Th) He and K-Ar in native Au and its fluid inclusions. Chem. Geol. 135 (3–4), 173–187.

Pinceaux, J.P., Maury, J.P., Besson, J.M., 1979. Solidification of helium, at room temperature under high pressure. J. Phys., Lett. 40 (13), 307–308.

Pinti, D.L., B. Marty, B., 1998. The origin of helium in deep sedimentary aquifers and the problem of dating very old groundwaters. Geol. Soc. 144 (1), 53–68. London, Special Publications.

Pinti, D.L., Marty, B., 1995. Noble gases in crude oils from the Paris Basin, France: Implications for the origin of fluids and constraints on oil-water-gas interactions. Geochimica et Cosmochimica Acta (59), 3389–3404.

Pinti, D.L., Marty, B., 2000. Chapter 7: noble gases in oil and gas fields: origins and processes. Minera. Assoc. Canada Short Course 28, 160–169.

Poreda, R.J., Jeffrey, A.W.A., Kaplan, I.R., Craig, H., 1988. Magmatic helium in subduction-zone natural gases. Chem. Geol. 71 (1–3), 199–210.

Poreda, R.J., Hunt, A.G., Lyons, W.B., Welch, K.A., 2004. The helium isotopic chemistry of Lake Bonney, Taylor Valley, Antarctica: timing of late Holocene climate change in Antarctica. Aquat. Geochem. 10 (3–4), 353–371.

Potuck, M., 2018. Helium Temporarily Bricks iPhones and Apple Watches in Chicago Hospital, Android Devices Immune. https://9to5mac.com/2018/10/31/helium-bricks-iphone-apple-watch/.

Power, M.J., Hattori, K., Pinti, D.L., Potter, E.G., 2013. Anomalous Abundance of He and Mobile Metals in Surface Media over the Deeply Buried Millennium U Deposit, Athabasca Basin, Canada, Goldschmidt Conference Abstracts, p. 1993.

Protin, M., Blard, P.-H., Marrocchi, Y., Mathon, F., 2016. Irreversible adsorption of atmospheric helium on olivine: a lobster pot analogy. Geochem. Cosmochim. Acta 179, 76–88.

Raitt, W.J., Schunk, R.W., Banks, P.M., 1978. Helium ion outflow from terrestrial ionosphere. Planet. Space Sci. 26, 255–268.

Ray, M.C., Hilton, D.R., Muñoz, J., Fischer, T.P., Shaw, A.M., 2009. The effects of volatile recycling, degassing and crustal contamination on the helium and carbon geochemistry of hydrothermal fluids from the Southern Volcanic Zone of Chile. Chem. Geol. 266 (1), 38–49.

Rogers, S., 1921. Helium-bearing Natural Gas. USGS. Professional Paper 121, 113pp.

Royston, M.K., Wheeler, H.P., 1956. Helium, Minerals Yearbook. USGS, p. 444.

Rumble, J.R., 2020. Handbook of Chemistry and Physics. CRC Press, p. 1572.

Sano, Y., Takahata, N., Igarashi, G., Koizumi, N., Sturchio, N.C., 1998. Helium degassing related to the Kobe earthquake. Chem. Geol. 150 (1–2), 171–179.

Sano, Y., Takahata, N., Kagoshima, T., Shibata, T., Onoue, T., Zhao, D., 2016. Groundwater helium anomaly reflects strain change during the 2016 Kumamoto earthquake in Southwest Japan. Sci. Rep. 6, 37939. https://doi.org/10.1038/srep37939, 2016.

Scharlin, P., Battino, R., Silla, E., Tuñón, I., Pascual-Ahuir, J.L., 1998. Solubility of gases in water: correlation between solubility and the number of water molecules in the first solvation shell. Pure Appl. Chem. 70 (10), 1895–1904.

Schmitt, H.H., 2013. Lunar Helium-3 energy resources. In: Ambrose, W.A., Reilly, J.F., Peters, D.C. (Eds.), Energy Resources for Human Settlement in the Solar System and Earth's Future in Space, AAPG Memoir 101, pp. 33–51.

Schmitt, H.H., 2014. Synthesis of Current Data on Helium and Hydrogen Concentration in Lunar Regolith, AAPG, Search and Discovery #70162, p. 26.

Sciencealert, 2021, www.sciencealert.com/china-s-nuclear-fusion-machine-just-smashed-germany-s-hydrogen-plasma-record.

Sears, W.M., 2015. Helium the Disappearing Element. Springer International Publishing, p. 138.

Seibel, C.W., Kennedy, H.S., 1934. Helium. USGS Minerals Yearbook, pp. 757–764, 1934.

Slyuta, E.N., Abdrakhimov, A.M., Galimov, E.M., 2007. Lunar and Planetary Science XXXVIII (1338), 2175.

Spacenews, 2021. Google to Shut Down Loon. https://spacenews.com/google-to-shut-down-loon/.

Stone, J.A., Stejskal, A., 2004. Using helium as a standard of refractive index: correcting errors in a gas refractometer. Metrologia 41 (3), 189–197.

Stroncik, N.A., Niedermann, S., Haase, K.M., 2007. Neon and helium isotopes as tracers of mantle reservoirs and mantle dynamics. Earth Planet Sci. Lett. 258 (1), 334–344.

Stuart, F.M., Burnard, P.G., Taylor, R.E.A., Turner, G., 1995. Resolving mantle and crustal contributions to ancient hydrothermal fluids: He-Ar isotopes in fluid inclusions from Dae Hwa W-Mo mineralization, South Korea. Geochem. Cosmochim. Acta 59 (22), 4663–4673.

Stute, M., Sonntag, C., Déak, J., Schlosser, P., 1992. Helium in deep circulating groundwater in the Great Hungarian Plain: flow dynamics and crustal and mantle helium fluxes. Geochem. Cosmochim. Acta 56 (5), 2051–2067.

Tardani, D., Reich, M., Roulleau, E., Takahata, N., Sano, Y., Perez-Flores, P., Sanchez-Alfaro, P., Cembrano, J., Arancibia, G., 2016. Exploring the structural controls on helium, nitrogen and carbon isotope signatures in hydrothermal fluids along an intra-arc fault system. Geochem. Cosmochim. Acta 184, 193–211.

Tolstikhin, I.N., Verchovsky, A.B., Kamensky, I.L., Skiba, V.I., Gannibal, M.A., Vetrin, V.R., V Tarakanov, S., 2016. Amphibole: a major carrier of helium isotopes in crustal rocks. Chem. Geol. 444, 187–198.

Tomonaga, Y., Marzocchi, R., Pera, S., Pfeifer, H.R., Kipfer, R., Decrouy, L., Vennemann, T., 2017. Using noble-gas and stable-isotope data to determine groundwater origin and flow regimes: application to the Ceneri Base Tunnel (Switzerland). J. Hydrol. 545, 395–409.

Torgersen, T., 1989. Terrestrial helium degassing fluxes and the atmospheric helium budget: implications with respect to the degassing processes of continental crust. Chem. Geol. Isot. Geosci. 79 (1), 1–14.

Torgersen, T., Clarke, W.B., 1985. Helium accumulation in groundwater, I: an evaluation of sources and the continental flux of crustal ^4He in the Great Artesian Basin, Australia. Geochem. Cosmochim. Acta 49 (5), 1211–1218.

Torgersen, T., Ivey, G.N., 1985. Helium accumulation in groundwater, II: a model for the accumulation of the crustal ^4He degassing flux. Geochem. Cosmochim. Acta 49 (11), 2445–2452.

Urban, P., Schmoranzer, D., Hanzelka, P., Sreenivasan, R., Srbek, L., 2013. Anomalous heat and transport and condensation in convection of cryogenic helium. Proc. Natl. Acad. Sci. U.S.A. 10 (20), 8036–8039.

Wakita, H., Fuiji, N., Matsuo, S., Notsu, K., Nagao, K., Takoka, N., 1978. Helium spots caused by a diapiric magma from the upper mantle. Science 200 (4340), 430–431.

Wakita, H., Nakamura, Y., Kita, I., Fuiji, N., Notsu, K., 1980. Hydrogen release, new indicator of fault activity. Science 210 (4466), 188–189.

Wang, X., Liu, W., Li, X., Li, Q., Tao, C., Xu, Y., 2020. Radiogenic helium concentration and isotopic variations in crustal pools from Sichuan Basin, China. Appl. Geochem. 117, 9.

Weiss, R.F., 1971. Solubility of helium and neon in water and seawater. J. Chem. Eng. Data 16 (2), 235–241.

Wikipedia, 2021, https://en.wikipedia.org/wiki/Helium.

Williams, M., 2007. Mining the Moon: Lab Experiments Suggest that Future Fusion Reactors Could Use Helium-3 Gathered from the Moon. MIT Technology Review. August 23rd.

Winckler, G., Severinghaus, J.P., 2013. Noble gases in ice cores: indicators of the earth's climate history. In: The Noble Gases as Geochemical Tracers. Springer, pp. 33–53.

Winckler, G., Suess, E., Wallmann, K., de Lange, G.J., Westbrook, G.K., Bayer, R., R., 1997. Excess helium and argon of radiogenic origin in Mediterranean brine basins. Earth Planet Sci. Lett. 151 (3–4), 225–231.

Wittenberg, L.J., Santarius, J.F., Kulcinski, G.L., 1986. Lunar source of ^3He for commercial fusion power. Fusion Technol. 10, 167–178.

Woods, A.D.B., Cowley, R.A., 1973. Structure and excitations of liquid Helium. Rep. Prog. Phys. 36 (9), 1135–1231.

Yang, S., Ellis, A.M., 2012. Helium droplets: a chemistry perspective. Chem. Soc. Rev. 42 (2), 472–484.

Yang, J.H., Wu, F.Y., Wilde, S.A., 2003. A review of the geodynamic setting of largescale Late Mesozoic gold mineralization in the North China Craton: an association with lithospheric thinning. Ore Geol. Rev. 23 (3), 125–152.

Zeitler, P.K., Enkelmann, E., Thomas, J.B., Watson, E.B., Ancuta, L.D., Idleman, B.D., 2017. Solubility and trapping of helium in apatite. Geochem. Cosmochim. Acta 209, 1–8.

Zephyrsolutions, 2015, https://zephyrsolutions.com/power-your-house-with-solar-helium-balloons/.

Zhou, Z., Ballentine, C.J., Schoell, M., Stevens, S.H., 2012. Identifying and quantifying natural CO_2 sequestration processes over geological timescales: the Jackson Dome CO_2 Deposit, USA. Geochem. Cosmochim. Acta 86, 257–275.

CHAPTER

2

Nitrogen, carbon dioxide, argon, neon, krypton, and xenon

Abbreviations

^{3}He ^{3}Helium
^{4}He ^{4}Helium
Al Aluminum
Sb Antimony
Ar Argon
As Arsenic
BCF Billions of cubic feet
Bi Bismuth
CO_2 Carbon dioxide
C Centigrade
m^3 Cubic meters
EARS East Africa Rift System
F Fahrenheit
ft Feet
He Helium
H Hydrogen
K Kelvin
Km Kilometers
Kr Krypton
Li Lithium
Mps Megapascal Pressure Unit
m Meters
CH_4 Methane
CH_4^+ Methane plus other hydrocarbons
Mi Mile
MMCF Millions of cubic feet of gas
MMCFHe Millions of cubic feet of Helium
MY Millions of years
Ne Neon
N_2 Nitrogen
O Oxygen
ppb Parts per billion
ppm Parts per million
ppt Parts per trillion
P Phosphorus

Geology and Production of Helium and Associated Gases
https://doi.org/10.1016/B978-0-323-90988-4.00004-9

psi Pounds per square inch
Si Silica
MCF Thousands of cubic feet of gas
KA Thousands of years
H_2O Water
Xe Xenon

Science ... never solves a problem without creating ten more. ***George Bernard Shaw.***

2.1 Introduction

Nitrogen (N_2) and carbon dioxide (CO_2) are commonly found in petroleum systems and are critical in terms of everyday life. Nitrogen is typically considered a waste product of and a contaminant in natural gas production and can significantly impact the economics of any project. Carbon dioxide in most cases is considered similar to N_2 in terms of being a contaminant, but in some cases, it is used as secondary recovery element that enhances oil production. Argon (Ar), neon (Ne), krypton (Kr), and xenon (Xe) are noble gases that are in high demand for industrial uses but they are not commonly found associated with petroleum or nonpetroleum gas deposits except in trace amounts. The noble gases are used routinely in evaluating petroleum and groundwater systems to determine both age and source of N_2, CO_2, and helium (He) with respect to petroleum systems. It is rare when noble gases, other than He, and in some cases Ar, are analyzed for in natural gas samples whether from the production stream, wellhead, or as part of a drill stem test (DST) during drilling. Presented here are short summaries of N_2, CO_2, and the noble gases characteristics, uses, and applications in the petroleum industry.

2.2 Nitrogen

Nitrogen is a common element as both a solid, liquid or gas phase or as part of various compounds in a petroleum accumulation. Nitrogen can have a significant impact when released through combustion environmentally. Characteristics of N_2 as described by Greenwood and Earnshaw (1997) and Rumble (2020) are:

1. Nitrogen is like a noble gas in many characteristics but does not conduct electricity;
2. Nitrogen has the atomic number 7;
3. Nitrogen was discovered by Daniel Rutherford in 1772 and it is the seventh most abundant element in the universe;
4. The name Nitrogène was suggested by French chemist Jean-Antoine-Claude Chaptal when it was found that N_2 was present in nitric acid and nitrates. The name Azote was suggested by Antoine Lavoisier but not adopted to worldwide use;
5. Nitrogen is an asphyxiant, part of N_2 compounds such as hydrazine, azides, and azo compounds.

6. Nitrogen has a density 1.2506 g/L at 0°C, 1013 mbar (STP);
7. Nitrogen's melting point is −210.0°C or −326°F and boiling point is 195.795°C (−320.431°F);
8. Nitrogen has three isotopes of which ^{14}N (99.6% of the atmosphere) and ^{15}N (0.4% of the atmosphere) are stable. The third isotope, ^{13}N has a half-life of 9.965 min and is very rare;
9. Nitrogen is at of the head of group 15 on the periodic table but its chemistry is very different than its group companions: phosphorus (P), arsenic (As), antimony (Sb), and bismuth (Bi).
10. Nitrogen forms dinitrogen which is identified as N_2, a colorless, odorless gas that constitutes 78% of the Earth's atmosphere and the most abundant uncombined element.
11. Nitrogen has a negative electron affinity which requires energy to add an electron. Nitrogen has three *p*-orbitals each occupied with one electron that causes an incoming electron to experience significant repulsion from these electrons in these orbitals.

Nitrogen occurs in all organisms, primarily in amino acids, in the nucleic acids, DNA and RNA, and in the energy transfer molecule: adenosine triphosphate (Nelson and Cox, 2017). The human body contains approximately 3% N_2 by mass and is the fourth most abundant element in the body after O, C, and H. Nitrogen is critically important to all life (Nelson and Cox, 2017). Nitrogen moves from the atmosphere into the biosphere and associated organic compounds and then at some point recycled back into the atmosphere.

Diagonal relationships are defined as relationships between elements of period 2 and period 3 in the Periodic Table (Fig. 2.1). Each period 2 element from lithium (Li) to oxygen (O) shows some similarities to the respective period 3 element; the degree of similarities drops off abruptly past the boron (B) – silicon (Si) pair (Greenwood and Earnshaw, 1997). The similarities of N_2 to sulfur (S) are mostly limited to sulfur nitride ring compounds when both elements are the only ones that are present. Nitrogen shares with both its horizontal neighbors a preference for forming multiple bonds, typically with Carbon (C), O, or other N atoms, through $p_\pi - p_\pi$ interactions. Nitrogen occurs as diatomic molecules and therefore has lower melting (−210°C) and boiling points (−196°C) than the rest of period 2 and 3. This effect is caused due to N_2 molecules being only held together by weak van der Waals interactions, so that only a few electrons are available to create significant instantaneous dipoles (Greenwood and Earnshaw, 1997; Rumble, 2020). This configuration is not possible for its vertical neighbors; thus, the nitrogen oxides (N_2O), nitrites (NO_2^-), nitrates (NO_3^-), nitro- (CH_3NO_2), azo-(R−N=N-R′), and diazo- ($R_2C=N^+=N^-$) compounds, azides (anion with formula $N^-{}_3$), cyanates (OCN^-), thiocyanates (SCN^-), and imino- (>C=NH) derivatives form easily with P, As, Sb, or Bi. Nitrogen and P form an extensive series of compounds with one another which have chain, ring, and cage structures.

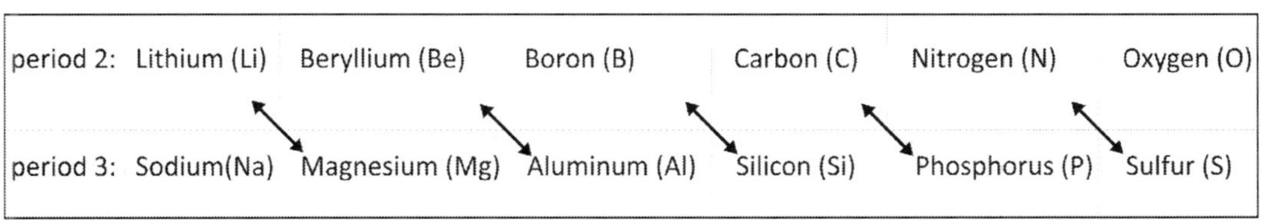

FIGURE 2.1 Diagonal relationships between elements of period 2 and period 3.

Carbon (C) and N tend to form ionic or metallic compounds with other metals. Nitrogen forms an extensive series of nitrides with C including those with chain structures. Nitrogen resembles O with its high electronegativity and concomitant capability for H bonding and the ability to form coordination complexes by donating its lone pairs of electrons (Greenwood and Earnshaw, 1997; Rumble, 2020). There are some parallels between the chemistry of NH_3 and H_2O. The capacity of both compounds to be pronated (add a proton to an atom, molecule, or ion) to give NH_4^+ and H_3O^+ or deprotonated (remove a proton from an atom, molecule, or ion) to give NH_2^- and OH^- with all of these able to be isolated in solid compounds.

2.3 Nitrogen uses

Nitrogen has multiple uses and it is a major component in modern day life. A short summary of it uses (Hatfield and Follett, 2008):

1. Synthetically produced N in the form of ammonia and nitrates are key industrial and private use fertilizers;
2. Breaking N bonds releases significant amount of energy that can be very useful;
3. Gunpowder and explosives—saltpeter (sodium nitrate or potassium nitrate);
4. Nitrogen is used in Kevlar for high-strength fabrics (bullet-proof vests) and in cyanoacrylate in superglue;
5. Nitrogen is an ingredient of every major pharmacological drug class, including antibiotics; In particular, N is an essential component of nucleic acids, amino acids, and thus proteins and the energy-carrying molecule, adenosine triphosphate, and is thus vital to all life on Earth.
6. Nitrogen gas is used as an industrial gas in a variety of applications where O is considered as risk for fire, explosion, or oxidizing hazard;
7. Nitrogen in a modified atmosphere, pure or mixed with CO_2, to nitrogenate and preserve the freshness of packaged or bulk foods (by delaying rancidity and other forms of oxidative damage);
8. Nitrogen is used in incandescent light bulbs as an inexpensive alternative to argon;
9. Nitrogen is used as a gas in inmate executions as an alternative to lethal injection;
10. Nireogen is used in fire suppression systems and for information technology (IT) systems;
11. Nitrogen is used in the manufacture of stainless steel;
12. Nitrogen is used in the case-hardening of steel by nitriding;
13. Nitrogen is used in some aircraft fuel systems to reduce fire hazard;
14. Nitrogen is used to inflate race car and aircraft tires, reducing the problems of inconsistent expansion and contraction caused by H_2O and O in natural air;
15. Nitrogen is commonly used during sample preparation in chemical analysis;
16. Nitrogen can be used as a replacement to, or in combination with, CO_2 to pressurize kegs of some beers, particularly stouts and British ales, due to the smaller bubbles it produces, which makes the dispensed beer smoother and headier;
17. Nitrogen tanks are also replacing CO_2 as the main power source for paintball guns.
18. Nitrogen must be kept at higher pressure than CO_2, making N_2 tanks heavier and more expensive;

19. The radioisotope ^{16}N is the primary radionuclide in the coolant of pressurized water reactors or boiling water reactors during normal operation. Nitrogen is a sensitive and immediate indicator of leaks from the primary coolant system to the secondary steam cycle, and is the primary means of detection for such leaks;
20. It is frequently used in nuclear magnetic resonance (NMR) spectroscopy to determine the structures of N-containing molecules, due to its fractional nuclear spin of one-half, which offers advantages for NMR such as narrower line width;
21. ^{15}N-enriched compounds have the advantage that under standard conditions they do not undergo chemical exchange of their N atoms with atmospheric N, unlike compounds with H, C, and O isotopes that must be kept away from the atmosphere;
22. ^{15}N:^{14}N ratio is commonly used in stable isotope analysis in the fields of geochemistry, hydrology, paleoclimatology, and paleoceanography, where it is called $\delta^{15}N$;
23. Nitrous oxide is used as an anesthetic for such procedures as heart surgery and dentistry.

2.4 Nitrogen toxicity

Nitrogen can be toxic to plants by overuse of fertilizers. Although very common, N_2 gas can be toxic to humans with the gas behaving as an asphyxiant that replaces oxygen. Nitrates and NH_4 can be deadly to humans, livestock, and other life and typically are delivered as toxins via drinking water, food, and air (Gupta et al., 2017). After ingestion, nitrates convert to nitrite by gastrointestinal microflora leading to methemoglobinemia, increasing free oxide radicals that predispose cells to having irreversible damage. Nitrates can cause cancer, increase infant mortality, abortions, birth defects, recurrent diarrhea, recurrent stomatitis, histopathological changes in cardiac muscles, alveoli of lungs and adrenal glands, and deterioration of immune system of the body (Gupta et al., 2017). Nitrates when inhaled can cause unconsciousness, vomiting, mental confusion, congestion and inflammation of respiratory tract, pulmonary edema, and genetic mutations thereby adversely affecting the development of fetus and decreasing fertility (Gupta et al., 2017). Ammonia toxicity occurs when the compound content in the blood supersedes the liver's capacity to eliminate it; this could be a result of either overproduction such as in congenital hyperammonemia or underelimination such as in liver cirrhosis (Padappayil and Borger, 2021).

2.5 Isotopes

The most common isotope of N is ^{14}N making up 99.634% of natural N present on Earth. ^{15}N, the second, slightly heavier than ^{14}N makes up the remaining 0.366% of N_2 present on the planet. Both of these stable isotopes are produced in the CNO cycle in stars. The relative abundance of ^{14}N and ^{15}N is constant in the atmosphere but can vary elsewhere, due to isotopic fractionation from biological redox reactions and the evaporation of NH_4 or HNO_3 acid. Laughery (2017) stated that N_2 dynamics are strongly controlled by assimilation, nitrification, and denitrification in the soil which typically result in ^{15}N enrichment of the substrate. N_2 stable isotope ($\delta^{15}N$) data cannot be used to quantify the contributions of different N_2 sources in subsurface fluids because of uncertainty in the isotopic range of various end members and the overlapping range of $\delta^{15}N$ for respective N_2 end members (Laughery, 2017). The combined use of $\delta^{15}N$ and noble gas systematics, however, effectively facilitates the interpretation of N_2 sources in petroleum reservoirs.

2.6 Sources of nitrogen

Natural gas in the Earth's crust typically contains 70%–100% CH_4, 1%–10% C_2H_6, lesser amounts of C_3H_8 through C_6H_{14}, and traces of higher hydrocarbons (Laughery, 2017). Nitrogen varies from traces to 100% and can affect the BTU value of any gas making it potentially uneconomic. Nitrogen has a variety of sources and is the most abundant nonhydrocarbon gas within most sedimentary basins. Ten percent of natural gases in the USA contain greater than 25% N_2 (Ballentine and Lollar, 2002). Rock N_2 concentrations range from trace levels (<200 mg N/kg) in granites to ecologically significant concentrations exceeding 1000 mg N/kg in some sedimentary and metasedimentary rocks (Holloway and Dahlgren, 2002). Fig. 2.2 is the N_2 cycle from a geological perspective (Holloway and Dahlgren, 2002). Fig. 2.3 is the relationship between mineral weathering and N_2 cycling. Nitrogen in natural gas reservoirs is generally thought to be derived from one or more of the following sources:

1. Volatilization of NH_4 and dinitrogen gas from the Earth's crust and mantle through hydrothermal activity and volcanism is thought to be the source of atmospheric N_2 (Holloway and Dahlgren, 2002). This volume of N_2 is 27%–30% of the N_2 on the planet (Bebout et al., 2015). The ultimate source of N_2 in the biosphere is outgassing from the mantle with N_2 incorporated into the biosphere by microbial N_2 fixation from the atmosphere (Holloway and Dahlgren, 2002);

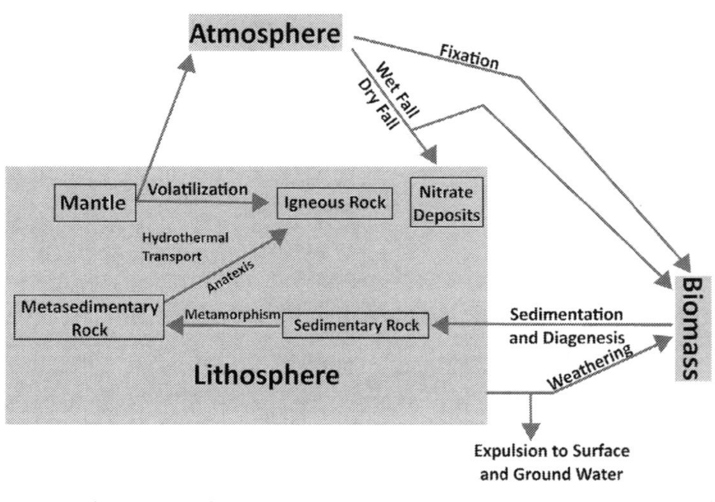

FIGURE 2.2 Nitrogen cycling in a geologic context (Holloway and Dahlgren, 2002). Major global pools of N_2 are shown in boxes with processes represented as arrows N_2. Larger arrows show processes that transfer nitrogen between global pools. This diagram represents a summary of processes that transfer N_2 between these pools, but is not inclusive of all processes (modified from Holloway and Dahlgren, 2002).

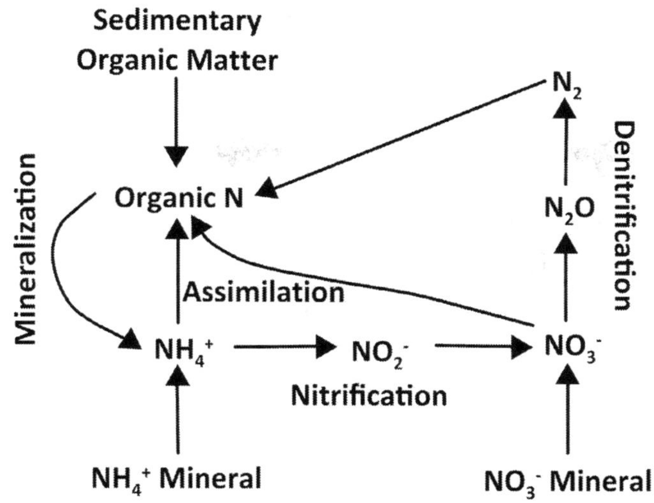

FIGURE 2.3 The relationship between nitrogen cycling and mineral weathering is represented as three forms of nitrogen in rock: (1) sedimentary organic matter, (2) ammonium minerals, and (3) nitrate minerals, entering soil microbial pathways for N_2 metabolism (modified from Holloway and Dahlgren, 2002).

2. Brown (2017), concluded that the major source of N_2 in natural, CO_2 and N_2 gas rich streams was from N_2 trapped in pore water that enters into the gas phase in the presence of CH_4.
3. Thermogenic degradation of sedimentary organic matter. An example of this is from Pironon and Grishina (1995) who found that NH ions were generated in early diagenesis from Bresse salt deposits, France, and the Gabon Cretaceous salt series of the Ogooue delta;
4. Oxidation of kerogen releases NH_4 in to pore waters;
5. Chemical or thermal release of N_2 from evaporites;
6. Nitrogen content of coals decreases during coalification. Nitrogen is formed from the thermal decomposition of N_2-containing organic compounds in the coal at temperatures above 600°C (Boudou and Espitalie, 1995) The denitrification process that occurs during coalification generally converts the N_2 in the coal to NH_4, which is effected into adjacent strata, groundwater, or brines (Ader et al., 1998);
7. High temperature release of N_2, NH_3, and NH_4 from various rocks;
8. Oxidation of clay minerals with NH_3 ions that are released then the NH_4 is oxidized in N_2 (Brown, 2017). An example of N_2 sourcing from clay minerals was reported in sediments in the Lena-Tunguska oil-bearing region of Siberian platform (Pironon and Grishina, 1995).
9. Release of mineral bound N_2 during metamorphism; and
10. Igneous or deep mantle sources (Ballentine and Sherwood Lollar, 2002).

Many N_2 accumulations can be related to mantle and crustal degassing and magmatic activity, but other accumulations are related to release or extraction of N_2 from clay minerals or interaction with pore water (Brown, 2017). Associated He is not necessarily related to similar processes but could be released from specific minerals, not necessarily the same that release N_2 or NH_4. The interpretation of the processes that form N_2 gas accumulations is based on isotopic interpretation of N_2 and noble gases, which is not always accurate. The concentration of N_2 in migrating gas is the balance between amounts of migrating hydrocarbons and N_2 in water exposed to the migrating gas (Brown, 2017). Typically, the huge amounts of thermogenic CH_4 swamp the N_2 contributed from water (Brown, 2017). Nitrogen source-rock volumes must be large to offset the low N_2 typically released from sediment. According to Brown (2017), even the largest N_2 accumulations can be explained by a relatively low volume of N_2 source rock.

Nitrogen's importance is not as a commercial commodity from natural gas production. Nitrogen's importance is in that the more it is present in a gas the more likely, a higher percentage of He is present. As a pathfinder tool for He or any other commodity N_2 is difficult to use. There are many sources of N_2, so its use in surface geochemistry for exploration has been nonexistent. Without extensive isotopic analysis, which is expensive, it would be impossible to separate biologic or near surface N_2 from subsurface generated N_2. The percentages of total N_2 to total percent of He present is also highly variable even within individual accumulations. It's presence in significant amounts in the subsurface brings hope to finding He accumulations. But hope always springs eternal in exploration.

The Panhandle-Hugoton Field (PHF), the largest gas field in the USA, contains significant N_2 but its He content has made it significantly economic and will be discussed in length in Chapter 3 along with several other N_2-He rich fields. Presented here will be a brief discussion of various deposits that contain partial or mainly all N_2. An example of a high N_2 reservoir that has been encountered but lacks He is the Robbins Gas Field, northern Sacramento Basin, California, USA, producing from the Magoon Sand member of the Forbes Formation, Upper Cretaceous in age from 6000 to 6800 ft (2000 to 2060 m) (Hector et al., 2013). The field is a stratigraphic trap composed of lenticular sands deposited in a deep-sea turbidite environment. The natural gas from the Robbins Gas Field is 86% N_2 and 14% CH_4, with a BTU of 140 and as of 2012 has produced almost 33 BCFG of which approximately 4.6 BCF was CH_4 (Hector et al., 2013). The Robbins Gas Field is part of a fairly large area, 23,000−43,000 acres, that is characterized by low BTU accumulations within the Forbes formation (Hector et al., 2013). The tectonic setting for the Sacromento Basin also would preclude the presence of anomalous He as the basement is young, active tectonically and is not underlain by Precambrian rocks, see Chapters 3, 4 and 6 for further discussion.

High-N_2 gas wells of greater than 10% N_2 occur throughout the Mid-Continent and in many cases the high N_2 content will cause them to be noncommercial (Ehinger, 2009). These isolated, high-N_2 wells occur at relatively shallow depths (<6000 ft or 1820 m), may be associated with red beds, structurally positive areas (arches), and/or deep, seated fault systems (Ehinger, 2009). Brown (2017) suggested that the high-N_2 gases common in the Mid-Continent and western USA could be due to N_2 generation from clays and organic matter that have undergone low thermal maturation. The relatively shallow upper Permian strata in Kansas could be a candidate for these processes. The primary stratigraphic unit holding this N_2 may be the widespread Cedar Hills Sandstone Leonardian Series, Nippewalla Group; and possibly stratigraphically nearby sandstones down to the underlying Runnymeade Sandstone (Leonardian Series, Sumner Group) (Newell, 2021). Nitrogen has proved to be a drilling hazard, as gas-charged Cedar Hills Sandstone was likely the source for what was probably the longest well blow-out (at least 2 weeks) in Kansas history (Lawrence Journal-World, 1984). The well was the Smith Cattle, Inc. #1 Helm (sec. 35-T.18S.-R.40W., at 2216 ft [675 m] depth), near the town of Tribune in Greeley County (Newell, 2021).

Another example of a N_2-rich reservoir is the Samson Oil and Gas Bluff 1-11 well drilled in the northern Denver-Julesburg Basin, southeast Wyoming, that flowed 7.8 MMCF at 2600 psi-ft from the Permian Hartville sandstone (Laughrey, 2019). The well lies on a four-way seismically defined structural closure with a seismic inversion anomaly at 7746 ft (2347 m) (Laughrey, 2019). The log neutron response through the reservoir was low suggesting no H, porosity of 29%, and a pressure transient test indicated approximately 500 md permeability (Laughrey, 2019). The gas composition was 97.58% N_2, 2.05% CH_4, and minor to trace amounts of higher hydrocarbons, CO_2, He, and Ar. This is an example of a seismically defined trap in a petroleum productive basin that unfortunately does not contain hydrocarbons. The cause of the N_2 accumulation is still unknown.

The Central European Basin and the North German Basin contains significant N_2 (90% of the gas) accumulations in the gas reservoirs especially in the Rotliegend reservoirs where thick, highly mature Paleozoic sedimentary sequences are present (Krooss et al., 2006; Plesen et al., 2007). Paleozoic sedimentary sequences are considered the most likely sources of N_2-rich natural gases of the Central European Basin and Northeast German Basin (Jurisch et al., 2007). The N_2 is sourced from sedimentary organic matter and NH_4-bearing minerals found in the Carboniferous sequences in the Northeast German Basin (Krooss et al., 2006; Jurisch et al., 2007; Pleseen et al., 2007). Part of the N_2 is fixed in the organic matter and the remaining is present as NH_4 in illites, indicating that illites may act as efficient N_2 storage sites during burial. Biological and thermal degradation by brine-induced K-NH_4 exchange reactions during diagenesis result in inorganic N compounds such as NH_4 which are substituted for K^+ in the interlayer sites of clay minerals. The process could be single or multistep, as shales vary in N_2 content, the thermal conditions in the basin vary laterally and vertically indicating varying amounts of N_2 release (Jurisch et al., 2007). More than 2000 ppm N_2 was detected in potential Carboniferous source rocks with an inorganic portion of up to 90%. The Upper Carboniferous Namurian shales from the Central European Basin from four deep wells at 14,520 to 23,100 ft (4400 to 7000 m) range between 520 ppm and 2680 ppm N_2 (Jurisch et al., 2007). Up to 90% of this N_2 occurs as NH_4 in minerals with ^{15}N values between +1‰ and +3.5‰. Nitrogen contents are as low as 460 ppm N_2. High ^{15}N values, up to +5.6‰, in one well in the basin center suggest a large-scale release of N_2 associated with isotopic fractionation. Thermal release of N_2 from pelagic Namurian A shales of NW and NE Germany occurred at significantly lower temperatures than from paralic Namurian B shales and terrestrial Westphalian rocks (Jurisch et al., 2007).

Brown (2017), concluded the setting and physical isolation of many high-N_2 gas accumulations in the USA Mid-Continent area indicate that N_2 associated with these accumulations was derived from thermally immature sedimentary rocks and not from basement or high-maturity sedimentary rock. The average N_2 concentration in older sedimentary rocks is approximately 0.1% (Brown, 2017). This conclusion would suggest that N_2 accumulations are commonly formed in the sedimentary section due to simple changes in chemistry and are devoid of He which has more of a deeper magmatic crustal origin. This conclusion further supports that N_2 is probably not a great guide in finding He deposits in some areas.

2.7 Carbon dioxide

Carbon dioxide (CO_2) is a critical compound in photosynthesis for plant growth and subsequently for the existence of animal life. Alternatively, it is an atmospheric poison that can inhibit or kill plant and animal life. Carbon dioxide's characteristics are (Karol, 2019; NETL, 2021):

1. Carbon dioxide is a colorless gas;
2. Carbon dioxide's density is 0.001977 g/cm^3 at STP compared to atmosphere 0.001225 g/cm^3;
3. Carbon dioxide consists of a single C atom covalently double bonded to two O atoms and occurs naturally in the Earth's atmosphere as a trace gas;
4. Carbon dioxide is the primary source of C for the cycle for life on Earth;
5. Carbon dioxide is regulated in the atmosphere and stratigraphic section by geologic process and photosynthesis since Precambrian times;
6. Carbon dioxide is produced by aerobic organisms and generally it is released into the environment through respiration;
7. Carbohydrates are photosynthesized using light, CO_2 and H_2O by plants, algae, and cyanobacteria. Anaerobic organisms decompose organic material producing CH_4 and CO_2 together with traces of other compounds. Regardless of the type of organic material, the production of gases follows a well-defined kinetic pattern.
8. Carbon dioxide is produced from anthropogenic activities of burning coal, natural gas, and wood; fermentation of beer, wine, and bread; a variety of chemical processes to make various chemicals.
9. Carbon dioxide produced from volcanoes has been at times affected global life in a negative way more than any of human industrial activity could do;
10. Carbon dioxide in the atmosphere provides protection for ultraviolet radiation bombarding the Earth which is what causes the greenhouse effect;
11. Through geological history CO_2 content has varied within the Earth's atmosphere and has been significantly higher in times in the past than today - at times reaching concentrations of close to 20%.

Uses of CO_2 (Karol, 2019; Aresta et al., 2020; NETL, 2021):

1. Carbon dioxide is used to enhance oil recovery;
2. Carbon dioxide is used in refrigeration as dry ice;
3. Carbon dioxide is used in the carbonation of beverages;
4. Carbon dioxide is a noncombustible gas used in fire extinguishers;
5. Carbon dioxide is used in the shredding of old tires;
6. Carbon dioxide is used in stripping of insulation from scrap wire;
7. Carbon dioxide is used in the removal of corn kernels from the cob during food processing;
8. Carbon dioxide is used to remove of hair from hogs in slaughterhouses;
9. Carbon dioxide is used in cooling of metal cutting tools, as it is cheaper than using He;
10. Carbon dioxide is used in stripping of paints;
11. The noncombustible atmosphere that CO_2 creates can be introduced into grain silos to prevent grain dust explosions;

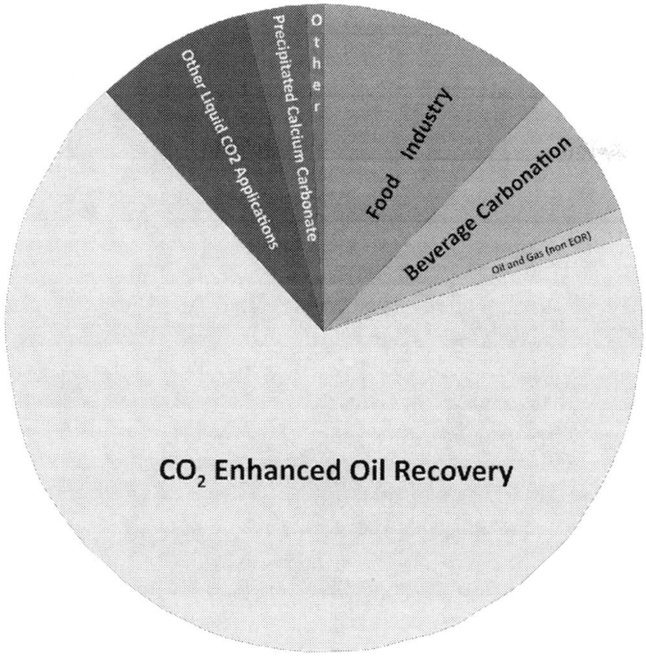

FIGURE 2.4 Approximate proportion of current CO_2 demand by end users (modified from NETL, 2021).

12. Carbon dioxide provides an inert, noncombustible atmosphere for welding;
13. Carbon dioxide is a nontoxic, aerosol propellant;
14. Carbon dioxide is used in the branding of livestock;
15. Carbon dioxide is used in the stimulation of plant growth in greenhouses, marijuana and hemp growers use this process to significantly enhance crop yield;
16. Carbon dioxide is an excellent solvent when in the supercritical state;
17. Carbon dioxide is a byproduct of the industrial production of H by steam reforming and the water gas shift reaction in ammonia production.

Fig. 2.4 is a graphic representation of the use of CO_2 industrially. The predominate use is in the petroleum industry (NETL, 2021).

2.8 Toxicity

Carbon dioxide is an asphyxiant gas and not classified as toxic or harmful in accordance with Globally Harmonized System of Classification and Labeling of Chemicals standards of United Nations Economic Commission for Europe by using the OECD Guidelines for the Testing of Chemicals (Wikipedia, 2021). In concentrations up to 1% (10,000 ppm), it will make some people feel drowsy and give the lungs a stuffy feeling. Concentrations of 7%–10% (70,000 to 100,000 ppm) may cause suffocation, even in the presence of sufficient O, manifesting as dizziness, headache, visual and hearing dysfunction, and unconsciousness within a few minutes to an hour (Wikipedia, 2021). The physiological effects of acute CO_2 exposure are grouped together under the term *hypercapnia*, a subset of asphyxiation.

Because it is heavier than air, in locations where there is subsurface volcanic or geothermal activity resulting in surface gas seeps in relatively high concentrations, without the dispersing effects of wind, it can collect in depressions below average ground level, causing animals located therein to be suffocated (Wikipedia, 2021). Carrion feeders attracted to the carcasses are then also killed. Children have been killed in the same way near the city of Goma, Democratic Republic of the Congo, by CO_2 emissions from the nearby volcano, Mt. Nyiragongo. The Swahili term for this phenomenon is 'mazuku' (Wikipedia, 2021).

2.9 Sources of carbon dioxide

Carbon dioxide is typically less than 1% in natural gas. The CO_2 in natural gases has been ascribed to many origins and the CO_2 in any one reservoir system may be derived from multiple sources, although a single source is usually dominant (MacDonald, 1983). Natural sources of CO_2 include volcanoes, hot springs, and geysers, and weathering. Also during chemical alteration of carbonate rocks it is released by dissolution in water and acids (MacDonald, 1983; Karol, 2019). Carbon dioxide is soluble in H_2O and found in groundwater, rivers and lakes, polar ice, glaciers, and seawater. Carbon dioxide content of the atmosphere has increased by industrial activity such as oil and gas production, mining, power generation, building of renewable facilities, transportation, to name a few. Carbon dioxide is critical to plant life and if the level in the atmosphere falls below a certain level, 180 ppm, it will cause most plants to suffocate. Carbon dioxide is vented from mantle and crustal rocks as part of the process of magma movement, heating of fluids, dissolution of carbonate rocks and at the surface vented from volcanoes and ground seeps such as hot springs (Karol, 2019). Carbon dioxide from these sources is typically unpredictable and can be extremely damaging to life locally if not globally based on the extent of the release of the gas, i.e., the eruption of Krakatoa in Indonesia in 1883. Carbon dioxide is found in natural gas and is rarely directly associated with He.

The source of CO_2 can be multiple or a single source depending upon the deposits. Chen et al. (2020) analyzed for CO_2 from 26 wells and 11 springs along active fault zones near Beijing, China, to determine if there were temporal variations. The CO_2 was found to be a combination of biogenic from oxidation of organic matter, mixtures of groundwater and carbonates at depth, and crustal and mantle-derived CO_2. Along the fault zones concentrations of CO_2 and He were found to vary through time and were related to earthquakes. Chen et al. (2020) work indicates that the process of generating CO_2 can be due to multiple processes and may not always reflect earthquake activity.

Sources of CO_2 in petroleum reservoirs include sedimentary organic matter and organic acids, microbial degradation or oxidation of hydrocarbons, thermochemical sulfate reduction (TSR), inorganic dissolution of carbonates, and magma (Laughery, 2017). Variations in the stable carbon isotope composition (^{13}C) of CO_2 and hydrocarbons might be diagnostic of gas origins, but the $^{13}CO_2$ of high- CO_2 reservoirs often fall within the overlapping range of carbonate breakdown and magmatic degassing, therefore making it difficult to distinguish between these sources (Laughery, 2017). The high solubility of CO_2 in water, and its high reactivity, complicates the interpretation of CO_2 sources because the extent of gas phase interaction with formation water is a critical parameter in determining CO_2 sinks (Laughery, 2017). The integration of noble gas systematics can be used to constrain CO_2 sources and interaction with formation H_2O (Laughery, 2017).

The following sources of CO_2 are currently thought to be the most common, found in natural gases (Wycherly et al., 1999; Whiticar, 1994; Hunt, 1996; Broadhead et al., 2009):

1. Degassing. Carbon dioxide degassing of the mantle and magmatic rocks is very common and is presumed to be of primordial origin. Mantle-derived CO_2 transports to the crust in the dissolved gas phase as part of rising magmas. Deep-seated faults augment migration and in CO_2 producing fields such as Big Piney-La Barge, Oakdale, Madden or Sheep Mountain is thought to be the main transport mechanism (see Chapter 3 for further discussion of these fields). As magma pressure decreases as it rises upward through the crust, the CO_2 is exsolved as a separate phase. The composition of volcanic gases is primarily water vapor and CO_2 gas (MacDonald, 1972; Baubron et al., 1990; Giggenbach, 1996; Giggenbach et al., 1991). Reservoirs with high percentages of CO_2 such as St. Johns, Des Moines, Big Piney-La Barge, Bravo Dome, Farnham Dome, Gordon Creek, Val Verde Basin fields, Jackson Dome, Doe Canyon, Sheep Mountain, and Oakdale to name a few are rare occurrences and seem to be related or near-magmatic, volcanic, and tectonic activity (see Chapter 3 for further discussions). At the Sheep Mountain and Oakdale fields, Huerfano Co., Colorado, magmas intrude into the sedimentary section forming an intrusive sill, dike, or laccolith that acts as a reservoir for the CO_2. Once in the reservoirs, it may be redissolved into the water system and migrate in water solution (Broadhead et al., 2009);

An example of mantle degassing is from Gilifillan et al. (2008) who analyzed 57 samples for noble gases concentrations from five fields, Bravo Dome, Sheep Mountain, and McCallum Dome (see Chapter 3 for field discussions) are located on the Colorado Plateau and the St. John's Dome, and McElmo Dome from within the uplift area. All fields have greater than 82% CO_2. Gilifillan et al. (2008) determined that the measured $CO_2/^3He$ ratios were consistent with a predominantly magmatic source. The oldest volcanic rocks in the area are associated with the McElmo Dome are from 70 MY to 42 MY ago, at the youngest with the Brazos Dome Field dated from 8 KA to 10 KA. The extent of contact of the CO_2 phase within the groundwater system is a critical parameter in understanding these systems. Gilifillan et al. (2008) indicated that coherent fractionation of groundwater $^{20}Ne/^{36}Ar$ with crustal radiogenic noble gases (^4He, ^{21}Ne, ^{40}Ar) is explained with a two-stage redissolution model:

Stage 1: Magmatic CO_2 injection into the groundwater system strips dissolved atmosphere derived noble gases and accumulated crustal/radiogenic noble gas by CO_2/H_2O phase partitioning. The CO_2 in the groundwater stripped gases that provides the first fluid and gas charge in the reservoir (Gilifillan et al., 2008). Subsequent charges of CO_2 did not provide additional atmosphere or crustal noble gases, and thus serve only to dilute the original atmosphere and crustal noble gas rich CO_2. Reservoir scale preservation of concentration gradients in atmosphere-derived noble gases indicate the direction of the CO_2 filling as found in the Bravo Dome and St. John's Dome fields (Gilifillan et al., 2008).

Stage 2: The noble gases redissolve into any available gas stripped groundwater (Gilifillan et al., 2008 This is modeled as a Rayleigh distillation process quantifies (1) the volume of groundwater originally "stripped" on reservoir filling; and (2) the volume of groundwater involved in subsequent interaction (Gilifillan et al., 2008). The original H_2O volume that is gas stripped varies from as low as 0.0005 cm^3 groundwater/cm^3

gas (STP) in one Bravo Dome sample, to 2.56 cm^3 groundwater/cm^3 gas (STP) in a St. John's Dome sample. Subsequent gas/groundwater equilibration varies within all fields. But the fields show a similar range, from zero to ~100 cm^3 water/cm^3 gas at reservoir pressure and temperature (Gilifillan et al., 2008).

2. Metamorphism. Regional metamorphism of basement rocks releases CO_2 gas and volumes are greatest when the these strata that contain large amounts of $CaCO_3$;
3. Magma Intrusion. Early investigators thought magmas intruding into the sedimentary strata rich in carbonate rocks and as a product of contact metamorphism at sufficiently high temperatures decompose and form Ca and CO_2 as products. Carbon dioxide can be generated by magmas or igneous rocks that intrude into a sedimentary section with coal seams, such as Spanish Peak, Colorado CO_2. Localized intrusive dikes as found in the Cherokee Basin in Kansas, central U.S.A. that produced CO_2 from thermal maturation of humic organic matter. Pironon and Grishina (1995) found that early diagenesis resulted in $CaCO_3$ dissolution in Lena-Tunguska oil bearing region in Siberia generating CO_2 under the influence of intrusive basalts. More recent studies do not favor this overall mechanism as the main source for many CO_2 accumulations (Broadhead et al., 2009). The generation of CO_2 from this process is probably very local and minimal in volume;

   Giggenbach et al. (1991) interpreted He (R_A = 3.0 to 5.6) and C ($\delta^{13}C$ from −7.2‰ to −3.4‰) isotopic compositions, and relative CO_2, CH_4, N_2, He, and Ar contents of CO_2-rich gases from Lakes Nyos and Monoun, Cameroon; Laacher See, Germany; Dieng Volcanic Plateau, Indonesia, and a well at Mt. Gambier, Australia, point to a common, essentially magmatic origin. The occurrence of these high-CO_2, low-temperature systems is likely to be favoured in tectonically active regions where mantle gases rise with sufficiently low regional heat flows. Many of these types of systems are generally not related to petroleum basins;
4. Groundwater. When groundwaters that are undersaturated with $CaCO_3$ move through a limestone, dissolution of the rock occurs, leaving behind porosity and additional CO_2 dissolved in the water. Limestone caves are the result of this process;
5. Thermal Degradation. Thermal degradation of organic material through the burial, compression and heating of organic rich sediment realeases CO_2. Carbon dioxide is formed relatively early in the maturation process along with CH_4 and humic nonmarine organic matter, such as coal, produce more CO_2 than sapropelic marine organic matter (Broadhead et al., 2009). Pironon and Grishina (1995) found the early generation of CO_2 from the Bresse salt deposits, France, by decarboxylation of biopolymers and catagenesis of kerogen in the Gabon Cretaceous salt series of the Ogooue delta. Maximum CO_2 production is attained prior to peak CH_4 production and both CH_4 and CO_2 need to be present for this process to be a major producer of CO_2 production in a sedimentary basin.

   The Yinggehai Basin, a Cenozoic rift that rests on Paleozoic and Mesozoic basement strata has numerous hydrocarbon accumulations that contain both inorganic and organic CO_2 (Huang et al., 2004). The Yinggehai Basin is filled with predominantly Eocene-Oligocene rift-stage sediments composed of fluvial, lake, and coal bearing facies with a thickness of 16,500–23,100 feet (5000–7000 m). The primary source of the hydrocarbons is from the carbonaceous mudstones of the Miocene age Meishan and Sanya formations that matured into the gas window 6.5–2.0 MY ago filling the reservoirs with mainly CH_4 (Huang et al., 2004). The only tectonic activity in the basin is a belt of shale diapirs that help develop CO_2 rich gas reservoirs. From approximately 4.5 MY to 0.01 MY ago volcanic activity in the Yinggehai Basin caused a high geothermal anomaly. Rapid subsidence occurred during this time with sediments reaching a maximum burial depth of greater than 21,300 ft (6500 m) (Huang et al., 2004). The lower Miocene calcareous shales under high pressure and temperature decomposed to form very large volumes of CO_2 gas. The geothermal anomaly and rapid subsidence caused by deformation of the shales forming diapir structures and associated faults which the CO_2 migrated (Huang et al., 2004). The CO_2 rich gases, along with hot fluids, moved upward through faults into the CH_4 reservoirs, destroying the hydrocarbons, to form CO_2 rich gas pools connected with the diapir structures.

6. Microbial degradation. Microbial degradation or oxidation of hydrocarbons as in an oil reservoir can result in CO_2 being produced when sufficient O is present. This biodegradation occurs with microbes were hydrocarbons are generally near the surface in proximity to groundwater;
7. Chemical alteration of clay minerals, calcites, and anhydrites. Anhydrite can provide sulfate anions that react with CH_4 to form CO_2 and H_2S (Hill, 1990). Carbon dioxide can be derived from calcite cement that were deposited within the pores rock and than partially dissolved by later fluids (Broadhead et al., 2009).

2.10 Other noble gases

2.10.1 Argon (Ar)

Argon is a noble gas with the symbol Ar, an atomic number 18 and the third-most abundant gas in the Earth's atmosphere at 0.934% (Crockett et al., 1973; Greenwood and Earnshaw, 1997; Rumble, 2020). The name "argon" is derived from the Greek word ἀργόν, neuter singular form of ἀργός meaning "lazy" or "inactive" as a reference to the fact that the element undergoes almost no chemical reactions. The complete eight electrons in the outer atomic shell makes Ar stable and resistant to bonding with other elements. Argon is more than twice as abundant as H_2O vapor, 23 times as abundant as CO_2, and more than 500 times as abundant as Ne (Crockett et al., 1973; Greenwood and Earnshaw, 1997). In the Earth's crust Ar is the most abundant noble gas at 0.00015% (Crockett et al., 1973). Argon in the Earth's atmosphere is radiogenic ^{40}Ar derived from the decay of ^{40}K in the Earth's crust but ^{36}Ar is more common in the universe where it is produced from nucleosynthesis in supernovas (Crockett et al., 1973). Argon is extracted industrially by the fractional distillation of liquid air.

Uses of Ar:

1. An Ar atmosphere is used in graphite electric furnaces to prevent the graphite from burning;
2. Argon is used in incandescent, fluorescent lighting, and other gas-discharge tubes;
3. Argon makes a distinctive blue–green colored gas laser;
4. Argon is used in fluorescent glow starters;
5. Liquid Ar is used as the target for neutrino experiments and direct dark matter research

6. Argon is used to displace O- and moisture-containing air in packaging material to extend the shelf-lives of the contents;
7. Argon is used in winemaking in a variety of activities to provide a barrier against O at the liquid surface, which can spoil wine by fueling both microbial metabolism (as with acetic acid bacteria) and standard redox chemistry;
8. Argon is sometimes used as the propellant in aerosol cans;
9. Argon is used as an inert shielding gas in welding and other high-temperature industrial processes where ordinarily unreactive substances become reactive;
10. Argon can be used as a preservative for such products as varnish, polyurethane, and paint, by displacing air to prepare a container for storage;
11. Argon may be used as the inert gas within Schlenk lines and gloveboxes;
12. Cryosurgery procedures such as cryoablation use liquid Ar to destroy tissue such as cancer cells;
13. Argon is used in a procedure called "argon-enhanced coagulation" a form of Ar plasma beam electrosurgery. The procedure carries a risk of producing gas embolism and has resulted in the death of at least one patient;
14. Blue Ar lasers are used in surgery to weld arteries, destroy tumors, and correct eye defects.
15. Argon has also been used experimentally to replace N_2 in the breathing or decompression mix known as *Argox*, to speed the elimination of dissolved N_2 from the blood. In scuba diving dissolved N_2 in the blood is released as the diver surfaces. If the decompression is done too rapidly, the N_2 will form bubbles in the blood, causing the bends. The bends is probably the most deadly aspect of scuba diving with compressed air;
16. Argon is used for thermal insulation in energy-efficient windows;
17. Argon is also used in technical scuba diving to inflate a dry suit because it is inert and has low thermal conductivity;
18. Argon is used as a propellant in the development of the Variable Specific Impulse Magnetoplasma Rocket (VASIMR). Compressed Ar gas is allowed to expand, to cool the seeker heads of some versions of the AIM-9 Sidewinder missile and other missiles that use cooled thermal seeker heads. The Ar gas is stored at high pressure;
19. ^{39}Ar, which has a half-life of 269 years and has been used for a number of applications, primarily ice core and ground water dating. Also, K-Ar dating and related Ar-Ar dating are used to age date sedimentary, metamorphic, and igneous rocks;
20. Argon has been used by athletes as a doping agent to simulate hypoxic conditions. In 2014, the World Anti-Doping Agency (WADA) added Ar and Xe to the list of prohibited substances and methods, although at this time there is no reliable test for the abuse.

The main isotopes of argon found on Earth are ^{40}Ar (99.6%), ^{36}Ar (0.34%), and ^{38}Ar (0.06%). Naturally occurring ^{40}K, with a half-life of 1.25×10^9 years, decays to stable ^{40}Ar (11.2%) by electron capture or positron emission, and also to stable ^{40}Ca (88.8%) by beta decay (Ozima and Podesek, 2002). These properties and ratios are used to determine the age of rocks by K-Ar dating. In the Earth's atmosphere, ^{39}Ar is made by cosmic ray activity, primarily by neutron capture of ^{40}Ar followed by two-neutron emission. In the subsurface environment, it is also produced through neutron capture by ^{39}K, followed by proton emission. ^{37}Ar is created from the neutron capture by ^{40}Ca followed by an alpha particle emission as a result of subsurface nuclear explosions (Ozima and Podesek, 2002). It has a half-life of 35 days.

2.10.2 Neon (Ne)

Neon is a noble gas element with the symbol Ne. Neon has the following characteristics (Crockett et al., 1973; Greenwood and Earnshaw, 1997; Rumble, 2020):

1. Neon has the atomic number 10;
2. Neon has the atomic weight of 20.179;
3. Neon is the fifth abundance in the universe but rare on Earth;
4. Neon is the second lightest noble gas, colorless, odorless, and chemically inert (Sherr, 2016);
5. Neon is an inert monatomic gas under standard conditions, a density of 0.90002 g/L at STP, a melting point of −248.59°C (−415.46°F), and a boiling point of −246.046°C (−410.883°F);
6. Neon is present in the atmosphere about 18.2 ppm by volume and a significantly smaller fraction in Earth's crust;
7. Neon's relative scarcity on Earth is due to its high volatility and it forms no compounds;
8. Neon has the narrowest liquid range of any element: from 24.55 to 27.05 K (−248.45 to −245.95°C or −415.21 to −410.71°F). It has over 40 times the refrigerating capacity (per unit volume) of liquid He and three times that of liquid H (Hammond, 2000);
9. Neon has three stable isotopes with abundances indicated in percentage: ^{20}Ne (90.48%), ^{21}Ne (0.27%), and ^{22}Ne (9.25%).

Neon was discovered in 1898 in conjunction with Kr and Xe as one of the three remaining residual rare inert elements in dry air, after N_2, O, Ar, and CO_2 (Crockett et al., 1973; Greenwood and Earnshaw, 1997; Rumble, 2020). In the universe the production of ^{20}Ne is primarily due to carbon-burning during stellar nucleosynthesis ($^{12}C + {}^{12}C\ {}^{20}Ne + {}^{4}He$) and was trapped by the Earth during accretion. Neon's most abundant isotope ^{20}Ne (90.48%) is created by the nuclear fusion of C in the C-burning process of stellar nucleosynthesis. This requires temperatures above 500 MK, which occur in the cores of stars of more than eight solar masses (Ryan and Norton, 2010). This accounts for the high abundance of ^{20}Ne in our atmosphere.

Uses of neon:

1. Neon has a distinct reddish–orange glow when used in low-voltage Ne, high-voltage discharge tubes and neon advertising signs. The red emission line from neon also causes the well-known red light of helium-neon lasers;
2. Neon is used in some plasma tube and refrigerant applications but has few other commercial uses;

Since the only economic source of neon is from the atmosphere, it is extracted by the fractional distillation. This makes Ne up to 55 times more expensive than He per MCF and this limits it uses.

Neon has three stable isotopes, and their abundances are indicated here in percentage: ^{20}Ne (90.48%), ^{21}Ne (0.27%), and ^{22}Ne (9.25%).

2.10.3 Krypton (Kr)

Krypton (from Ancient Greek: κρυπτός, romanized: *kryptos* "the hidden one") is a chemical element with the symbol Kr and atomic number 36 (Crockett et al., 1973; Greenwood and

Earnshaw, 1997; Rumble, 2020). It is a colorless, odorless, tasteless noble gas that occurs in trace amounts in the atmosphere and is often used with other rare gases in fluorescent lamps. With rare exceptions, Kr is chemically inert. Krypton's concentration in the atmosphere is approximately 1 ppm and is extracted from liquid air by fractional distillation (Crockett et al., 1973; Greenwood and Earnshaw, 1997; Rumble, 2020). The amount of Kr in space is uncertain, but present measurements suggest an abundance.

Uses of Kr:

1. Krypton's multiple emission lines make ionized krypton which is used in photography as a white light source especially for when high speeds are needed;
2. Krypton is combined with mercury and glows with a bright greenish–blue light;
3. Krypton mixed with Ar makes energy efficient fluorescent lamps but also reduces the light output and raises the cost;
4. Krypton along with Xe is used to fill incandescent lamps to reduce filament evaporation—the ability to operate under higher temperatures;
5. Krypton has higher light power than Ne in the red spectral line region, and for this reason, red lasers for high-power laser light shows are often Kr lasers with mirrors that select the red spectral line for laser amplification and emission. Helium-Ne lasers cannot achieve the same multi-watt outputs as Kr lasers;
6. In experimental particle physics, liquid Kr is used to construct quasihomogeneous electromagnetic calorimeters;
7. Krypton is used in insulating window panes;
8. Krypton is used as part of a propellant for electric propulsion system.
9. The sealed spark gap assemblies in ignition exciters in some older jet engines contain a small amount of ^{85}Kr that allows consistent ionization levels and uniform operation;
10. ^{83}Krypton has application in magnetic resonance imaging (MRI) for imaging airways allow it to distinguish between hydrophobic and hydrophilic surfaces;
11. Krypton is mixed with Xe to avoid the unwanted effects of a high partial pressure of Xe gas only systems during computed tomography (CT) scans;
12. ^{81}Krytpon is a metastable isotope used in nuclear medicine for lung ventilation/perfusion scans, where it is inhaled and imaged with a gamma camera;
13. ^{85}Krypton in the atmosphere has been used to detect clandestine nuclear fuel reprocessing facilities in North Korea and Pakistan. Those facilities were detected in the early 2000s and were believed to be producing weapons-grade plutonium;
14. Krypton is used in isotope studies to determine the history of groundwater, and natural gases specifically ^{81}Kr. ^{81}Kr has been used for dating 50,000–800,000-year-old groundwater.

There are five stable isotopes of Kr in the Earth's atmosphere. There is a sixth isotope that is considered relatively stable, ^{78}Kr, with a half-life of 9.2×10^{21} years. There are 30 unstable isotopes and isomers that have been identified. Traces of ^{81}Kr, a cosmogenic nuclide produced by the cosmic ray irradiation of ^{80}Kr, occur in nature which has a half-life of 230,000 years. Krypton is highly volatile and does not stay in solution in near-surface water. ^{85}Kr is an inert radioactive noble gas with a half-life of 10.76 years that is produced from the fission of U and Pu as a product of nuclear bomb testing and is released during the reprocessing of fuel rods from nuclear reactors.

2.10.4 Xenon (Xe)

Xenon has the symbol Xe, atomic number 54, is a colorless, dense, odorless noble gas found in Earth's atmosphere in trace amounts (Crockett et al., 1973; Greenwood and Earnshaw, 1997; Rumble, 2020). Xenon can undergo a few chemical reactions such as the formation of xenon hexafluoroplatinate which was the first noble gas compound to be synthesized. Xenon was discovered in England by the Scottish chemist William Ramsay and English chemist Morris Travers in September 1898, shortly after the discovery of Kr and Ne as a residue left over from evaporating components of liquid air (Crockett et al., 1973; Greenwood and Earnshaw, 1997; Rumble, 2020). The name *xenon* for this gas is from the Greek word ξένον *xénon*, neuter singular form of ξένος *xénos*, meaning "foreign(er)," "strange(r)," or "guest." In 1902, Ramsay estimated that Xe was present in the Earth's atmosphere at a concentration of one part in 20 million.

Uses of Xe:

1. Xenon is used in flash and arc lamps;
2. Xenon can be used as a general anesthetic;
3. Xenon induces robust cardioprotection and neuroprotection through a variety of mechanisms;
4. Xenon, particularly hyperpolarized ^{129}Xe, is a useful contrast agent for magnetic resonance imaging (MRI). In the gas phase, it can image cavities in a porous sample, alveoli in lungs, or the flow of gases within the lungs;
5. Xenon is soluble both in water and in hydrophobic solvents, so it can be used to image various soft living tissues;
6. Inhaling a Xe/O mixture activates production of the transcription factor HIF-1-alpha, which may lead to increased production of erythropoietin. The latter hormone is known to increase red blood cell production and athletic performance and therefore is illegal in sports;
7. The first excimer laser design used Xe as the lasing medium, and the earliest laser designs used Xe flash lamps as pumps;
8. Xenon is used to search for hypothetical weakly interacting massive particles and as the propellant for ion thrusters in spacecraft;
9. In nuclear energy studies, Xe is used in bubble chambers, probes, and in other areas where a high molecular weight and inert chemistry is desirable. A byproduct of nuclear weapon testing is the release of radioactive ^{133}Xe and ^{135}Xe. These isotopes are monitored to ensure compliance with nuclear test ban treaties and to confirm nuclear tests by states such as North Korea;

Naturally occurring xenon consists of seven stable isotopes and two long-lived radioactive isotopes. More than 40 unstable xenon isotopes undergo radioactive decay, and the isotope ratios of xenon are an important tool for studying the early history of the Solar System. Radioactive ^{135}Xe is produced by beta decay from ^{135}Iodine as a product of nuclear fission; it is the most significant and unwanted neutron absorber in nuclear reactors.

2.11 Isotope ratios in earth systems

The following is presented as a short summary of isotopes that are used in evaluating the sources of various elements in terms of whether they are from the mantle, crust, or atmosphere. This summary is not designed to present in detail discussions but to present to the reader an overview of how isotopes are used in oil and gas exploration. Ozima and Podosek (2002) discussed how many isotopes have exceptions and biases that can potentially be resolved but must be addressed in the context of any study. Isotopes unto themselves are not a stand-alone exploration tool. They are more of a look back at how and why a specific deposit may have formed or evaluation of the potential make-up of the fluids and gases in a reservoir. Why many deposits form, even with isotopic analysis, is merely a matter of conjecture based on assumptions that seem to make sense. The database for isotopes for He, N_2, and CO_2 in relation to hydrocarbon, He, and CO_2 deposits is limited, as it is typically concentrated in some very specific areas. It can be argued that the number of samples that are available in the literature are insignificant and not representative of an area. This is typically related to the lack of access to many sampling sites and the cost of analysis. However, many of these papers are a good start in understanding the history of hydrocarbon, He, N_2, and CO_2 migration, formation, and accumulation.

A key to evaluation of isotopes is isotope fractionation. The isotope fractionation processes focus on the relative abundance of isotopes of the same element which allows interpretation of the origin of an element (mantle, crustal, sedimentary, and atmospheric) and its relationship to migration of fluids and gases in the subsurface. Isotopic fractionation is measured by isotope analysis, using isotope-ratio tool to understand geochemical as well as biological systems.

There are four types of isotope fractionation, of which the first two are normally most important (Crockett et al., 1973; Hoefs, 2004; Sharp, 2006; Holland and Turekian, 2011):

1. Equilibrium fractionation is the partial separation of isotopes between two or more substances in chemical equilibrium which is strongest at low temperatures;
2. Kinetic fractionation is an isotopic fractionation process that separates stable isotopes from each other by their mass during unidirectional processes;
3. Mass-independent fractionation (or non-mass-dependent fractionation) is not as common as the first two listed above. This type of fractionation occurs mainly in photochemical and spin-forbidden reactions;
4. Transient kinetic isotope fractionation occurs when the reaction leading to isotope fractionation does not follow pure first-order kinetics and therefore isotopic effects cannot be described with the classical equilibrium fractionation equations or steady-state kinetic fractionation equations (also known as the *Rayleigh equation*).

Nitrogen from the air defines the origin of isotopic scale at zero ‰ (Schimmelmann, 2019). Nitrogen stable isotope ratios $^{15}N/^{14}N$ are more difficult to measure than those of C and according to Schimmelmann (2019) (Fig. 2.5):

1. the heavier isotope ^{15}N is relatively less abundant than ^{13}C;
2. N_2 is more difficult to ionize in the ion source of a mass spectrometer than CO_2;
3. N_2 in general is much less abundant in organic matter and minerals than carbon;
4. It is sometimes difficult to separate different oxidation levels of N, organic versus nitrate versus NH_3, etc.

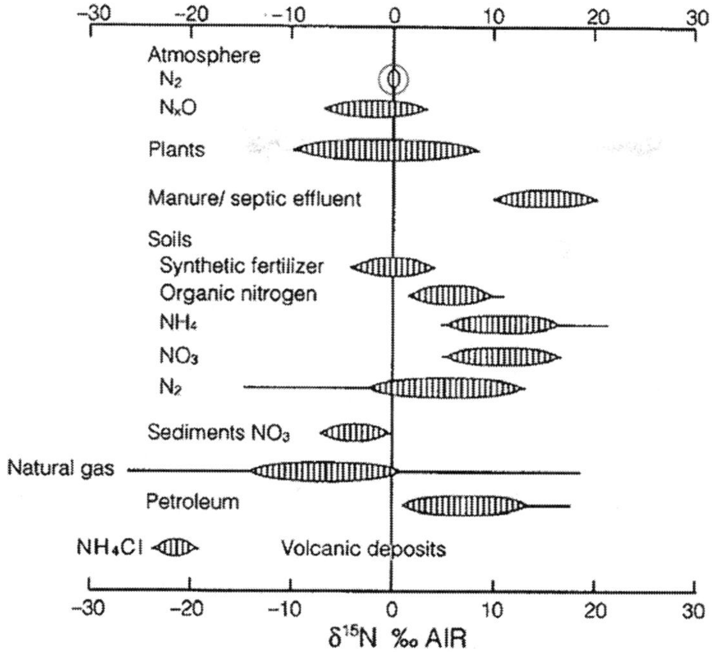

FIGURE 2.5 The range for ^{15}N in natural materials (Schimmelmann, 2019).

5. High-performance liquid chromatography and other separation methods have been used to prepare pure biochemical compounds to determine compound-specific $\delta^{15}N$ values (Schimmelmann, 2019).

The N isotope system is different than the C isotope system, where discrimination against ^{13}C occurs early during assimilation of inorganic C due to photosynthesis and chemolithotrophy. The biochemical fixation by bacteria in the soil and of N_2 in some N_2-deficient marine environments by bacteria and algae involves no significant isotope fractionation, resulting in newly fixed organic N_2 with $\delta^{15}N$ values of $\sim -3‰$ to $+1‰$ (Schimmelmann, 2019). Important and large fractionations of N_2 isotopes occur during conversions ("recycling") of N_2 compounds, mostly among different oxidation states (NH_4, NO_3^-, NO_2^-, organic N_2, loss to elemental N) that favor conversion/loss of ^{14}N. Nitrogen according to Schimmelmann (2019):

1. Biological nitrification (= oxidation of NH_3 to nitrite and nitrate, which can then be assimilated to form organic N_2) by nitrifying organisms fractionates kinetically by up to 20‰;
2. The direct uptake of dissolved NH_4 by biota to form organic nitrogen favors $^{14}NH_3$, depending on the concentration of NH_3, with a kinetic fractionation of up to 27‰;
3. Phytoplankton discriminates against ^{15}N in its uptake of DIN (DIN is short for dissolved inorganic N = Σ [NO_3^-, NO_2^-, NH_3, NH_4^+], but not counting elemental N_2) and synthesis of organic N_2 in oceans with high concentrations of DIN: as a result, the $\delta^{15}N$ value of residual DIN tends to increase as DIN is depleted by phytoplankton (Schimmelmann, 2019);

4. Biological denitrification, a reduction of NO_2^- to elemental N_2 that occurs in O-deficient environments. The kinetic fractionation depends on the reaction rate, with values up to 30‰ favoring the loss of ^{14}N to N_2 and a corresponding enrichment of any remaining NO_2^- in ^{15}N;
5. Volatilization of NH_3 out of an aqueous phase, such as soil pore water, favors the loss of $^{14}NH_3$, with a fractionation up to 35‰. This fractionation can be based on equilibrium conditions but may also have a kinetic component. The strong NH_3 odor in some types of wet manures is derived from NH_3 volatilization (Schimmelmann, 2019).

Diagenesis of particular organic nitrogen (PON) in soils and N reaching the sediment/water interface in lakes and oceans is lost during early diagenesis to "remineralization" by microbiota. The remaining organic N in refractory organic matter is often believed to reflect the isotopic composition of the original PON, and thus also to reflect the organic N of the original biomass. With increasing temperature or maturity of organic matter, organic N is lost from this material to form inorganic NH_3. Ammonia may become fixed into clay minerals where its N_2 isotopic composition would resemble that of the original organic N. However, NH_4 may also migrate together with formation waters and may at elevated temperatures, undergo chemical reactions with organic matter elsewhere. This would represent abiotic generation of new organic N.

Nitrogen isotopes are fractionated substantially during nitrification/denitrification by anthropogenic/bacterial processes. Increasing $\delta^{15}N$ values as a result of denitrification processes in recharge in the clay layer (enrichment).

Schimmelmann (2019) indicated that $^{15}N/N_2$ values indicate origin of N:

1. 19‰ to −19‰ maybe from immature and early mature sedimentary organic matter;
2. 10‰ to −2‰ may indicate origin from mature sedimentary organic matter;
3. 2‰ to +1‰ may imply an origin from the deep crust or mantle;
4. 0‰ and N_2/Ar ratios of 38−84 suggest an atmospheric origin;
5. + 1‰ to +4‰ characterize N_2 from NH_4 clay minerals during metamorphism;
6. +4‰ to +18‰ indicate origin from post-mature sedimentary organic matter.

Fig. 2.6 is the plotting of N_2/Ar ratios versus $^{15}N/_{N2}$ that originated from the West Siberian (Prasolov et al., 1982), Yinggehai Basin and California Great Valley (Jenden et al., 1988; Zhu et al., 2000) and Mid-European Basin (Schimmelmann, 2019). Fig. 2.5 clearly indicates different sources of N_2 for each area.

The example presented previously in Section 2.6, of the Samson Oil and Gas Bluff 1−11 well drilled in the northern Denver-Julesburg Basin, southeast Wyoming that flowed 7.8 MMCF of almost pure N_2 from the Permian Hartville was analyzed by Laughrey (2019) to determine the sources of N_2. Laughrey (2019) determined that N_2 analysis of ^{15}N of the produced N_2 is +20.4‰, an enriched value compatible with a crustal or a magmatic source. However, the $^3He/^4He$, $CH_4/^3He$, $^{20}Ne/^{22}Ne$, and $^{21}Ne/^{22}Ne$ ratios of the gas eliminate magmatic sources from consideration and indicate a crustal origin for the N_2 (Laughrey, 2019). The ^{15}N-enriched isotope composition of the gas, low $^4He/N_2$, and high $N_2/^{20}Ne$ indicate that the N_2 was generated by denitrification of post-mature organic matter from downdip Desmoinesian carbonaceous and marlstone source rocks. Combined $^4He/CH_4$ and $^{20}Ne/^{36}Ar$ systematics suggest diffusive gas loss from the reservoir under low gas volume/water volume conditions and significant solubility fractionation through H_2O-saturated crust (Laughrey, 2019). High $^{20}Ne/^{36}Ar$ and low $^{132}Xe/^{36}Ar$ ratios in the gas samples indicate gas-water

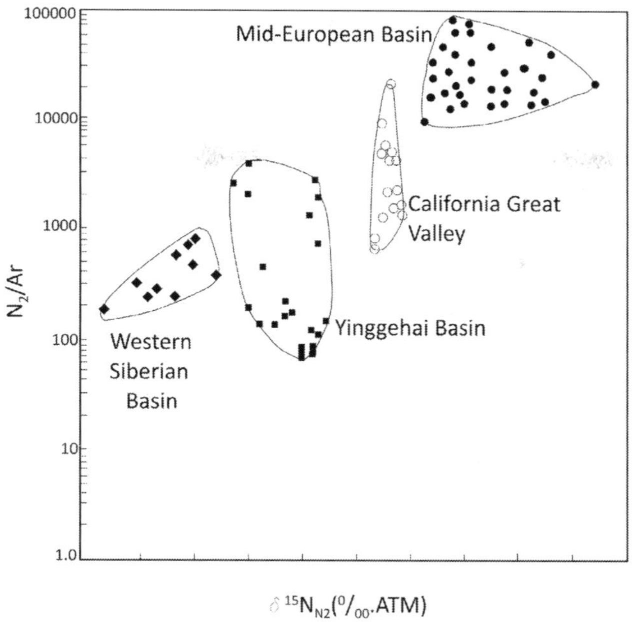

FIGURE 2.6 Semilogarithmic plot of N_2/Ar ratios versus $^{15}N/_{N2}$ of natural gases from the West Siberian (Prasolov et al., 1982), Yinggehai Basin and California Great Valley (modified from Jenden et al., 1988; Zhu et al., 2000), and Mid-European Basin (Schimmelmann, 2019, reprinted by permission of the AAPG whose permission is required for future use).

equilibrium in the Hartville sandstone reservoir; the less soluble Ne was preferentially partitioned in the gas phase compared to Ar, while more soluble Xe remained in the H_2O phase that indicated a gas/water contact downdip from the gas cap (Laughrey, 2019).

2.11.1 Neon

Origin of ^{21}Ne and ^{22}Ne are nucleogenic Ne observed in the volcanic gases (^{20}Ne) as well as derived from U-series decay chain (^{21}Ne and ^{22}Ne). Diamonds have higher amounts of ^{20}Ne. By analyzing all three isotopes, the cosmogenic component can be resolved from magmatic and nucleonic Ne. Lower ratio of $^{20}Ne/^{22}Ne$ and higher ratio of $^{21}Ne/^{22}Ne$ found in U rich rocks such as granite Ne isotopes will be useful tool in determining cosmic exposure ages of surficial rocks.

The isotopes of He and Ne define the nature of the mantle source regions and the degassing history of the planet. The main source of Ne isotope systematics of the oceanic mantle comes from arc- and back-arc basalts. Since the production of ^{20}Ne in the crust is negligible compared to ^{20}Ne sourced from the atmosphere, any ^{20}Ne detected in shallow reservoirs is determined to have been introduced into the system by groundwater (Ballentine et al., 1991; Ozima and Podosek, 2002; Ballentine and Burnard, 2002). Ratios of Ne isotopes, $^{20}Ne/^{22}Ne$, have a strong correlation in some cases with high $^{3}He/^{4}He$ ratios indicating a mantle source or a remnant of early Earth (Honda et al., 1991). In several cases $^{4}He/^{20}Ne$ ratios indicate that atmospheric ratios should be approximately 0.288. Various studies have found that with changes in pressure and temperature the ratio and calculated radiogenic production value of 1.0×10^7 (Kyser and Rison, 1982; Ozima and Podosek, 1983; Greenwood and Earnshaw, 1997) is inaccurate. The use of Ne isotopes therefore should be used with caution.

Ballentine and Lollar (2002), investigated the PHF using He/N_2 ratios to identify the transport mechanism from source to trap. The PHF contains an average of 15% N_2 and the Hugoton part of the field has a vary uniform CH_4 to C_3H_8 composition. Ballentine and Lollar (2002) collected 31 samples analyzed for noble gases and N_2 across a 210 mi (350 km) north-south traverse in the PHF. The PHF has and He/N_2 ratios ranging from ~0.025 in the Kansas Hugoton to ~0.05 in the Texas Panhandle. Ballentine and Lollar (2002) used $^3He/N_2$ and $^{20}Ne/N_2$ ratios to rule out a magmatic or air source for the N_2 in the PHF. ^4He is derived from crustal metamorphic rocks. In a plot of $^4He/N_2$ versus $^{20}Ne/^4He$ all data lie on a single mixing line between two N_2 end members, one that has neither ^4He nor ^{20}Ne and a N_2 component with a near constant $^4He/^{20}Ne/N_2$ value. The mixing between these endmember components is also reflected in the N_2 isotopes, which vary between +10‰ and +3‰ respectively. ^{20}Ne is an unambiguous tracer of groundwater involvement and shows that between 26% and 41% of the N_2 in the gas field has been transported in solution by the regional groundwater system. Ballentine and Lollar (2002) found no observed correlation between noble gases and the hydrocarbon phase. Fig. 2.7 is the plot the $\delta^{15}N(N_2)$ (0/00) across the field from south to north showing how the ratios change over distance implying how the migration history and chemistry for PHF is variable.

In contrast to earlier reports, they found those to be consistent with oil-associated gases produced from a moderately mature, marine source rock with a vitrinite reflectance of approximately 0.8%. The mantle $^3He/N_2$ and groundwater-recharge $^{36}Ar/N_2$ ratios enable us to rule out significant magmatic or atmosphere contributions to the gas field N_2, which is dominantly crustal in origin. We have used a combination of noble gas and stable isotopes

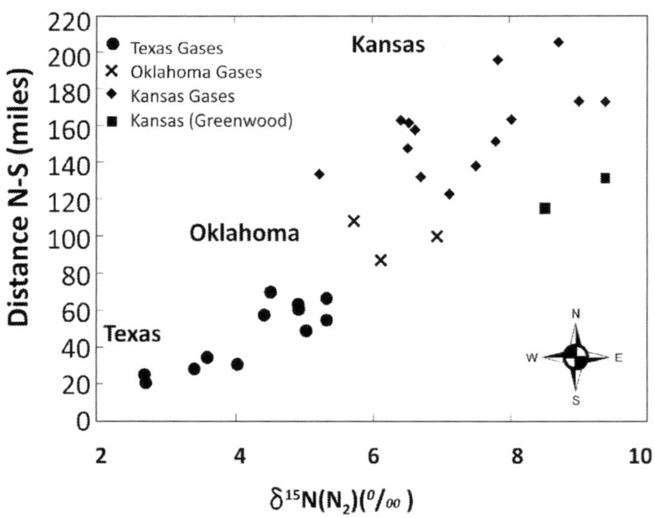

FIGURE 2.7 The change in δ15NN2 in a north-south traverse along the field. The "distance" is the perpendicular translation of the well location onto the north-south axis of the field with the reference point or "zero distance" being the southern gas water contact of the main field. Modified from Ballentine and Lollar, 2002.

to identify two crustal nitrogen components in the PHF. One N_2 component accounts for up to 60% of the N_2 in the Panhandle portion of the field, dropping to as little as 25% in the Kansas/Oklahoma section. This N_2 component is characterized by $^{15}NN_2 = -3‰$, $^4He/N_2 = 0.077$ and $N_2/^{20}Ne = 4.4 \times 10^5$. This "He-associated" N_2 is primarily from the devolatilization of low-grade metamorphic rocks, dissolved in the groundwater system and transported to the Texas Panhandle portion of the field. The hydrocarbons and related gases were already in place, and provide the mechanism with which to quantitatively partition the dissolved gas species into the reservoir gas phase. Magmatic ^3He is also identified within the reservoir in very high concentrations and from reasonably uniform $^3He/^4He$ is therefore linked with the ^4He source. A combination of magmatic ^3He-source requirement and ^4He mass balance limit the source of the ^4He-, ^3He-, and He associated N_2 to the great plains W–E flowing groundwater system to the west of the PHF. The second N_2 component accounts for between 40% of the N_2 in the Panhandle part of the field and as much as 75% of the N_2 in the Hugoton part of the field. This N_2 component is characterized by $^{15}N_{N2}$ 13+‰, but is associated with neither resolvable ^4He nor ^{20}Ne. The isotopic composition of this N_2 is consistent with denitrification of a relatively mature marine source rock. The location of the highest concentrations of this N_2, both isolated from the regional groundwater system and furthest away from the hydrocarbon producing areas, suggests that formation of this N_2 component predates the dominant hydrocarbon filling phase. Hydrocarbon filling in turn is prior to regional groundwater degassing of ^4He-, ^3He-, ^{20}Ne-, and He associated N_2.

2.11.2 Xenon

Xenon is a tracer for two extinct nuclides. Isotopic ratios in meteorites are powerful tools for studying the condensation of the Solar System. Excess amounts of $^{129}Xe/^{130}Xe$ found in CO_2 well gases from New Mexico is believed to be derived by decay of mantle-derived gases soon after the formation of the Earth or attributed to meteoritic contamination. Xe isotopic is useful to study the evolution of mantle reservoirs from Mid-Ocean Ridge Basalts (MORB) and different conditions of diamonds.

2.12 Summary

Carbon dioxide, N_2, and the other noble gases are part of petroleum systems and in the case of the first two, can create their own gas systems such as Harley Dome in Utah which is N_2 rich and Big Piney-La Barge Field, Wyoming which is rich in CO_2. Understanding these elements or compounds in relation to mainly petroleum systems can help lead to finding new He accumulations. Many of the examples in the literature from old wells or fields that contain CO_2 and N_2 in significant quantities generally are presented without any detailed analysis to determine how they may have been introduced to the reservoir or their source. While knowing the isotopic nature of noble gases in any deposit is of interest it does not generally lead to finding economic accumulations of He. Knowing the isotopic history of a petroleum system can lead to an understanding, assuming the interpretation is correct, of whether there is a possibility of He accumulations but not where they are or if they still exist in the trap. While N_2 is not related to He chemically, its presence in significant amount can be an indicator of increased He potential.

References

Ader, M., Boudou, J.P., Javoy, M., Goffe, B., Daniels, E., 1998. Isotope study on organic nitrogen of Westphalian anthracites from the Western Middle field of Pennsylvania (U.S.A.) and from the Bramsche Massif (Germany). Org. Geochem. 29, 315–323.

Aresta, M., Karimi, I., Kawi, S., 2020. An Economy Based on Carbon Dioxide and Water: Potential of Large Scale Carbon Dioxide Utilization. Springer Nature, p. 632.

Ballentine, C.J., Burnard, P.G., 2002. Production, release, and transport of noble gases in the continental crust. Rev. Mineral. Geochem. 47, 481–538.

Ballentine, C.J., Lollar, B.S., 2002. Regional groundwater focusing of N and noble gases into Hugoton-Panhandle giant gas field, USA. Geochem et Cosmo Acta 66 (14), 2483–2497.

Ballentine, C.J., O'Nions, R.K., Oxburgh, E.R., Horvath, F., Deak, J., 1991. Rare gas constraints on hydrocarbon accumulation, crustal degassing and groundwater flow in the Pannonian Basin. Earth Planet Sci. Lett. 105 (1–3), 229–246.

Baubron, J.C., Appard, P., Toutain, J.P., 1990. Diffuse volcanic emissions of carbon dioxide from Vulcano Island, Italy. Nature 344, 51–53.

Bebout, G.E., Lazzeri, K.E., Geiger, C.A., 2015. Pathways for nitrogen cycling in Earth's crust and upper mantle: a review and new results for microporous beryl and cordierite. Am. Mineral. 61.

Boudou, J., Espitalie, J., 1995. Molecular nitrogen from coal pyrolysis: kinetic modelling. Chem. Geol. 126, 319–333.

Broadhead, R.F., Mansell, M., Jones, G., 2009. Carbon Dioxide in New Mexico: Geologic Distribution of Natural Occurrences, vol. 514. NMBGMR, OFR, p. 131.

Brown, A.A., 2017. Possible Origins for Low Thermal Maturity, High-Nitrogen Natural Gases. American Association of Petroleum Geologists, Search and Discovery. Article 42065, Available online: http://www.searchanddiscovery.com/pdfz/documents/2017/42065brown/ndx_brown.pdf.html.

Chen, Z., Li, Y., Martinelli, G., Lu, C., Zhao, X., 2020. Spatial and temporal variations of CO_2 emissions from the active fault zones in the capital area of China. Appl. Geochem. 112, 10.

Crockett, A.H., Smith, K.C., Bartlett, N., 1973. The Chemistry of Monatomic Gases. Pergamon, p. 208.

Ehinger, R., 2009. High Nitrogen Gas Wells – a Mid-continent Mystery. AAPG, Search and Discovery. Article 90097, Abstract.

Giggenbach, W.F., 1996. Chemical composition of volcanic gases. In: Scarpa, R., Tilling, R.I. (Eds.), Monitoring and Mitigation of Volcanic Hazards. Springer- Verlag, Berlin, pp. 221–256.

Giggenbach, W.F., Sano, Y., Schminke, H.U., 1991. CO_2-rich gases from Lakes Nyos and Monoun, Cameroon, Laacher See, Germany: Dieng, Indonesia; and Mt. Gambier, Australia – variations on a common theme. J. Volcanol. Geoth. Res. 45 (3–4), 311–323.

Gilifillan, C., Ballentine, G., Holland, D., Lollar, B.S., Steven, S., Cassidy, M., 2008. The noble gas geochem of natural CO2 gas reservoirs from the Colorado Plateau, and Rocky Mountain Provinces, USA. Geochim. Cosmochim. Acta 72, 1174–1198.

Greenwood, N.N., Earnshaw, A., 1997. Chemistry of the Elements. Butterworth-Heinemann, p. 1364.

Gupta, S.K., Gupta, A.B., Gupta, R., 2017. Pathophysiology of nitrate toxicity in humans in view of the changing trends of the global nitrogen cycle with special reference to India. In: The Indian Nitrogen Assessment, Sources of Reactive Nitrogen, Environmental and Climate Effects, Management Options, and Policies, pp. 459–468.

Hammond, C.R., 2000. The Elements. In: Handbook of Chemistry and Physics, 81st ed. CRC Press, p. 19.

Hatfield, J.L., Follett, R.F., 2008. Nitrogen in the Environment. Academic Press, p. 720.

Hector, S., Conrad, G., Blake, K.E., April 19–25, 2013. Robbins gas field: a low BTU, nitrogen rich gas accumulation. In: 2013 Pacific Section AAPG, SPE and SEPM Joint Technical Conference, Monterey, California. AAPG, Search and Discovery. Article #90162.

Hill, C.A., 1990. Sulfuric Acid Speleogenenesis of Carlsbad Cavern and its Relationship to Hydrocarbons, Delaware Basin, New Mexico and Texas, vol. 74. AAPG, Bull v, pp. 1685–1694 (11).

Hoefs, J., 2004. Stable Isotope Geochemistry. Springer Verlag, pp. 1–244.

Holland, H.D., Turekian, K.K., 2011. Radioactivity Geochronometry: from the Treatise on Geochemistry. Academic Press, pp. 1–267.

Holloway, J.M., Dahlgren, R.A., 2002. Nitrogen in rock: occurrences and biochemical implications. Glob. Biochem. Cycl. 16 (14), 17p.

Honda, M., McDougall, I., Patterson, D.B., Doulgeris, A., Clague, D.A., 1991. Possible Solar Noble-gas Component in Hawaiian Basalts, vol. 349, pp. 149–151.

Huang, B., Xiao, X., Zhu, W., 2004. Geochemistry, Origin, and Accumulations of CO_2 in Natural Gases of the Yinggehai Basin, Offshore S. China Sea, vol. 88. AAPG, pp. 1277–1293 (9).

Hunt, J.M., 1996. Petroleum Geochemistry and Geology, second ed. W.H. Freeman and Company, San Francisco, p. 743.

Jenden, P.D., Kaplan, I.R., Poreda, R.J., Craig, H., 1988. Origin of nitrogen-rich natural gases in the California Great Valley—evidence from helium, carbon and nitrogen isotope ratios. Geochem. Cosmochim. Acta 52 (4), 851–861.

Jurisch, A., Kross, B.M., Littke, R., Plessen, B., Lüders, V., Hoth, P., 2007. Concepts and models for the formation of N_2-rich natural gas reservoirs. In: AAPG Hedberg Conference: Basin Modeling Perspectives: Innovative Developments and Novel Applications, May 6–9, Search and Discovery, p. 90066.

Karol, P.J., 2019. The Legacy of Carbon Dioxide: Past and Present Impacts. Taylor & Francis Group, CRC Press, p. 241.

Krooss, B.M., Jurisch, A., Plessen, B., 2006. Investigation of the fate of nitrogen in Palaeozoic Shales of Central European Basin. J. Geochem. Explor. 89, 191–194.

Kyser, T.K., Rison, W., 1982. Systematics of rare gases in basaltic lavas and ultramafic xenoliths. J. Geophys. Res. 87, 5611–5630.

Laughery, C.H., 2017. Origin of Non-hydrocarbon Gases in Petroleum Systems - a Review. AAPG, Search and Discovery. #90308.

Laughrey, C., May 19–22, 2019. Stable isotope and noble gas geochemistry of a high-nitrogen permian sandstone reservoir, Northern Denver-Julesburg (D-J) Basin, Southeastern Wyoming, U.S.A. In: 2019 AAPG Annual Convention and Exhibition, San Antonio, Texas. AAPG.

Lawrence Journal-World, 1984. Gas well blowout longest in Kansas, p. 18. June 20.

MacDonald, G.A., 1972. Volcanoes. Prentice-hall, Englewood Cliffs NJ, p. 510.

MacDonald, G.J., 1983. The many origins of natural gas. J. Petrol. Geol. 5 (4), 341–362.

National Energy Technology Laboratory (NETL), 2021. Commercial Carbon Dioxide Uses. www.netl.doe.gov/research/coal/energy-systems/gasification/gasifipedia/eor.

Nelson, D.L., Cox, M.M., 2017. Lehninger Principles of Biochemistry, vol. 1. W.H. Freeman and Company, p. 130.

Newell, K.D., 2021. Stratigraphic and Regional Trends in Chemistry and Quality of Natural Gas in Central and Western Kansas. Mid-Continent, USA, KGS (in press).

Ozima, M., Podosek, F., 1983. Noble Gas Geochemistry. Cambridge University Press, p. 367.

Ozima, M., Podosek, F., 2002. Noble gas geochemistry. Cambridge University Press, pp. 1–300.

Padappayil, R.P., Borger, J., 2021. Ammonia Toxicity. https://pubmed.ncbi.nlm.nih.gov/31536283/.

Pironon, J., Grishina, S., 1995. Hydrocarbon and non-hydrocarbon gas in salt environments, a contribution to gas genesis understanding. In: 1995 International Convention and Exposition Meeting, Nice, France. AAPG, Search and Discovery. Article #90956.

Plessen, B., Hoth, P., Lüders, V., 2007. Formation of nitrogen-rich gas accumulations in the Central European Basin. I. The significance of fixed ammonium in Palaeozoic Rocks. In: Annual Convention, Long Beach, California. AAPG Search and Discover. Article #90063.

Prasolov, E.M., Kamenskiy, I.L., Meshik, A.P., Subbotin, Y.S., Surovtseva, L.N., Yakovkev, O.N., 1982. Formation of gas fields of the North of West Siberia from isotope data. Petrol. Geol. 20 (7), 316–319.

Rumble, J.R., 2020. Handbook of Chemistry and Physics, 101st ed. CRC Press, p. 1572.

Ryan, S.G., Norton, A.J., 2010. Stellar Evolution and Nucleosynthesis. Cambridge University Press, p. 135.

Schimmelmann, A., September 7-11 2019. Isotopes of Unconventional Systems, Part I, the Society for Organic Petrology, Annual Meeting, Bloomington, Indiana, p. 56.

Sharp, Z., 2006. Principles of Stable Isotope Geochemistry. Prentice Hall, pp. 1–344.

Sherr, J., 2016. Neon. Saltire Books, Glasgow, Scotland, pp. 1–228.

Whiticar, M.J., 1994. Correlation of natural gases with their sources. In: Magoon, L.B., Dow, W.G. (Eds.), The Petroleum System from Source to Trap: American Association of Petroleum Geologists, Memoir, vol. 60, pp. 261–283.

Wikipedia, 2021. https://en.wikipedia.org/wiki/Carbon_dioxide.

Wycherly, H., Fleet, A., Shaw, H., 1999. Some observations on the origins of large volumes of carbon dioxide accumulations in sedimentary basins. Mar. Petrol. Geol. 16, 489–494.

Zhu, Y., Shi, B., Fang, C., 2000. The isotopic compositions of molecular: nitrogen implications on their origins in natural gas accumulations. Chem. Geol. 164 (3–4), 321–330.

CHAPTER

3

Geology of helium, carbon dioxide, and nitrogen case histories and sources

3.1 Introduction

Examples of helium (He), carbon dioxide (CO_2), and nitrogen (N_2) deposits are found throughout the literature. This chapter discusses the classification of He, CO_2, and N_2 deposits in general terms and presents summaries of many of these fields found in the literature. One of the great difficulties in exploring He and CO_2 is the lack of coherent sources that bring together all of the many divergent examples of these deposits and theories of their formation. The majority of He production is from just a small group of fields that display very unique characteristics or from processing facilities that extract the element from high-volume gas production. Carbon dioxide production is also very localized but, because of the high volumes and use in enhanced oil extraction, is commonly transported by pipeline to its destination. Helium's multiple uses and scarcity has caused the elements exploration to become a worldwide hunt for reserves. Helium's uniqueness of origin, habitat, and associations, having a single document to present and evaluate the many examples, makes the process easier. This chapter was written with this goal in mind.

Helium and associated CO_2 and N_2 are generated and released from underlying crustal or mantle rocks. Helium is generated from decaying U and Th and, because of its low density and small size, easily migrates outward from the planet unless it is temporarily trapped below very efficient seals such as salt, anhydrite, or low permeability shale. In compiling this book there are numerous histories for He and CO_2 reservoirs that are scattered through various publications, bulletins, company presentations, articles, and newsletters. This dispersion of information makes it difficult for the explorationist to efficiently evaluate the various types of reservoirs and habitat of accumulations that exist with He. Only the federal government in the United States, a few US States, and Canadian Provinces keep any sort of records related to He production. In the US federal government, the collection of data is due largely to the various Helium Acts enacted from World War I through the 1970s. In Alberta, Canada, the government has been keeping records due to a requirement of submitting gas analysis for

https://doi.org/10.1016/B978-0-323-90988-4.00007-4

all gas wells. In many cases, the analyses include He, N_2, and CO_2 data. A small number of companies may have He data because they purchase and sell He and CO_2 as a commodity but rarely release them publicly. For CO_2 there are even less fields being exploited and they tend to be super large accumulations and generally have very little public data to evaluate from. The compilation in this chapter is a combination of various published field histories with updated maps where possible.

3.2 Helium, nitrogen, and carbon dioxide field and reservoir classifications

Helium, N_2, and CO_2 reservoirs can be classified in terms of areal extent, quality of gas, geologic and tectonic setting, and reservoir and production characteristics. Danabalan (2017) laid out a classification scheme which is presented below and modified based on additional data. A He-rich gas field is classified as one with 0.3% by volume or greater of said noble gas (Danabalan, 2017; Halford, 2018). An economic He-producing field can be classified as anywhere from 0.1% or greater and whose financial success is directly related to price, total volume of helium in the reservoir, and cost of processing, royalties, taxes, lease operating, and disposing of any waste products. Carbon dioxide, CH_4+, and H_2S represent significant problems in processing and disposal when not extracted in economic quantities. Nitrogen can be vented to the atmosphere with little if any environmental issues.

Helium-rich fields can be classified according to their primary gas component as one of three main types:

3.2.1 Methane + rich He reservoirs

Helium productive reservoirs are associated with CH_4 plus other hydrocarbons. Hydrocarbons are generated through the heating and pressurization of sedimentary rocks with high organic content. Helium and other noble gases require higher temperatures and pressures and much longer time periods to produce significant quantities of gas. However, although the sources are different, the gasses are trapped in the same conventional traps, i.e., with a porous reservoir overlain by an efficient seal. Examples of such CH4+-rich fields' accumulations with economic He include the Hugoton-Panhandle (US), the North Dome-South Pars (Qatar and Iran), and Hassi R'Mel (Algeria) fields. The TPH is a classic example of a He-rich field where the element is found in high enough concentration to be economic on its own, i.e., it has at least 0.3% He by volume of gas in the reservoir. However, the Hassi R'Mel and North Dome–South Pars fields are also classified as He-rich although the He concentration in these fields is much less. However, in the latter fields, the production of the He is economic as part of the overall production stream. Indeed, the South Pars part of this combined field contains one of the largest helium reserves on Earth, dwarfing any currently producing in the United States. Estimates of the helium reserve in this field run to around 10.1 billion m^3—a quarter of the current world helium reserves. Despite this, under current economic considerations, the North Dome–South Pars field would not be considered as a viable He-rich field since this field only contains approximately 0.04% He. Hassi R'Mel also contains subeconomic concentrations of around 0.05% He. However, due to the size of the natural gas reserves in this reservoir, this diluted concentration still equates to a large overall volume of economically viable He production.

3.2.2 Nitrogen-rich He reservoirs

Nitrogen is always found in He-producing fields, but He is not always found with N_2 production. This paradox derives from the fact that N_2 is produced when He is produced but it is also produced in other ways as well. Thus, He is not found in such N_2-rich fields in Germany (Barnard and Cooper, 1983; Gerling et al., 1983; Danabalan, 2017; Halford, 2018). Nitrogen-rich fields can contain up to 10% He, but the amount of He in relation to N_2 is variable and not predictable. Usually the ratio of He/N_2 varies from 0.02 to 0.20 (2%–20%). In CH_4+-rich fields with He there is always some amount of N_2. The reason very few N_2-rich He accumulations have not been found outside the United States is probably the lack of exploration focusing on finding them. This lack of focus is because most countries outside the United States, Canada, and Australia do not support viable small and midsize independent companies that explore for medium-sized hydrocarbon accumulations. Danabalan (2017) indicated that radiogenic He associated with a N_2 component showed a light ^{15}N signature (−3.00‰). The noble gas and other stable isotopes indicated that low-temperature metamorphism of basement rocks both in situ and from the surrounding region sourced both the He and N_2 (Ballentine and Lollar, 2002).

3.2.3 Carbon dioxide-rich reservoirs

Carbon dioxide–rich reservoirs in rare cases can also be rich in N_2 and He such as in the Colorado Plateau (Gilfillan, 2006). In rare cases, gas fields such as Big Piney–La Barge and Doe Run gas fields, located on the Colorado Plateau, USA, are CO_2- and He-rich (Gilfillan, 2006; Merrill et al., 2014). As will be discussed below, the majority of CO_2-rich reservoirs usually originate from a purely magmatic provenance. These reservoirs can contain economic He not because they have high percentage of the element but by the sheer volume of the CO_2 produced with reserves in the hundreds of BCF to over a TCF in reserves. Helium is generally obtained from shallow crustal sources which are probably stripped from shallow crustal rocks during the CO_2 migration upwards toward the surface (Weinlich et al., 1999; Barry et al., 2013; Danabalan, 2017).

3.3 Texas Panhandle–Hugoton Field

The Panhandle–Hugoton Field (PHF) discovered in 1918 covers 8500 square mi (13,600 km), 5.5 million acres, and 10,500 wells, is the largest natural gas field in the United States (Fig. 3.1), with a potential recovery of 75 TCF of gas, and is located in Hamilton, Kearney, Finney, Stanton, Grant, Haswell, Morton Stevens, and Seward counties, Kansas, Texas County, Oklahoma, and Hansford, Sherman, Hartley, Moore, Hutchinson, Potter, Carson, Gray, Wheeler, and Collinsworth counties, Texas (Fig. 3.1) (Mason, 1968; Pippin, 1968;

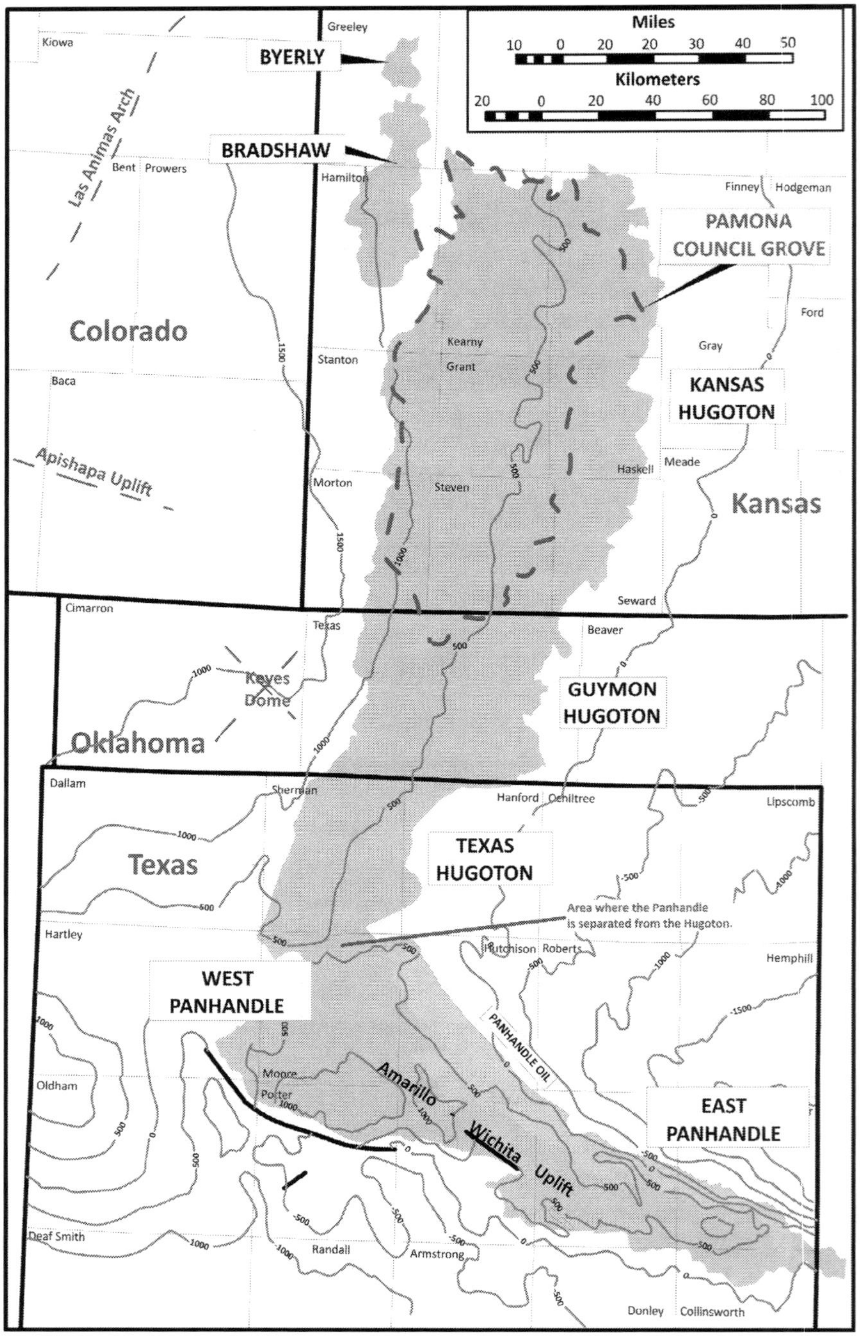

FIGURE 3.1 Location of the Texas Panhandle-Hugoton Field in Texas–Oklahoma–Kansas in the Mid-Continent USA with Wolfcamp Structure (Permian) from Sorenson (2003 and *reprinted by permission of the AAPG whose permission is required for future use*) and Pippin (1970).

Sorenson, 2005). The Panhandle portion of the PHF Field was found by mapping of a surface feature of the John Ray Dome in Potter County, Texas, in December 1918. To date, this part of the field has made over nine TCF of natural gas. The Hugoton part of the PHF was discovered in Seward County, Kansas, in 1922, but its significance was not realized until the 1930s (Pippin, 1968). Brown (2010) has estimated a recovery of 400 BCF of He from the PHF. The PHF is a unique geologic anomaly with a unique geologic history. The PHF produces from Permian Chase, Council Grove, and Admire carbonates that are sealed by interbedded anhydrites and shale (Fig. 3.2). The field at time of discovery contained 109 TCF (3.1 Tcm^3) with a relative uniform pressure of 450 psi (Brown, 2019).

The PHF is separated by a saddle in the north part of Moore County, Texas (Fig. 3.1). This structural separation does not have an effect on the He concentration. However, the He percentage of the PHF is, as indicated in Fig. 3.1, higher on the margins of the field. This variation represents the hydrodynamic nature of the trap as water on the margins of the field is highly saturated with He. The present-day setting of the PHF area and Mid-Continent USA is indicated in Fig. 3.3A. Fig. 3.3B indicates the nature of the hydrodynamic trap. Fig. 3.4 is a cross section through the field showing the structural/stratigraphic nature of the trap.

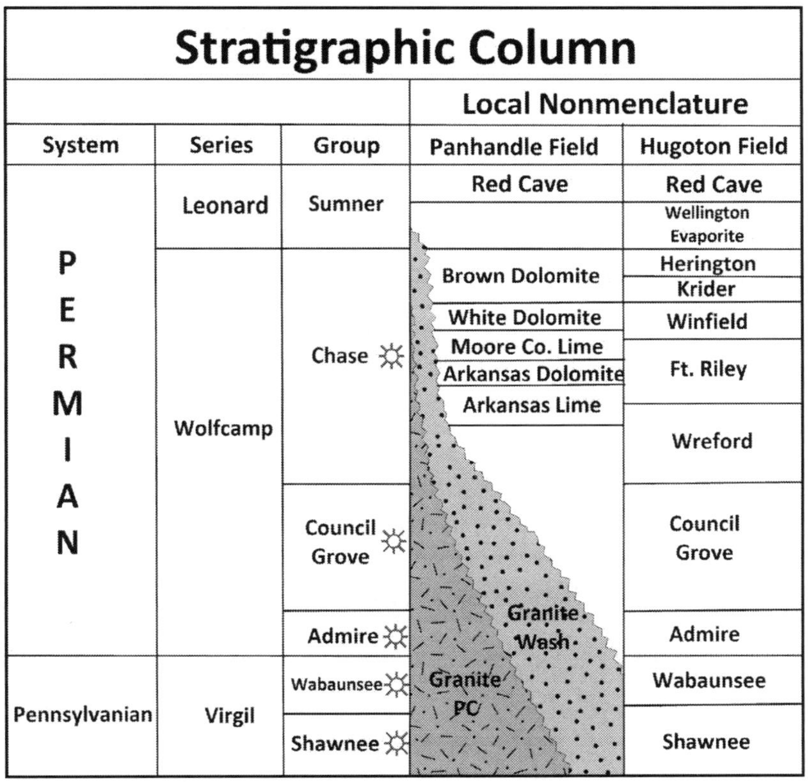

FIGURE 3.2 Stratigraphic column for the Texas Panhandle and Hugoton Fields (Pippin, 1970).

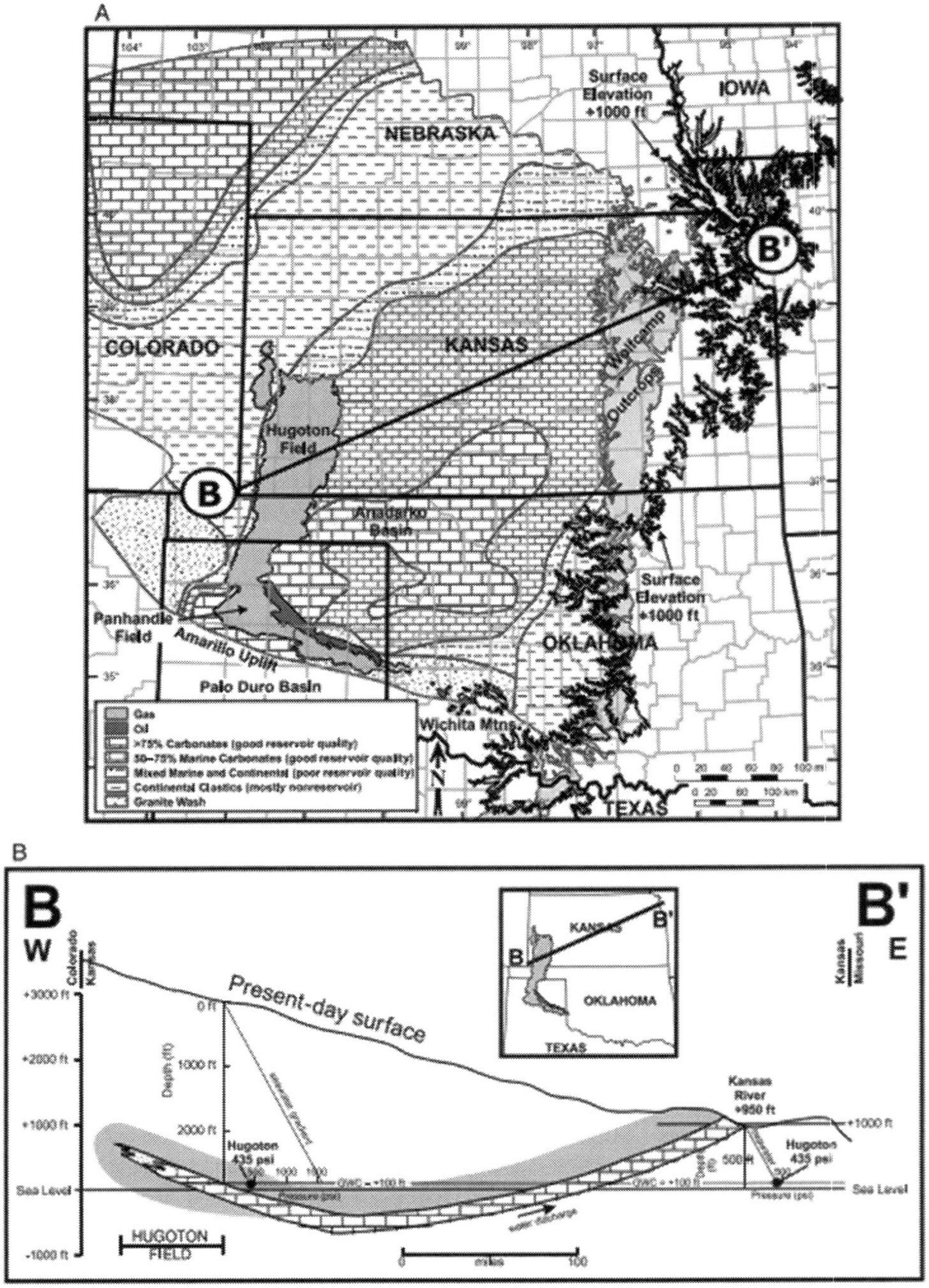

FIGURE 3.3 Sorenson (2005) indicates (A) mid-continent distribution of upper Wolfcampian carbonate facies (after Rascoe, 1968, 1988; Rascoe and Adler, 1983), relative to outcrops at surface elevations near or below +1000 ft (+300 m), primarily in northeast Kansas. Outcrops of Wolfcampian continental facies below +1000 ft (+300 m) are common in central Oklahoma. Surface elevations increase to the west. Upper Wolfcampian facies distribution from Rascoe (1968). (B) Simplified west-east cross section B to B′. Pressure–depth diagrams show the calculated hydrostatic pressure at sea level +100 ft (+30 m) for the Hugoton Field and the Wolfcampian outcrop area, assuming a seawater pressure gradient and vertical communication with the surface (Sorenson, 2005). The observed Hugoton discovery reservoir pressure of 435 psi (3 MPa) yields a hydraulic head approximating the reservoir outcrop elevation, instead of the field surface elevation, indicating possible aquifer communication between the gas field and Wolfcampian outcrops in northeast Kansas (Sorenson, 2005).

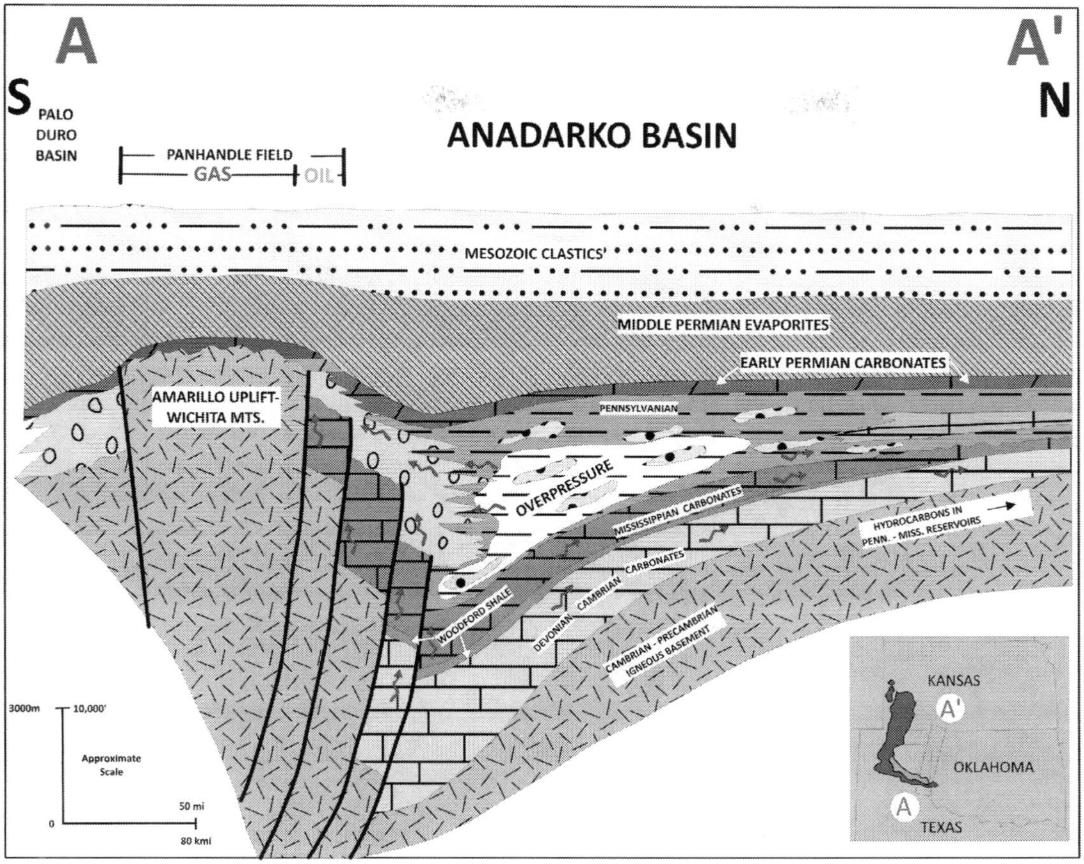

FIGURE 3.4 Cross section through the Panhandle–Hugoton Field after Pippin (1970), Johnson (1989), Dutton and Garnett (1989) and Sorenson (2005).

The formation of the PHF trap and reservoir began in Early Permian Wolfcampian time with deposition of carbonates across an extensive margin along the northern portion of the Permian Basin and into the western Anadarko Basin (Fig. 3.3A). The developing carbonate reservoirs were bounded by impermeable continental red beds and overlain by Middle Permian Leonardian evaporites that formed the top seal (Mason, 1968; Pippin, 1968; Sorenson, 2005; Brown, 2010). Early traps formed drape structures on the erosional topography of the Amarillo Uplift and Wichita Mountains. The Anadarko Basin underwent rapid burial during the Permian, expelling hydrocarbons and brines from the Late Devonian–Early Mississippian Woodford carbonaceous mudstones into the Amarillo Uplift area. By the end of the Permian, the reservoirs in the Amarillo Uplift area were filled with petroleum and normally pressured at 1000 to 1500 psi at 3000 to 4000 ft (910–1220 m). Throughout Mesozoic time, the burial of the Anadarko Basin continued until it reached its maximum depth in Late Cretaceous into Early Tertiary time (Mason, 1968; Pippin, 1968; Sorenson, 2005; Brown, 2010). The Permian strata were regionally tilted to the west. Paleozoic shales and carbonaceous mudstones within the center of the Anadarko Basin overmatured and began generating more gas that migrated into the PHF. The present-day axis of the field represented an eastward shift that aligned parallel to the Cretaceous Western Interior Seaway axis. At that time PHF had a normal pressure gradient of 1500 to 2500 psi at 4000 to 6000 ft (1220 to 1820 m). Fig. 3.4 is a cross section by Sorenson (2005) that layouts the present configuration of the PHF Field in association with the Wichita Uplift and Anadarko Basin.

The Laramide orogeny of the Rocky Mountains of Early Tertiary time caused regional uplift and eastward tilting in the mid-continent area (Mason, 1968; Pippin, 1968; Sorenson, 2005; Brown, 2010). The PHF maintained its internal pressure acting like a sealed container, not releasing any hydrocarbons. Internally though, the eastward tilt caused redistribution of oil and gas westward. In Later Tertiary time the margins of the Permian strata, specifically porous carbonates and sandstone, in Eastern Kansas eroded away causing depressurization of the field and allowing the shifting of hydrocarbons in the Texas Panhandle portion of the PHF northward into the Hugoton Field (Fig. 3.5) (Mason, 1968; Pippin, 1968; Sorenson, 2005; Brown, 2010). Because of the depressurization, dissolved gases in the groundwater were released, causing the composition of the gas to become enriched in N_2 and He, thus reducing the BTU content. Lower BTU contents are found along the northern and western margins of the field.

In the Quaternary, continental glaciation reached part of the Permian outcrops in northeastern Kansas. Melting of the glacier increased erosion (Fig. 3.6), which resulted in increased exposure of the Permian carbonates and further depressurizing the PHF Field to its present pressure of 435 psi at the time of discovery (Mason, 1968; Pippin, 1968; Sorenson, 2005; Brown, 2010). This erosion resulted in the displacement of part of the gas field with 500 billion barrels of aquifer water, filling the PHF to its current limit (Brown, 2010). Fig. 3.7 is the present-day He percentage in the PHF.

One of the concerns in the above model is the numerous structural features between the location of the PHF and the area where depressurization would occur. The Nemaha Ridge and Central Kansas Uplift to the east and the Las Animas Arch to the west would present significant barriers to migration of fluids and hydrocarbons and, therefore, is not adequately explained in the current model.

Brown (2010), 2017, 2019) indicated the PHF is one of the largest N_2-hydrocarbon accumulations in the world. Ballentine and Lollar (2002) identified three N_2 end members:

(1) Mantle N_2 forms a minor end member (<1.5% of the in-place N_2) not discussed further here.

(2) Light (−3‰ d15Natm) N_2 end member has $^4He/N_2$ of 0.077 and elevated ^{20}Ne. Ballentine and Lollar (2002) interpret this as a low-grade metamorphic source for He with transport to the field dissolved in groundwater.

(3) Relatively heavy (+13‰ d15Natm) N_2 end member has no ^{20}Ne or ^4He. Ballentine and Lollar (2002) interpret this as a relatively mature marine source rock in the Anadarko Basin, with charge predating filling by hydrocarbons.

Brown (2010) integrated the field charge history reported by Sorenson (2005) with additional geochemical data. Two high-N_2 and one low-N_2 end members with different He/N_2 ratios were identified. Both high-N_2 end members were exsolved from H_2O from beneath the field, which was originally charged with the low-N_2 gas end member. Water flowed from two sources, the Palo Duro Basin and an area west of Hugoton field—the Las Animas

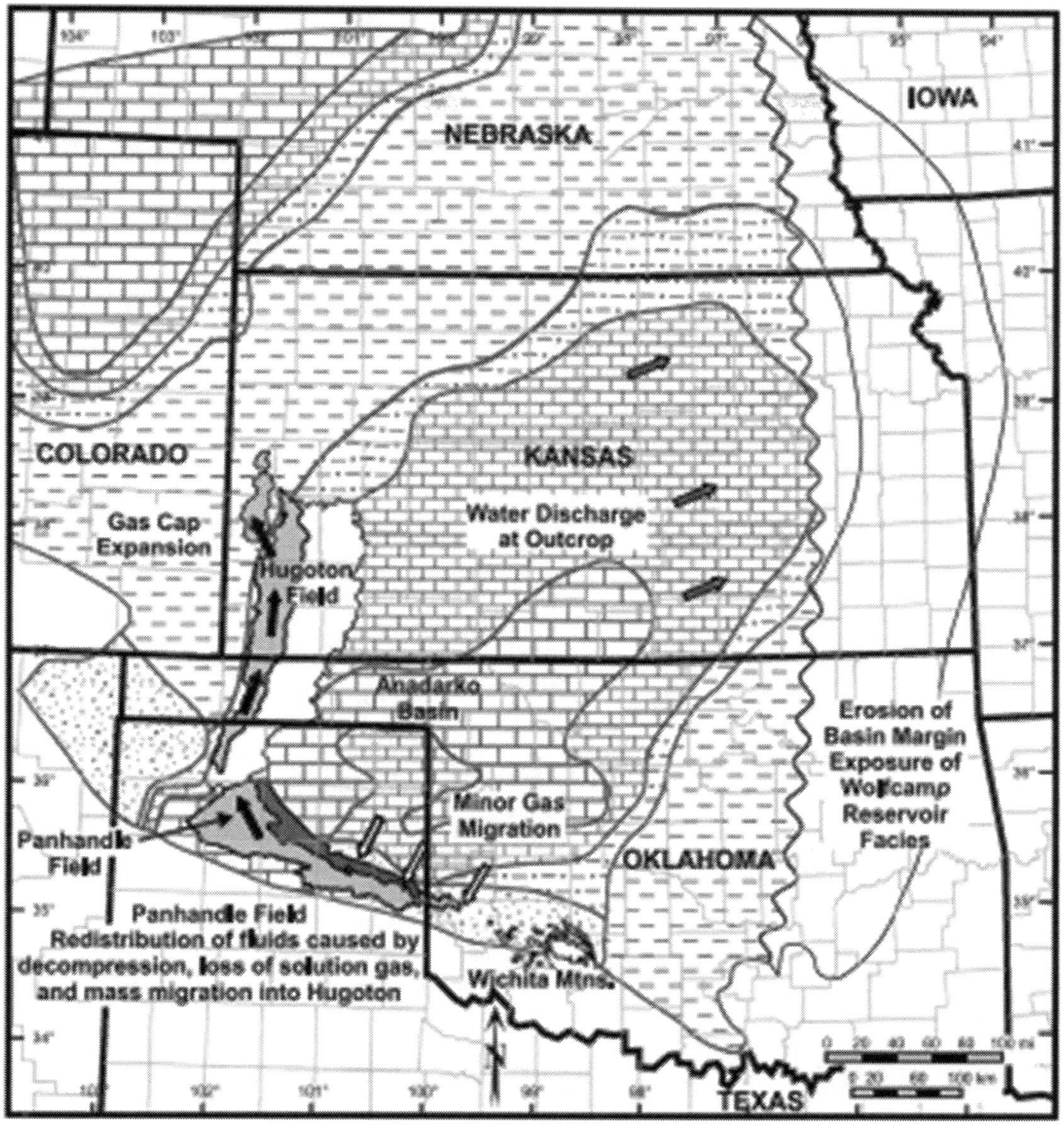

FIGURE 3.5 Sorenson (2005) displays the Late Tertiary oil and gas distribution for Wolfcampian reservoirs, showing rapid volumetric expansion of the Panhandle Field gas cap, mass movement of gas into Oklahoma and Kansas, and the charging of reservoirs updip to Hugoton at pressures above discovery reservoir conditions (Sorenson, 2005). Wolfcampian reservoirs become exposed in northeast Kansas, allowing discharge of aquifer fluids and regional loss of reservoir pressure. Upper Wolfcampian facies distribution from Rascoe (1968).

Arch (Brown, 2010, 2017, 2019). The different He/N_2 and the dominance of the two-component mixing trends are the results of the initial gas charge from the two aquifers. Brown (2010, 2017, 2019) interpret He and N_2 to be generated from the sedimentary rocks and possibly shallow, fractured basement.

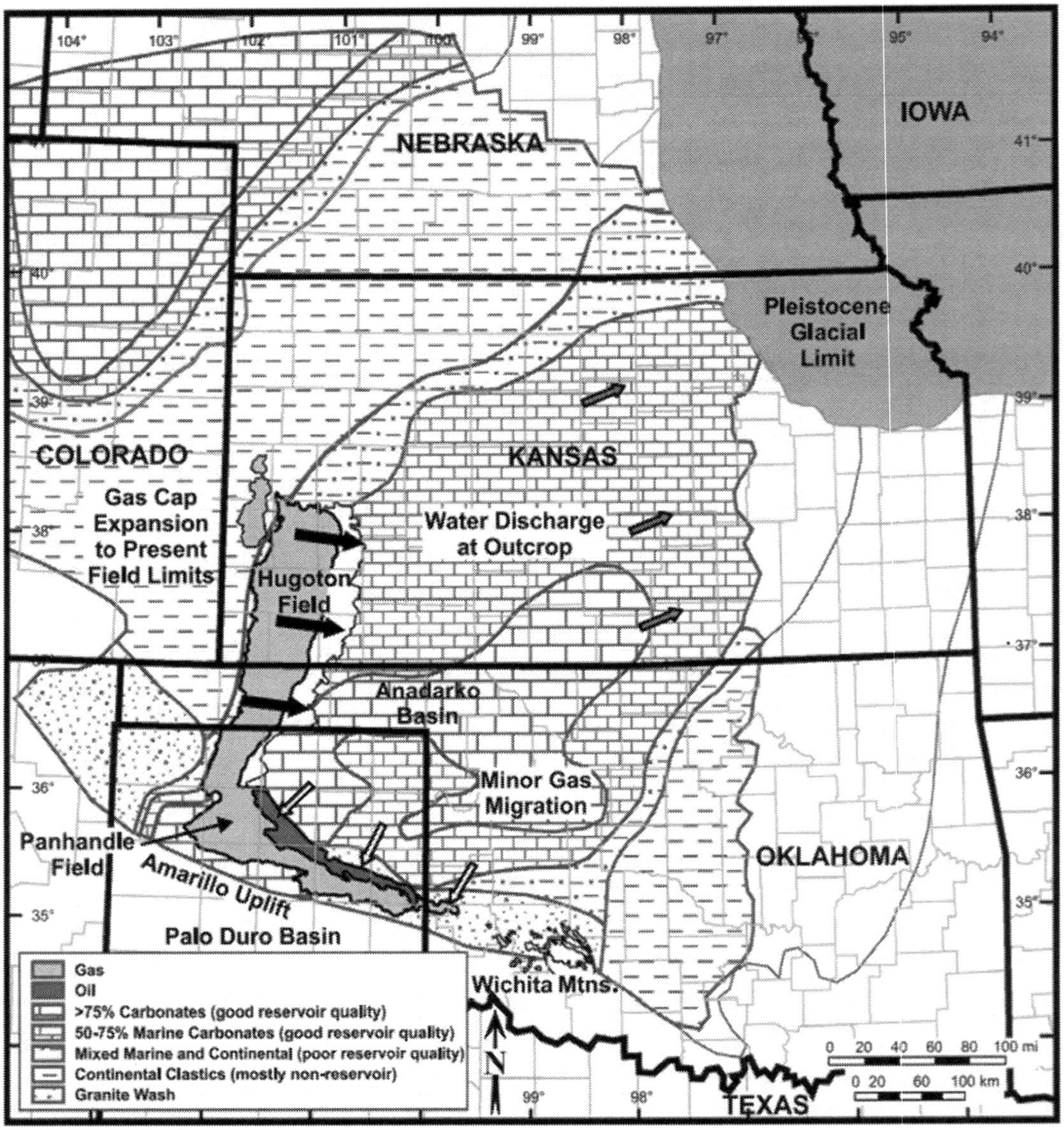

FIGURE 3.6 Accelerated quaternary erosion of Wolfcampian outcrops in northeast Kansas caused by continental glaciation and increased water flow. Continued discharge of Wolfcampian water allows regional pressure to drop to discovery levels and expansion of the Panhandle–Hugoton gas oil and gas accumulation to its present configuration (Sorenson, 2005). Glaciation limits after Beck (1959) and Frye and Leonard (1952). Upper Wolfcampian facies distribution from Rascoe (1968), used with permission from the Rocky Mountain Association of Geologists (Sorenson, 2005).

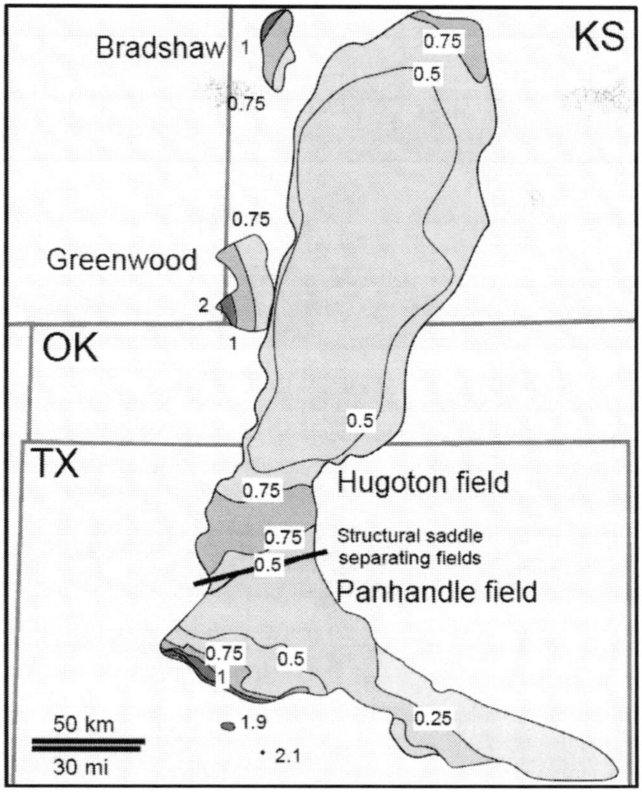

FIGURE 3.7 Helium percentage concentration across the Panhandle–Hugoton Field (Brown 2010 and *reprinted by permission of the AAPG whose permission is required for future use*).

The two different models can be compared by integrating bulk gas composition with other data sets (Brown, 2010, 2017, 2019). The Ballentine and Lollar (2002) end members are ratios that plot as lines on He–N_2 plots; therefore, unique end member contributions to each gas sample cannot be determined. The Ballentine and Lollar (2002) model does not explain the absence of gases with He/N_2 ratios less than about 0.02 or why gases with high He/N2 ratios do not contain more than 2% He (Brown, 2010, 2017, 2019). Most Anadarko Basin gases have low N_2 and He, inconsistent with the high N_2, zero He end member proposed by Ballentine and Lollar (2002). Almost all PHF gas analyses can be explained by mixing between the end members Brown (2010). Gas compositions from different parts of the field form different linear trends indicating predominantly two-component mixing of the two high He, high-N_2 end members with the low-He end member (Brown, 2010, 2017, 2019). Southeast Colorado gases, based on Brown (2010, 2017, 2019), show mixing between PHF gases and higher He gases of the Las Animas Arch area. Nitrogen isotopic compositions of almost all gas samples are also explained by the three end members.

The quantity of N_2 in the PHF is excessively large, at approximately 460 billion m^3. Of the greater than 75 TCF OGIP in PHF, about 16 TCF is N_2. According to Brown (2010, 2017, 2019), an increase in N_2 corresponds to increase in He. Thus, determining the source of He indicates the likely source of N_2. If insufficient He could be generated in sediments near the field, a larger or additional source, such as the Precambrian basement crust, must be considered for both N_2 and He. Brown (2010, 2017, 2019) conclusions are contrary to Ballentine and Lollar (2002), and in agreement with Pierce et al. (1964), sufficient sedimentary rock is available to generate all He in the PHF. The aquifers potentially supplying He and N_2 to the PHF extend to the south into the Dalhart Basin in eastern New Mexico and across the northern part of the Las Animas Arch. Source area for dissolved He and N_2 could be significantly larger than modeled because N_2 and He could have been transported as dissolved components in hydrodynamic water. Brown (2010, 2017, 2019) suggested if N_2 and He are co-sourced, N_2 could have been generated in the sediment column. If the modeled sediment is the source, an average of 0.0025 mg N_2/g rock (25 wt. ppm) must be released from the total rock volume to achieve the amount of N_2 present in the field (Brown, 2010, 2017, 2019).

Trace Ar and Ne concentrations follow the three-component mixing trends evident in the N_2—He data. Neon isotopic ratios do not show major systematic variation with N_2 concentration (Brown, 2010, 2017, 2019). The ratio of $^{20}Ne/^{21}Ne$ is lower than atmosphere, indicating ^{21}Ne generation from relatively old water. Together, these data indicate that trace gas concentrations differ by dilution and mixing between gases with approximately the same $^{20}Ne/^{21}Ne$ age (Brown, 2010, 2017, 2019).

Brown (2010, 2017, 2019) determined that isotopic ratios of the radiogenic gas components can be used to assess the duration of He generation. Radiogenic ^{20}Ne and ^{21}Ne generations are mainly controlled by the concentrations of U, Th, and O with smaller contribution from Mg and Na (Ballentine and Burnard 2002). Because O concentration is relatively constant in rocks, the generation rates of these two isotopes are proportional to the rate of He generation. Brown (2010, 2017, 2019) stated that $^{20}Ne/^{4}He$, $^{21}Ne/^{4}He$, and $^{20}Ne/^{21}Ne$ all change systematically with age due to mixing of atmospheric Ne and radiogenic production of He and Ne. If porosity, temperature, and pressure of gas equilibrium with H_2O can be estimated, and if all He and Ne go into the water and gas phases, then the He and Ne concentrations in gas equilibrated with the water will change systematically with duration of generation (Brown, 2010, 2017, 2019).

Models using conditions in the Palo Duro Basin (for generation) and PHF (for exsolution) show that the Ne forms a dilution trend extrapolating to an He generation duration between about 180 and 300 MY (Brown, 2010, 2017, 2019). Ratios of ^{20}Ne to ^{21}Ne are consistent with a source from sediment with generation duration over about 150—420 MY, depending on assumptions. The $^{20}Ne/^{21}Ne$ is well below that of the atmosphere but far greater than those of metamorphic fluid inclusions (Brown, 2010, 2017, 2019).

Brown (2010, 2017, 2019) concluded that the gas components of the PHF were the product of mixing of three sources, two of which are associated with aquifers. The Palo Duro and western Hugoton embayments provided the volume of He and N_2 needed to fill the PHF. Both Ne/He ratios and Ne isotopic ratios point toward a generation duration of about 200—300 million years, the expected duration of sedimentary generation prior to Cenozoic hydrodynamics Brown (2010, 2017, 2019). The ratio of $^{20}Ne/^{21}Ne$ excludes both a young water age and a basement origin Brown (2010, 2017, 2019).

Reservoir Characteristics of Panhandle–Hugoton Field (Dubois et al., 2007; Sorenson, 2005; Brown, 2010, 2019):

Productive area:	Approximately 7800 mi^2 (20,000 km^2)
Producing formations:	Wolfcampian series of Permian and Virgilian series of upper Pennsylvanian
Depth to pay:	2500 to 3300 ft (780–900 m)
Net pay:	200–600 ft (61–180 m) gross, 65 to 88 net ft (20–27 m), from 17 different intervals
Lithology:	Carbonates
Porosity:	Reservoir rocks within the Hugoton Field (Chase and Council Grove strata) have a median 6%–8%. Average 9.2%, standard deviation 5.1%, median of 8.7%, mode of 7.0%, Range of 1%–34%
Permeability:	Reservoir rock within the Hugoton (Chase and Council Grove strata) nearly 80% is less than 1 md, 56% is 0.1 md and 35% has less than 0.01md
Water saturation:	30%–70%
Initial field pressure:	435 psi average, 300 to 550 psi
Type of drive:	Gas depletion
Type of trap:	Hydrodynamic, structure, and stratigraphic
Estimated ultimate recovery:	>75 TCF, 1.4 BBO; 400 BCF He
Gas characteristics and analysis:	He in the Hugoton Field: Minimum 0.110%; average −0.441%; median −0.411%; maximum −0.97% (USGS). He in the Panhandle Field: Minimum −0.10%; average −0.629%; median −0.5%; maximum −82.7% (USGS). Fig. 3.7 is the He indicates the distribution of the element throughout the PHF (Brown, 2010).

The Hugoton Field is located in Hamilton, Kearney, Finney, Stanton, Grant, Haswell, Morton Stevens and Seward counties, Kansas and Texas County, Oklahoma. The field lies on the western side of the Anadarko Basin, bounded by the Las Animas Arch to the west and the Central Kansas Uplift to the north (Fig. 3.1). The pay zones are Wolfcampian marine carbonate reservoirs in the Chase and Council Grove groups that thin toward the updip margin and generally pinch out at, or just west of, the margin of the Hugoton and Panoma fields (Dubois et al., 2007). Red continental rocks, primarily very fine to coarse siltstones, are thickest at the western field margin and thin basin ward across the shelf (Dubois et al., 2007). These red beds are thought to be the lateral seal, which in conjunction with Leonardian-age evaporite top seal created a giant stratigraphic trap (Dubois et al., 2007). Laterally continuous high porosity and permeability marine and continental sandstones are found at the updip margin in the northwest part of the field (Dubois et al., 2007). These sandstones are gas productive inside the field boundaries but are water saturated outside the field despite being structurally higher with no identified physical barrier (Dubois et al., 2007). This discrepancy has been used as an argument against the red beds acting solely as the seal for hydrocarbon entrapment. The Hugoton Field was deposited in a shelf setting in shallow waters as evidenced by the lack of carbonaceous shales found outside the field representing deeper water (Dubois et al., 2007).

The Greenwood, Vilas, Vilas North, Playa, Stonington, Walsh, and Windmill fields are known as the Greenwood Topeka Trend (Beams, 1982) that is part of the Hugoton Field and its western most extension located in Baca County, Colorado. The Greenwood Topeka Trend is part of the Hugoton Field and is located in Morton County Kansas, Texas County, Oklahoma, and in Baca County, Colorado (Fig. 3.8). The field produces from the Upper Pennsylvanian Virgil limestones, covering 214,000 acres with a projected recovery of over 1 TCF of natural gas. The field was discovered in 1952 producing from the Virgil Greenwood limestone zones at 2988 to 3158 ft (905–960 m) at 16 MMCFGPD (Wingerter, 1968). The field is located on a gentle monocline that is part of the Keyes Uplift where folding occurred in Early Pennsylvanian times. Movement in Permian time was titling down to basin eastward followed by northward titling in Cretaceous time. There is natural gas and oil production from the Morrowan and Desmoinesian of Pennsylvanian age, limestones in anticlinal traps of Missourian age, all of which are minor compared to oil and gas production from the Virgil age reservoirs.

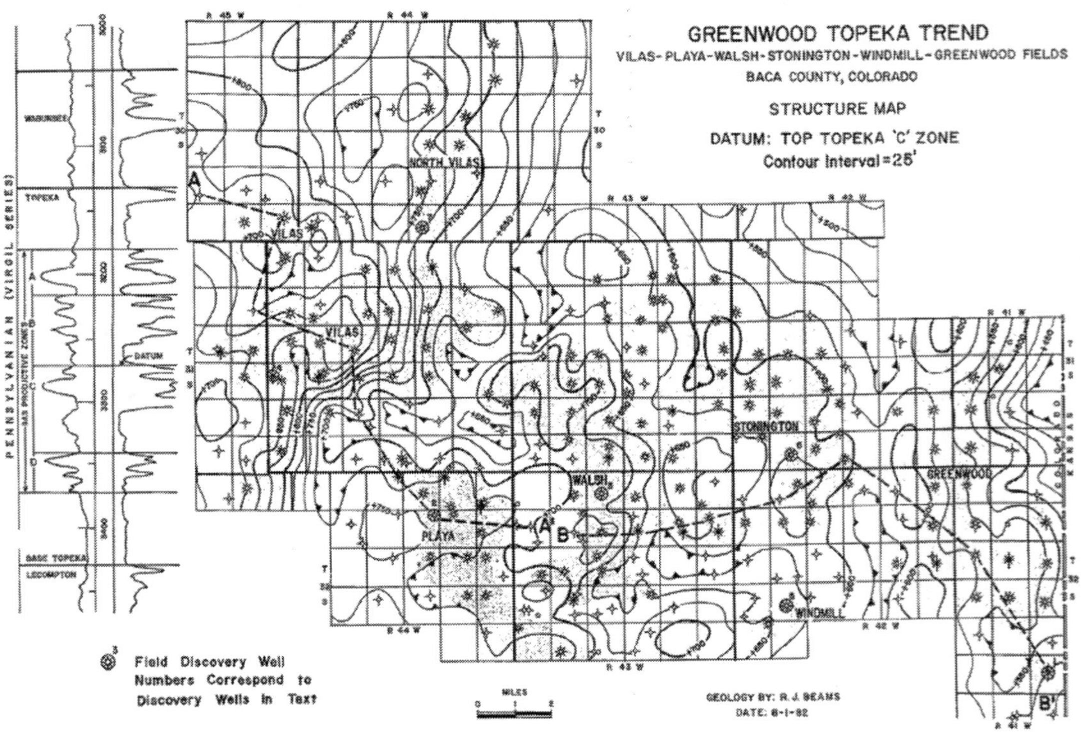

FIGURE 3.8 Greenwood Topeka Trend Field (Beams, 1982), structure on top of the Topeka "C" Zone, contour interval 25 ft (7.6 m).

Reservoir Characteristics of Greenwood Topeka Trend Gas Field (Wingerter, 1968; Beams, 1982):

Productive area:	80,000 to 214,000 approximately (area varies based on reference)
Depth to pay:	2988 to 3158 ft (905–960 m)
Net pay:	200–600 ft (61–180 m) gross, 65 to 88 net ft (20–27m), from 17 different intervals.
Porosity:	8%–22% plus in fractures
Permeability:	1 to 250 millidarcies
Water saturation:	A zone less than 65%, B zone less than 50%, C zone less than 40%
Initial field pressure:	490 psi
Type of drive:	Depletion
Type of trap:	Stratigraphic with a hydrodynamic component
Producing formations:	Virgil limestone (Pennsylvanian)
Gas characteristics and analysis:	He 0.6%, CO_2 0.03, N_2 21.09%, CH_4 57.15%, C_2H_6 5.86%, C_3H_8 3.48%, iC_4H_{10} 0.44%, nC_4H_{10} 1.10%, iC_5H_{12} 0.24%, nC_5H_{12} 0.34%, C_6H_{14} 0.27% and He 0.659%, CO_2 0.46, N_2 33.27%, CH_4 53.368%, C_2H_6 5.645%, C_3H_8 4.267%, iC_4H_{10} 0.5%, nC_4H_{10} 1.398%, $nC_5H_{12}+$ 0.847%. BTU is 500–920, average 850
Production:	Kansas only: 1.2 TCF of natural gas; 100 BCF He

3.4 Wyoming

Wyoming has several world class CO_2 accumulations that are associated with He and CH_4+. These are primarily located in the Greater Green River Basin and Moxa Arch area (Fig. 3.9). The largest of these fields is the Big Piney–La Barge complex, which is part of the Overthrust Belt and is currently the major producer of He for the United States (De Bruin, 1991 and 1995). The lone field that is CO_2 productive outside of the Moxa Arch area is the Madden Field in the Wind River Basin (Fig. 3.9). The majority of CO_2 and He production in Wyoming is from the Madison carbonates which is equivalent to the Leadville Limestone in other areas such as Utah (De Bruin, 1991 and 1995). Fig. 3.10 is a general stratigraphic column for the Overthrust Belt, Wind River, and Green River basins. There are 666 samples that have been recorded by the USBM and BLM for He content of which the majority fall below 0.1% He. The data set has an average of 0.11%, median of 0.02% and a standard deviation of 0.23%. Most anomalous data points from the database are isolated areas of high He, except for the Big Piney–La Barge area. The Yellowstone Volcanic and Grand Teton regions in the northwestern corner of the state has very low He content (Merrill and Hunt, 2017). Therefore, based on existing sampling, Wyoming has very limited potential for future He reserves except for the Greater Green River Basin in conjunction with CO_2 fields that are present or yet to be discovered.

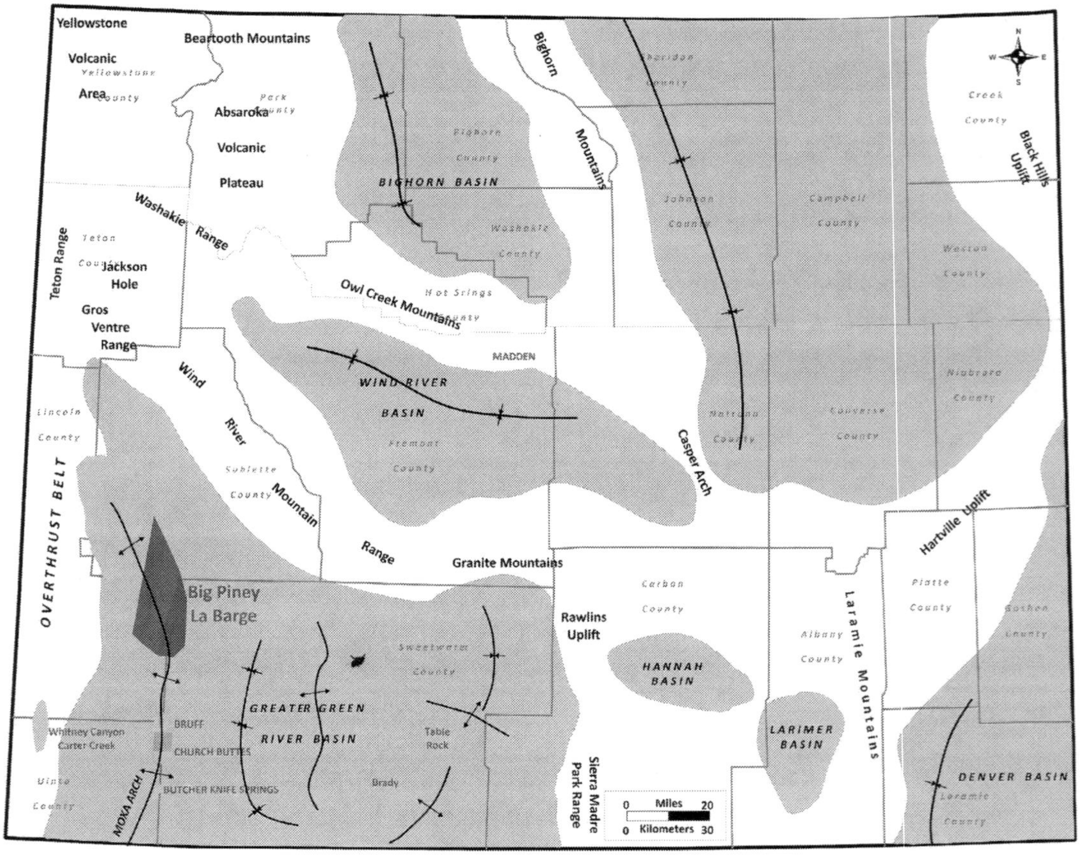

FIGURE 3.9 Location of the Big Piney–La Barge and Madison Fields in Wyoming. Also denoted are the structural elements for the State of Wyoming. Production of the existing CO_2 fields correlates with the color of the zone labeled in Fig. 3.9.

3.4.1 Big Piney–La Barge area

The State of Wyoming produces 55% of He in the United States from the Madison Limestone (Mississippian) at the Big Piney–La Barge Field, in the northern end of the Moxa Arch, Greater Green River Basin, Sublette, Sweetwater, and Lincoln counties (Fig. 3.11). The field is 21 mi by 65 mi (34 km 105 km), was discovered in 1952, and has several Paleozoic zones capable of producing CO_2 (Stilwell, 1989). There is shallower production from the Pennsylvanian strata but is not related to the CO_2 reservoir in the Madison Formation. In 2014, the field produced an average of 45 MMCFGPD which was sold to the Shute Creek gas facility, which has a capability of 600 MMCFGPD (De Bruin, 1991; Eppink et al., 2014). The gas is typically two-thirds CO_2. The H_2S and He are removed and the CO_2 is piped to various enhanced oil recovery projects in Wyoming and adjacent states (Eppink et al., 2014).

The present-day structural configuration of the Big Piney–La Barge Field is a large anticline trending north-northwest to south-southeast that was formed during Cretaceous times

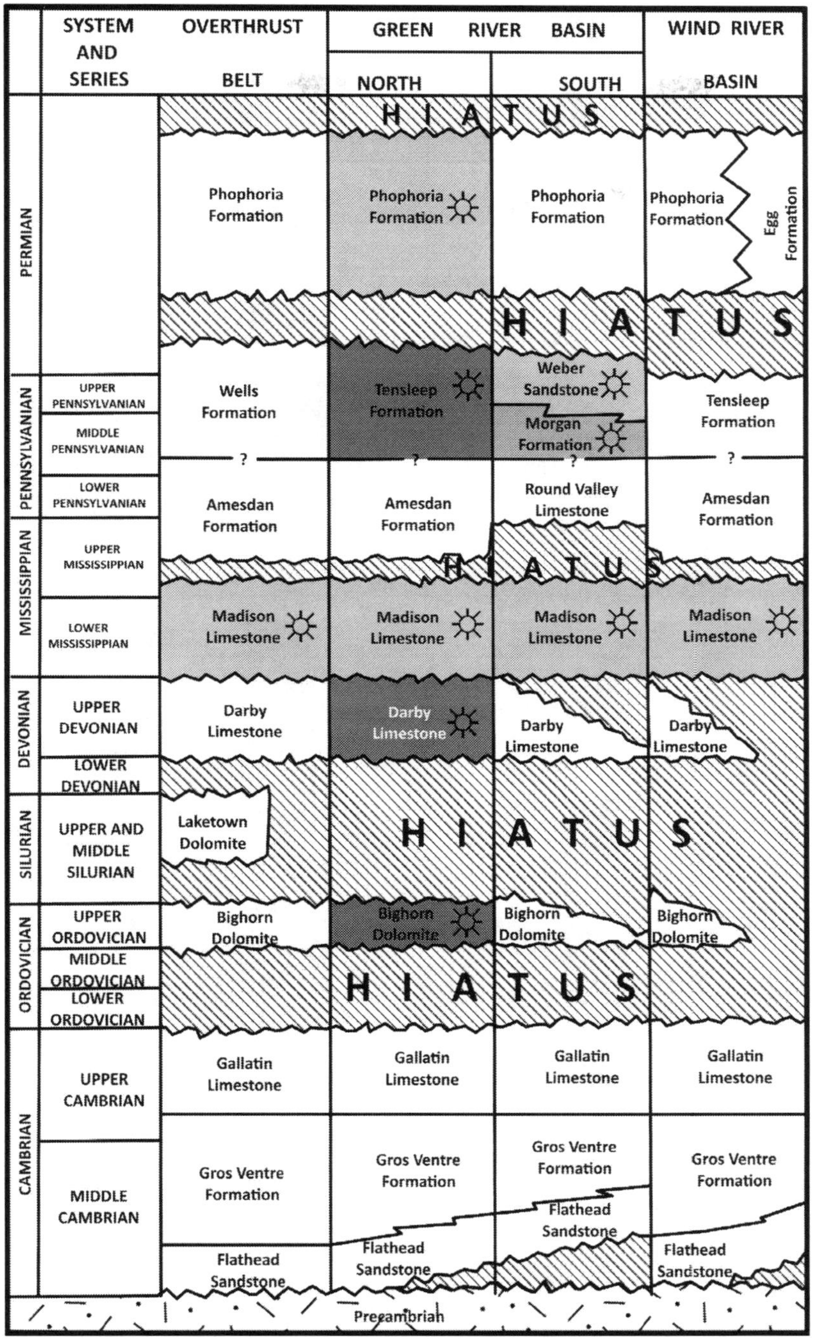

FIGURE 3.10 Stratigraphic column for the Overthrust Belt, Green and Wind River basins, modified from DeBruin (1991). Colors associated with certain formation names are CO_2-producing zones as indicated in Fig. 3.9.

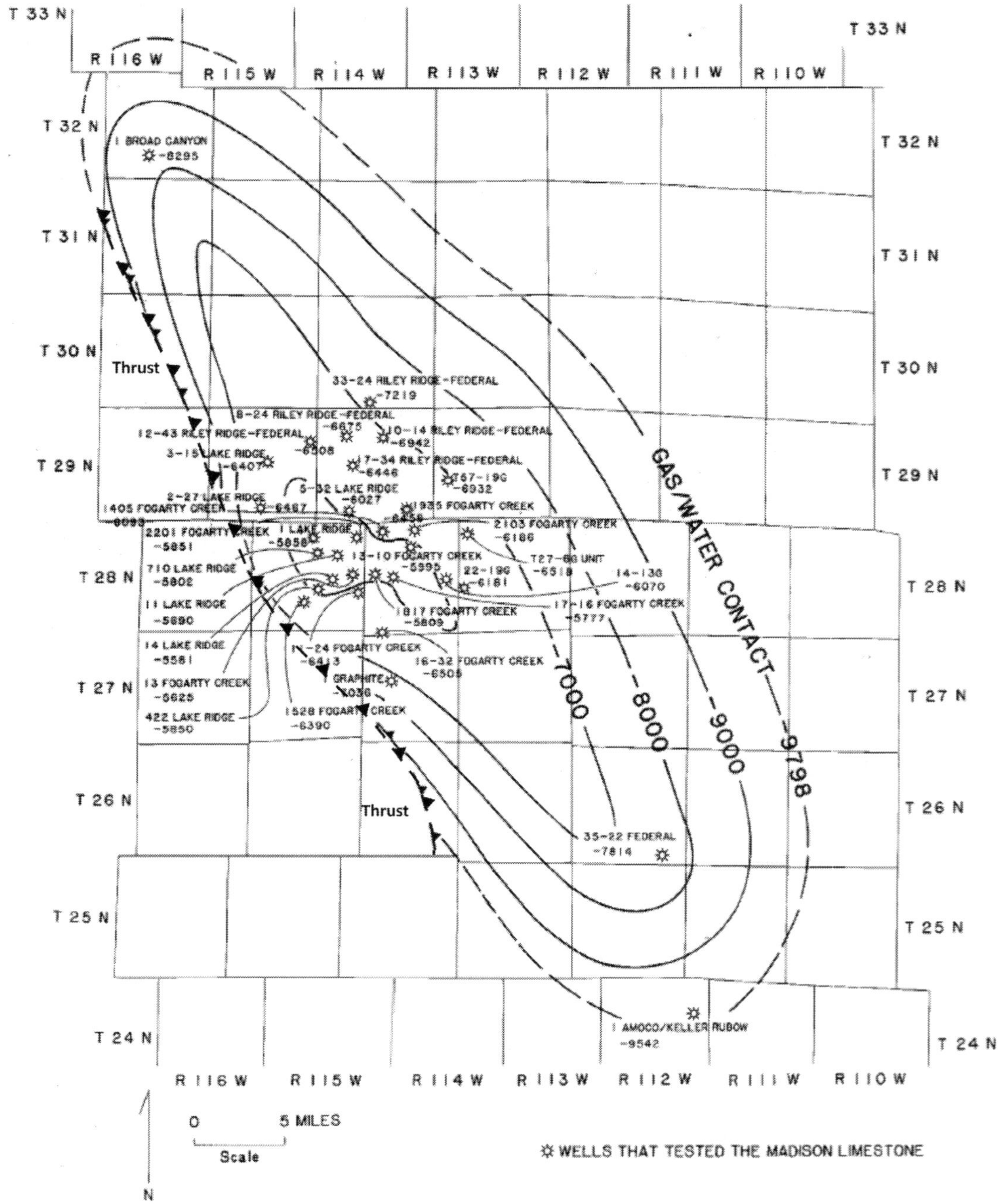

FIGURE 3.11 Map of the Big Piney–La Barge Field showing the structural outline on top of the Madison reservoir, with the oil–water contact and the concentrations CO_2 (De Bruin, 1991).

(Fig. 3.11). The Big Piney–La Barge tectonic feature first appeared as a broad rolling structure that through time morphed into its present-day configuration (Becker and Lynds, 2012). The structure was deformed during the Sevier–Laramide orogeny and then Early Tertiary phase of the larger Moxa Arch deformation (Becker and Lynds, 2012). The structure was uplifted over 3 mi (4.8 km) and folded over the Moxa Arch and any Paleozoic faulting tends to lose displacement and was flattened in the Triassic evaporite-rich interval. Thrusting from the east in Tertiary times affected the west side of the structure causing high-angle reverse and tear faults and anticlines across the field and associated area.

Other than the Mississippian Madison Limestone, only the Ordovician Bighorn dolomite is considered prospective for CO_2 (Becker and Lynds, 2012). The lower Madison Formation is composed of dolomitized limestones and dolostones which are overlain by the upper Madison Formation which consists of sabkha and karst deposits that form the seal. Eppink et al. (2014) defined three different producing reservoir terrains within the lower Madison Formation: basinal, highland, and foreland.

Two periods of migration seem to have occurred in the field: the first trapping hydrocarbons followed by migration of CO_2 into the Madison reservoir (Stilwell, 1989). The source of the CO_2 has the following suggested origin:

(1) From the adjacent Tertiary volcanic activity that contains CO_2 or from hot waters rising from depth on the down side of the Big Piney–La Barge structure that would have released the compound from Paleozoic carbonates (Stilwell, 1989).
(2) The source of CO_2 from hot waters from the basement would allow migration of He from crustal rocks rather than volcanic rocks which tend to be lacking in He.
(3) Inorganic thermochemical sulfate reduction involving the anhydrite seals, methane, and other hydrocarbons within the reservoir (Becker and Lynds, 2012).

The Big Piney–La Barge Field produces primarily CO_2 and hydrocarbons with secondary production of He. He is believed to be radiogenically produced from sedimentary rocks and was mobilized by deep basinal groundwater flow or possibly hydrocarbon migration and concentrated by diffusion into stable traps (Merrill and Hunt, 2017). Water and gas samples were collected from six gas wells producing from the Madison Limestone in the Big Piney–La Barge Field along with 17 regional gas-effusive surface springs within a 200-km radius of the field (Merrill and Hunt, 2017). Noble gas isotopes analysis from the study identifies three distinct regional He source groups that approximately match the three major geologic provinces in the area. The Big Piney–La Barge Field lies in the first province, composed of tectonically stable sedimentary strata with concentrations of He of 0.5%–1.0% that is primarily of crustal origins (Merrill and Hunt, 2017). There was a minor population of He within this group that is not purely crustal and its source is unknown. The second He source group is to the west of Big Piney–La Barge Field, in the Basin and Range province of southeastern Idaho where He is from the mantle with a concentration range of 0.0004%–0.008%. The third He source group is in the Yellowstone Volcanic province to the north and is sourced from the mantle (Merrill and Hunt, 2017). Geochemical analyses of gas from La Barge Field show increase of CO_2 and He and decrease of hydrocarbons and H_2S with structural depth (Merrill and Hunt, 2017). These trends suggest a well-connected stable and stratified Madison Limestone reservoir, where the gas constituents have been able to concentrate by density through a permeable reservoir. Studies of He isotopes suggest there are two different pools of gas:

(1) one to the southwest on the steeper side of the structural crest of the platform, and
(2) one to the east on the more shallowly dipping side of the platform (Merrill and Hunt, 2017).

Because of the complex structural nature of the Big Piney–La Barge Field, there are sufficient migration pathways that allowed accumulation of He and CO_2. Limits of production development have been imposed by the US Bridger-Teton National Forest and the Bureau of Land Management concerns for animal life, biodiversity, sensitive environmental habitats, steep slopes, and many other unknown stipulations.

Reservoir Characteristics of Big Piney–La Barge Area (Stilwell, 1989; Eppink, 2014):

Productive area:	Basinal: 138,000 acres
	Foreland: 125,000 acres
	Highland: 388,000 acres
Producing formations:	Madison Limestone (Mississippian)
Average net pay:	275 ft (85 m)
Depth to pay:	Basinal: 15,680 ft (4750 m)
	Foreland: 16,356 ft (4955 m)
	Highland: 18,154 acres ft (5500 m)
Lithology:	Sandstone and dolomites
Average porosity:	9%, with a range of 6%–12%
Average permeability:	10 to 50 millidarcies
Water saturation:	15%
Initial field pressure:	Basinal: 6585 psi
	Foreland: 6870 psi
	Highland: 7625 psi
Type of drive:	Gas expansion
Type of trap:	Structural
Formation temperature:	Basinal: 275°F (135°C)
	Foreland: 324°F (162°C)
	Highland: 349°F (176°C)
Producing formations:	Madison Limestone (Mississippian)
Gas characteristics and analysis:	Basinal: CO_2 85%, CH_4+ 9.0%, N_2 3.0%, He 0.5%, and H_2S 2.4%
	Foreland: CO_2 74%, CH_4+ 15.0%, N_2 6.0%, He 0.5%, and H_2S 4.4%
	Highland: CO_2 81%, CH_4+ 11.0%, N_2 4.0%, He 0.5%, and H_2S 3.0%

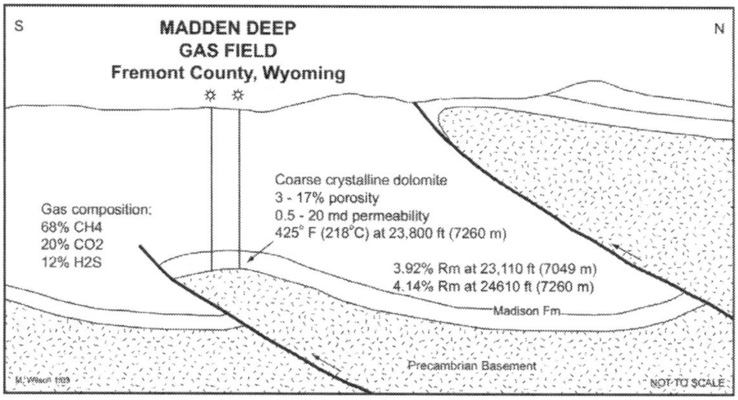

FIGURE 3.12 Schematic cross section of the Madden Field (Wilson et al., 2003).

3.4.2 Madden Field

The Madden Field is located in Townships 38 and 39 North, Ranges 88 to 91 West, Fremont County, Wyoming, northern part of the Wind River Basin (Fig. 3.12). The field produces CO_2 from the Madison Dolomite which is sealed by the Madison strata itself. The field produces from an east-west trending anticline, 20 mi (34 km) long and 10 mi (17 km) wide (Fig. 3.12). The field has a smaller structural closure, East Madden Field that lies to the east and is bordered by the Owl Creek Fault Thrust in the Wind River Basin (Barrett and Hubley, 1969). The reservoir is heavily fractured across the structure but not off structure in wells that were dry in the Madison (Hawkins and Baugh, 2006). Fig. 3.12 is a schematic cross section of the Madden Field by Wilson et al. (2003).

Reservoir Characteristics of the Madden Field (Barrett and Hubley, 1969; Hawkins and Baugh, 2006; Eppnik et al., 2014):

Productive area:	80,000 acres
Producing formations:	Mississippian Madden Formation
Reservoir:	Madison
Depth to pay	24,000 ft (7275 m)
Net pay:	85 ft (25.8 m)
Lithology type:	Dolomite
Porosity:	3%–17%, average 16.2%
Permeability:	5–20 md, average 10 md
Water saturation:	42%
Initial field pressure:	4265 psi
Type of drive:	Gas expansion

(Continued)

Type of trap:	Structure
Temperature:	425°F (218°C)
Gas characteristics and analysis:	CO_2 20%, CH_4+ 67%, and H_2S 12.0%

3.5 Kansas

Helium production in Kansas is either associated with the low-grade He deposit in the Byerly–Bradshaw–Hugoton Field or high-grade He fields (greater than 1.0%) as in the Ryersee, Reichel, and Otis-Albert fields. Carbon dioxide fields have not been found to date in Kansas because of the lack of igneous activity typically associated with CO_2 accumulations found in Wyoming, Arizona, Colorado, and New Mexico related to venting from mantle or crustal rocks. Fig. 3.13 is a generalized stratigraphic column for Kansas from Newell (2021)

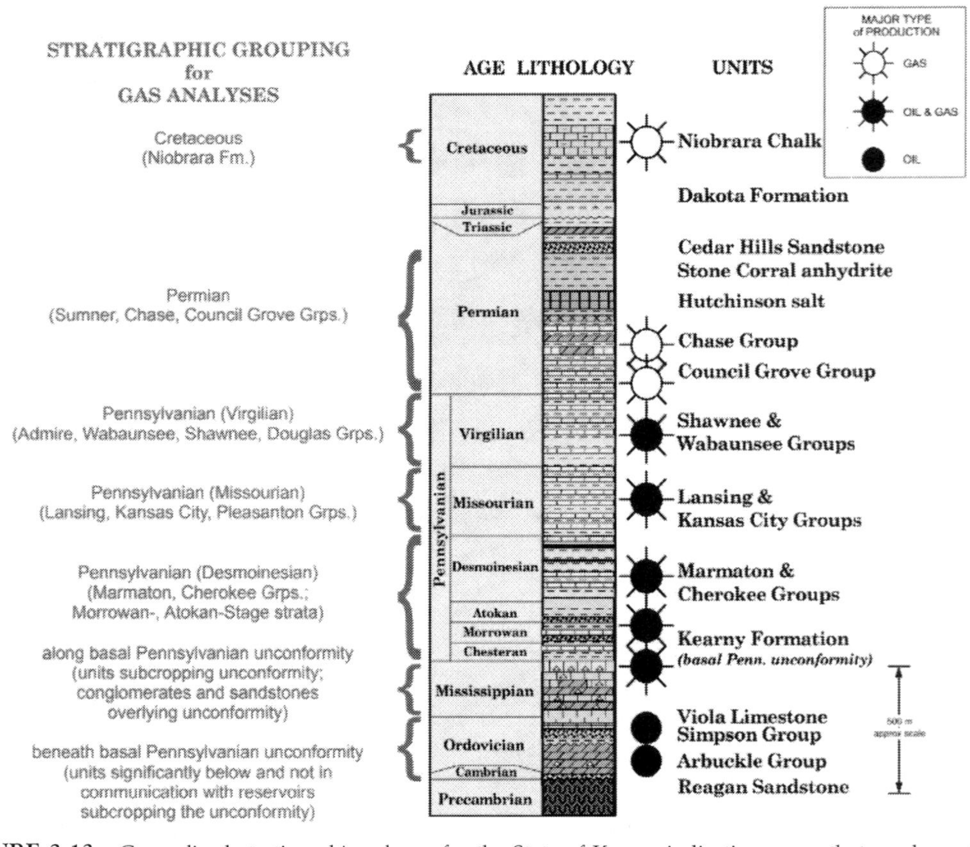

FIGURE 3.13 Generalized stratigraphic column for the State of Kansas indicating zones that produce oil, gas, and He.

showing the production of oil, gas, and He zones. The stratigraphy in Kansas and extending into Oklahoma consists primarily of Paleozoic rocks. Structurally, Kansas is underlain by the Cherokee, Forest City, Salina, Sedgwick, northeastern edge of the Denver, and northern part of the Anadarko Basins. The Salina Basin has marginal oil production along its southern edge. The Cambridge and Bourbon arches, Nemaha Ridge, and Central Kansas Uplift are major positive features that have had a significant impact on the petroleum migration and accumulations. Fig. 3.14 is the He database from the United States Geological Survey (USGS) and United States Bureau of Mines (USBM).

The top of the Precambrian, major structural trends, and He data are indicated in Fig. 3.15 along with the fields discussed in the text. The Hugoton Basin is associated with the northwestern flank of the Anadarko Basin, the Ryersee—Reichel—Otis-Albert fields are associated with the southwestern flank of the Central Uplift, and the Elm—Elmsdale—Vetter fields are associated with a junction of the Bourbon Arch, a Precambrian high, and the Nemaha Ridge. Fig. 3.16 is modified from Gerhard (2004) and shows petroleum migration from the Anadarko, Ardmore, and Arkoma basins in Oklahoma. Along with the petroleum migration, N_2, CO_2, He, and brines also migrated. The Hugoton—Texas Panhandle is considered a hydrodynamic trap that terminates before the Colorado—Kansas state line. Helium in the Reichel—Otis-Albert fields forms a structural trap along the "Russell Rib." The Ryersee is a field that produces He from a structural closure adjacent to the Central Kansas Uplift that lies along migration pathways. The Elm-Elmsdale-Vetter fields lie on a junction of the Bourbon Arch and the Nemaha Ridge and overlie a major Precambrian high where at its center, all pre-Lansing—Kansas City strata have been removed. Helium may be derived both

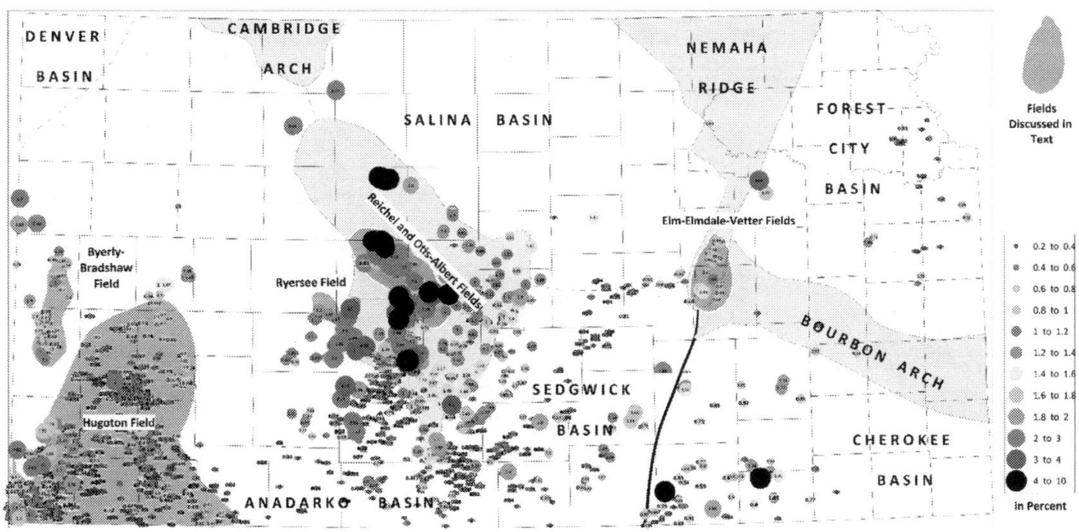

FIGURE 3.14 Classed post map for He in the State of Kansas based on the USGS and USBM database and additional data from Running Foxes Petroleum in relation to structural elements and the basins in Kansas. Helium data is from Rogers, 1921; Anderson and Hines, 1951; Moore, 1978; Moore, 1981; Moore and Sigler, 1985; Wescott, 1989; Hamal and Driskill, 1996; Gage and Driskill, 2003; Gage and Driskill, 2005.

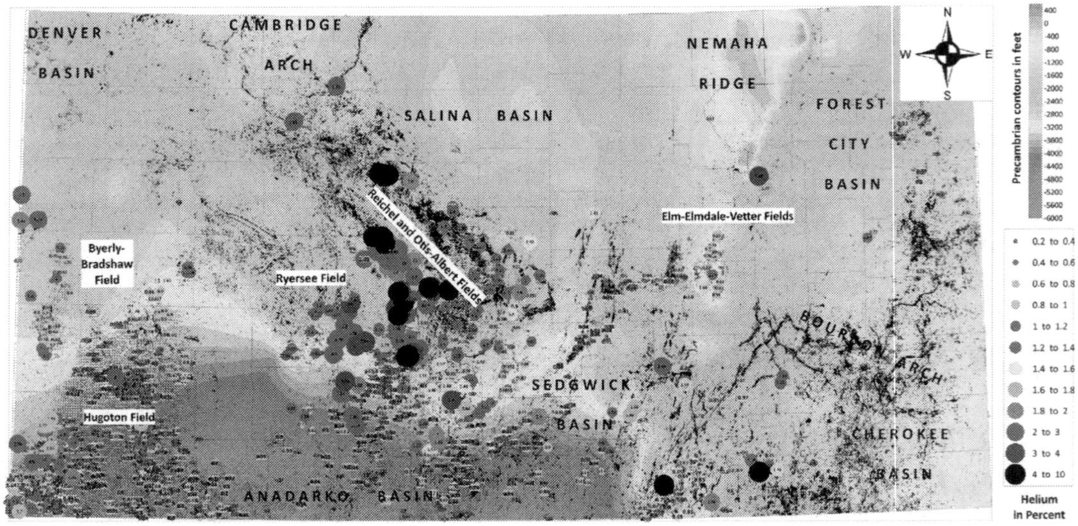

FIGURE 3.15 Structure on top of the Precambrian for the State of Kansas indicating major structural/tectonic features and location of productive or anomalous areas containing He. Helium data are from Rogers (1921); Anderson and Hines (1951); Moore (1978); Moore (1981); Moore and Sigler (1985); Wescott (1989); Hamal and Driskill (1996, 2003, 2005).

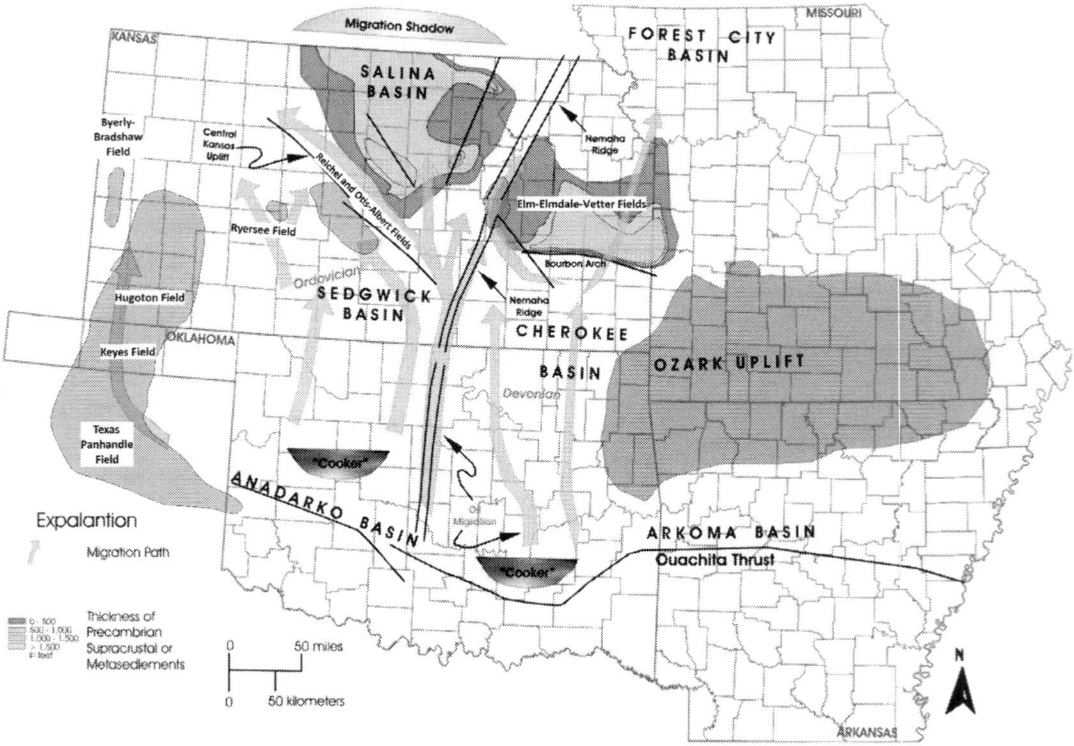

FIGURE 3.16 Migration pathways as suggested by Gerhard (2004) with the areas of productive or anomalous He gas fields in Kansas, Oklahoma and the Panhandle of Texas. The age of that oil began migrating and their pathways is indicated.

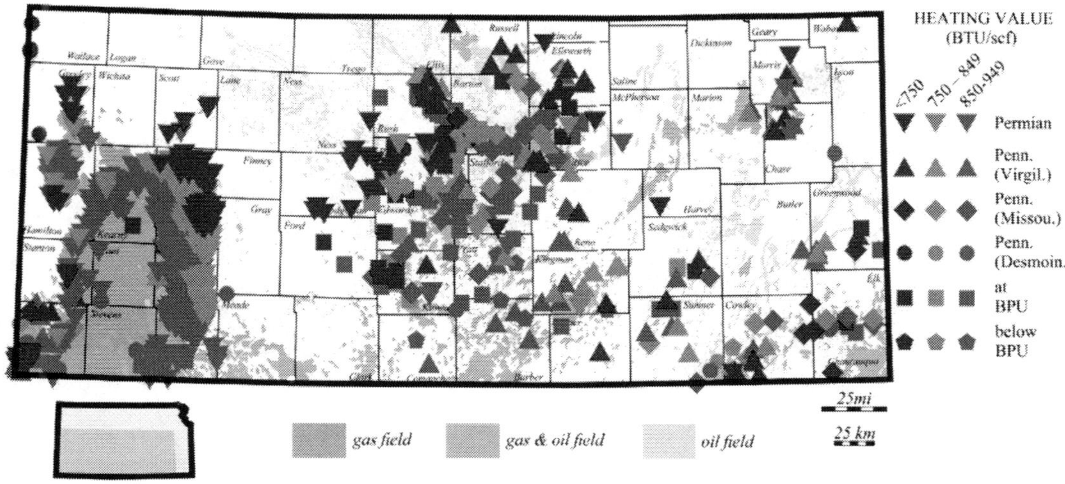

FIGURE 3.17 BTU map from Newell (2021) indicating changes in calorific values that reflect changes in He and N_2. Helium data are from Rogers (1921), Anderson and Hines (1951), Moore (1978), Moore (1981), Moore and Sigler (1985), Wescott (1989), Hamal and Driskill (1996, 2003, 2005).

from migrating brines which terminated at this structural closure and from Precambrian clastic rocks underlying the area.

Newell (2021) mapped He and N_2 across Kansas and found that in the Hugoton–Byerly–Bradshaw area that BTU content decreased toward the edges of the field with corresponding increase in He and N_2 (Fig. 3.17). Newell (2021) identified that He content increased from west to east with a decrease in N_2/He ratios in the area of the Central Kansas Uplift (Fig. 3.18). The economic He for the Central Kansas Uplift occurred due to migrating with low-temperature hydrothermal brines that also carried petroleum from the Anadarko Basin to the south (Jenden et al., 1988). There is probably a crustal and mantle contribution locally to many of the reservoirs as the expelled brines from the Anadarko or locally generated basement fluids stripped, He, and N_2 during migration.

Newell et al. (2003) defined the He and N_2 ratio for gases from Mississippian rocks across several basins versus rocks productive in the Hugoton Embayment (Fig. 3.19). The gases from Mississippian and older rocks clearly indicate higher He and lower N_2 then in the Hugoton Embayment area. In addition, the gas from the Ryersee Field along with other fields in the Central Uplift and Pratt Anticline is different from the gases in the Hugoton Embayment and Mississippian gas (Newell et al., 2003). This suggests there are differences in source of origin and migration history for all three groups of gases.

There is a lack of data in Eastern Kansas but it is sufficed to say that localized high-He percentages in the Nemaha Ridge and Cherokee Basin area can be related to localized basement features that accumulated He during fluid migration from the Ardmore and Arkoma basins to the south. The localized areas of high He could also be sourced from the upwelling Precambrian fluids during tectonic activity. Newell (2021) further concluded that there was no specific decrease or increase through the Paleozoic section for He. The

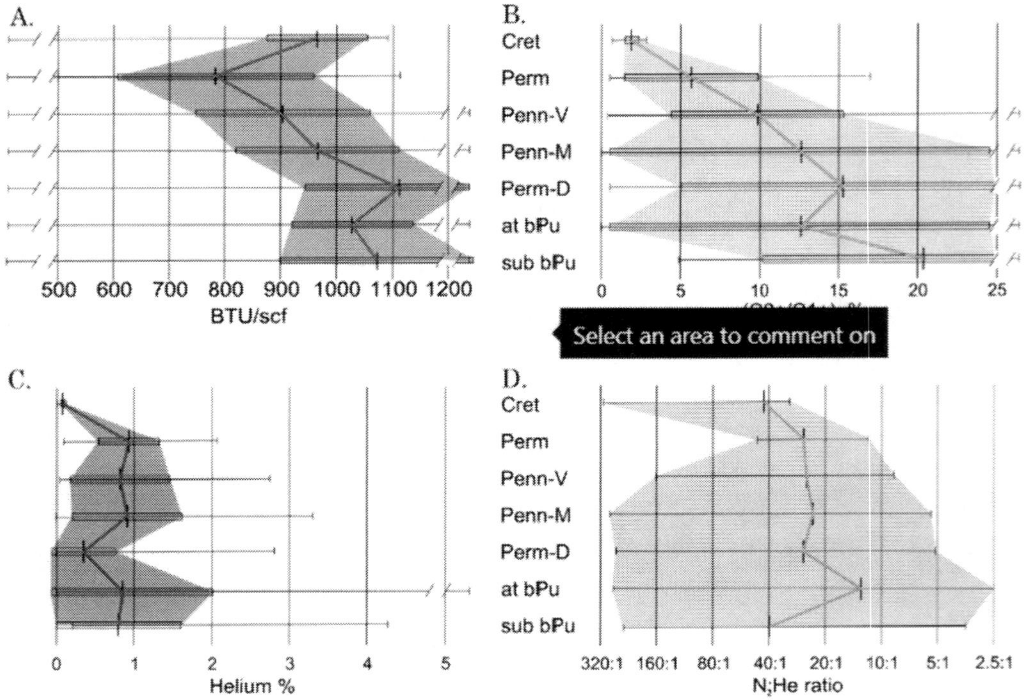

FIGURE 3.18 Variations in natural-gas characteristics by stratigraphic unit: (A) BTU content, (B) hydrocarbon wetness, (C) He %, (D) N_2/He ratio. The dark line between each stratigraphic interval follows changes in the average or median; the lighter areas marks +/− standard deviation for (A), (B), and (C) and upper and lower ranges for (D) from Newell, 2021.

results may be skewed in terms of sampling which is at its greatest in areas of He production and leanest in several areas that have never been sampled for He and N_2 let alone natural gas content.

3.5.1 Ryersee Field

The Ryersee Field is located at Township 18 South, Range 21 West, Ness County, Kansas (Fig. 3.19), and produces He and low BTU gas (520 BTU/scf) from porous dolomites (Fig. 3.20) of the Permian Chase Group. Fig. 3.20 is the structure on top of the Mississippian carbonates at the Ryersee Field (Newell et al., 2003). The gas from the field is 48.6% N_2, 1.4% He, and 50% CH_4+ that is structurally trapped at 2250 ft (682 m) depth along the Bazine Anticline (Fig. 3.21). The Bazine Anticline is a south-southwest plunging fold on the northeastern flank of the Hugoton Embayment. The gas in the Chase Group at Ryersee Field is chemically different from the gas at the Hugoton Embayment and is more similar to the typical fields along the Pratt Anticline in the Central Kansas Uplift and the Sedgwick Basin. These characteristics make it more akin to widely separated Chase Group fields on the Central Kansas Uplift, Pratt Anticline, and Sedgwick Basin.

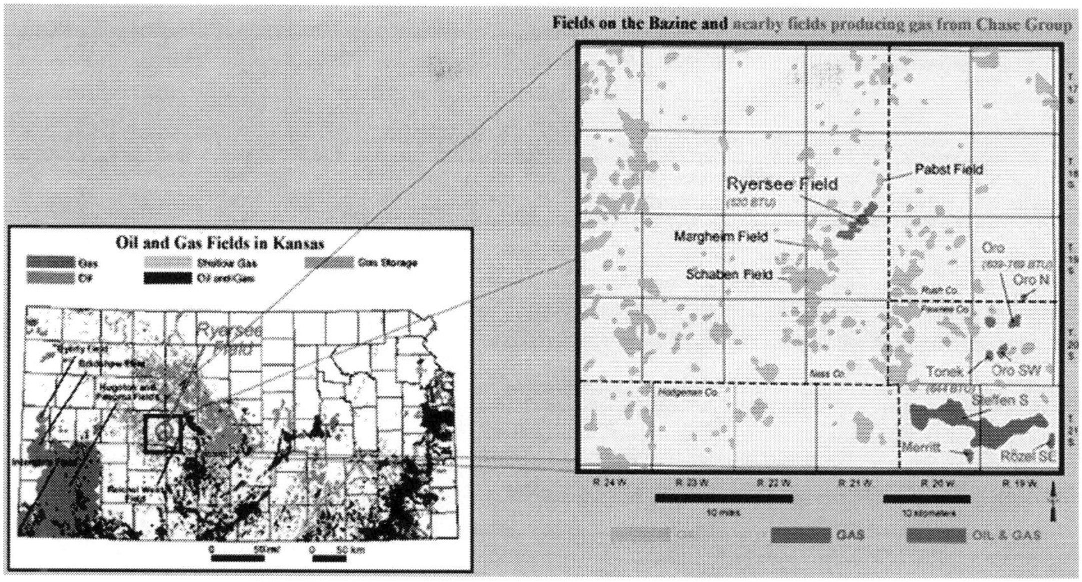

FIGURE 3.19 Location of the Ryersee Field in Ness County, USA (Newell et al., 2003).

The field was discovered in 1928–29 with shows of oil in the Cherokee and gas in the Chase Group. The oil was not produced until 1956. The field has produced 1.45 MMBO from the Cherokee, Marmaton, and Mississippian rocks 3800 to 4250 ft (1150 to 1290 m). The delay in extracting He from the gas flow was the lack of available technology until the later 1990s when microprocessing and cryotechnology began to be available for smaller accumulations (Newell et al., 2003).

The operator at the Ryersee Field has a cryogenic N_2-rejection plant with a design capacity of 5 MMCFD of feed gas. Raw He is cryogenically enriched to He quality of 50%–60% pure and then upgraded via conventional membrane technology to a level of 90+% for shipping purposes (Newell et al., 2003). The hydrocarbon gas is then sold to the market.

Reservoir Characteristics of the Ryersee Field (Newell et al., 2003):

Producing formations:	Chase Group (Permian)
Productive area:	3840 acres
Reservoir:	Chase Group (Permian)
Depth to pay	2190 to 2510 ft (663–760 m)
Net pay:	>10 ft (>3.3 m)
Porosity:	Variable
Permeability:	Unknown
Water saturation:	Unknown
Initial field pressure:	Unknown

(Continued)

Type of drive:	Gas expansion
Type of trap:	Structure
Gas characteristics and analysis:	CH_4 48.52%, C_2H_6 0.34%, CO_2 0.04%, N_2 48.62%, He 1.44%, BTU 520
Production:	None indicated, CH_4+ may be injected into the lower oil reservoir for pressure maintenance

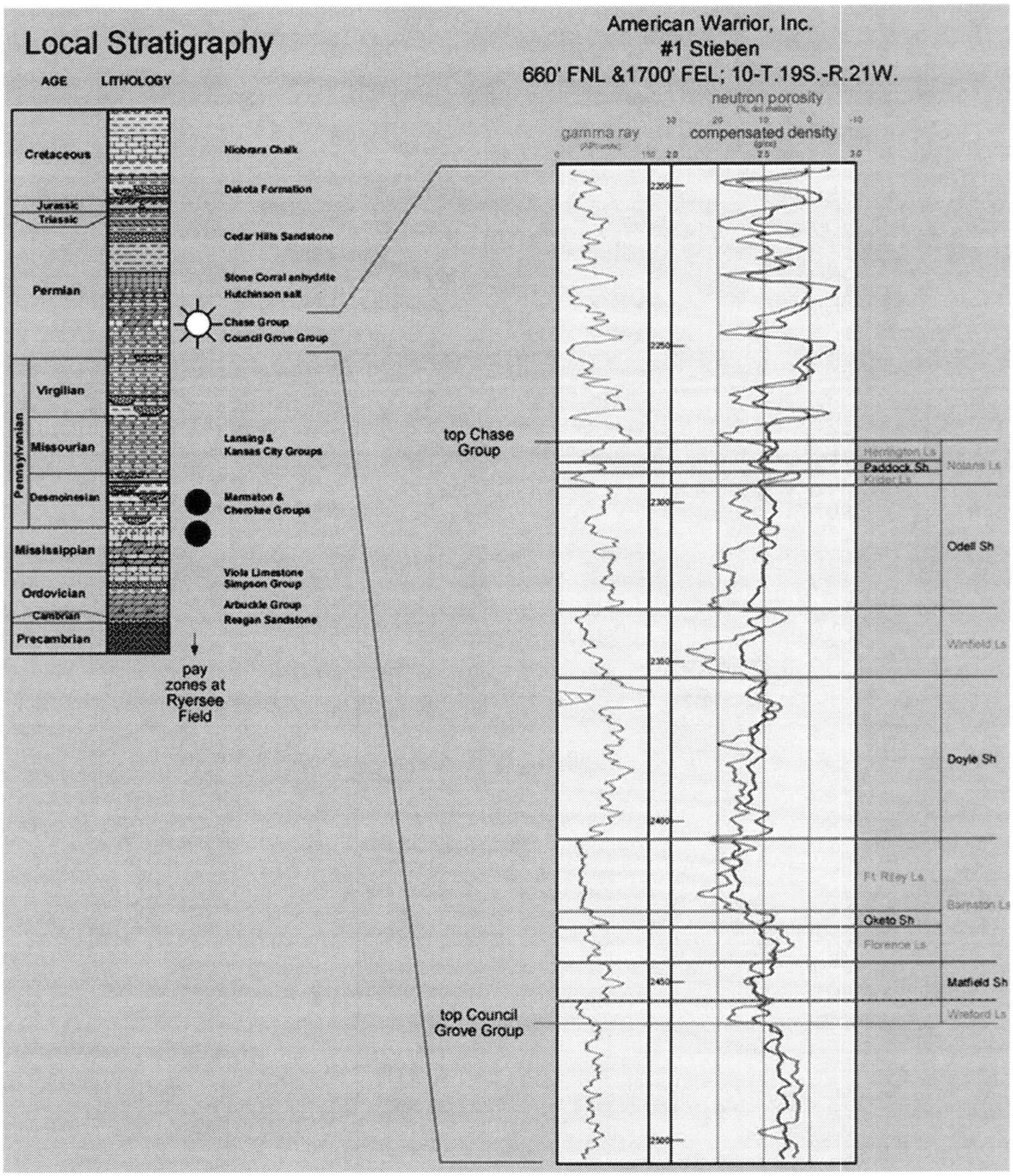

FIGURE 3.20 Geological column and mechanical log from the for the Ryersee Field, Ness County, Kansas (Newell et al., 2003).

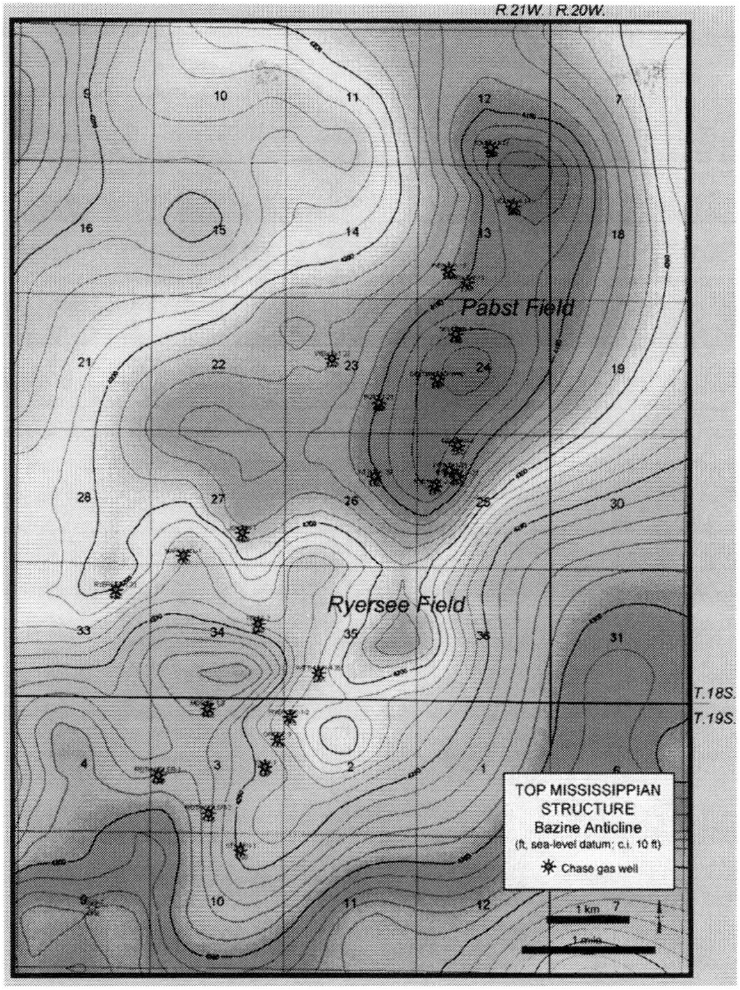

FIGURE 3.21 Structure on top of the Mississippian Limestone strata at the Ryersee Field, Ness County, Kansas (Newell et al., 2003). Structural contour interval is 10 ft (3 m).

3.5.2 Otis-Albert Field

Otis-Albert Field discovered in 1930 is located in Township 18 South, Ranges 15, 16, and 17 West, Township 19 South, Range 15 West, Rush and Barton counties Kansas and produces oil and gas from Reagan, Arbuckle, Pennsylvanian conglomerate, Lansing–Kansas City, and Topeka and Chase reservoirs (Miller, 1968). The production is extracted from 1848 to 3527 ft (560–1069 m) with present production records indicating 15.8 MMBO and 10.3 BCF from Kansas Corporation Commission records. These records did not start until the 1970s, however. These records did not start until the 1970s, however. Miller (1968) indicated that

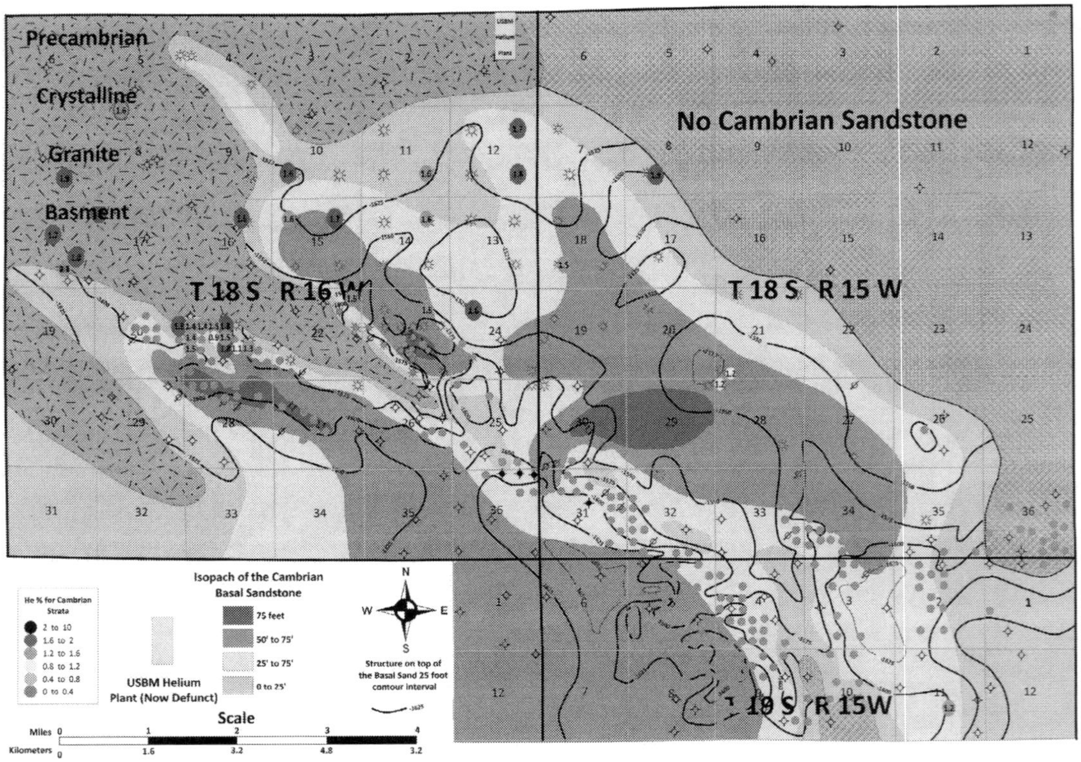

FIGURE 3.22 Otis-Albert Field in Rush and Barton counties, Kansas with structure on top of the Reagan or Basal sandstone (Cambrian), isopach of the Reagan or Basal sandstone (Cambrian) and He % for the Cambrian sandstone (USGS and USBM database). The structure and isopach are modified from Miller (1968).

the field had produced 173 BCF of gas but was depleted by 1960 and 20 BCF of gas had been vented. The gas processing plant processed 32.3 BCF of gas producing 321 MMCF He from March 1, 1945 to January 1, 1960.

Fig. 3.22 is a structure and isopach map on top of the Reagan Sandstone (Cambrian) across the Otis-Albert Field with associated He% values. The Reagan pinches out up against a basement high, the Shaffer Ridge (Miller, 1968). The Reagan Sandstone and Arbuckle carbonates were originally deposited across the entire area but were eroded off topographical highs such as the Shaffer Ridge prior to Pennsylvanian times. The Reagan Sandstone is the primary He-producing reservoir and generally is overlain Pennsylvanian sediments. The Arbuckle Group (Middle Ordovician) is found in localized outliers that are productive on the southeast corner of the field. Helium content is approximately 1.9%. The high He and N_2 percentages that have been localized in the Otis-Albert Field area may be the result of mixing of migrating fluids from the Hugoton Field to the west, localized fluids expelled from the basement along the Central Uplift and fluids expelled from the Anadarko Basin to the south (Newell, 2021).

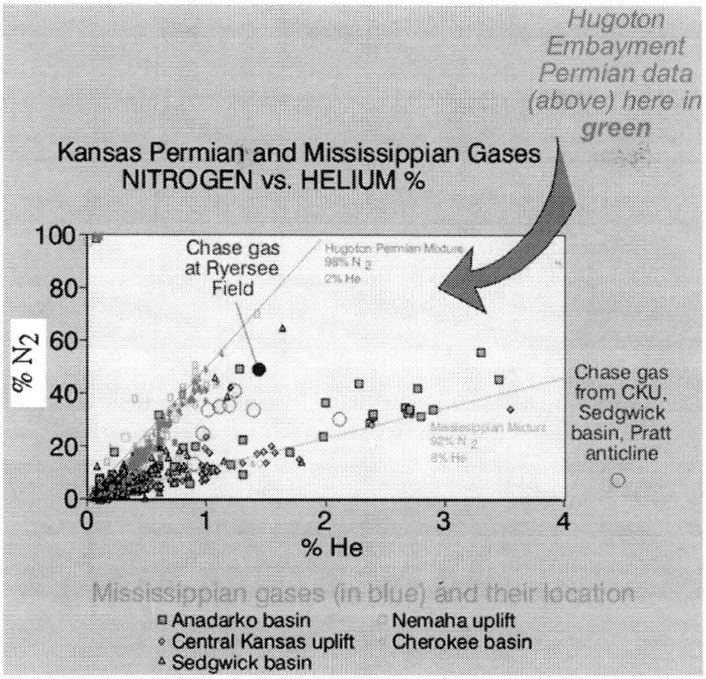

FIGURE 3.23 Comparison of N_2 to He ratios for gas from the Hugoton Embayment, Ryersee Field, and other tests in the Central Kansas Uplift are of Kansas (Newell et al., 2003).

The He plant at the Otis-Albert Field also processes gas from the adjacent Reichel Field to the north-northwest. The Reichel Field lies on trend with the Otis-Albert Field and the reservoirs are stratigraphic associated with the Shaffer Ridge or "Russell Rib." Isotopic analysis by Jenden et al. (1988) concluded that all the gases in the Permian and Pennsylvanian reservoirs are nearly identical in composition which suggests that migration occurred through fault and fracture systems along the bounding horst to fill the reservoirs from lower strata (Newell, 2021). Fig. 3.23 is a comparison of N_2 to He ratios for gas from the Comparison of N_2 to He ratios for gas from the Hugoton Embayment and the Mississippian rocks in several basins that demonstrate variability in sources of the gas (Newell et al., 2003).

Reservoir Characteristics of Otis-Albert Field (Miller, 1968):

Productive area:	28,800 acres
Producing formations:	Reagan Sandstone (Cambrian), Arbuckle group (Upper Cambrian–Lower Ordovician), Simpson Sandstone (Middle Ordovician), Pennsylvanian Conglomerate, Lansing–Kansas City carbonates (Pennsylvanian), Topeka Group (Pennsylvanian) and Chase Group (Permian).

(Continued)

Depth of pay:	3527 ft (1070 m)
Net pay:	1–80 ft (1–24 m), average 30 ft (9 m)
Porosity:	12%, range from 5% to 22%
Permeability:	Unknown
Water saturation:	Unknown
Initial field pressure:	1068 psi
Type of drive:	Gas expansion
Type of trap:	Primarily stratigraphic with associated Precambrian structure ridge
Gas characteristics and analysis:	Reagan Sandstone: CH_4 76.4%, CH_2H_6 4.3%, C_3H_8 1.5%, iC_4H_8 0.5%, nC_4H_8 0.8%, iC_5H_{12} 0.6%, nC_5H_{12} 0.2%, C_6H_{14} 0.2%, N_2 12.8%, He 1.6% with a range 1.17%–1.9%, 979 BTU; Chase Group-Pennsylvanian Conglomerate-Kansas City: He 1.58%–2.31%
Production:	15.8 MMBO, 10,282,264 MCFG, estimated He in MCF 169,143 of He based on the Reagan/Simpson sandstone

3.6 Colorado Plateau

The Colorado Plateau contains numerous CO_2 and He deposits in association with the Black Mesa, Holbrook, San Juan, Piceance, Unita, and Henry basins, and associated Uncompahgre and Defiance Uplifts. The Colorado Plateau is a thick crustal block located to the east of the Basin and Range Province and to the west of the Rocky Mountains, encompasses 140,000 square mi (233,000 square km), and is relatively stable geologically and relatively undeformed (Foos, 1999). The Colorado Plateau is located at the intersection of the northwestern-southeastern trending Olympic-Wichita Lineament, the northeast-southwest trending Colorado Lineament (Fig. 3.24), and the margins are marked by major volcanic accumulations and active mountain building by thrusting of the Rocky Mountains. There has been a lack of significant folding and faulting within the Plateau over the last 600 million years. The Colorado Plateau has been like a raft of crustal strata floating in and surrounded by a sea of chaos in the lower crust and mantle. The Colorado Plateau is a large area that represents a wide variety of depositional settings with a similarly wide variety of stratigraphic names, interpretations, and correlations. Presented here is a general summary of the Colorado Plateau. Fig. 3.25 is the present-day structural configuration on top of the Precambrian from Craddock et al. (2017).

The Colorado Plateau is underlain by the Precambrian age Vishnu igneous and metamorphic rocks that are from 1950 to 1680 million years old. The erosional surface on the top of the Vishnu rocks is overlain by sedimentary rocks and basalt flows that extend from 1250 to 750 million years ago. These strata were uplifted into a basin and range type mountain province. This mountain range was eroded and beveled, and then Early Cambrian ocean swept across parts of the plateau. From Cambrian to Permian times, sediments were deposited across the plateau, and toward the end of the Mississippian, the Uncompahgre Mountains were formed.

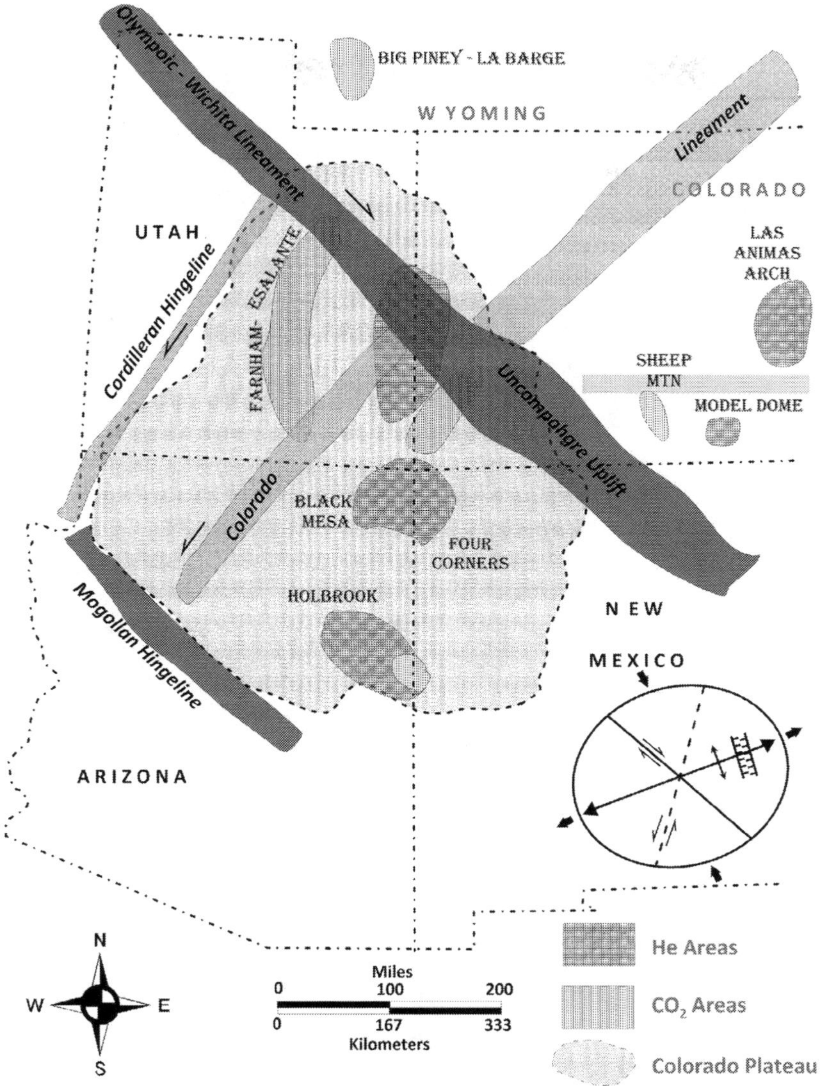

FIGURE 3.24 Map showing location of Colorado Plateau and relationship to a major orthogonal set of lineaments modified from Baars and Stevenson (1981). Northwest-trending lineaments are right lateral, and northeast-trending lineaments are left lateral. Stress—strain ellipsoid is oriented such that maximum compressive stress is directed from the north.

These mountains lay where the present-day Uncompahgre Uplift is located. The Uncompahgre Mountains shed sediment into the adjacent Paradox and Central Colorado basins. The adjacent Paradox Basin accumulated almost four miles (6.4 km) of sediment and evaporites. Most of the strata were deposited in warm shallow seas and near-shore environments with multiple transgression and regression events occurring. During most of the Mesozoic, the plateau was exposed and eroded.

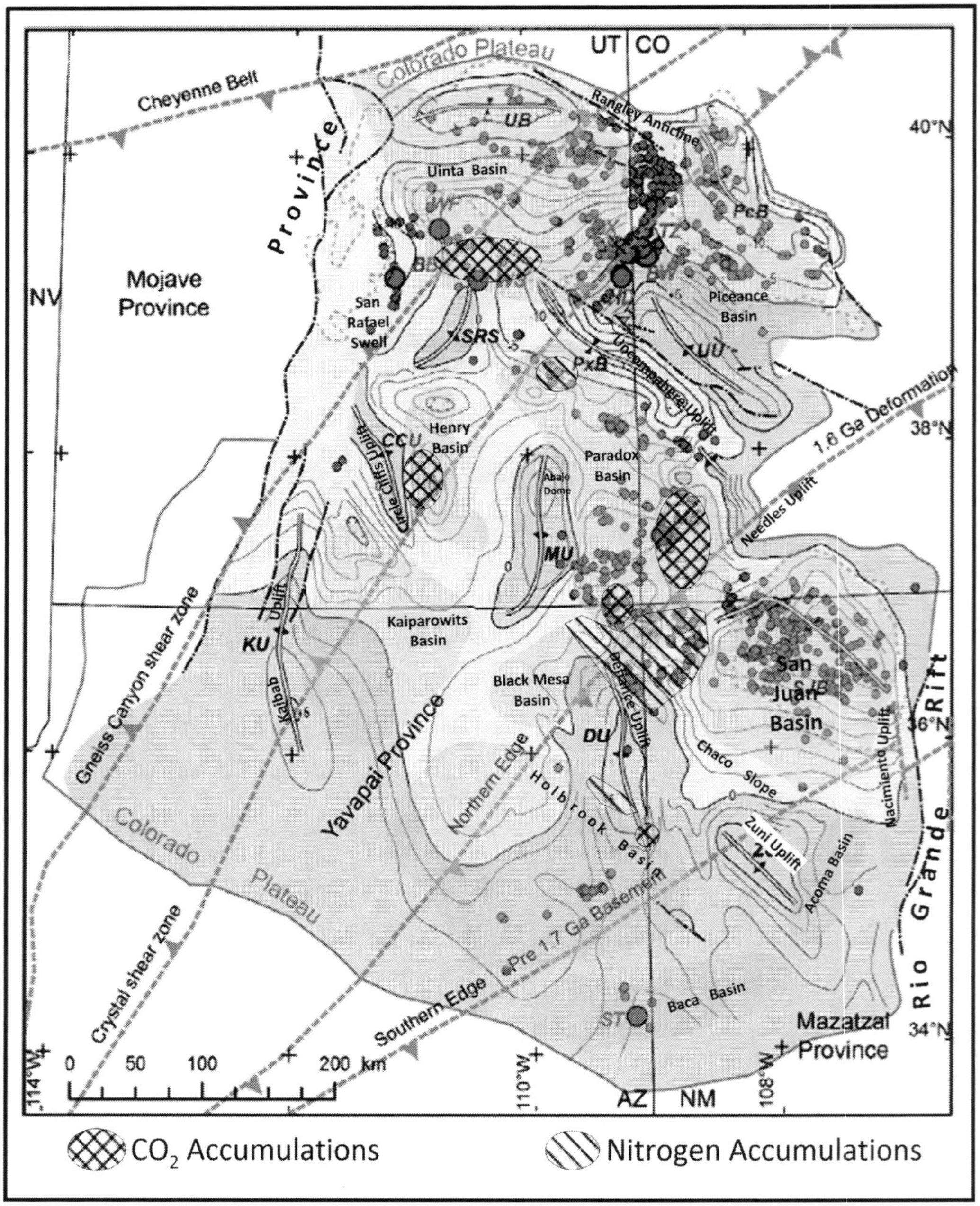

FIGURE 3.25 Map of Arizona (AZ), Colorado (CO), Nevada (NV), New Mexico (NM), and Utah (UT) showing province boundaries, basement structure, and locations of wells which were sampled for gas composition (Craddock et al., 2017). Datum is sea level, contour interval is 1 km. Areas where basement is above sea level are shaded tan and areas where basement is below sea level are shaded light blue. Selected structural highs and lows are labeled as follows: Circle Cliffs Upwarp (CCU), Defiance Upwarp (DU), Kaibab Upwarp (KU), Monument Upwarp (MU), Piceance Basin (PcB), Paradox Basin (PxB), San Juan Basin (SB), San Rafael Swell (SRS), Uncompahgre Uplift (UU), Uinta Basin (UB), and Zuni Upwarp (ZU). Areas of major Cenozoic sedimentation outlined in red Craddock et al. (2017).

The Upper Cambrian (Fig. 3.26) marine rocks comprise the oldest rock unit in the area and are relatively uniform and contiguous through the Paradox, Holbrook, and Black Mesa basins but were not deposited in the Defiance Uplift and Four Corners Platform regions (Fig. 3.26). The Defiance Uplift was a positive feature throughout most of the Paleozoic time. The

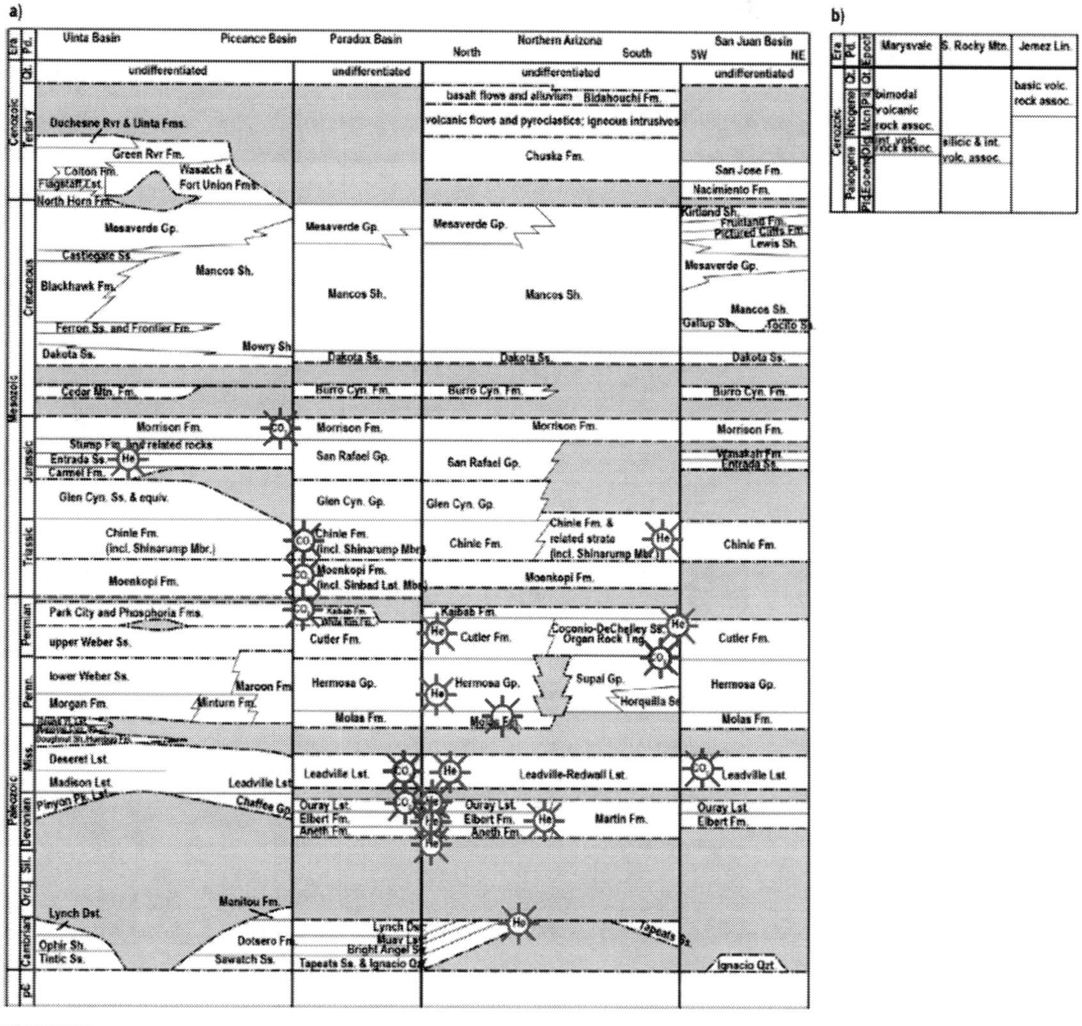

FIGURE 3.26 (A) Phanerozoic stratigraphy of the Colorado Plateau and (B) Cenozoic igneous associations of selected volcanic fields along the margins of the region (Craddock et al., 2017). Reservoirs containing >75 mol percent (%) CO_2 or >5% He are shown by reservoir symbols (circles with spikes) and their locations in the column are based exclusively on the compilation of gas composition data by the US Bureau of Mines and US Bureau of Land Management (see text for details). Dashed lines indicate unconformities and solid lines indicate conformable contacts. Abbreviations are as follows: *assoc.*, association; *Fm.*, Formation; *Fms.*, Formations; *Gp.*, Group; *int.*, intermediate; *Mbr.*, Member; *SS*, Sandstone; *volc.* = volcanic. Purple gas wells are He-producing zones and blue gas wells are CO_2 producing zones, Craddock et al., (2017).

Cambrian strata consist of a basal clastic section overlain by limestone and dolostone that thickens from west and north across the plateau (Fig. 3.26). Ordovician and Silurian rocks are generally absent throughout the area either through nondeposition or erosion.

The Devonian system (Fig. 3.26) is represented by the Aneth Formation overlain by the Elbert Formation throughout most of the area. However, the latter was not deposited or eroded off of the Defiance area at the southern part of the Four Corners Platform. The Devonian system (Gorham, 1975) lies unconformably on Cambrian or Precambrian rocks and seems to have been deposited in localized depressions on the Cambrian surface. The Mississippian strata (Fig. 3.26) overlie the Devonian rocks unconformably and are similar in thickness to previous rock units, but their areal extent is greater (Gorham, 1975). The Mississippian rock units, commonly known as the Madison, Redwall, or Leadville strata, consist predominantly of limestone and dolomites and developed karst features on its surface, indicating that they were subaerially exposed through Early Pennsylvanian times.

Pennsylvanian strata (Fig. 3.26) are composed of clastic, carbonate, and salt sediments, depending upon which basin they were deposited, and are found across most of the plateau except for the Emery, Defiance, and Uncompahgre uplifts. The Pennsylvanian Hermosa Group encompasses all strata of Later Morrowan to early Virgil time (Fig. 3.26). Deposition of Pennsylvanian sediments began in Atokan time with the Sandia, Molas, and Pinkerton Trail formations, which were deposited along the southern edge of the Paradox Basin, the northern edge of the Mesa Basin, and into the Four Corners Platform. The area of the present-day Black Mesa and Holbrook basins remained subaerially exposed in the Pennsylvanian and is known as part of the Piute—Zuni—Defiance Platform. The Atokan rocks are a mixture of carbonates, limey carbonate shales, plus silty, and shaley sandstones. Desmoinesian rocks (Fig. 3.26) had similar lithologies and were deposited across the Mesa and Holbrook basin's area, thus burying the central portion of the Piute-Zuni-Defiance Platform. The Paradox Formation consists of up to 4000 ft (1200 m) evaporites with associated extensive marine carbonates across the Paradox Basin. In adjacent areas, the Paradox Formation consists of carbonates and clastic strata but lacks the evaporites. The overlying Permian rocks were deposited across the plateau, have an average thickness of 2000 ft (666 m), and drowned the Defiance uplift specifically during Wolfcamp and Lower Leonardian times. In the Upper Leonardian and Guadalupe times, the area was subaerially exposed. The Coconino Sandstone of Lower Leonardian age is a major reservoir for He, oil, and gas in Arizona and New Mexico.

During the Triassic (Fig. 3.26), the Colorado Plateau became a basin where predominantly shale, shaley sandstone, and sandstones were deposited as represented by the Moenkopi, Chinle, and Dolores formations. Jurassic rocks are represented by the Navajo, Entrada, and Morrison formations (Fig. 3.26) which are dominated by eolian sandstones that were deposited across the northern part of the area. Volcanic material was sourced from the south and west during Chinle time. The Lower Cretaceous rocks are missing in most of the area and are not until the Upper Cretaceous that numerous carbonaceous mudstones (Mancos, Niobrara, Claggett, Eagle, and Judith units) were deposited. The west southern boundaries of the plateau and the Defiance Uplift had an intermittent effect on deposition.

The Colorado Plateau was affected by tectonic activity related to the formation of the Pangea supercontinent with associated orogenic events such as the Marathon-Ouachita; sedimentation was dominated by terrestrial environments such as the Entrada, Morrison, and Dakota through Cretaceous times. During the Mesozoic, volcanic mountain ranges to the west inundated periodically large areas of the Colorado Plateau with ash debris. In Cretaceous times, the Western Interior seaway opened and deposited the Mancos and Niobrara marls and shales across the plateau.

The Laramide orogeny uplifted the Colorado Plateau as part of a large belt of crustal rocks from Mexico to Montana. The plateau was uplifted generally as a single, rigid block, with adjacent thrust faults causing a slight clockwise movement of the block. Compressional forces were transmitted through the block instead of deforming it. During this orogeny, preexisting weaknesses in Precambrian rocks were generally reactivated by the compression, causing much of the province's relatively small and gentle anticlines, synclines, and monoclines to form. Several mountain ranges of the Plateau have igneous rocks as cores that were emplaced about 70 million years ago. During Upper Cretaceous to Tertiary times, the Colorado Plateau was subject to emplacement of deep plutons, volcanoes, and multiple sills and laccoliths throughout the area. This igneous activity also caused degassing of mantle and basement rocks allowing the migration of He, N_2, and CO_2 which in certain areas accumulated in sedimentary reservoirs or in some cases igneous ones.

The southern part of the Paradox, the San Juan, the Black Mesa, and the Holbrook basins, the Defiance Uplift, and the Mogollon Slope form a complex stratigraphic, structural, and tectonic area (Fig. 3.27). The area also lies across the intersection of the Transcontinental Arch and Uncompahgre Uplift. Some of the highest He values have been measured in this area along with elevated CO_2 accumulations. The State of Arizona, USA, is one of the richest areas in terms of He production, outside of the Texas PHF, and was a major producing region in the 1960 and 1970s (Rauzi, 2003). There are three main tectonic features in Eastern Arizona where He has been produced from the Holbrook, the San Juan, and the Mesa basins, Defiance Uplift, and Mogollon Slope.

Igneous and volcanic activity in the southern part of the Paradox, the norther part of the San Juan, the Black Mesa, and Holbrook basins, the Defiance Uplift, and the Mogollon Slope were generally absent throughout Paleozoic time and did not start impacting the area until Upper Cretaceous time (Craddock et al., 2017). During the Eocene, Oligocene, and Miocene of the mid-Tertiary, there was a mountain building event tied to the underlying Farallon Tectonic Plate steeply subducting, causing intense volcanic activity, and venting huge amounts of ash that created the Superstition, Galiuro, Chiricahua, and Tumacacori mountains. During the Miocene, Pliocene, Pleistocene, and into the current Holocene, the Farallon Tectonic Plate was cut off by the San Andreas Fault's strike-slip motion and creation of a transform boundary that caused significant volcanic activity (Craddock et al., 2017). This resulted in the Farallon Plate subducting more slowly, creating steep normal faults, horst and grabens, and creation of the San Francisco Volcanic Field. The area is still technically active suggesting that there is an ongoing venting of He, N_2, and CO_2 from the mantle.

There are multiple tectonic features within the Colorado Plateau that were caused by the subduction of the Farallon Plate and subsequent volcanism (Craddock et al., 2017). This activity has created numerous structures both as the result of extension (horst and graben development) and creating structures locally within the stratigraphy by intrusion of igneous material. So far all of the He or CO_2 fields are structure related. The He or CO_2 accumulations with no specific preference seem to be in a variety of reservoirs. The accumulations seem to be

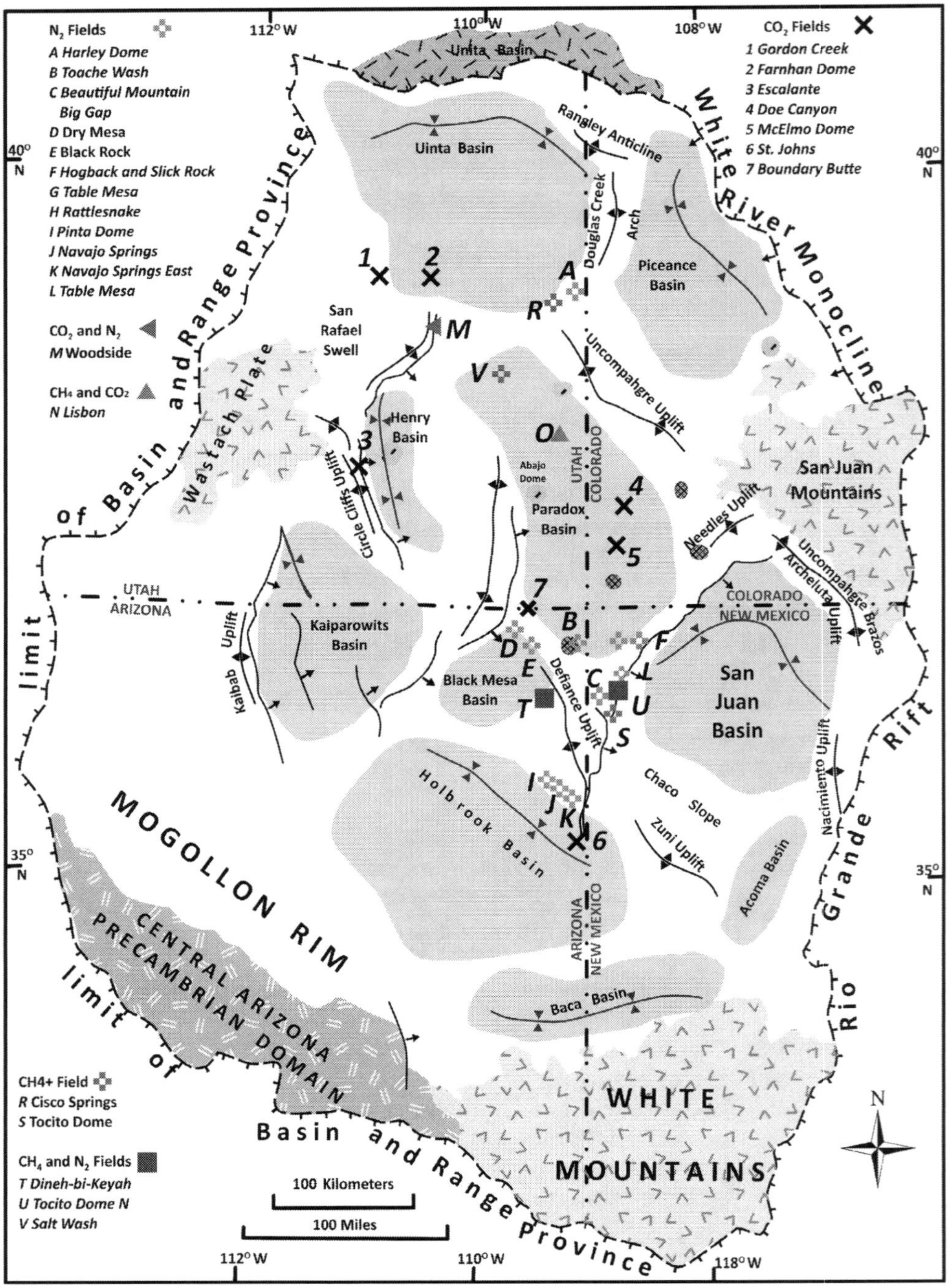

FIGURE 3.27 Location of the southern Paradox, San Juan, Black Mesa, and Holbrook basins, the Defiance Uplift, and Mogollon Slope in Southeast Utah, Southwestern Colorado, Eastern Arizona, and eastern New Mexico. The existing He, CO_2, and N_2 fields have been productive or discovered in the Colorado Plateau.

related to faulted structures which suggest that accumulations rely on a fault and fracture system that extends to basement (Craddock et al., 2017). Groundwater movement also aids in the migration of He and CO_2 to the trap laterally and vertically.

Helium values in the Colorado Plateau area are some of the highest in the United States, Canada, and maybe the world. Fig. 3.28 is a map of the He values from different sources, and as indicated, the greater He values lie in a boundary area between the San Juan and southern Paradox basins, the east side of the Black Mesa Basin, and the Holbrook Basin. The known uranium (U) and thorium (Th) lie away from the anomalous He gas productive areas. The shallow U and Th deposits have clearly not contributed any significant He to local reservoirs. The proximity of the anomalous He values lies in known areas of subsurface Miocene, Pliocene, Pleistocene, and Holocene volcanic activity that is technically ongoing. Uranium and Th deposits at the surface are exhumed potential sources of He but are no longer active. It is more likely the He, N_2, and CO_2 that is anomalous are related to ongoing degassing from

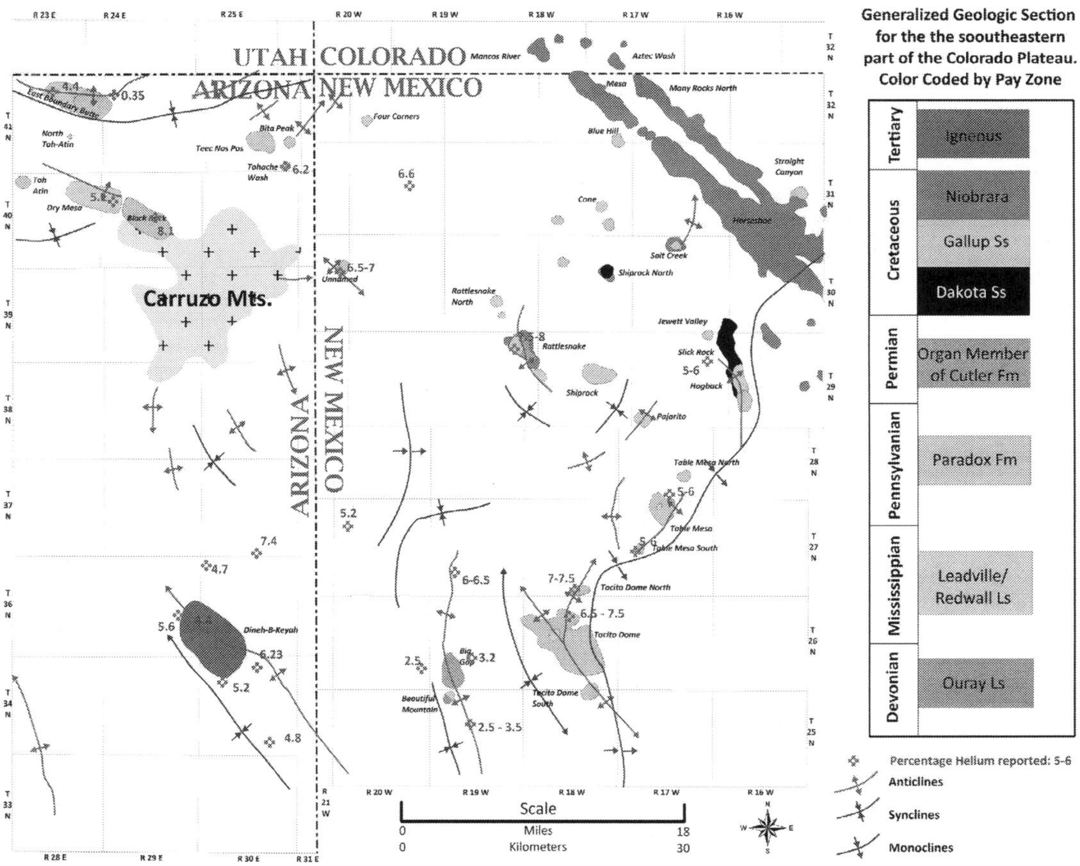

FIGURE 3.28 Structural elements and oil and gas fields of the Black Mesa, Defiant Uplift, and Four Corners Platform that form the He-producing area of the southern portion of the Colorado Plateau. The public He data are indicated where available.

the mantle due to the Farallon Plate subduction that has caused volcanic activity and provides pathways for degassing of noble and associated gases.

3.6.1 Black Mesa—four corners platform, Arizona, and New Mexico

The southeastern portion of the Colorado Plateau (Figs. 3.27 and 3.28) located in Apache County, northeastern Arizona and San Juan County, and northwestern New Mexico has been the focus of extensive and intense exploration for He since the 1940s (Broadhead and Gillard, 2004) and 1960s from Arizona (Razzi, 2003). Helium in the area is exceptionally high and in the Black Mesa, Defiant Uplift, and Four Corners Platform exceeds 7% (Fig. 3.28). A number of fields have been developed such as the Dineh-bi-Keyah (DBK), Rattlesnake, Beautiful Mountain, Hogback, Tocito Dome, and Tocito Dome North to name a few. Several anticlines have had excellent He numbers but limited reservoir, low gas flows, water encroachment, or poor reservoir characteristics. Within some reservoirs such as East Boundary Butte, the He percentages have high variability suggesting compartmentalization.

The Holbrook Basin to the south of the Black Mesa Basin is essentially a single field that has been compartmentalized based on more recently developed structural boundaries into three subfields. The three subfields are Pinta Dome, Navajo Springs, and East Navajo Springs, all of which have produced He (Fig. 3.28). Carbon dioxide reservoirs are also present, such as the St. Johns Field, but they are not as common as compared to other parts of the Colorado Plateau. The rest of the Holbrook Basin lacked exploration or seemed to be barren of any similar economic He or CO_2 accumulations until recently when new discoveries have been made. At time of the printing of this book, those discoveries have not been put into production.

3.6.1.1 Dineh-Bi-Keyah Field

The Dineh-Bi-Keyah (DBK) Field, which in Navajo means "People's or Pasture," is located in Townships 35 and 36 North, Ranges 29 and 30 East, Apache County, Arizona, USA, on the northern edge of the Defiance Uplift adjacent to the Black Mesa Basin (Fig. 3.28). The DBK Field lies on the north end of the Toadlena anticline where it produces oil from an igneous syenite sill that has been intruded into the Lower Pennsylvanian strata (McKenny and Masters, 1968; Danie, 1978), at less than 3864 ft (1,171m). The syenite is a coarse-grained intrusive igneous rock of Tertiary age whose composition is similar to granite but deficient in quartz. The field did not start producing He until 2003. Fig. 3.29 is a structure on top of the sill reservoir in the DBK and enhances the closure on the strata above. Fig. 3.30 is an isopach of the sill which thickens to the south (Danie, 1978). The reservoir creates its own structure which is caused by an increase in the thickening of the sill along the anticline. Fig. 3.31 is a structure on top of the Barker Creek subzone of the Hermosa Group (Pennsylvanian) and defines the northwest-trending anticline. The Barker Creek substage reservoirs of the Hermosa Group He have not been developed (Halford, 2018). Fig. 3.32 is a cross section through the DBK Field indicating the location of the sill within the Hermosa Group and modified from Danie (1978). Helium is likely from a shallow crustal source, based on N_2 and noble gases (Halford, 2018). Note the closure to the southeast is seemingly barren of hydrocarbons and He.

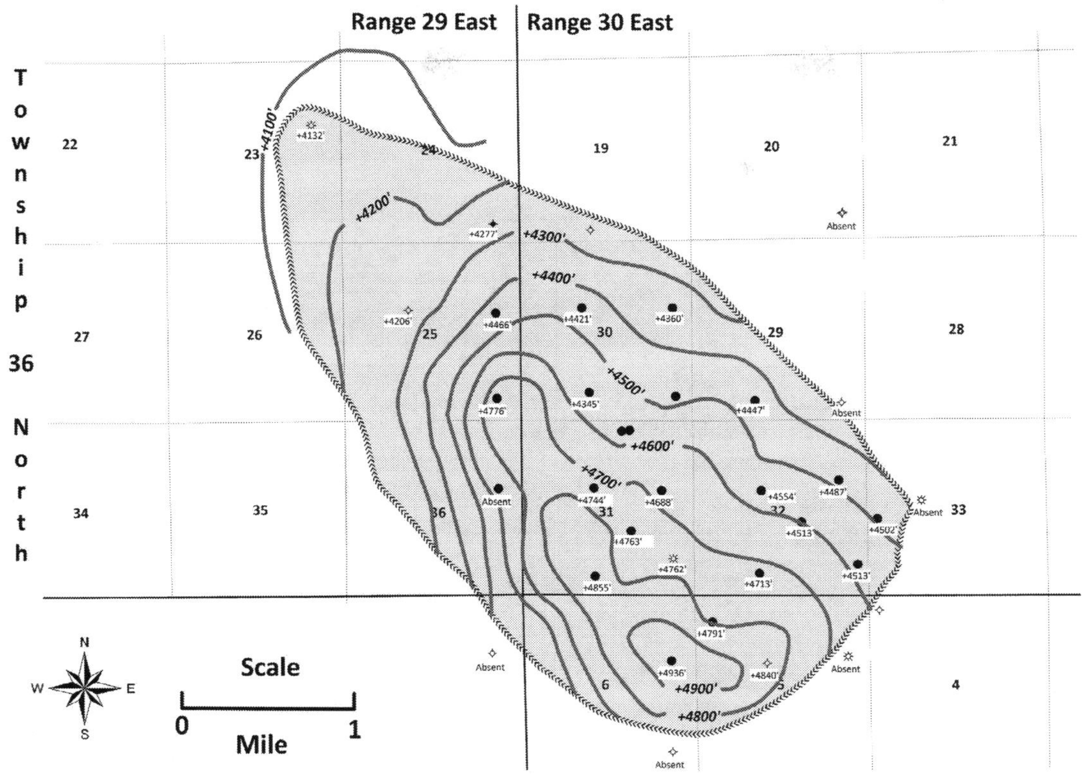

FIGURE 3.29 Structure on top of the sill (reservoir) at the Dineh-bi-Keyah Field. Drilling is as of 1978. Contour interval is 100 ft (33 m) and the structure dips to the northwest and is modified from Danie (1978).

Reservoir Characteristics of Dineh-Bi-Keyah Field (based on oil reservoir characteristics) (Danie, 1978; Rauzi, 2003):

Productive area:	Approximately 3500 acres
Producing formations:	Syenite Sill (Tertiary); Pennsylvanian Paradox Barker Creek substage, and Devonian Elbert Formation McCracken
Depth to pay:	Syenite 2860 to 2,885 ft (866–874 m)
Net pay:	Syenite: 80 ft (26.5 m), range of 60–166 ft (20–50 m)
Porosity:	Syenite: 10% (average), ranges from 5% to 17% (based on core analysis)
Permeability:	Syenite: 0.01 to 25 millidarcies
Lithology:	Tertiary: Syenite
	Paradox Barker Creek substage: carbonates
	McCracken: Sandstone

(Continued)

Water saturation:	Syenite: 35%
Initial field pressure:	Syenite: 729 psi
Type of drive:	Syenite: Solution gas
Type of trap:	Syenite: Primarily stratigraphic but with associated structure
Formation temperature	Syenite sill: 102°F (39°C)
Gas characteristics and analysis:	Syenite: He 4.39%, CH_4+ 30.23%, N_2 63.5%, CO_2 0.1% Paradox Barker Creek substage: 5% He McCracken Sandstone: He 3.49%, CH_4+ 2.5%, N_2 78.3%, CO_2 9.9%
Production:	18,812,252 BO; 6,684,348 MCFG; and at 5.5% He and if extracted from all gas produced (which did not happen) may have yielded: 376,639 MCF He from the syenite sill. The other two reservoirs have not produced.

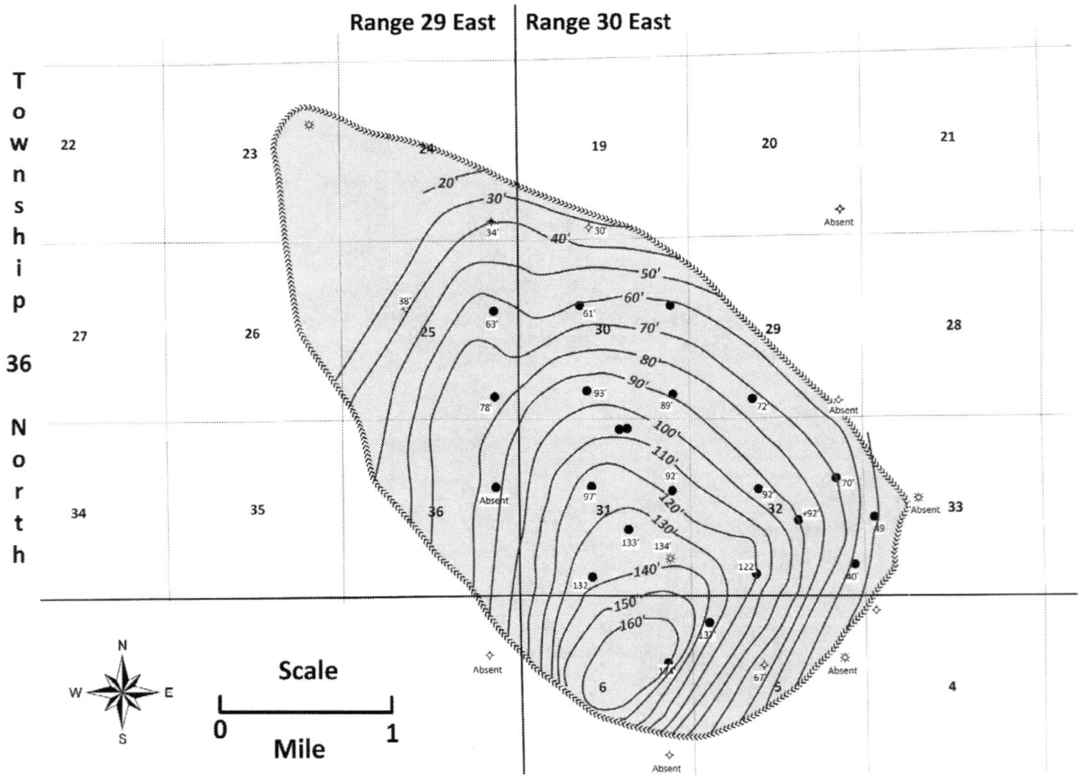

FIGURE 3.30 Isopach on top of the sill (reservoir) at the Dineh-bi-Keyah Field. Contour interval is 10 ft (3.3 m) and thickens to the southwest. *Modified from Danie, T.C., 1978, Dinah-Bi-Keyah, In: Fassett, J.E., Thomaidis, N.D., Matheny, M.L. (Eds.), Oil and Gas Fields of the Four Corners Area, Four Corners Geological Society, vol. 1, pp. 73–76.*

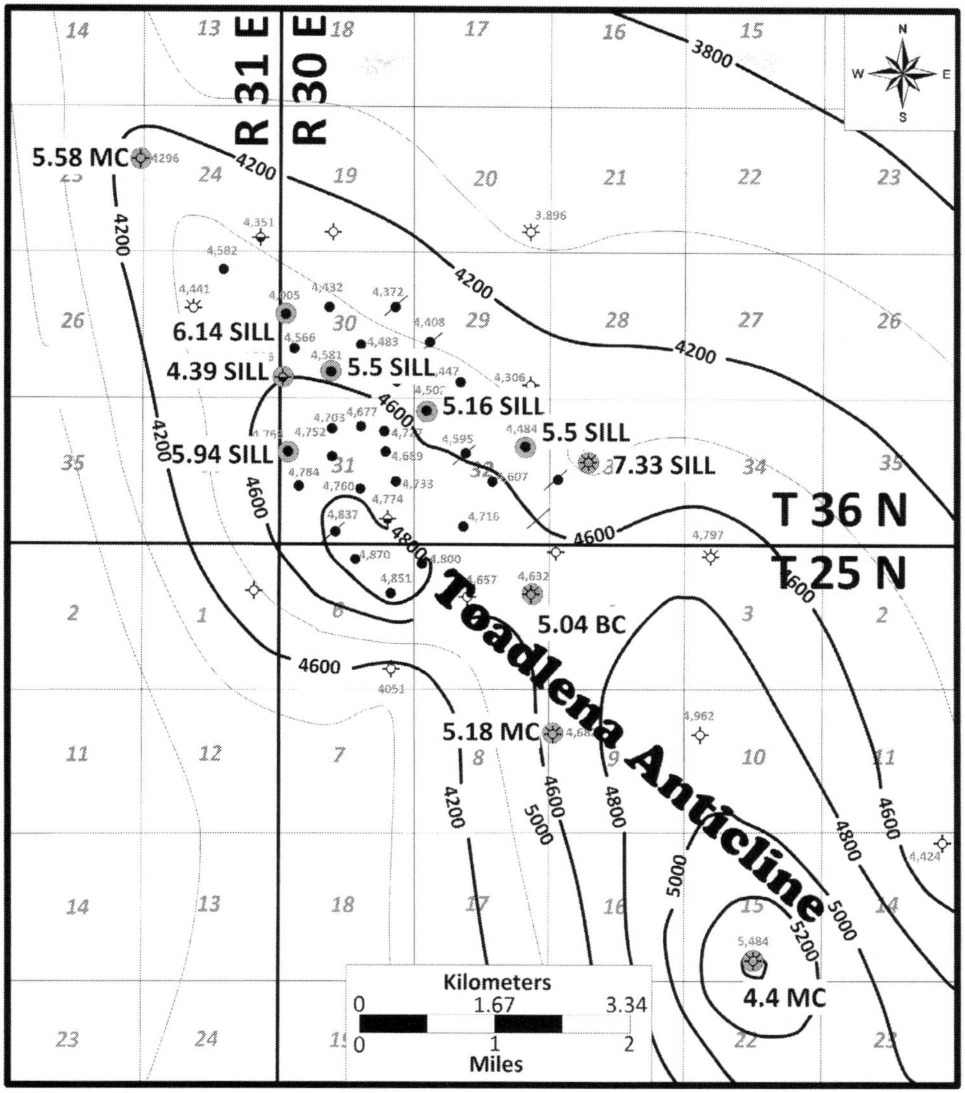

FIGURE 3.31 Structure map on the top of the Barker Creek in the Dineh-bi-Keyah Field (contour interval: 100 ft, 30 m). Barker Creek sample is on a structural high area. Barker Creek is the same reservoir producing hydrocarbons and barely economic levels of He in the Tocito Dome. Seals are likely overlying black shale.

3.6.1.2 Tohache Wash Area

The Tohache Wash Area is located in Township 41 North, Range 30 East, Apache County, Arizona (Fig. 3.28), and produces from the Aneth Formation (Devonian) from 6270 ft (1900 m) and the Leadville Formation (Mississippian) from 6758 ft (2045 m) (Spencer, 1978a). The field produced oil from the Aneth Formation until depleted and then the operator moved

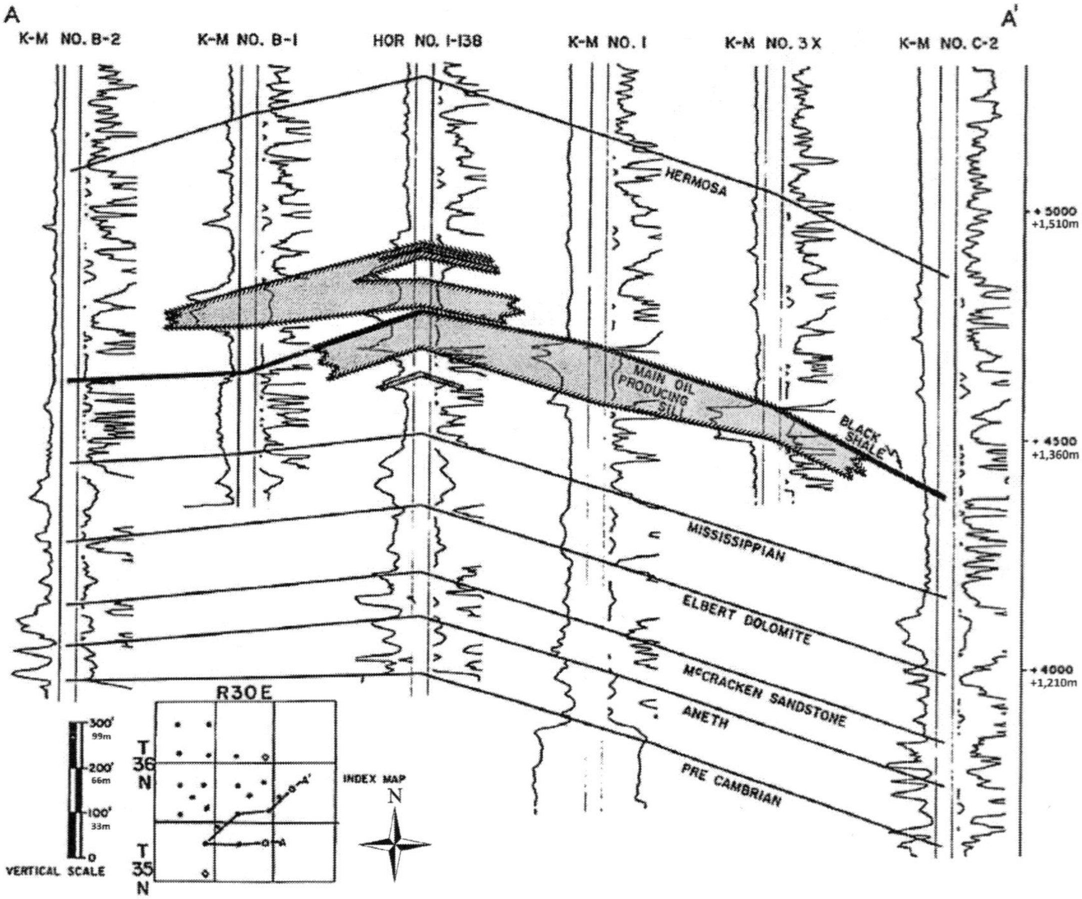

FIGURE 3.32 Cross section through the Dineh-bi-Keyah Field indicating the location of the sill within the Hermosa Group and modified from Danie (1978).

up hole to the Leadville Formation to produce gas for approximately 17 months until low He prices caused plugging of the well. Other oil and natural gas shows were encountered in the Devonian Elbert Formation and Pennsylvanian Hermosa Formation.

Reservoir Characteristics of Tohache Wash Area (Spencer, 1978a):

Productive area:	Unknown
Producing formations:	Leadville (Mississippian) Formation and Aneth (Devonian) Formation
Depth to pay:	Leadville: 6270 to 6320 ft (1900 to 1915 m)
	Aneth: 6758 to 6793 (2047 to 2,058 m)

Pay:	Leadville: 31 ft (9.4 m)
	Aneth: 8 ft (2.4 m)
Porosity:	Leadville: 5%–9%, with average of 7%
	Aneth; 4%–6% matrix porosity with fractures
Permeability:	Unknown
Water saturation:	29%
Initial field pressure:	2395 psi (Aneth), Leadville is unknown
Type of drive:	Aneth: Water
	Leadville: Gas expansion and water
Type of trap:	Structural closure
Formation temperature:	136°F (57°C)
Gas characteristics and analysis:	Aneth gas flammable, no analysis
	Leadville gas: He 5.91%, 54.1% N_2, O_2 trace, Ar 0.7%, CO_2 22.1%, CH_4+ 17.1%
Production:	384,774 MCF of total gas, field is plugged and abandoned

3.6.1.3 *Beautiful Mountain Field*

The Beautiful Mountain Field is located in Townships 26 and 27 North, Range 19 West, San Juan County, New Mexico (Fig. 3.28), and produced oil and gas from the Mississippian Leadville Formation at 5960 to 6040 ft (1806 to 1830 m). The trap is a structural nose that is producing from densely fractured limestone. Also, it is reported that the Devonian age Ouray Limestone is contributing to production (Broadhead and Gillard, 2004).

Reservoir Characteristics for the Beautiful Mountain Field (Brown, 1978, Broadhead and Gillard, 2004):

Productive area:	Approximately 320 acres
Depth to pay:	5960 to 6040 ft (1806 to 1830 m)
Reservoir:	Leadville (Mississippian) and Ouray Formation (Devonian)
Lithology:	Carbonates
Net pay:	8 ft (2.4 m) net pay
Porosity:	Unknown
Permeability:	Unknown
Water saturation:	Unknown
Initial field pressure:	Unknown

(*Continued*)

Type of drive:	Water and solution gas
Type of trap:	Fractures on well-developed structural nose
API oil gravity:	41 degrees
Gas characteristics and analysis:	He 7.14%, CH_4+ 3.41%, N_2 87.83%, CO_2 1.62%
Production:	Leadville/Ouray 2,455,230 MCFG, 175,303 He

3.6.1.4 *Big Gap Field*

The Big Gap Field is located in sections 27, 28, 33, and 34, Township 27 North, Range 19 West, San Juan County, New Mexico (Fig. 3.28), and produced He from the Permian age Organ Rock member of the Cutler Formation at 3900 ft (1,181m). The Organ Rock member of the Cutler Formation consists of poorly indurated sandstones that are interbedded with red shales (Broadhead and Gillard, 2004). The field is a four-way structural closure. The Mississippian Leadville also had significant He shows (Walker, 1978).

Reservoir Characteristics for the Beautiful Mountain Field (Walker, 1978; Broadhead and Gillard, 2004):

Productive area:	Unknown
Depth to pay:	3900 ft (1181 m)
Reservoir:	Permian Organ Rock member, Cutler Formation
Lithology:	Sandstone
Net pay:	Unknown
Porosity:	Unknown
Permeability:	Unknown
Water saturation:	Unknown
Initial field pressure:	Unknown
Type of drive:	Solution gas
Type of trap:	Structural
Gas characteristics and analysis:	He 5.5%
Production:	Permian Organ Rock member, Cutler Formation: 3,260,416 gas, 179,323 He

3.6.1.5 *Rattlesnake Field*

The Rattlesnake Field is located in sections 35 and 36, Township 30 North, Range 19 West; sections 1, 2, 11, 12, and 13, Township 29 North, Range 19 West, San Juan County, New Mexico (Fig. 3.28), and produced gas from the Leadville (Mississippian) Limestone and the Ouray (Devonian) Formation at 7030 ft (2,130m). The pay zones consist of 85 ft (26 m) of porous limestones and dolostones. The Rattlesnake Field has only produced He from the Leadville (Mississippian) Limestone and the Ouray (Devonian) Formation. Other productive zones in the field, the Hermosa Group (Pennsylvanian), Dakota (Cretaceous) Sandstone, and the Gallup (Cretaceous) Sandstone have only trace amounts to 0.3% He (Broadhead and Gillard, 2004).

Reservoir Characteristics of Rattlesnake Field (Baars, 1983; Broadhead and Gillard, 2004):

Productive area:	320 acres
Producing formations:	Leadville (Mississippian) Formation and Ouray (Devonian) Formation
Depth to pay:	6948 ft (2105 m) Open hole
Pay:	Leadville: 60 ft (19 m) or 85 ft (26 m)
Porosity:	Unknown
Permeability:	Unknown
Water saturation:	Unknown
Initial field pressure:	3605 psi (Leadville Formation)
Type of drive:	Unknown
Type of trap:	Structural closure
Gas characteristics and analysis:	Leadville gas: He 7%–8%
Production:	Mississippian Leadville/Ouray Limestone 150 MCF He

3.6.1.6 *Hogback and Slick Rock fields*

The Hogback and Slick Rock fields are located in sections 6, 7, 18, 19, and 30, Township 29 North, Range 16 West; sections 1 and 12, Township 29 North, Range 17 West, San Juan County, New Mexico (Fig. 3.28) (Garvin, 1978; Maynard, 1978a and b; Broadhead and Gillard, 2004), and produced oil from the Dakota, significant He from the Pennsylvanian Paradox Formation from a gross interval at 6166–6632 ft (1868–2009 m), and the Mississippian Leadville Limestone at 6930–7030 ft (2100–2130 m). The Hogback and Slick Rock fields lie on the Hogback structure which is a north-northwest to south-southeast trending anticline with multiple closures along its crest as defined by mapping of the Mississippian strata and Barker Creek member of the Hermosa Group (Pennsylvanian) (Figs. 3.33 and 3.34). The Hogback structure is bounded to the east by the Hogback Monocline (Fig. 3.28) (Maynard, 1978a and b; Broadhead and Gillard, 2004).

The Hogback and Slick Rock fields have a long history of oil and gas exploration and production that began with drilling for the Dakota Sandstone in the 1920s. In the early 1950s, drilling expanded with deeper tests of the Pennsylvanian, Mississippian, and Devonian strata (Maynard, 1978a and b; Broadhead and Gillard, 2004).

The field was initially found by surface mapping and the Dakota produces from 800 to 900 ft (242–275 m). In 1952, Stanolind drilled the 19-13 well to 7,470 ft (2263 m) on the crest of the anticline. The well reached total depth in the Precambrian and had minor amounts of

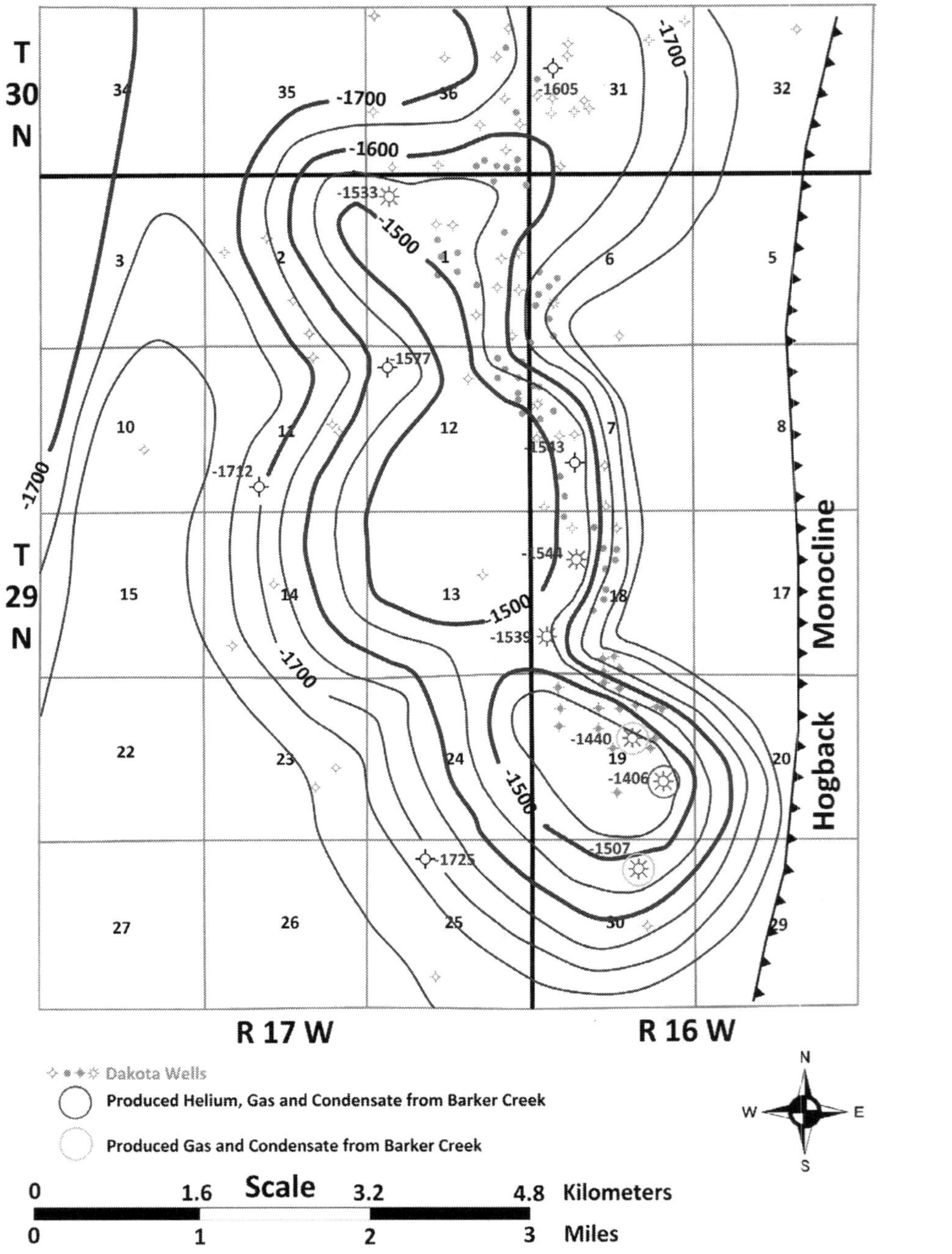

FIGURE 3.33 Structure on top of the Paradox Barker Creek member (Pennsylvanian) at the Hogback and Slick Rock fields. Contour interval is 50 ft (17.5 m).

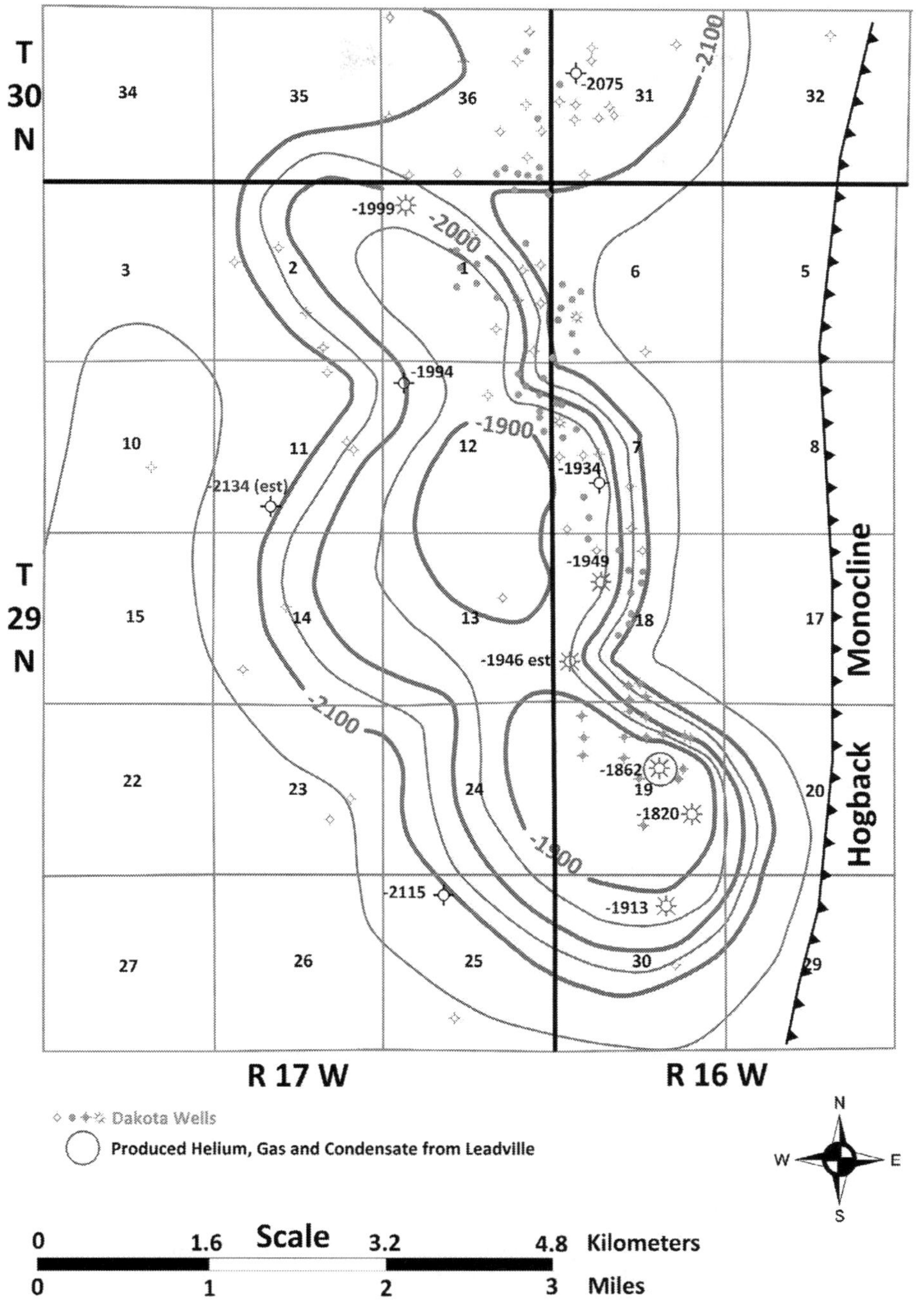

FIGURE 3.34 Structure on top of the Mississippian Leadville Formation at the Hogback and Slick Rock fields. Contour interval is 50 ft (17.5 m).

gas with 5% He produced from the Mississippian strata. The well was eventually plugged back and gas was produced from the Pennsylvanian Paradox Formation (Maynard, 1978a and b; Broadhead and Gillard, 2004). In 1958, Pan American (formerly Stanolind) drilled the 19-13 well and completed as a gas producer from the Paradox Formation with 7% He, 55% N_2, and 29% CH_4+ gas content that expanded the areal extent of the reservoir.

The untested northern part of the Hogback structure was drilled and completed in 1959 at the Navajo C-1 by Pan American in Paradox Formation with 5.7% He. This part of the structure became to be known as the Slick Rock Field (Maynard, 1978a and b; Broadhead and Gillard, 2004). Three additional wells were subsequently drilled, of which the 18-47 well was the structurally highest but had poor well tests and was plugged and abandoned. This result suggests there is a stratigraphic component to the porosity and permeability zones. The amount of He produced from the Hogback structure is unknown.

Reservoir characteristics of Hogback and Slick Rock fields (Garvin, 1978; Lauth, 1978a; Maynard, 1978a and b; Broadhead and Gillard, 2004):

Productive area (both reservoirs):	Approximately 160 acres
Producing formations:	Leadville (Mississippian) and Paradox (Pennsylvanian) Formations
Depth to Paradox pay:	6166 to 6632 ft (1868 to 2009 m)
Depth to Leadville pay:	6930 to 7030 ft (2100 to 2130 m)
Paradox net pay:	16 ft (4.8m)
Leadville net pay:	48 ft (14.5 m)
Paradox porosity:	Unknown
Leadville porosity:	8%–20%
Average permeability:	Unknown
Paradox water saturation:	43%–100%
Leadville water saturation:	Unknown
Paradox initial field pressure:	3175 psi (estimated)
Leadville initial field pressure:	3615 psi (estimated)
Type of drive (both):	Volumetric
Type of trap:	Faulted anticline
Gas characteristics and analysis:	Paradox Formation: He 7.17%, CH4+ 35.4%, N_2 54.6%, Ar 0.6%, CO_2 2%, BTU 454
	Leadville Formation: CH_4+ 10.9%, N_2 63.5%, Ar 0.5%, CO_2 20%, He 5.15%, BTU 214
Production:	Pennsylvanian Paradox Formation: 666,714 gas MCFG; 47,803 MCFHe

3.6.1.7 Table Mesa Field

The Table Mesa Field is located in sections 3, 4, 9, and 10, Township 27 North, Range 17 West, San Juan County, New Mexico (Fig. 3.28), and produces from the Paradox member of the Pennsylvanian Hermosa Group at approximately 7000 ft (2121 m) and the Mississippian Leadville Formation at 7500 ft (2272 m) (Hoppe, 1978; Broadhead and Gillard, 2004). There is a structural closure with a northwest-west and south-southeast trending fault on the north-side of the field (Lauth, 1978b). The field is a combination of structural and stratigraphic trap. Both the Paradox and Leadville pay zones have been abandoned as of 2004.

Reservoir characteristics of the Table Mesa Field (Hoppe, 1978; Broadhead and Gillard, 2004):

Productive area:	Paradox member: 2640 acres
	Leadville: 160 acres
Producing formations:	Leadville (Mississippian) and Paradox (Pennsylvanian) Formations
Depth to Paradox pay:	7010 to 7060 ft (1868 to 2009 m)
Depth to Leadville pay:	7.450 to 7533 ft (2100 to 2130 m)
Paradox net pay:	40 ft (12 m)
Leadville net pay:	17 ft (5 m)
Lithology:	Paradox Barker Creek Stage vugular coral and algal plates carbonate
	Leadville: fracture, intercrystalline fossiliferous limestone
Paradox porosity:	Unknown
Leadville porosity:	Unknown
Average permeability:	Unknown
Paradox water saturation:	Unknown
Leadville water saturation:	Unknown
Paradox initial field pressure:	4140 psi
Leadville initial field pressure:	4295 psi
Type of drive (both):	Water and gas solution drive
Type of trap:	Structure and stratigraphic (porosity trends internally)
Gas characteristics and analysis:	Paradox Member: He 5.37%, CH4+ 11.9%, N_2 77.83%, CO_2 0.19%, BTU 420
	Leadville Formation: He 5.7%, CH_4+ 8.3%, N_2 83.8%, O_2 0.1%, Ar 0.7%, CO_2 1.4%, BTU 131
Production:	Paradox Barker Creek Stage: 7,100,076 MCFG; 381,274 MCFHe
	Leadville: 1,193,006 MCFG; 68.001 MCFHe

(*Continued*)

Gas analysis:	Paradox Formation (Pennsylvanian), He 5.37%, N_2 77.83%, CH_4 16.61%, and CO_2 0.19% (Hoppe, 1978; Broadhead and Gillard, 2004)
	Leadville Formation (Mississippian), He 5.7%, N_2 83.83%, CH_4 8.3%, and CO_2 1.0%
Production:	Pennsylvanian Paradox Member: 7,100,076 MCFG, 381,274 MCFHe
	Mississippian Leadville Limestone: 1,193,006 gas; 68,001 He (Broadhead and Gillard, 2004)

3.6.1.8 Tocito Dome

The Tocito Dome is located in sections 7, 8, 14, 15, 16, 17, 20, 21, 22, 23, 26, 27, 28, and 34, Township 26 North, Range 18 West, San Juan County, New Mexico (Fig. 3.28), and produces from the Paradox "D" member of the Pennsylvanian Hermosa Group at 6350–6500 ft (1925–1970 m). The reservoir consists of fossiliferous, calcareous boundstone, packstone, and grainstone with interbeds of varicolored claystone, siltstone, containing phosphatic nodules (Spencer, 1978b). The field is a northwest-southwest to south-trending structural dome bounded to the east a north-south fault with over 200 ft (60 m) of throw (Fig. 3.35) (modified from Spencer, 1978b; Halford, 2018). The field was found by surface mapping and seismic by Amoco and Texaco. The initial test flowed 5077 MCFGPD from the Amoco Navajo Tribal No. 1 (Spencer, 1978b). Production has also been established in the Mississippian Leadville Formation.

Based on isotopic analysis by Halford (2018), He is derived primarily from shallow crustal rocks that migrated upwards into the Leadville Limestone and Barker Creek member of the Paradox Formation via faulting and leaky seals. There are indications of some secondary migration via groundwater based on ^{20}Ne and ^{36}Ar isotopes (Halford, 2018).

Reservoir Characteristics of the Tocito Dome (Spencer, 1978b; Halford, 2018):

Productive area:	6380 acres
Producing formations:	Pennsylvanian Paradox Formation D member
	Mississippian Leadville Formation
Depth to pay:	Paradox: 6350–6500 ft (1925–1970 m)
	Leadville: 6812 to 6828 ft (2064 to 2070 m)
Net pay:	Paradox: 17 ft (5.1m) net pay, gross thickness of 120 ft (36 m)
Average porosity:	Paradox: 8.6%
Average permeability:	Paradox: Variable, 94 millidarcies
Water saturation:	Paradox: 20%–30%
Initial field pressure:	Paradox: 3217 psi
Type of drive:	Combination water, solution gas, and gravity drainage
Type of trap:	Structural
Gas characteristics and analysis:	Paradox Formation D member: Amoco 3 Navajo: He 0.51%, CH_4+ 79.9%, C_2H_6 7.4%, C_3H_8 2.3%, C_4H_{10} 0.7%, C_5H_{12} trace, N_2 7.9%, CO_2 0.8%, O_2 0.3%, Ar 0.1%, BTU 1030
Production:	Estimated recovery of 15,000,000 BO and 28,000,000 MCFG

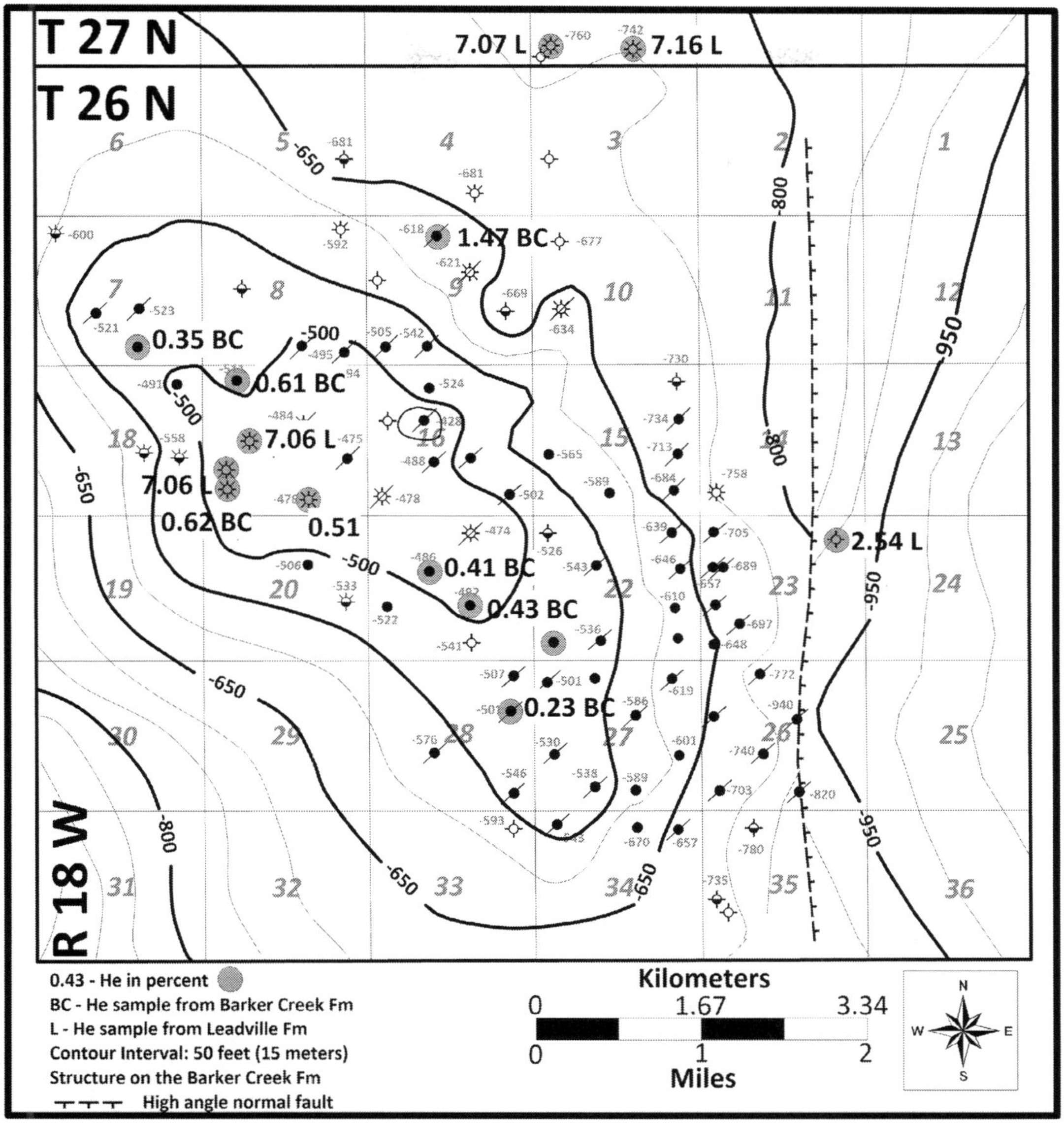

FIGURE 3.35 Structure map on the top of the Barker Creek, contour interval equals 50 ft (15 m) for the Tocito Dome and modified from Halford, 2018. He in percent is indicated next to the well with analysis.

3.6.1.9 Tocito Dome North

The Tocito Dome North is located in section 9, Township 26 North, Range 18 West, San Juan County, New Mexico (Fig. 3.28). The field produces from the Paradox "D" member of the Pennsylvanian Hermosa Group at 6322 ft (1916 m) and the Mississippian Leadville Formation at 6812 ft (2065 m) (Spencer, 1978b). The Pennsylvanian Paradox D Formation reservoir consists of fossiliferous, calcareous boundstone, packstone, and grainstone with interbeds of varicolored claystone and siltstone (Spencer, 1978b). The Mississippian Leadville Limestone has a gross thickness of 180 ft (55 m) and is overpressured. The field is located to the northeast of the northwest-southeast trending Tocito Dome structural anticline that is bound to the east by a north-south fault. The field was found by surface mapping and seismic by Amoco and Texaco. The initial test flowed 384 BOPD and 8275 MCFGPD from the Campbell, Kiel, and Rothwell, Navajo Tocito No. 1 (Spencer, 1978b). The Tocito Dome also produces oil and gas from other Pennsylvanian strata, but the He is 0.51%, N_2 7.9%, and CO_2 0.8%, and the oil API gravity is 47° to 49° API gravity. The Pennsylvanian reservoir rocks are very similar in lithology to the Paradox member of the Hermosa Group.

Reservoir Characteristics of the Tocito Dome North (Riggs, 1978; Spencer, 1978b; Broadhead and Gillard, 2004):

Productive area:	Paradox: 550 acres
	Leadville: 640 acres
Producing formations:	Paradox Barker Creek substage: fossiliferous, calcareous boundstone, and packstone
	Leadville: Carbonates
Depth to pay:	Paradox Barker Creek substage: 6322 to 6328 ft (1915–1917 m) based on perforations
	Leadville: 6812–6828 ft (2065–2069 m)
Average net pay:	Paradox Barker Creek substage: 10 ft (3 m)
	Leadville: 64 ft (19 m)
Average porosity:	Paradox Barker Creek substage: 6%–14%
	Leadville: 10% (estimate)
Average permeability:	Unknown
Water saturation:	Paradox Barker Creek substage: 32%–43%
	Leadville: minor amounts of water produced
Initial field pressure:	Paradox Barker Creek substage: 3155 psi
	Leadville: 3537 psi (wellhead)
Type of drive:	Paradox Barker Creek substage: combination of fluid expansion, solution gas, and water
	Leadville: Gas expansion
Type of trap:	Structural
Gas characteristics and analysis:	Paradox Barker Creek substage: He 3.26%, CH_4+ 60.0%, N_2 34.2%, CO_2 2.1%, O_2 0.1%, Ar 0.3%, BTU 752
	Mississippian Leadville: He 7.0%, CH_4+ 1.2%, N_2 90.3%, CO_2 0.1%, O_2 trace, Ar 1.0%, BTU 21 (sample no. 2)
Production:	Paradox Barker Creek substage: 532,856 MCFG; 17,371 MCFHe
	Leadville: 1,104,668 MCFG; 79426 MCFHe

3.6.2 Holbrook Basin

The Holbrook Basin (Figs. 3.28 and 3.36) produces He primarily from the Permian Ft. Apache limestone and Coconino sandstones. A total of 9.23 BCF of He has been produced from 22 wells from 1961 to 1976. The production is from three fields: Pinta Dome, Navajo Springs, and East Navajo Springs (Razzi, 2003). Exploration in the basin has recently been expanding but historically only these three existing fields have provided any long-term production. Most of the high He percentages have proven disappointing, due to low volume gas flows or limited reserves. The St. Johns Field produces CO_2 from the Supai Formation and Precambrian fractured rocks. The thick salt section (Fig. 3.36) overlying the potentially productive reservoirs acts as an effective seal for underlying reservoirs, but to date a few reported discoveries via press release have been made but no production has been established.

3.6.2.1 *Pinta Dome, Navajo Springs, and East Springs fields*

The Pinta Dome Field is located in sections 28, 32, 33, 34, and 35 of Township 20 North, Range 26 East, and sections 2, 4, and 10, Township 19 South, Range 26 East, Apache County,

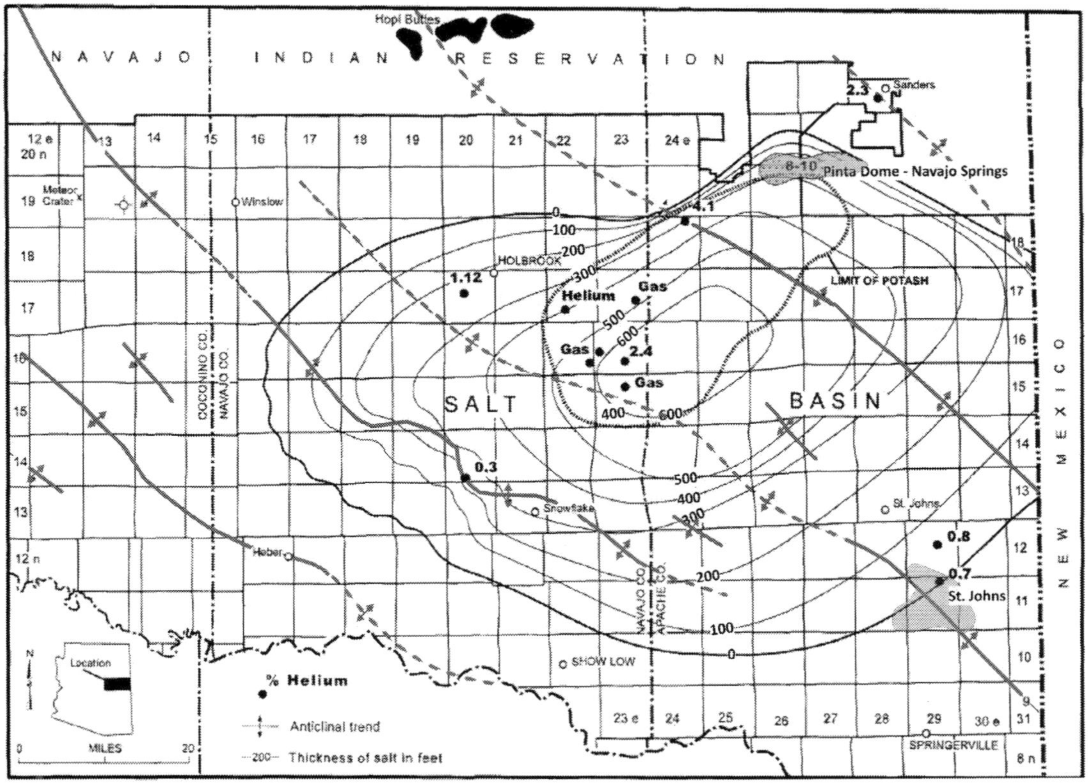

FIGURE 3.36 The Holbrook Basin modified from Rauzi (2003) indicates the anticlines present and the two main fields, one producing He and one producing CO_2. The basin has a thick section of salt which can act as an effective seal.

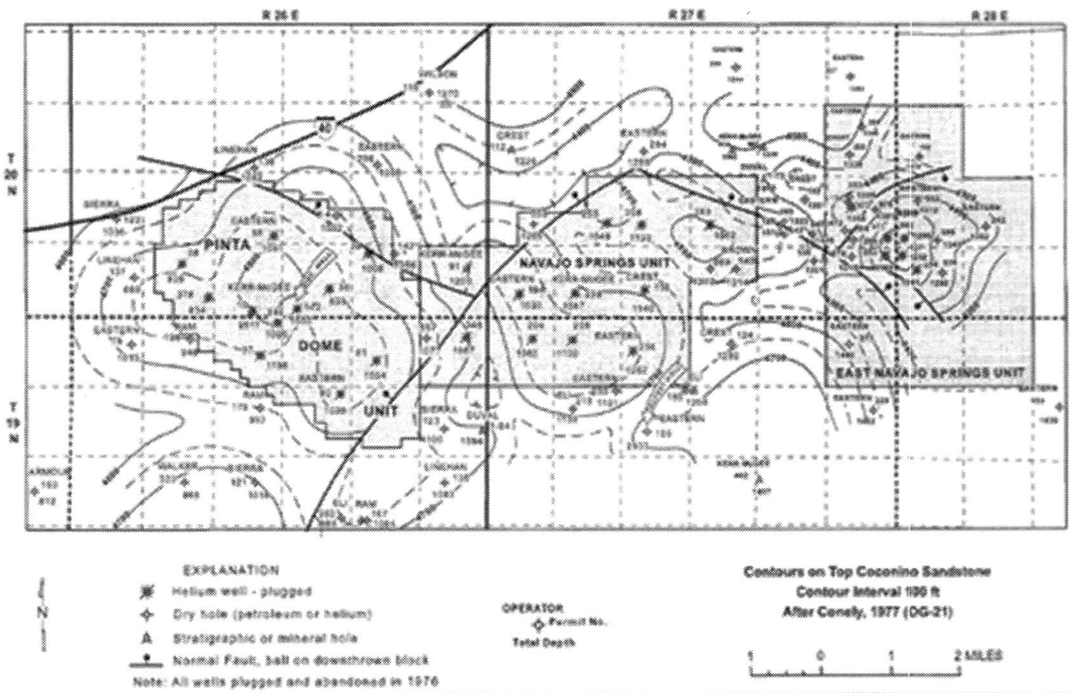

FIGURE 3.37 Pinta Dome, Navajo Springs and East Navajo Springs Units, Holbrook Basin, Apache County, Arizona. Structure on top of the Coconino sandstone (Rauzi, 2003).

Arizona, USA (Figs 3.28 and 3.37), and produces from the Coconino Sandstone of Permian age from 1026 to 1098 ft (311–333 m). Navajo Springs and East Navajo Springs fields are located in sections 30 and 31 in Township 19 South, Range 28 East; sections 25, 28, 29, 31, 32, 33, and 36 of Township 19 North, Range 27 East; and sections 4, 5, and 6 in Township 20 South, Range 27 East, Apache County, Arizona, USA. These two fields have produced He from the Permian age Coconino sandstone (Allen, 1978a,b) from 1140 ft (315 m). The Pinta Dome, adjacent Navajo Springs, and East Navajo Springs fields are productive structural closures defined by horst and graben formation (Fig. 3.38). The pay zone, the Coconino sandstone, is continuous across all three fields. Based on Halford (2018), using N_2 and noble gases, the He originates from the shallow crust. The isotopes further indicated there is significant groundwater–gas interaction, large amount of fractionation, and the He has migrated over a significant distance from its source.

Reservoir Characteristics of the Pinta Dome Field (Allen, 1978a; Rauzi, 2003):

Productive area:	Approximately 4926.25 acres
Producing formations:	Coconino sandstone (Permian)
Depth to pay:	1026 to 1100 ft (311–334 m)
Net pay:	61 ft (21 m)
Porosity:	14%
Permeability:	110 millidarcies (average)

Water saturation:	29%
Initial field pressure:	102 psi (estimated)
Type of drive:	Gas expansion with northward tilted gas expansion
Type of trap:	Faulted anticline
Gas characteristics and analysis:	He 8.2%, N_2 90.0%, CH_4 0.2%, and other 1.6%
Production:	6,533,976 MCFG; field is plugged and abandoned

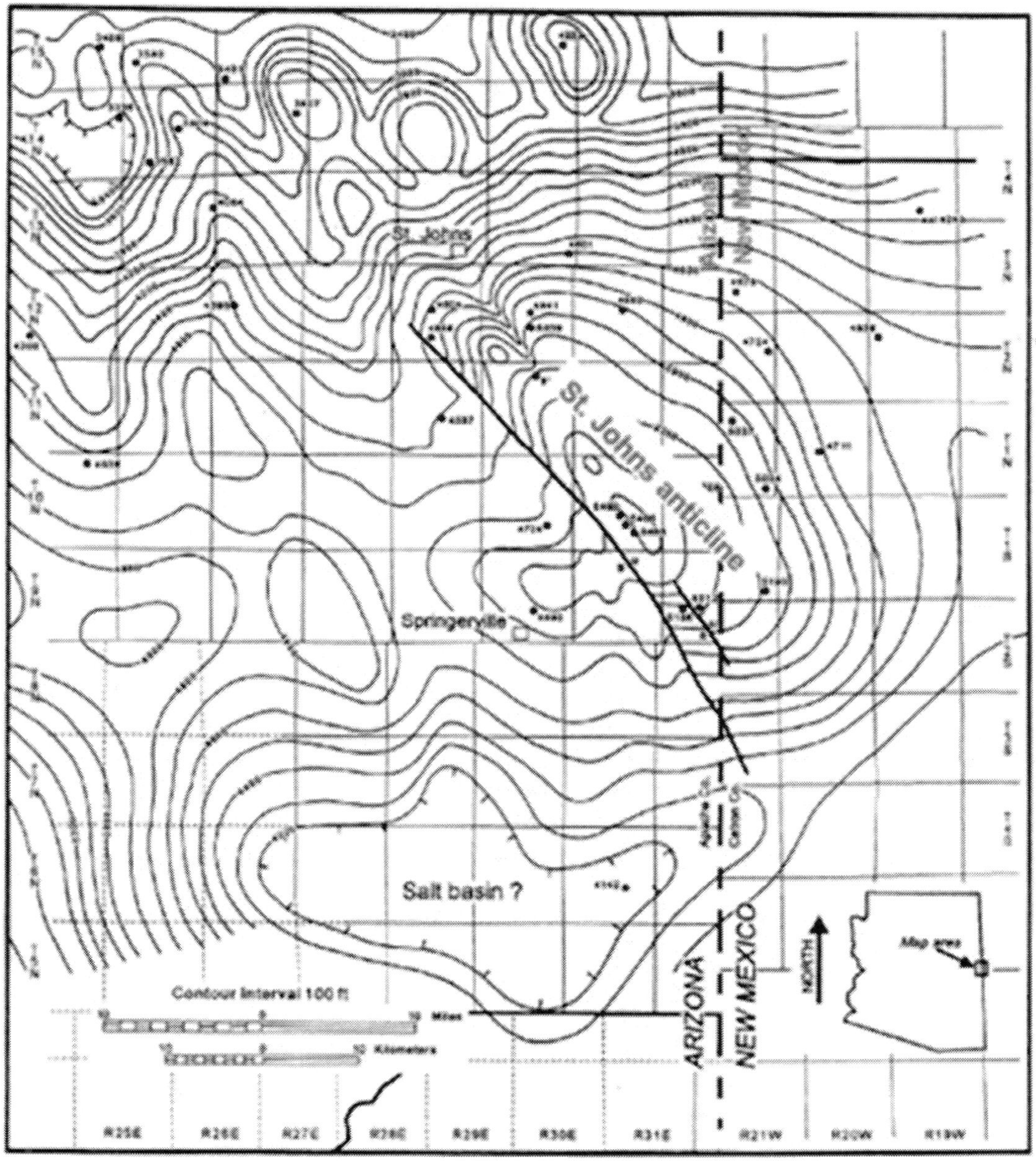

FIGURE 3.38 Structure on the base of Fort Apache Limestone Member of Supai Formation (Permian, equivalent to part of middle part of Yeso Formation), Apache County Arizona and western most Catron County, New Mexico (Rauzi, 1999).

Reservoir Characteristics of the Navajo Springs and East Navajo Springs fields (Allen, 1978b; Rauzi, 2003):

Productive area:	Approximately 4425 acres
Producing formations:	Coconino sandstone (Permian)
Depth to pay:	1062 to 1072 ft (322–325 m)
Net pay:	32 ft (10 m)
Porosity:	16%
Permeability:	164 millidarcies (average)
Water saturation:	32%
Initial field pressure:	124 psi (estimated)
Type of drive:	Gas expansion with northward tilted gas–water contact
Type of trap:	Faulted anticline
Gas characteristics and analysis:	He 8.2%, N_2 89.9%, CH_4 0.2%, and other 1.7%
Production:	2,308,403 MCFG; field is plugged and abandoned

3.6.2.2 St. Johns and Springerville Fields

The St. John's Gas Field is located in Township 9 North, Range 31 East; Township 10 North, Ranges 30 and 31 East; Township 11 North, Ranges 29, 30, and 31 East; Township 12 North, Ranges 28, 29, 30, and 31 East, Apache County, Arizona; Township 2 South, Ranges 19, 20, and 21 West; Township 1 South, Ranges 19, 20, and 21 West; Township 1 North, Ranges 19, 20, and 21 West; Township 2 North, Ranges 19, 20, and 21 West; and Township 3 North, Ranges 20 and 21 West, Catron County, New Mexico, and was discovered in 1959 (Figs. 3.28 and 3.38). The St. John's Field is located in the transitional area between the Colorado Plateau and the Basin and Range and Rio Grande Rift along the Mogollon Slope (Gilfillan et al., 2008). The field produces CO_2 and He from the Supai Formation (Rauzi, 1999). The reservoir is dissected by the major Coyote Wash northwest-southeast trending fault and has seals of impermeable anhydrites that separate the individual sandstone reservoirs (Gilfillan et al., 2008). There has been minor He production from the field (Broadhead, 2009). On the New Mexico side of the field, He has been measured at around 0.2% but increases to 8% along with 6%–10% N_2 on the Arizona side of the basin (Broadhead et al., 2009). The Springerville Volcanic Field lies approximately 30 mi (50 km) to the southwest of the St. John's Field that was emplaced in the Pleistocene to Holocene time. The volcanic field is a combination of a mantle and ocean–island type source with an early 5–6 Ma olivine pyretic basalt and a later (0.3–2.1 MY) alkali basalt (Rauzi, 1999; Gilfillan et al., 2008).

Reservoir Characteristics of St. Johns and Springerville fields (Rauzi, 2003; Gilfillan et al., 2008; Broadhead et al., 2009; Eppink et al., 2014):

Productive area:	220,000 acres
Depth to pay:	1526 ft (462 m)
Pay zone:	Supai Formation arkosic sandstone (Permian), fractured Precambrian basement
Net pay:	Unknown
Porosity:	10%–15%
Permeability:	10 md
Water saturation:	Unknown
Initial field pressure:	Unknown
Type of drive:	Gas expansion
Type of trap:	Doubly plunging anticline
Gas characteristics and analysis:	CO_2 99.42%, CH_4+ 0.13%, and He 0.2%–8+%
Production:	377,896 MCFCO_2

3.6.3 Utah and Southwestern Colorado

The State of Utah has at least two fields that have historically produced He, the Salt Wash and Harley Dome fields. There are several other fields in the area with anomalous high He. These include the Boundary Butte and Lisbon fields (San Juan County); Cisco Springs Field, San Arroyo Area, and Big Flat Area (Grand County); Woodside Dome Field, Temple Spring Prospect, and Bow Knot Prospect (Emery County); Clay Basin Field (Daggett County); and Grassy Trail Area (Carbon and Emery Counties) (Fig. 3.39) (Wiseman and Eckels, 2020). Harley Dome and Temple Spring Prospect are the only two areas with reported He concentration above 2%. All the rest are 1.8% or lower (Wiseman and Eckels, 2020).

There are active faults within the Colorado Plateau leaking CO_2. Examples include the Salt Wash and Little Grand Wash faults near the town of Green River Grand County, Utah (Fig. 3.40) (Heath et al., 2009; Dockrill and Shipton, 2010; Urquhart, 2011; Han et al., 2019). The Little Grand Wash Fault was evaluated for modeling to define what happens when CO_2 is injected into an underground reservoir. The result is a model that defines CO_2 migration to the surface for the area (Fig. 3.41). Urquhart (2011) determined at the Little Grand Wash Fault that the expected mineralization resulting from CO_2 migration and alteration in the fault zone and host rock did not occur. They determined that the chemistry of the CO_2 system and host rock dictates what alteration and mineralization will occur. However, the chemical alteration may have happened but, as the measurements were made at the

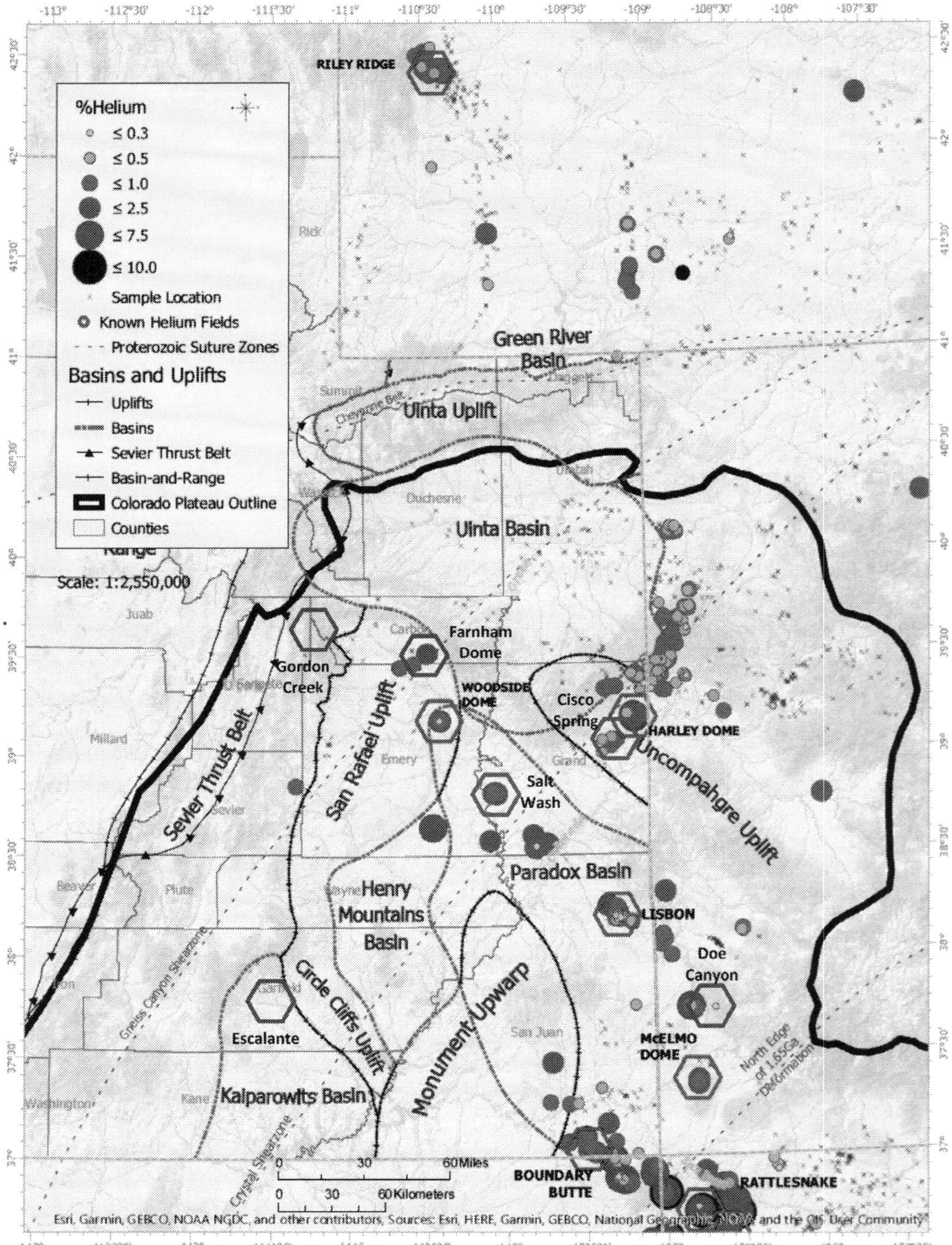

FIGURE 3.39 Location of He and CO_2 fields in Utah and modified from Wiseman and Eckels, 2020. Purple hexagon indicates CO_2 fields discussed in this text.

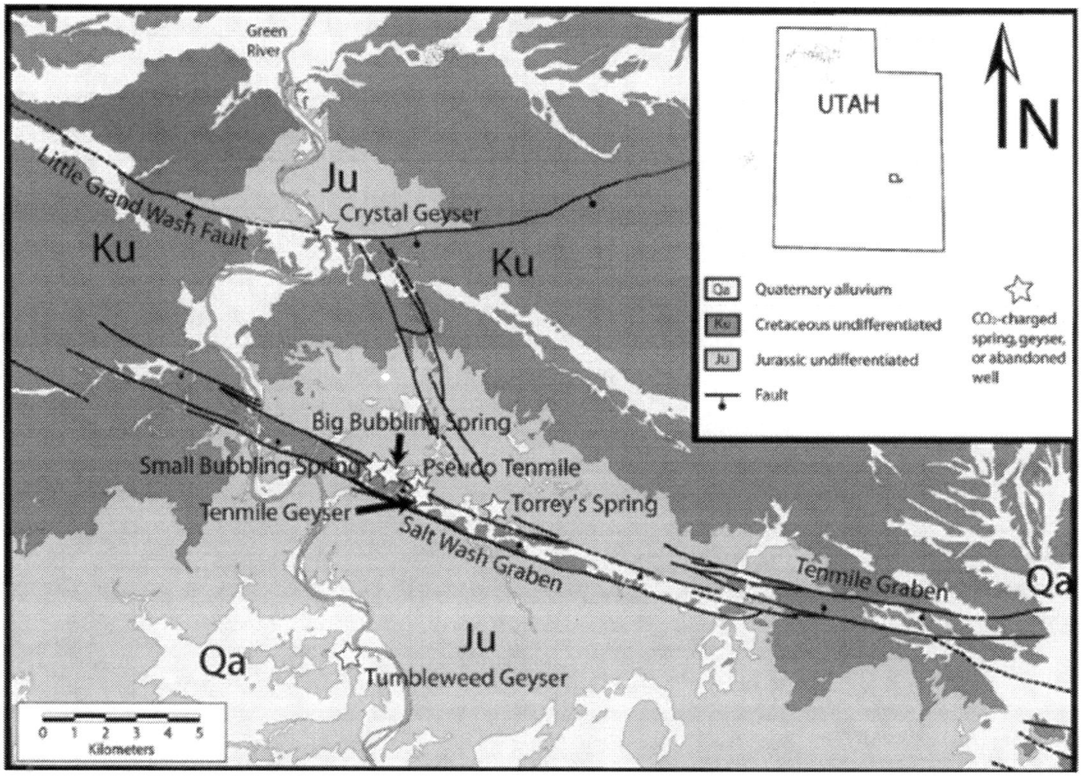

FIGURE 3.40 Geologic map showing the Little Grand Wash and Salt Wash Fault zones and locations of springs, geysers, and groundwater from abandoned wells modified from Heath et al., 2009.

surface (Urquhart, 2011), may have been removed by surface processes. The result of these studies is that there is an active system moving CO_2 from a deep source, whether from magmatic activity in the deep crust, clay–carbonate–metamorphic reactions and/or groundwater recharge, filling various reservoirs during migration and at the same time destroying existing hydrocarbon and He-rich accumulations (Fig. 3.42). The fault system present in Figs. 3.40 and 3.41 lies close to the Salt Wash, Cisco Dome, and Cisco Springs that are discussed below.

3.6.3.1 Harley Dome

The Haley Dome Field was discovered in 1925 and is located in section 8 of Township 19 South, Range 25 East, Grand County, Utah (Bon, 1999) (Figs. 3.39 and 3.43). The geology of the area is dominated by Cretaceous to Triassic Jurassic rocks that are less than 3500 ft (1150 m thick), underlain by Precambrian basement (Fig. 3.44). The Mancos is at the surface

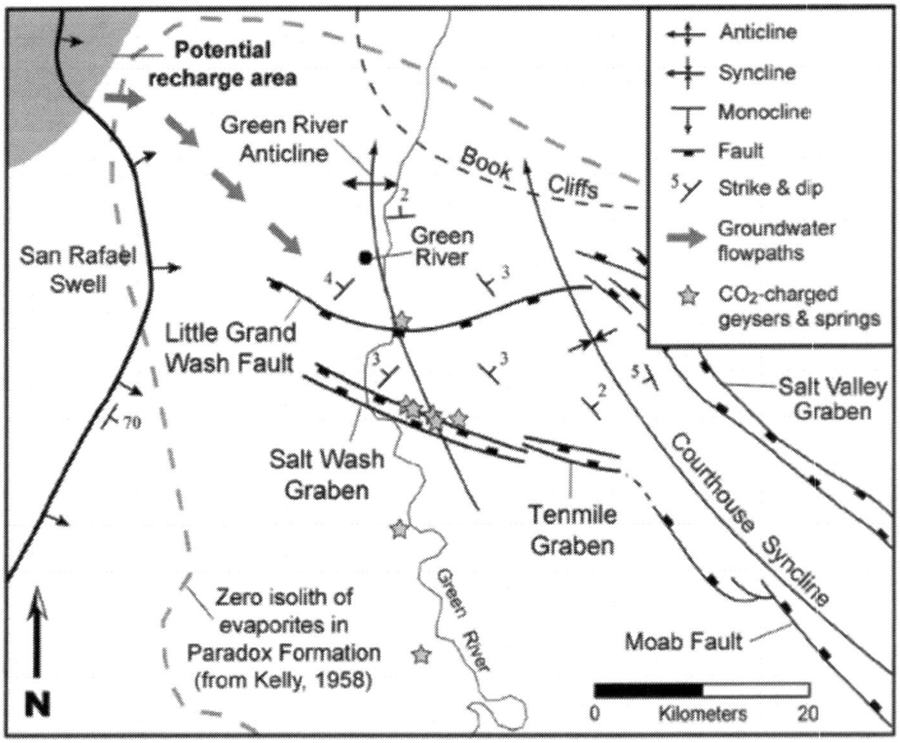

FIGURE 3.41 Map showing direction of groundwater flow near and in the area of potential recharge. Gray arrows represent groundwater flow paths determined from the potentiometric surface of the Navajo sandstone (Heath et al., 2009).

and is overlain by the Dakota Sandstone of Cretaceous age (Fig. 3.44). The Dakota in the area consists of fluvial deposits and can be divided into three sandstone benches across the field (Bon, 1999). The trapping mechanism for petroleum or He in this reservoir is closure of fluvial channels across structural highs, change in direction of channels, and permeability/porosity pinch-outs. The underlying Cedar Mountain Formation (Fig. 3.44) has similar reservoir and depositional attributes as the Dakota Sandstone. Underlying the Dakota is the Morrison Formation of Late Jurassic age (Fig. 3.44) which is divided into the Brushy Basin and Salt Wash members. Both members are similar in trapping and reservoir configuration to the overlying Dakota Sandstone. The Entrada zone of Middle Jurassic age (Fig. 3.44) is the main producing zone at Harley Dome. The Entrada is regional sheet sandstone composed of very fine to coarse-grained sandstones of mostly eolian origin. These sands trap hydrocarbons, He, N_2, and CO_2 when there is a structural closure present. There is no oil in any of the reservoirs at Harley Dome and the amount of natural gas in the producing area is less than 15% from the main producing zone (Fig. 3.44).

The Harley Dome structure is a northwest-southeast trending, down to the NW nose (Fig. 3.45). Fig. 3.46 is a structural cross section through the Harley Dome Field (Bon, 1999). The Harley Dome Field is part of the Greater Cisco Field that includes a series of

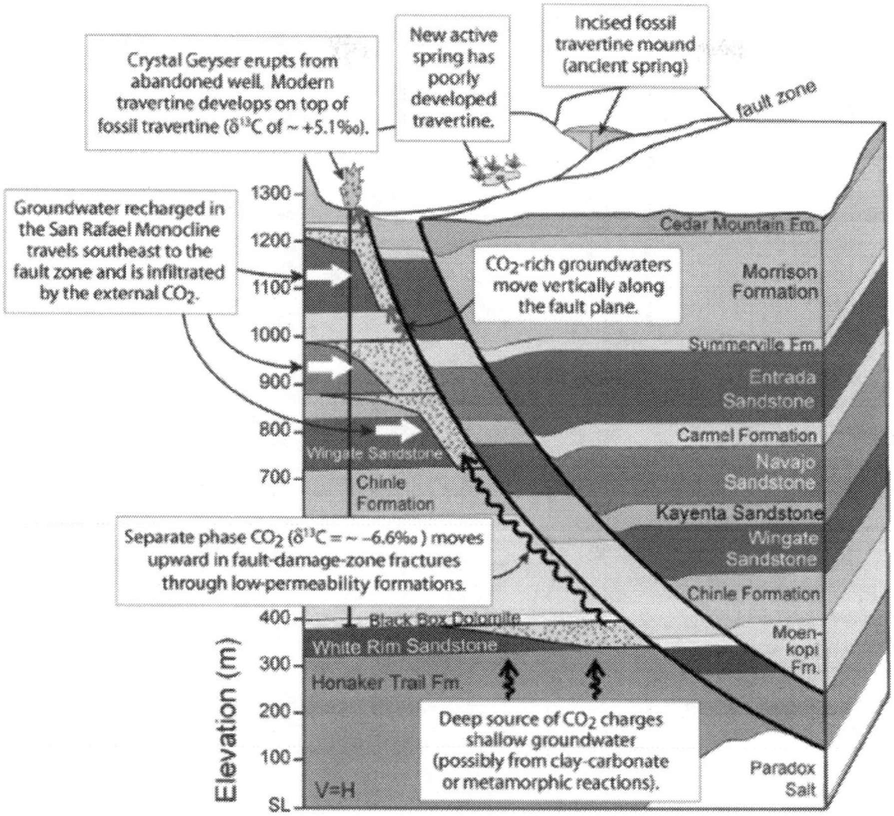

FIGURE 3.42 Graphical presentation of the flow system superimposed on a north-south cross section perpendicular to the Little Grand Wash Fault zone (Heath et al., 2009). North is to the left of the cross section.

northwest-southeast–trending structural domes, anticlines, monoclines, and closures along the Uncompahgre Uplift. The Uncompahgre Uplift is an arch that forms the boundary between the Paradox, Piceance, and Uinta basins. The State of Utah has designated 4920 acres for the field but the productive area is less than 320 acres. Table 3.1 indicates the gas quality from some of the wells at Harley Dome (Bon, 1999). Table 3.2 is the recent production from the field (Wiseman, 2019; Wiseman and Eckels, 2020). The gas stream is dominated by 38.7%–86.1% N_2 (Wiseman and Eckels, 2020).

Production from the Harley Dome from December 2014 to March 2018 is in Table 3.2. The Harley Dome Field has produced 42 MCF He at a rate of 50 MCFHEPD over a 30-month period. The low recovery in He despite its richness could be related to the shallow nature of the reservoir and trap and low pressure associated with the area.

3.6.3.2 *Salt Wash Field*

The Salt Wash Field is located in sections 16 and 17 East, Township 23 South, Range 17 East, Grand County, Utah (Fig. 3.39), and produced from the Leadville Limestone

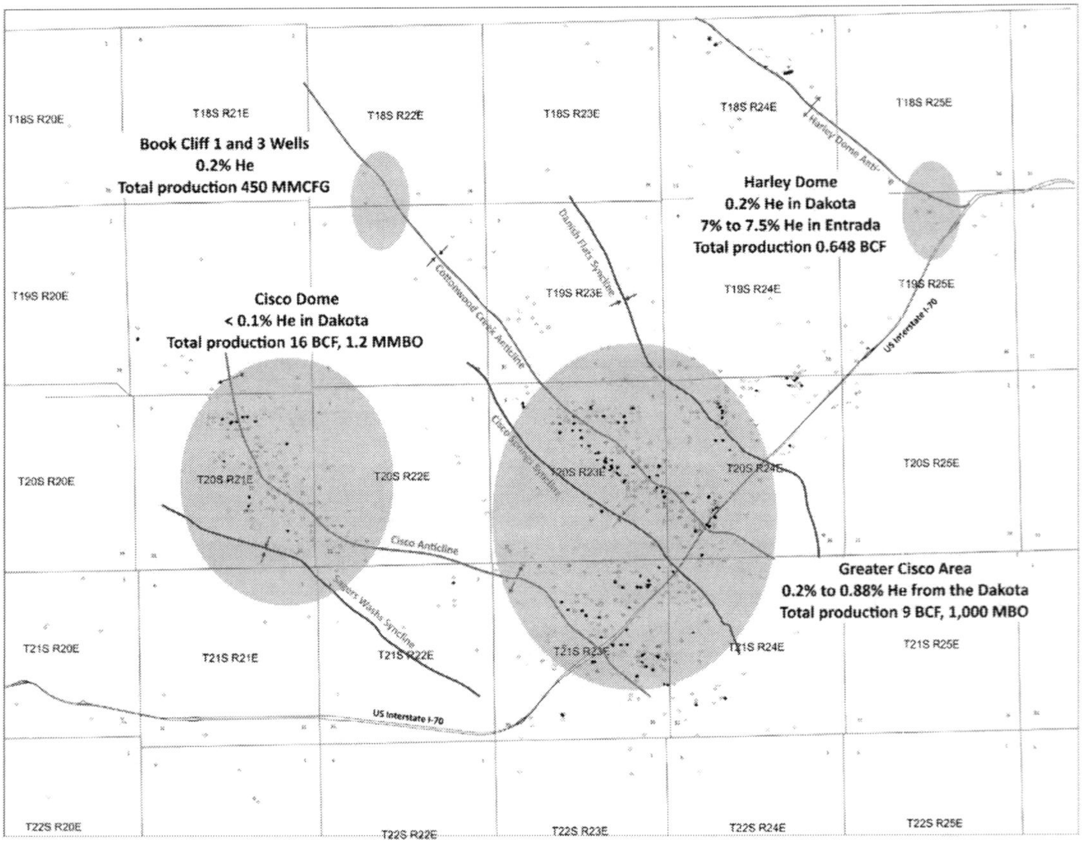

FIGURE 3.43 Map of the Harley Dome, Cisco Springs, and Cisco Dome Fields, Grand County, Utah. The Cisco Dome Field (yellow in color) lacks any significant He which is consistent with the lack of granitic intrusions (Chapter 4).

(Mississippian) from 8400 to 8700 ft (2545–2635 m). The field has been plugged and abandoned but has produced 1,331,512 BO and 10,739,012 MCFG according to the State of Utah records. The field contained significant He assets that were vented through its productive life.

The field produced the majority of oil and gas from two wells. The Devonian Elbert Formation contributed small amounts of oil (Morgan, 1994). The structure on top of the Mississippian Leadville Limestone in Fig. 3.47 indicates a north-northwest plunging structural nose. Production is associated with isolated porosity zones that consist of 5–10 ft (1.7–3.3 m) stringers across a structural closure in the Leadville Formation. According to Morgan (1994), the Salt Wash structure is associated with a high angle east-west fault, down to the south, that moved during Late Mississippian to Early Pennsylvanian times. The fault has been identified on seismic profiles (Morgan, 1994). Salt movement underlying the structure has caused a considerable amount of fault movement in and around the field which juxtaposed organic rich shales of the Paradox Formation (Pennsylvanian) against the

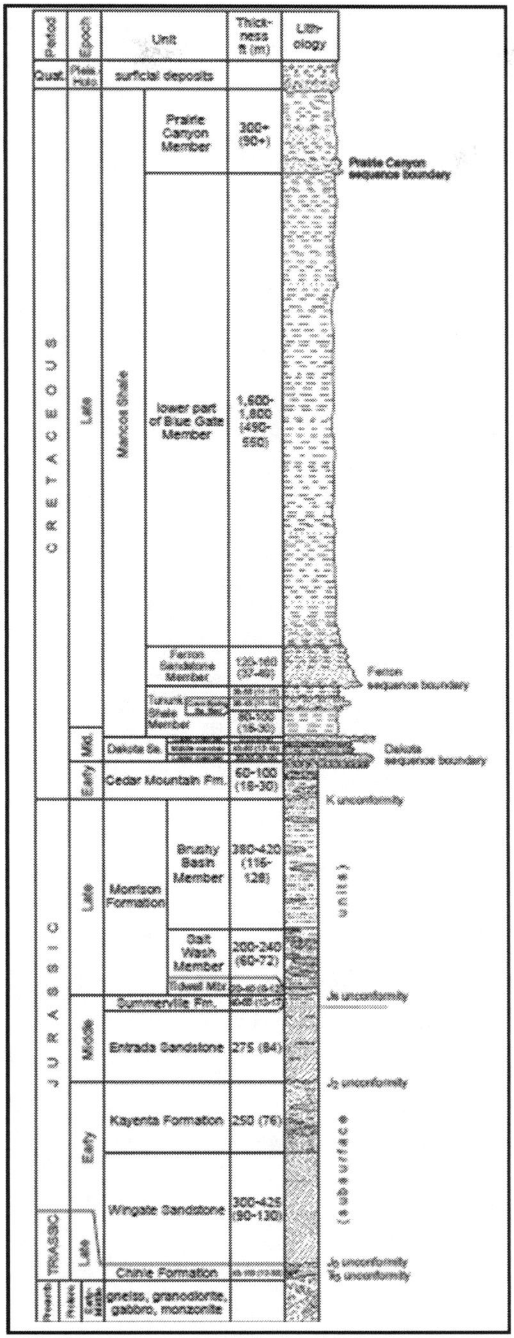

FIGURE 3.44 Stratigraphic column for the Harley Dome Field, Grand County, Utah (Bon, 1999).

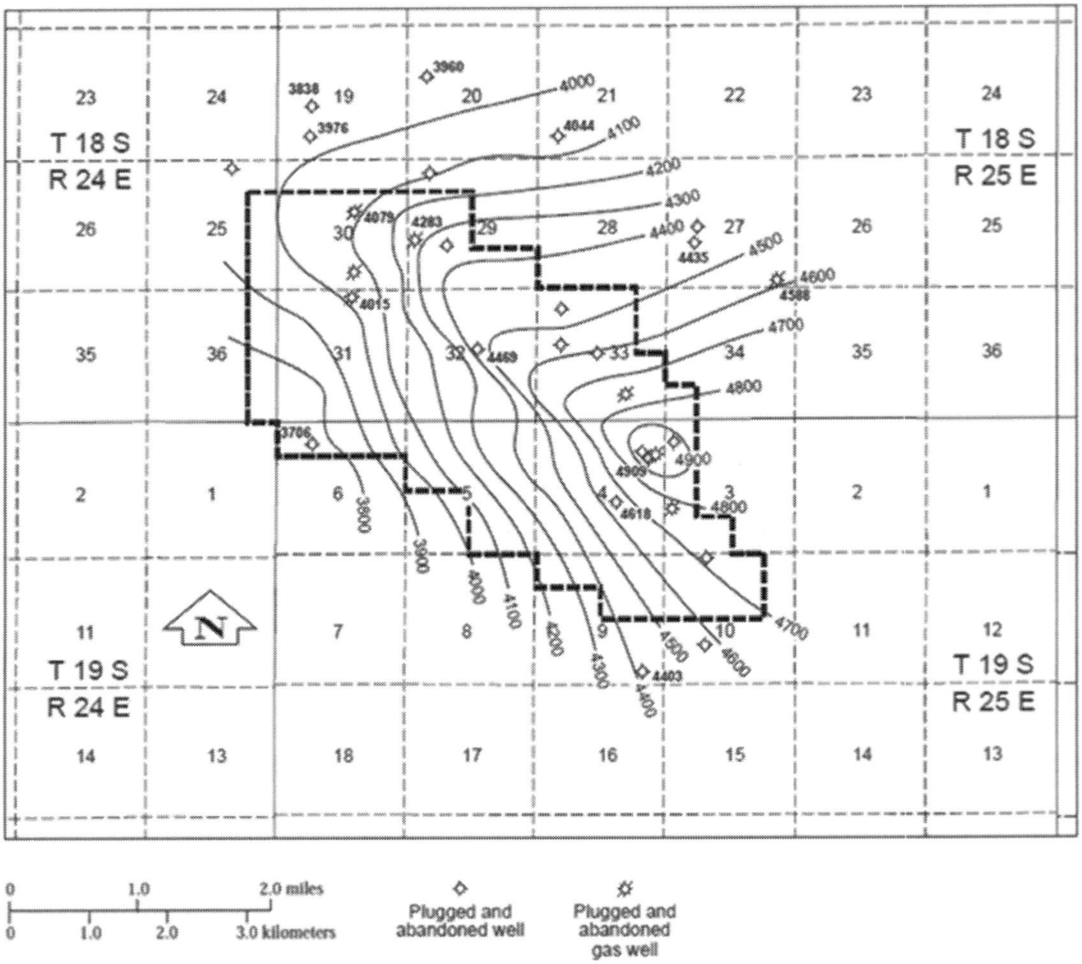

FIGURE 3.45 Structure on top of the Dakota Formation, Harley Dome Field, Grand County, Utah (Bon, 1999). Contour interval is 100 ft (30 m). The Harley Dome is a plunging to the northwest anticline that produces He from the Entrada Sandstone.

Leadville Limestone (Morgan, 1994). Fig. 3.47 indicates a classed post map of oil and gas production which suggests that the trap is a combination of structure and stratigraphic elements. The gas composition of the Salt Wash Field is indicated in Table 3.3 (Wiseman and Eckels, 2020). There was no known He production but Pan American (1964) indicates in filed documents with the State of Utah that they recognized the loss of several millions of dollars of He revenue. The sealing mechanism for the hydrocarbon trap is the encasing dense limestone and dolostones by the overlying thick salts of the Paradox Formation. The reservoir is reportedly heavily fractured which is likely the cause of the excessive water production (Smouse, 1993). The field has been subsequently abandoned with no He production. Many of the wells suffered casing collapse sometime due to salt movement (Morgan, 1994).

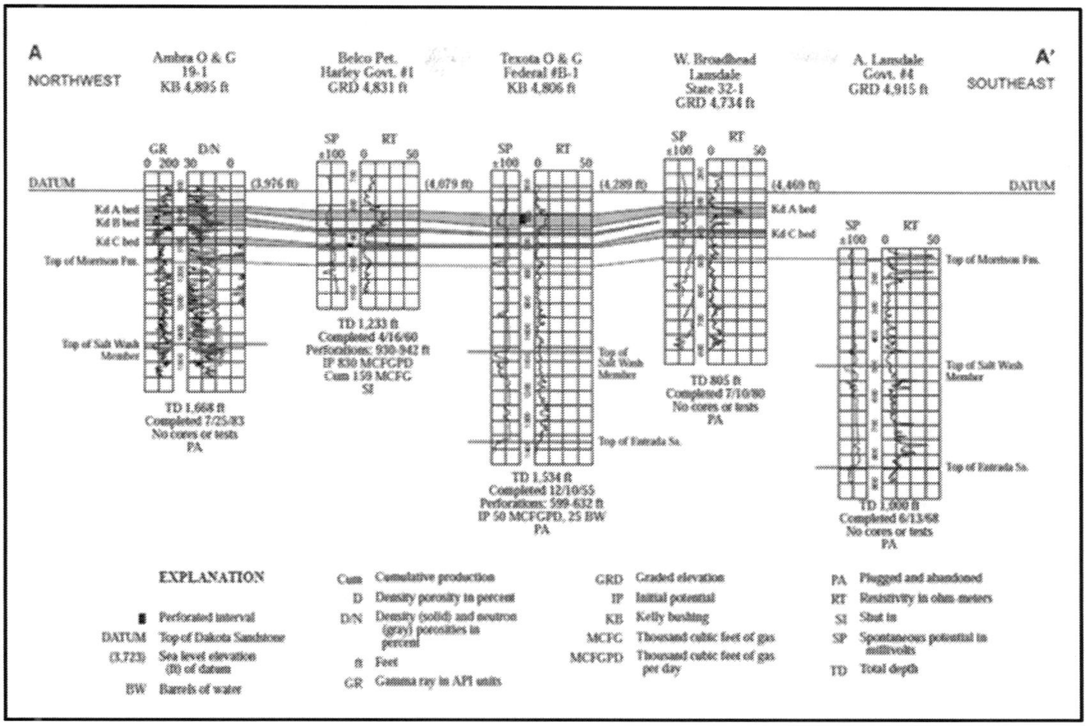

FIGURE 3.46 Cross section through the Harley Dome Field, Grand County, Utah (Bon, 1999).

Reservoir Characteristics of the Salt Wash Field (Norton, 1978; Smouse, 1993; Morgan, 1994):

Productive area:	800 acres
Pay zone:	Leadville (Mississippian)
Depth to pay:	8693 to 8707 ft (2650 to 2654 m)
Pay thickness:	10–27 ft (3–8.2 m)
Porosity:	7.8% average
Permeability:	Unknown
Water saturation:	25%
Type of drive:	Gas cap with encroaching water drive
Type of trap:	Structure and stratigraphic
Bottom hole temperature:	143°F
Initial field pressure:	4075 psi
Oil characteristics:	50 degrees API gravity, pourpoint 40°F
Gas characteristics:	He 1.2%, CH_4 43%, N_2 56%–80%
Oil and gas column:	370 ft (112 m)
Production:	11 BCF, 1.4 MMBO; and a loss of 13.2 MMCF He

TABLE 3.1 Gas constituents of the various wells at the Harley Dome Field, Utah (Bon, 1999).

Well	Harley Government 1	Lansdale Government 4	Harley Dome 2	Flatirons Fed 1–4	Lansdale Government 1	Lansdale Government 15	Lansdale Government 13	Lansdale Government 13	Bashor 1/Home Oil 1
Formation	Dakota (Cretaceous)	Entrada (Jurassic)	Entrada (Jurassic)	Entrada (Jurassic)	Entrada (Jurassic)	Entrada (Jurassic)	Morrison–Entrada (Jurassic)	Morrison (Jurassic)	Morrison (Jurassic)
Depth (ft)	930	836	860	868	836	915	890–946	635	538
O	ND						0.0%	0.0%	
N_2	ND	86.1%	84.6%	83.64%	46.5%	38.7	83.8%	81.3%	43.8
CH_4	84.2%	5.56%	5.5%	7.55%	49%	57.1%	8.5%	14.8%	50%
C_2H_6	0.7%						0.3%	0.5%	
C_3H_8	0.3%						0.1%	0.2%	
CO_2	0.5%	0.91%	0.8%	0.8%	0.1%	0%	0.8%	0.4%	0.3%
He	<0.1%	7.31%	7.06%	6.82%	2.51%	2.43%	6.5%	2.8%	2.25%
BTU/Ft^3	877	58	92	-	95	610	NA	163	571

NA, not applicable; *ND*, no data.

TABLE 3.2 Helium production from Harley Dome from 2014 to March 2018 (Wiseman, 2019; Wiseman and Eckels, 2020).

Well	Location	Gas production in MCF	He Transported in MCF
Lansdale Government 1	Section 33 T19S R25E	86,978	5563
Lansdale Government 4	Section 3 T18S R25E	209,915	14,271
Flatirons Federal 1–4	Section 4 T18S R25E	348,515	22,286
Lansdale Government 15	Section 35 T19S R25E	Shut-in	–

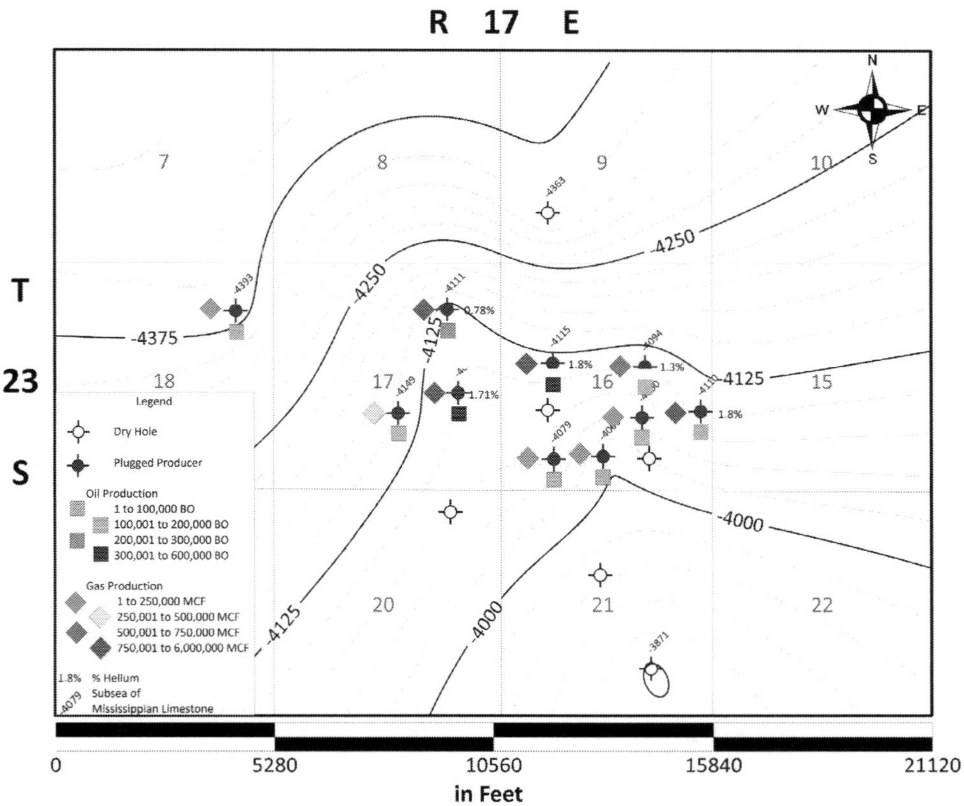

FIGURE 3.47 Structure on top of the Leadville Formation (Mississippian) at the Salt Wash Field, Grand County, Utah. Contour interval is 25 ft (8 m). The Salt Wash Field is a plunging to the northwest anticline that has produced He from the Mississippian Limestone. Production of oil and gas is indicated with a classed post.

3.6.3.3 *Lisbon Field*

The Lisbon Oil and Gas Field is located in sections 2, 3, 9, 10, 11, 12, 13, 14, 15, and 16, Township 30 South, Range 24 East, San Juan County, Utah, USA (Fig. 3.39). The field produces from the McCracken Sandstone Member (Pennsylvanian) and Leadville Limestone

TABLE 3.3 Gas content for wells within the Salt Wash Field, Grand County, Utah (Wiseman and Eckels, 2020).

Well	Salt Wash 1	CF & I 22–16	Government smoot 3	CF & I 42–16	Government smoot 1
Formation	Leadville (Mississippian)	Leadville (Mississippian)	Leadville (Mississippian)	Leadville (Mississippian)	Leadville (Mississippian)
Depth (ft):	8553	8793	8630	8912	8732
N_2	77.3%	77.8%	84.6%	70.4%	55.9%
CH_4	11.8%	11.8%	12.1%	12.2%	16.1%
CO_2	2.2%	2.3%	3.2%	2.9%	5.3%
He	1.8%	1.8%	1.71%	1.3%	0.78%
BTU/Ft^3	319	297	322	484	794

(Mississippian) and is a faulted anticlinal closure trending northwest-southeast on the larger Lisbon Anticline. The Lisbon reservoir is unique as it is completely encased in salt (Parker, 1981). The reservoir fluids have been noted to be very unique compared to other fields in the area (Parker, 1981). The amount of recoverable He from the field has been estimated at 750 MMCF He from the Leadville Limestone (Mississippian). However, because the gas is sour, the field has not been a major source of He, as processing it out of the gas stream is difficult. Wiseman and Eckels (2020) indicate there are indications of other potential He targets based on limited surface geochemical sampling. The sour nature of gas in the area may degrade the potential He resource due to unfavorable economics. Fig. 3.48 is a schematic cross section of the Lisbon Field from Wilson et al. (2003).

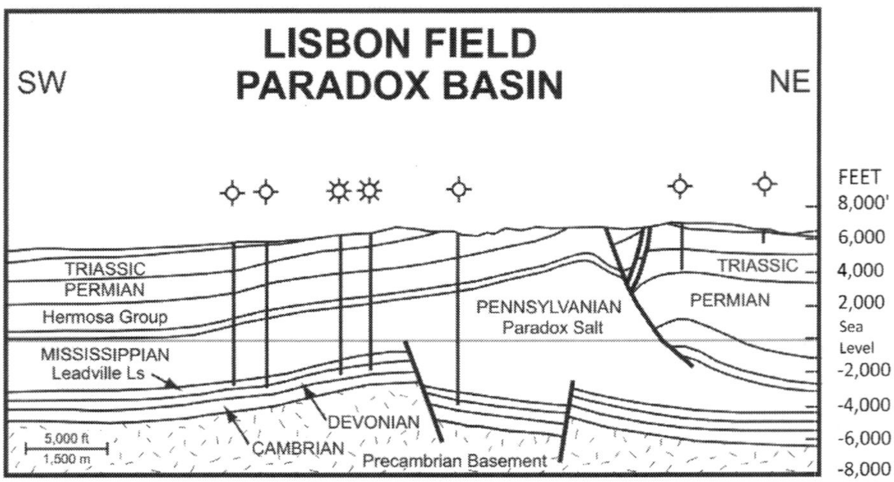

FIGURE 3.48 Schematic cross section of the Mississippian Lisbon Field, San Juan County, Utah (Wilson et al., 2003).

Reservoir Characteristics of the Lisbon Field (Clark, 1978; Parker, 1981; Wiseman and Eckels, 2020):

Productive area:	Leadville Limestone: 5120 acres
	McCracken Sandstone: 1050 acres
Depth to pay:	Leadville Limestone: 7567 to 7970 ft (2293 to 2415 m)
	McCracken Sandstone: 8261 to 8293 ft (2503 to 2513 m)
Pay zone:	Leadville Limestone (Mississippian) and McCracken Sandstone (Devonian)
Net pay thickness:	Leadville Limestone: Net 225 ft (68 m)
	McCracken Sandstone: Net 25 ft (7.5 m)
Type of trap:	Structural closure with variations caused by karsting and dolomitization
Porosity:	Leadville Limestone: 5.5% average
	McCracken Sandstone: 8%
Permeability:	Leadville Limestone: 22 millidarcies but highly variable
	McCracken Sandstone: 2.6 millidarcies
Water saturation:	Leadville Limestone: 39%
	McCracken Sandstone: 43%
Type of drive:	Leadville Limestone: Expanding gas cap and gravity drainage.
	McCracken Sandstone: Solution gas
Bottom hole temperature:	Unknown
Initial field pressure:	Leadville Limestone: 2788 psi at 7,773 ft (2,355m)
	McCracken Sandstone: 2721 psi at 8271 ft (2506 m)
Gas characteristics:	Leadville Limestone: He 0.72%, CH_4 38.75%, N_2 6.7%, CO_2 26.2%, 864 BTU average
	McCracken Sandstone: He 0.5%, CH_4 59.86%, N_2 11.3%, CO_2 2.043%, 1249 BTU average
Oil and gas column:	−1800 ft (545 m) from sea level

The State of Utah has several areas with potential He production (Wiseman and Eckels, 2020). One of the concerns for many of these leads is the remoteness of many of the potential areas (Wiseman and Eckels, 2020). Grand County, Utah, is an excellent area to explore for He because the natural gas can be sold into existing pipelines, access to the I-70 interstate highway corridor for services and access to markets, several wells have tested significant He results, geologically its generally shallow and along areas of tectonic activity with associated intrusions at depth.

3.6.3.4 Woodside Field

The Woodside Field is located in E/2 Township, 19 South Range 13 East, W/2 Township, 19 South Range 14 East, sections 24, 25, 35, and 36 of Township 18 South, Range 13 East, sections 19, 30, and 31, Township 18 South, Range 14 East, Emery County, Utah, and produces from the Kaibab (Permian) sandstone (Fig. 3.39) (Morgan and Chidsey, 1991; Eppnick et al., 2014). The trap is a double plunging anticline trending north-south.

Reservoir Characteristics of the Woodside Field (Morgan and Chidsey, 1991; Eppnick et al., 2014):

Productive area:	8000 acres
Reservoir:	Kaibab Sandstone (Permian)
Depth to pay:	3120 ft (945 m)
Net pay:	135 ft (52 m)
Lithology:	Silty limestone
Average porosity:	9%
Permeability:	Unknown
Water saturation:	Unknown
Gas column:	200+ ft (61 m)
Initial field pressure:	924 psi
Bottom hole temperature:	96°F
Type of drive:	Gas expansion
Type of trap:	Anticline
Gas characteristics and analysis:	CO_2 31.7%, N_2 61%, C_2H_6 5.7%, O_2 0.3%, H_2S 0.0%, and He 1.31%

3.6.3.5 Gordon Creek Field

The Gordon Creek Field is located in sections 24 and 25, Township 14 South, Range 7 East, and sections 7, 17, 18, 19, 20, 29, and 30, Township 14 South, Range 7 East, Emery County, Utah, and produces CO_2 from the Coconino sandstone (Permian), Sinbad Limestone member of the Moenkopi Formation (Triassic), and Ferron Sandstone and the Blue Gate Shale (Cretaceous) of the Mancos Formation (Cretaceous) (Fig. 3.39) (Morgan and Chidsey, 1991; Eppnick et al., 2014). The Coconino sandstone has excellent reservoir continuity but the other three reservoirs are limited in lateral extent (Morgan and Chidsey, 1991).

Reservoir Characteristics of the Gordon Creek Field (Morgan and Chidsey, 1991; Eppnick et al., 2014):

Productive area:	8320 acres
Depth to pay:	10,890 ft (3300 m)
Pay zone:	Coconino Sandstone (Permian), Sinbad Limestone member (Triassic) of the Limestone of the Moenkopi Formation, Ferron Sandstone (Cretaceous), and the Blue Gate Shale (Cretaceous)
Net pay:	Coconino: 175 ft (53 m);
	Sinbad limestone: 23 ft (7 m)

Porosity:	8%–12% estimate for Coconino, Sinbad 6%, Ferron Sandstone 10%–14%, Blue Gate Shale is fracture porosity
Permeability:	Unknown
Water saturation:	Unknown
Gas column:	Coconino Sandstone: 560 ft (170 m)
Initial field pressure:	4200 psi
Type of drive:	Gas expansion
Type of trap:	Faulted anticline
Bottom hole temperature:	198°F for the Coconino Sandstone, 98°F for the Ferron Sandstone
Gas characteristics and analysis:	Coconino Sandstone: CO_2 98.82%, N_2 1.03%, CH_4+ 0.14%, O_2 0.01%, and H_2S 0.0%; Sinbad Limestone CO_2 99.5%, N_2 0.0%, CH_4+ 0.03%, O_2 0.01%, H_2S 0.0%, and He trace%

3.6.3.6 *Farnham Dome Field*

The Farnham Dome Field is located in sections 11 and 12, Township 15 South, Range 11 East, sections 7, 8, 17, and 18 of Township 15 South, Range 12 East, Carbon County, Utah, and produces CO_2 from the Navajo Sandstone (Jurassic) (Fig. 3.39) (Morgan and Chidsey, 1991). Fig. 3.49 is a structure map on top of the Navajo Sandstone.

Reservoir Characteristics of the Farnham Dome Field (Morgan and Chidsey, 1991):

Productive area:	2370 acres
Pay type:	Navajo Sandstone (Jurassic)
Depth to pay:	3093 to 3105 ft (937–940 m)
Average net pay:	40 ft (12 m)
Lithology:	Sandstone
Porosity:	12%
Permeability:	Unknown
Water saturation:	Unknown
Gas column:	330 ft (100 m)
Initial field pressure:	1008? psi
Bottom hole temperature:	125°F
Type of trap:	Faulted anticline
Type of drive:	Gas expansion
Gas characteristics and analysis:	CO_2 98.9%, N_2 0.9%, and O_2 0.02%

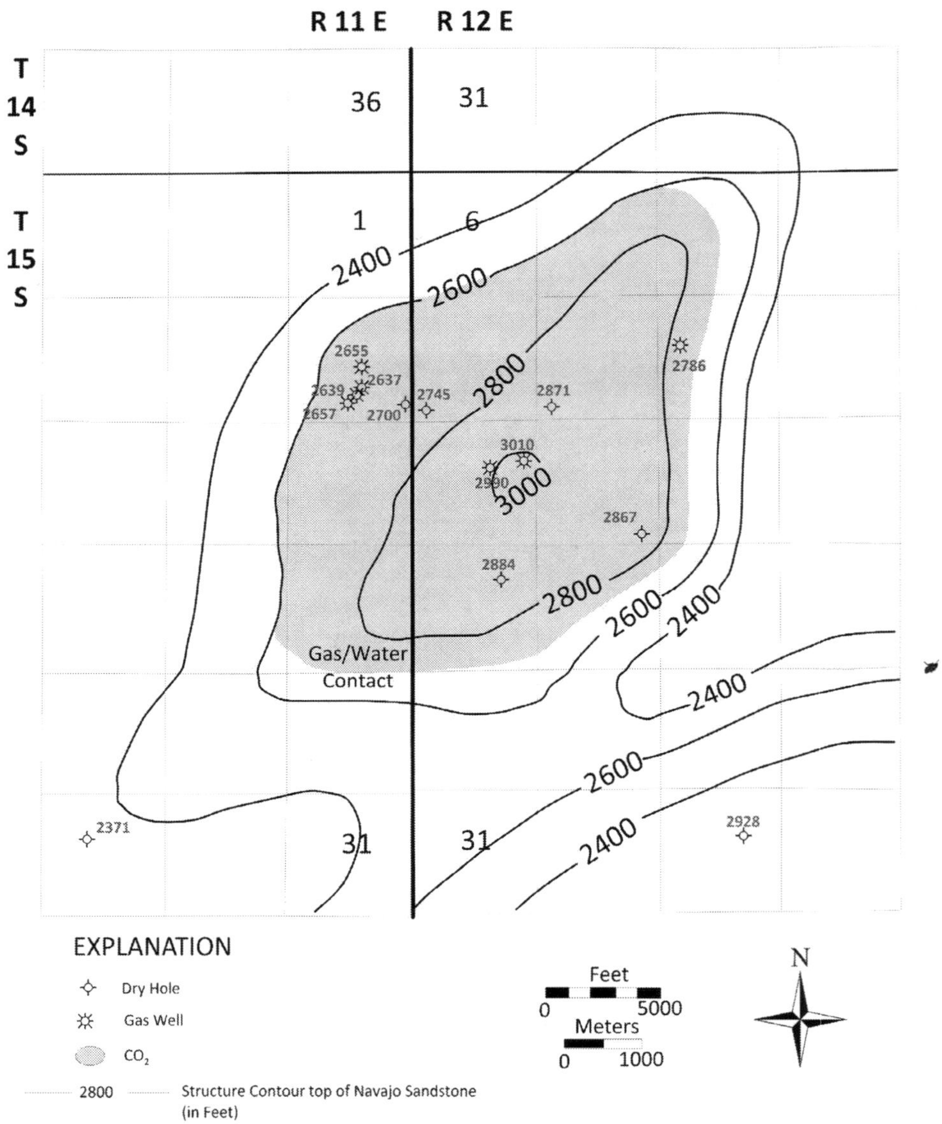

FIGURE 3.49 Structure on top of the Navajo Sandstone for the Farnham Dome Field, Carbon County, Utah (Morgan, 2007). The gas—water contact is indicated.

3.6.3.7 Escalante Anticline

The Escalante Anticline is located is Township 32 South, Range 3 East, Garfield County, Utah, USA. Significant CO_2 resources were identified in the Cedar Mesa Sandstone, White Rim Sandstone, Toroweap Formation, and karst features in the Kaibab Limestone (Permian) (Fig. 3.39) (Eppnick et al., 2014; Wiseman and Eckels, 2020). The anticline is poorly defined by drilling as it is in a remote area and constrained by National Forests. The initial well flowed 72 MMCFGPD of nonflammable gas with oil before being plugged and abandoned (Shannon, 1962).

Reservoir Characteristics of Escalante Anticline (Eppnick et al., 2014; Wiseman and Eckels, 2020):

Productive area:	37,000 acres (est.)
Reservoir:	Cedar Mesa Sandstone, White Rim Sandstone, Toroweap Formation (Leonardian), and karst features in the Kaibab Limestone (Permian)
Average net pay:	172 ft (40 m)
Depth to pay:	2272 ft (690 m)
Lithology:	Sandstone and limestone
Average porosity:	7%
Average permeability:	Unknown
Average water saturation:	Unknown
Initial field pressure:	1057 psi
Type of drive:	Gas expansion
Type of traps:	Anticline but it is not well defined
Gas characteristics and analysis:	CO_2 95%, N_2 4%, and He 0.01% (Eppink et al., 2014) and CO_2 93.1%, N_2 5.5%, Ar 0.1%, CH_4+ 0.4%, and He 0.01% (Wiseman and Eckels, 2020)

3.6.3.8 McElmo Dome

The McElmo Dome Field is located in Townships 37 and 38 North, Ranges 18, 19, and 20 West, Dolores and Montezuma counties, Colorado, USA (Fig. 3.39), and produces from the Leadville Formation at 6600 to 8400 ft (2000–2545 m) (Gerling, 1983). The field was discovered in 1948 but it was not until 1984 that the field began delivering CO_2 to market (Gerling, 1983). The field, including the adjacent Doe Canyon Field to the north-northeast, is one of the largest sources of CO_2 in the Rocky Mountain Region. The field is actually a series of compartments or isolated reservoirs that have been given an overall field name. The field is a plunging anticline with the reservoir being hydrodynamically controlled and is underpressured (Eppnick et al., 2014).

Early studies of the McElmo Field suggested that the CO_2 present was either the result of thermal in situ decomposition of carbonate–sulfate assemblages in the Leadville Limestone or magmatic gas release (Cappa and Rice, 1995; Gilfillan et al., 2008; Adams et al., 2015). The fundamental source of the gases, however, remained poorly understood. Adams and Gonzales (2015) used noble gas isotope signatures in an attempt to fingerprint the source of the CO_2 gas and test competing hypotheses on its origin, migration, and evolution. Analyses of noble gas isotopes, stable isotopes, and major gas compositions across the McElmo Dome–Doe Canyon fields reveal variable and mixed mantle-crust signatures which are dominated by the addition of radiogenic crustal signatures (^4He, ^{21}Ne, ^{40}Ar). A comparison of $CO_2/^3He$ against CO_2 concentrations is consistent with a magmatic ^3He source that is mixed with crustal contributions (Adams et al., 2015). The contribution of crustal rocks is indicated by He isotope ratios $^3He/^4He$ (where the ratio of RAIR = 1) from 0.057 to 0.215 (R/Ra), nucleogenic (following U and Th decay) $^{20}Ne/^{20}Ne$ (<8.5), $^{21}Ne/^{22}Ne$ (>0.10), and highly elevated radiogenic Ar with $^{40}Ar/^{36}Ar^*$ >15,000 (Adams et al., 2015). CO_2 gas was likely sourced from Cenozoic magmatic activity in the area that filled traps during that time. Magmatic events spanned the period from 75 to 5 Ma and involved melting of the Proterozoic lithospheric mantle that was a key source of carbonated mantle melts in the Oligocene, potentially related to the Ute Mountain intrusive (Cappa and Rice, 1995; Gilfillan et al., 2008; Adams et al., 2015). Mafic rocks generated from these melts have elevated K, U, Th, and F, and these magmas could have been a major source of the exceptionally high nucleogenic (^{21}Ne, ^{22}Ne) and radiogenic (^4He,^{40}Ar) signatures of noble gases in the McElmo Dome and Doe Canyon CO_2 fields (Adams et al., 2015).

Reservoir Characteristics of McElmo Dome Field (Krivanek, 1978; Eppink et al., 2014):

Productive area:	203,714 acres (est.)
Reservoir:	Leadville Formation (Mississippian)
Depth:	6600 to 8400 ft (2000 to 2545 m)
Net pay:	70–150 ft (21–45 m)
Lithology:	Dolomites
Porosity:	3.5% to over 21%
Average permeability:	Highly variable with locally over 200 millidarcies
Average water saturation:	20%
Basin:	Paradox Basin
Initial field pressure:	2580 psi at 1640 ft (497 m), which is underpressured
	Eppink et al. (2014)
Type of trap:	Northwest trending anticlinal structure that is hydrodynamically controlled.
Type of drive:	Pressure depletion with minor water-drive
Gas characteristics and analysis:	CO_2 98.2%, N_2 1.6%, He 0.7% and CH_4+ 0.2%. He varies from the 0.014% to 0.28% in the Leadville. The Shinarump conglomerate of Triassic age 1.03%–1.12% He.

3.6.3.9 Doe Canyon Field

The Doe Canyon Field is located on Township 40 North, Ranges 17 and 18 West, Dolores County, Colorado, USA (Fig. 3.39). The productive Mississippian Leadville reservoir is trapped on the upthrown side of an east-west trending fault that borders the southside of an east-west trending anticline (Krivanek, 1981). The Mississippian Leadville is truncated at the top of the structure and the accumulation has a 500-ft (150 m) gas column (Krivanek, 1981). The reservoir is sealed by overlying salts and anhydrites of the Paradox Formation.

Reservoir Characteristics of Doe Canyon Field (Eppink et al., 2014):

Productive area:	82,000 acres
Reservoir:	Leadville Limestone (Mississippian)
Depth to pay:	9000 ft (2730 m)
Average net pay:	60 ft (18 m)
Average porosity:	10%
Average permeability:	Unknown
Water saturation:	Unknown
Initial field pressure:	3960 psi
Type of trap:	Structural trap that is faulted
Type of drive:	Unknown
Gas characteristics and analysis:	95% CO_2, He varies from 0.190% to 2.622%

3.7 Las Animas Arch helium region

The Las Animas Arch Helium Region (LAAHR) is in Lincoln, Kit Carson, Kiowa, Cheyenne, Baca, Prowers, Otero, and Cowley counties, Colorado; Cimarron County and west half of Texas County, Oklahoma; and Sherman, Wallace, Greeley, Hamilton, Stanton, and Morton counties, Kansas (Fig. 3.50). The region is centered around the Las Animas Arch which is a northeast-southwest trending positive feature that has been present since the end of Precambrian times. The arch forms the boundary between the Denver Basin to the west and the Hugoton Embayment to the east and southeast. The southern boundary of the LAAHR is the Apishapa and Sierra Grande uplifts. The northern boundary is the Cambridge Arch and northern Denver basin. The unique nature of the LAAHR is that the majority of the He accumulations are found in the Pennsylvanian Morrow Sandstone reservoir with lesser gas accumulations in the Mississippian, Atoka, Cherokee, Marmaton, Wabaunsee, and Topeka strata (Fig. 3.51). The LAAHR has a deceptively complex history being a double closed positive feature since at least the Mississippian (Merewether, 1987). The Las Animas Arch sedimentary rocks range in thickness from approximately 2000 ft (600 m) at the southern end of the arch to approximately 8000 ft (2400 m) along its western margin.

The Granite Wash and Reagan sandstones of Precambrian to Cambrian age (Fig. 3.51) are the oldest rocks in the LAAHR and are derived from weathered Precambrian strata that form sheet-like to lenticular deposits across the area (Ross and Tweto, 1980). The Granite Wash and Reagan sandstones are porous and permeable and can provide fairways for migrating fluids and gases from both the Anadarko Basin to the east and southeast and the Denver Basin from the west (Ross and Tweto, 1980).

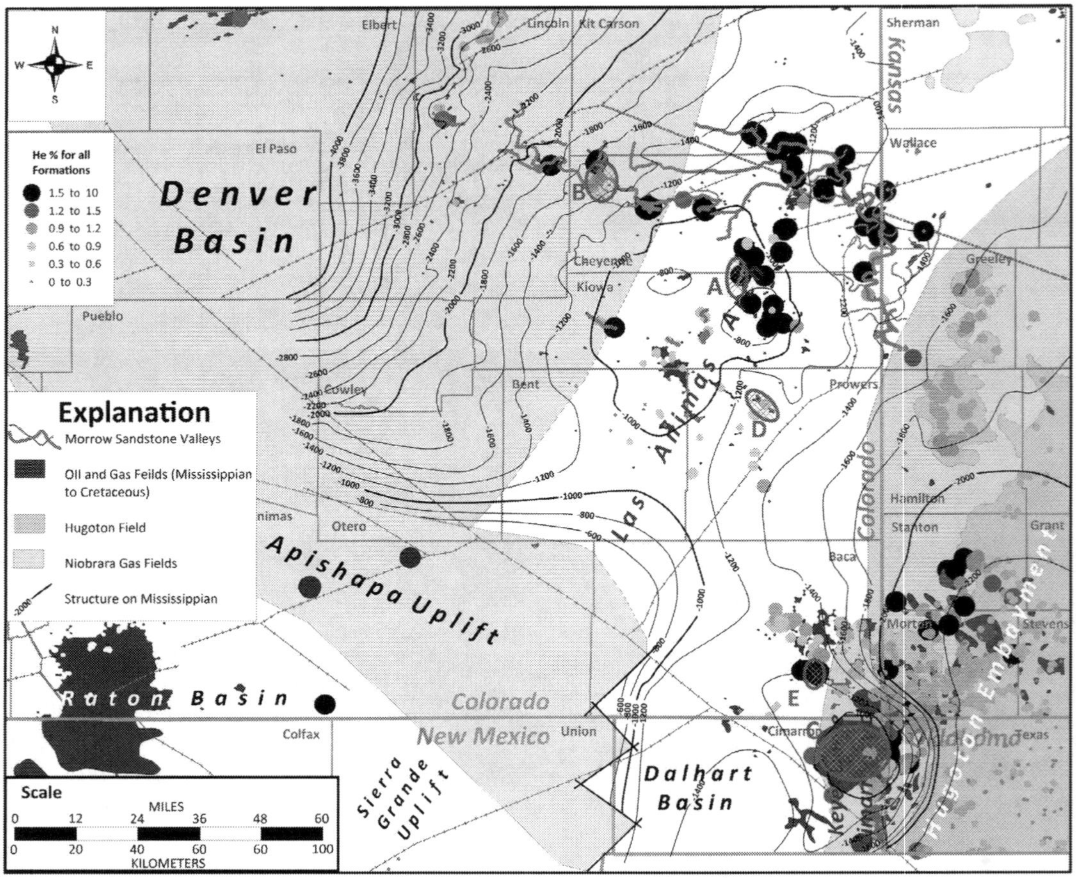

FIGURE 3.50 The LAAHR region in Southeast Colorado, western Kansas, and western Oklahoma with structure on top of the Mississippian strata and all He data are from Rogers, 1921; Anderson and Hines, 1951; Moore, 1978; Moore, 1981; Moore and Sigler, 1985; Wescott, 1989; Hamal and Driskill, 1996; Gage and Driskill, 2003; Gage and Driskill, 2005. The list of fields discussed in the text are: A—Arrowhead Field, B—Bledsoe and Castle Peak fields, C—Keyes Field, D—Clyde Field, and E—Spelunker Field.

Overlying the Granite Wash and Reagan sandstones is the Arbuckle Group (Late Cambrian to Lower Ordovician) that is composed of dense to porous, highly fractured limestone and dolostone with minor shale and sandstone units (Fig. 3.51) which thin from east to west (Ross and Tweto, 1980). Overlying the Arbuckle, especially in the eastern part of the region, is the Simpson Group (sandstone–shale–carbonate) and Viola limestone (Middle Ordovician age) which disappear west of the Las Animas Arch. Devonian–Silurian or Hunton Group strata are missing in the region so that the Mississippian strata lie directly on Ordovician age rocks. Mississippian rocks (Fig. 3.51) are a series of carbonates that generally have minor shale and sandstone interbeds (De Voto, 1980a). The Mississippian surface is eroded and thins to the west where it has been removed on the westside of the Las Animas Arch (Fig. 3.51). Many of the petroleum fields in the Mississippian are structurally controlled

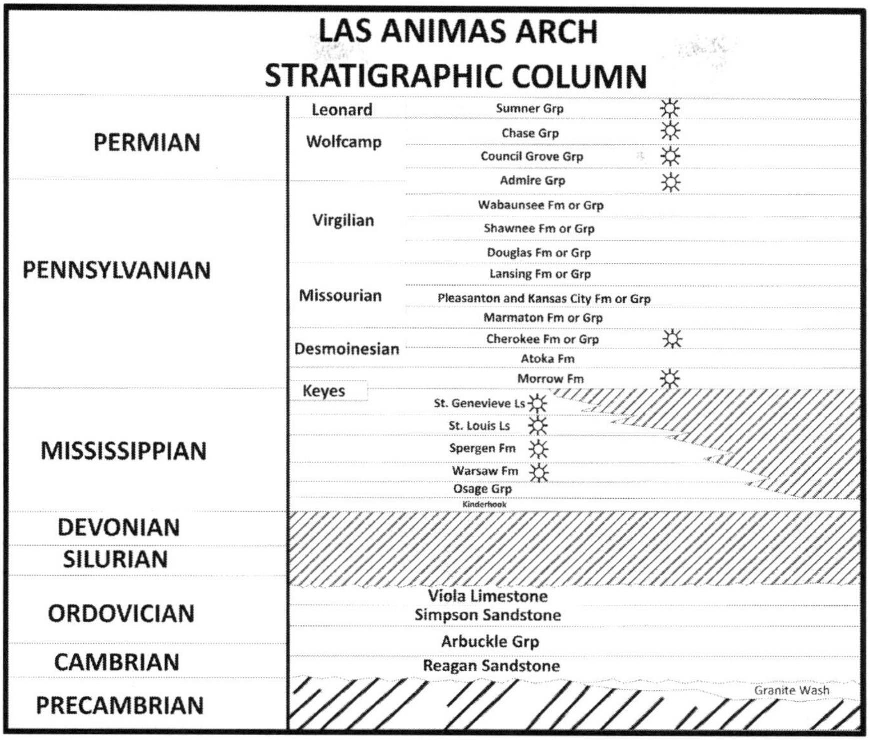

FIGURE 3.51 Generalized Stratigraphy for the LAAHR in southeast Colorado and west central Kansas.

accumulations such as Brandon, Ladder Creek, Arikaree Creek, and Golden Spike to name a few. The reservoirs within several of these Mississippian fields are internally controlled by karsting, depositional porosity such as oolites, low-temperature, hydrothermal-altered limestone to dolomitization, which increases porosity. The majority of the fields are related to significant faulting whether normal or wrench, that is critical conduits for migrating petroleum, He, N_2, and CO_2.

Overlying the Mississippian strata locally is the Keyes Formation (De Voto, 1980b), potentially an upper Mississippian or a lower Pennsylvanian weathered zone with reworked clastic material deposited on the eroded Mississippian surface (Fig. 3.51). The Morrow Formation overlies the Keyes Formation or Mississippian strata (Fig. 3.51) and is a series of fluvial channel sandstones encased in shale, deposited during a series of low-stand events that extend from the northern boundary of the LAAHR into Cimarron and Texas counties, Oklahoma (Bowen and Weimer, 2003). In Morrowan time, the southern part of the Denver Basin adjacent to the Apishapa Uplift began downwarping and became the depocenter for all Pennsylvanian and Permian strata (De Voto, 1980b). Thick alluvial sandstones developed from the erosion of the Apishapa interfinger with Morrowan to Permian marine sediments (De Voto, 1980b). There are a series of Morrowan valleys that developed from the northwest to the southeast draining into the Anadarko Basin (Bowen and Weimer, 2003). These incised valleys are partly filled with high porosity and permeable sandstones that are generally oil productive on the north and east side of the Las Animas Arch (De Voto, 1980b; Bowen and Weimer, 2003). The Morrowan sandstones transgressing south across the arch tend to be gas productive. Helium tends to be prolific along the northern part of the Arch but then diminishes to the south until absent in the area of the Keyes Dome (Fig. 3.51). There are thin discontinuous carbonaceous mudstones present in the Morrow Formation that have been invoked as a source of petroleum and He. However, this hypothesis is highly unlikely, as based on the literature, most of the oils in the LAAHR are derived from organic material in the Mississippian strata and the thin Atokan and Cherokee carbonaceous mudstones.

The Morrow Formation is overlain by the Atoka Formation (Fig. 3.51), which is composed of dense limestone, multiple thin, mappable, and aerially extensive carbonaceous mudstones that extend across the LAAHR into the Anadarko Basin (De Voto, 1980b). During Atoka time, the greatest amount of sediment was shed into the Denver Basin off of the ancestral Rockies to the west and the Apishapa Uplift to the south (De Voto, 1980b). The strata between the mappable carbonaceous mudstones thins across the arch and into the Anadarko Basin whereas the thickness of individual units shows relatively little thinning. The Atokan carbonaceous mudstones interbed in the Texas Panhandle–Hugoton area with dense carbonates and are referred to as the 13 finger "limestone." The carbonaceous shale within the Atoka averages over 11% total organic carbon (TOC) and displays the typical high uranium profile on gamma logs (Bagley et al., 2011; Tedesco, 2020). The Atokan carbonaceous mudstones in the Denver Basin and along the arch are of terrestrial origin and generate oil with a high pour-point, greater than 60°F (Tedesco, 2020).

The Atoka Formation is overlain by the Cherokee Formation of Lower Desmoinesian age (Fig. 3.51) and contain four to six thin carbonaceous mudstone layers that cover the region, have upwards of 25%+ TOC, are very uranium rich, and are marine in origin. This formation can be mapped across the entire region. The formation is composed mainly of gray shales with minor carbonates and carbonaceous mudstones. The Cherokee depocenter is coincident with that of the underlying Morrow and Atoka formations. The main carbonaceous mudstones in the Cherokee are designated in ascending order as Tebo, B, Tebo, V, and Excello which do not thin across the arch, but the strata between does. The Cherokee Formation reservoirs are the Cherokee "A" and "C" carbonates that represent lenticular porosity zones formed during deposition (oolites) or created later during secondary alteration (Tedesco, 2020). The Marmaton Formation of Desmoinesian age (Fig. 3.51) overlies the Cherokee Formation and is predominantly a thick carbonate that internally contains porosity zones known as the Marmaton "A" and "B" with interbedded gray shales and thin carbonaceous mudstones (De Voto, 1980b). The Upper Pennsylvanian rocks in the Las Animas Arch region consist of alternating sequences of carbonates, shales, and minor sandstones.

The Permian rocks in the LAAHR are similar to those in the Hugoton Embayment and have been previously described. The rocks are rarely petroleum and He productive except on the east and southeastern edge of the LAAHR. Lincoln and Elbert counties on the west side of the Las Animas Arch contain Permian rocks that have generally elevated He levels but have no significant accumulations.

3.7.1 Arrowhead Field

The Arrowhead Field is located in sections 26 and 34, Township 16 South, Range 46 West, Cheyenne County, and sections 2, 3, 10, and 15, Township 17 South, Range 46 West, Kiowa County, Colorado, USA, and produces natural gas with He from the Morrow Sandstone (Pennsylvanian) from 4800 to 4950 ft (1450 to 1500 m) (Fig. 3.50, field is labeled A) (Rogers, 2014). The gas contains approximately 2.554% He (from one well), the sandstones are fluvial in natural, and the reservoirs are isolated with their own distinct fluid and gas properties and rock characteristics (Rogers, 2014). Fig. 3.52 is a structure map on top of the Morrow Sandstone and an isopach of the Morrow Sandstone. Note the production figures and that the productive wells are located generally on the structural highs. The cross section in Fig. 3.53 indicates the difficulty in mapping Morrow fluvial sandstones and determining if the sandstones are connected.

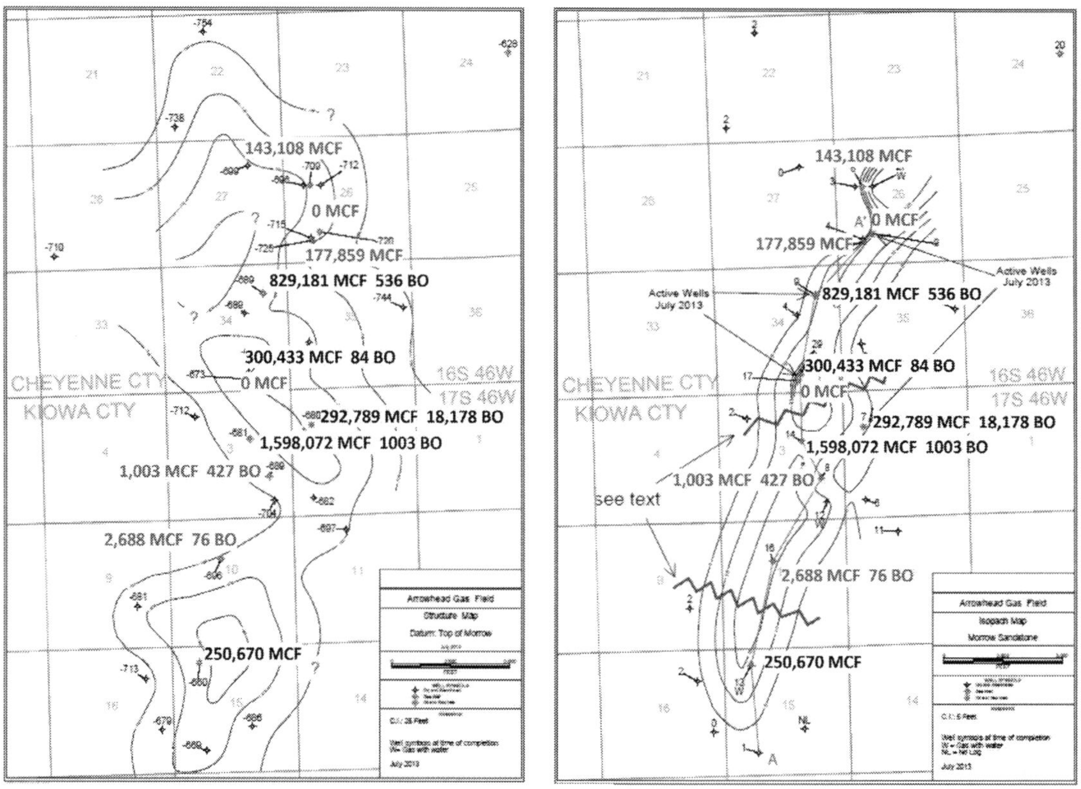

FIGURE 3.52 Isopach of fluvial sandstones in the Morrow Formation (Pennsylvanian) and structure map on top of the Morrow Formation for the Arrowhead Field located in Cheyenne and Kiowa counties, Colorado, Las Animas Arch (Rogers, 2014).

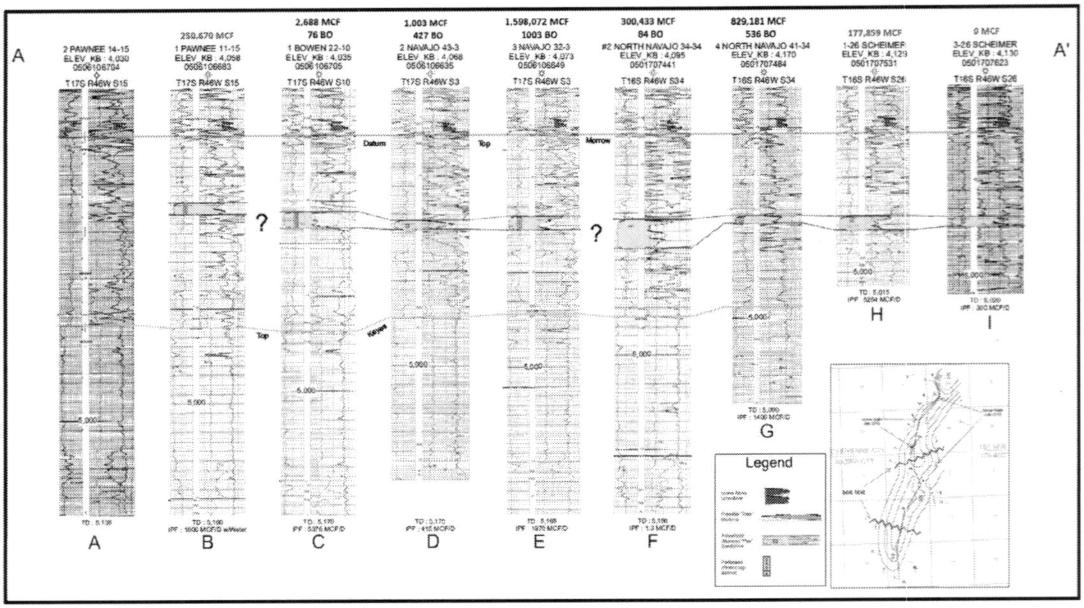

FIGURE 3.53 Cross section through the Arrowhead Field located in Cheyenne and Kiowa counties, Colorado, Las Animas Arch (Rogers, 2014).

Reservoir Characteristics of Arrowhead Field (Rogers, 2014):

Productive area:	320 acres
Pay zone:	Morrow Sandstone fluvial channels (Pennsylvanian)
Depth to pay:	4800 to 4900 ft (1450 to 1500 m)
Average net pay:	10 ft (3.3 m)
Average porosity:	15%–30%
Average permeability:	Unknown, but many of the wells are at the end of their life or have been plugged suggesting high permeability.
Average water saturation:	Unknown
Initial field pressure:	Unknown
Type of drive:	Water
Type of traps:	Stratigraphic–structural, isolated Morrow channels along a north-south trend
Gas characteristics and analysis:	CH_1 47.781%, C_2H_4 7.242%, C_3H_6 4.648%, iC_4H_{10} 0.587, nC_4H_{10} 1.373%, iC_5H_{12} 0.295%, nC_5H_{12} 0.339%, C_6H_{14} 0.449% CO_2 0.747%, N_2 33.985%, and He 2.554% from the Scheimer 1-24
Production:	Projected 4.5 BCFG, 200MBO; presently 3,595,803 MMCFG.

3.7.2 Bledsoe Ranch and Castle Peak Fields

The Bledsoe Ranch Field is located in sections 25 and 36, Township 12 South, Range 51 West, sections 30 and 31, Township Range 50 West, and Castle Peak Field is in sections 1, 10, 11, and 12, Township 13 South, Range 51 West, Cheyenne County, Colorado, USA (Fig. 3.50, field is labeled B). Both fields produce from the Morrow Sandstone at 6000 ft (1820 m). Together, the fields have produced 4.585 MBO of oil and 7.202 BCFG (Sellars, 2014). Fig. 3.54 (Sellars, 2014) is a structure map on the top of the Morrow Sandstone and indicates a lack of any closure. Fig. 3.55 is an isopach on top of the Morrow showing a clear channel system that is productive along most of its trend and where dry holes represent tight reservoir (Sellars, 2014). There are wells within the field that produce from the Mississippian, Cherokee, Marmaton, and Codell (Cretaceous) zones. The Morrow channel lies along a channel system defined by Bowman and Weimar (2003) and lies at the intersection of two merging tributary meanders. The gas characteristics are 1.046% He for Bledsoe Ranch, 720 BTU, and 1.199% He for Castle Peak.

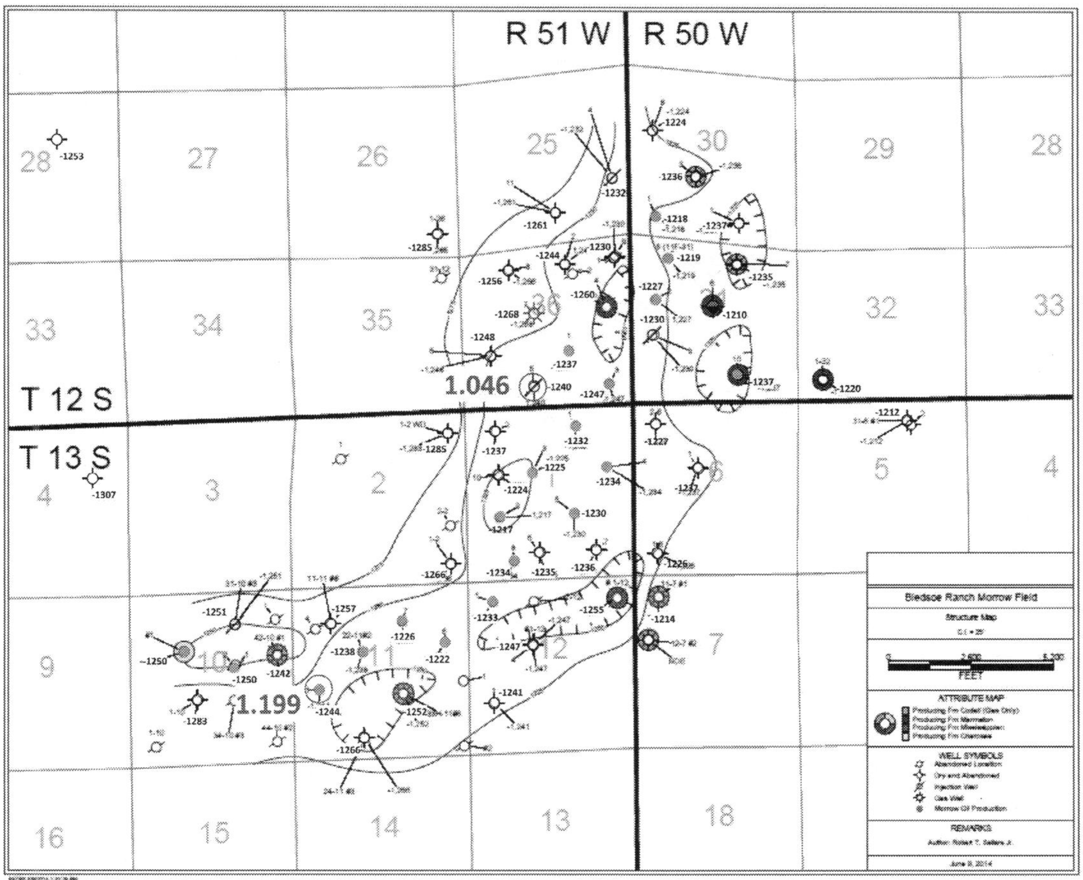

FIGURE 3.54 Structure map on top of the Morrow Sandstone (Pennsylvanian) at the Bledsoe Ranch and Castle Peak fields from Sellars (2014).

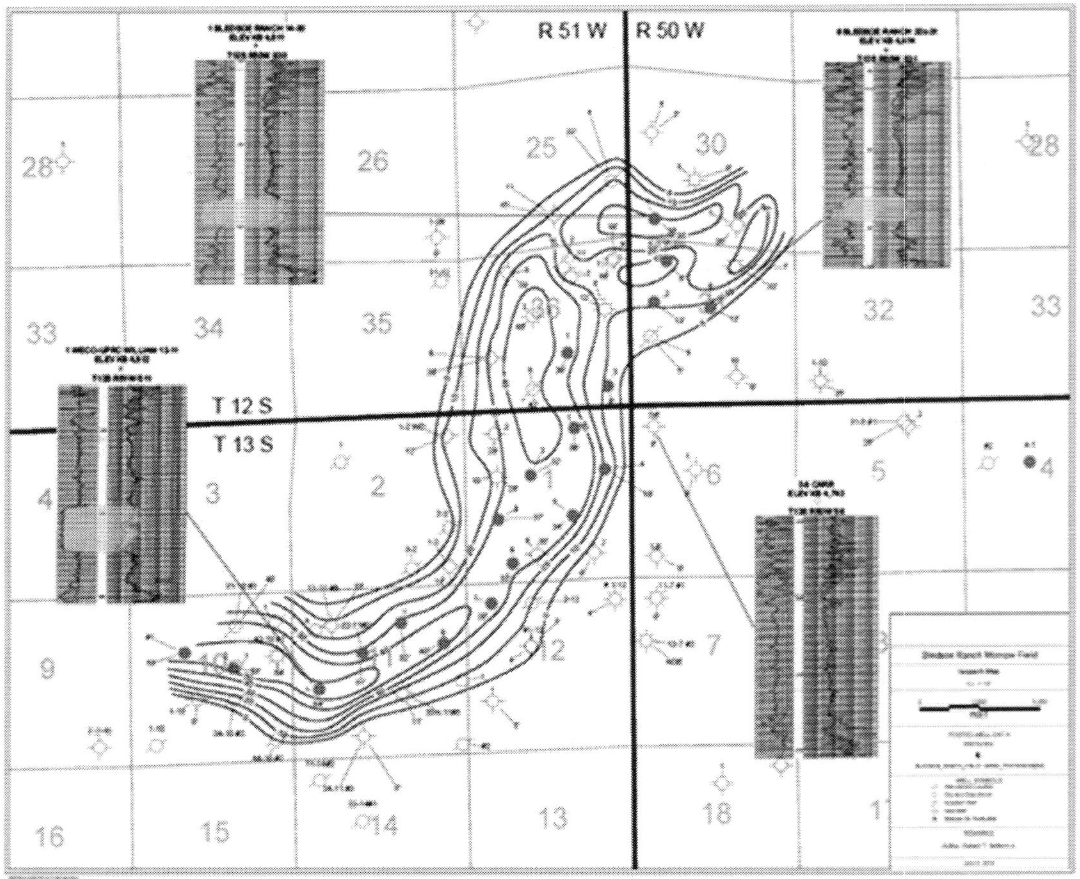

FIGURE 3.55 Isopach of the Morrow Sandstone (Pennsylvanian) of the Bledsoe and Castle Peak Field from Sellars (2014).

Reservoir Characteristics of Bledsoe Ranch Field:

Productive area:	320 acres
Pay zone:	Morrow Sandstone fluvial channels (Pennsylvanian)
Depth to pay:	4800 to 4900 ft (1450 to 1500 m)
Average net pay:	10 ft (3.3 m)
Average porosity:	15%–30%
Average permeability:	Unknown, but many of the wells are at the end of their life or have been plugged suggesting high permeability.
Average water saturation:	Unknown

Initial field pressure:	Unknown
Type of drive:	Water
Type of traps:	Stratigraphic–structural, isolated Morrow channels along a north-south trend
Gas characteristics and analysis:	1.046 for Bledsoe Ranch, 720 BTU. 1.199% He for Castle Peak (Sellars, 2014); 1.9% He, 37.7 N_2, 0.72% CO_2, and 59.3% CH_4 (Sonnenberg and von Drehle, 1990)
Production:	Projected 7.3025 BCFG, 4.585 MMBO (Sellars, 2014)

3.7.3 Keyes Field

The Keyes Field is located in Townships 3, 4, 5, and 6 North, Ranges 8, 9, and 10 East, Cimarron County, Oklahoma, USA, and produces from the Cherokee, Purdy, Morrow, and Keyes sandstones from 4070 to 4751 ft (1230 to 1440 m) (Fig. 3.50, field is labeled C) (Wagner, 1961). The Keyes Field was discovered in 1943, is located to the west of the Hugoton Embayment and at the southern end of a series of gas productive fields that extend into Baca County, Colorado, and Morton County, Kansas (Wagner, 1961). The Keyes Field is typically assigned to the Hugoton–Texas Panhandle Field. However, in this discussion it is made part of the LAAHR as its production is from the Pennsylvanian Morrow and Late Mississippian or Early Pennsylvanian Keyes reservoirs (Carver, 1956). Fig. 3.57 is a structure map by Carver (1956) on top of the Keyes across the field. The productive sandstone reservoirs in the Keyes Field are structurally controlled sandstone lens draped across a structural high (Fig. 3.56) encompassing 90,000 acres (Wagner, 1961). The sandstones thin across the structural high that is formed by the Mississippian topography. Helium content of 2% within the field is restricted to the Lower Morrow and Keyes sandstones and declines in the older Paleozoic strata to 0.1%–0.3% He (Moore, 1981). The Keyes Field He processing plant began operation in 1959, and in that year produced 24% of the country's He which was shipped to the cliffside facility for underground storage. By 1968, it was the Bureau of Land Management's largest plant (Weaver, 2021). The plant was dismantled in 1982 due to low He prices and depletion of gas from the Keyes and other fields in the surrounding area (Weaver, 2021).

Reservoir Characteristics of Keyes Field (Wagner, 1961):

Approximately area:	90,000 acres
Producing formations:	Cherokee, Purdy, Keyes, and Morrow sandstones
Depth:	4500 to 4600 ft (1360 to 1400 m)
Net pay:	Cherokee: 10 ft (3.3 m)
Purdy:	20 ft (6.6 m)
Lower Morrow:	10 ft (3.3 m)
Keyes:	100 ft (33 m)
Porosity:	Cherokee: 17% (based on core)

(Continued)

Purdy:	16.9% (based on core)
Keyes:	15% (based on core)
Permeability:	Cherokee: 880 md (based on core)
Purdy:	72 md (based on core)
Keyes:	100 md (based on core)
Water saturation:	Unknown
Initial field pressure:	710 psi Cherokee, 1040 psi Purdy, 820 psi Lower Morrow, 840 psi Keyes
Type of trap:	Varies by pay from structural to combination stratigraphic–structural
Type of drive:	Gas
Gas characteristics and analysis:	Keyes sandstone: He 2%, N_2 24.82%, CO_2 0.3%, CH_4+ 67.15%, CH_2H_6 2.47%, C_3H_8 1.8%, iC_4H_{10} 0.29%, nC_4H_{10}+ 1.17%, Cherokee 790 BTU, Purdy 1040 BTU, Lower Morrow 1014 BTU, Keyes 820 BTU

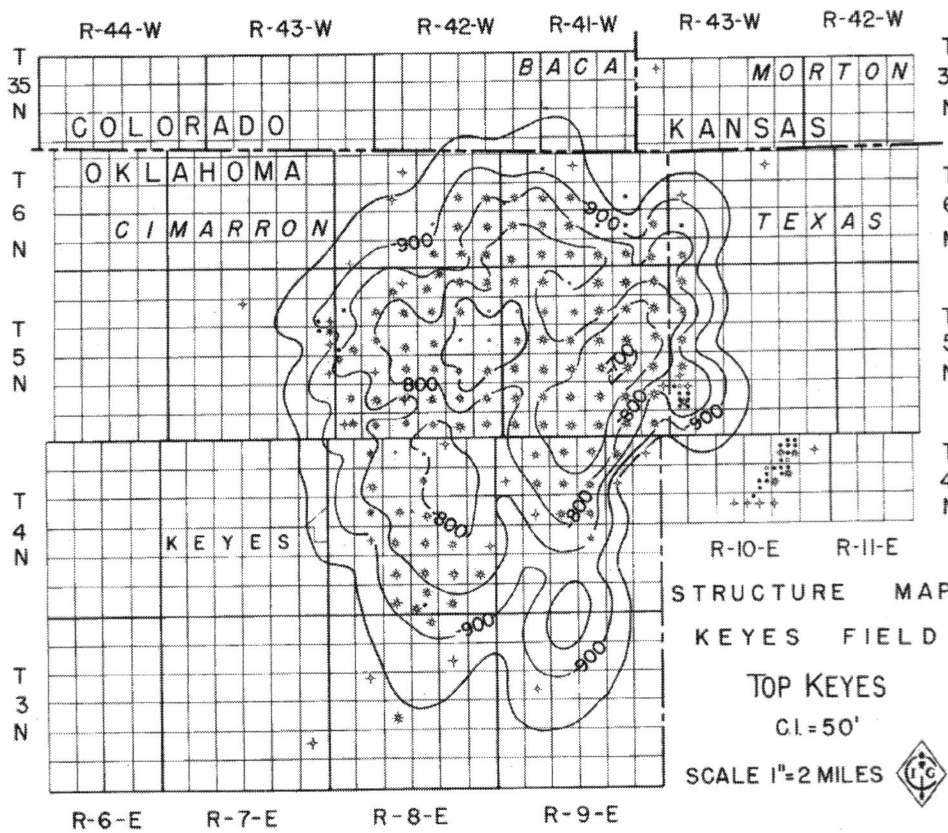

FIGURE 3.56 Keyes Field, structure on top of the Keyes strata (Carver, 1956).

3.7.4 Clyde Field

The Clyde Field is located in sections 5, 8, 9, 15, 16, 22, 23, and 26, Townships 22 South, Range 45 West, Prowers County, Colorado, USA, on the east flank of the Las Animas Basin and produces from Clyde or Morrow Sandstone of Pennsylvanian age (Fig. 3.50, field is labeled D) (Oty, 2014). Gas from within the field has He content in the range of 0.55%–0.7%, which is relatively low compared to He productive areas further north. The field, while not a significant He producer, does lend itself as an example of a productive field in a Morrow channel whose productive wells are defined by an incised valley channel system that forms a stratigraphic trap (Fig. 3.57). Relative production rates and cumulative production are then internally dependent on structural closure for effective production (Fig. 3.57A and B). The Clyde sandstone is found at approximately 4600 ft (1400 m) and the pay zone at the center of the channel is 40 ft (12 m) thick (Oty, 2014). There is shallower production in the northwestern part of the field from the Marmaton carbonates (Pennsylvanian). The field has made 7.27 BCF of gas from the Clyde sandstone (Oty, 2014) and, based on reported percentages of He, would have potentially made 40 to 50.9 MMCF He. If we use the price of natural gas at $2.50 an MCF and He at $300 an MCF, the lost value of the vented He is approximately $11M (US) and the natural gas value is $18M (US). While this does not include lease operating expenses, cost of a cryogenic unit, royalties, etc., it demonstrates the economic attraction of extracting He from a reservoir along with the natural gas.

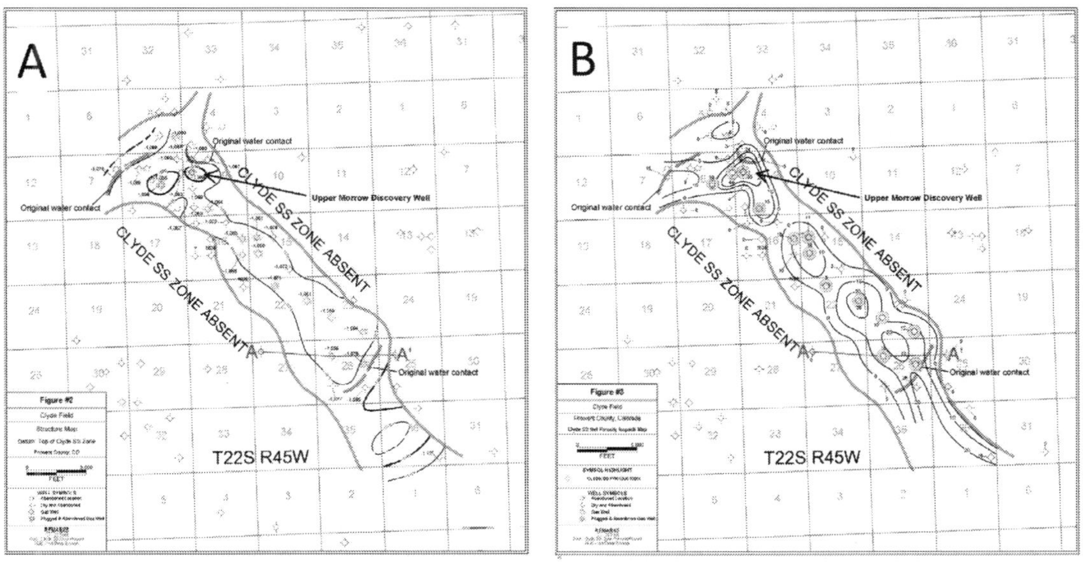

FIGURE 3.57 The Clyde Field located in Prowers County, Colorado, with (A) structure on top of the Clyde Sandstone within the Morrow Formation and (B) isopach of the Clyde Sandstone porosity (Oty, 2014).

Reservoir Characteristics of Clyde Field (Oty, 2014):

Approximately area:	520 acres
Producing formations:	Morrow Sandstone; Keyes, and Marmaton
Depth:	4600 ft (1400 m)
Net pay:	40 ft (12m)
Porosity:	Unknown
Permeability:	Unknown
Water saturation:	Unknown
Initial field pressure:	Unknown
Type of drive:	Gas
Type of trap:	Primarily structural
Gas characteristics and analysis:	0.55%–0.7% He

3.7.5 Spelunker Field

The Spelunker Field is located in sections 7, 8, 9, 16, 17, 18, 19,20, 21, 29, and 30, Township 33 South, Range 44 West, Baca County, Colorado, USA, along the Cimarron Arch (Fig. 3.50, field is labeled E) (Goff and Goff, 2014). The field was discovered in 1996 and has produced He and CH_4 from the Red Cave formation from 1684 ft (510 m) with minor production from the Wabaunsee and Topeka zones. The reservoir is very fine-grained dolomite-cemented feldspathic sandstone that was deposited in a braided stream system in a very arid environment overlain by Stone Corral Anhydrite as a seal (Goff and Goff, 2014). The field is defined as a northwest-southeast trending structural closure with the reservoir trapped across. The trap is bounded by a steeply dipping fault on the west side (not shown) that is the updip edge of the reservoir and wet sandstone going downdip to the east (Goff and Goff, 2014).

Reservoir Characteristics of Spelunker Field (Goff and Goff, 2014):

Approximately area:	6400 acres approximately
Pay zone:	Permian Red Cave; Pennsylvanian Wabaunsee, and Topeka
Lithology:	Sandstones
Depth:	1684 ft (510m)
Net pay:	17 ft (5.1 m)
Porosity:	20%–29%
Permeability:	100–250 md
Water saturation:	Unknown

Type of trap:	Structure–stratigraphic
Initial field pressure:	345 psi
Production:	As of May 2013, 7.8 BCF (Goff and Goff, 2014), He 1.6%, N_2 59%, 440 BTU. Approximately 124.8 MMCF He may have been produced based on volumetrics.

Jacobs (1998) suggested that the anomalous He accumulations in the LAAHR were generated from metasediments and biotite-rich granites in the Precambrian basement. Oil and natural gases were derived from Mississippian, Atokan, and Cherokee organic-rich marine and terrestrial carbonaceous mudstones and are unrelated to the He and most of the N_2 present in the LAAHR. Mantle-derived He seems to be a minor component that is found along the freeze-out fault system associated with the southern part of the LAAHR region (Jacobs, 1998).

3.7.6 General discussion of LAAHR

The Morrowan, Atokan, Cherokee, and Marmaton strata in the LAAHR are typically low porosity and permeability carbonates, carbonaceous mudstones, and shale that act as a seal to retain not only hydrocarbons in Morrowan reservoirs but He too. Fig. 3.58 is a cross section from west to east across the northern end of the Las Animas Arch. Generally, the underlying Mississippian and older strata are porous, permeable, and highly fractured (Tedesco, 2021) that only in a few cases allow trapping of He.

Fig. 3.59 is a basement map (where available) of the LAAHR with associated basement faults as defined by Sims et al. (2001). The Morrow valley and channel system as defined by Bowen and Weimer (2003) lies along a series of Precambrian faults and erosional boundaries. The presence of the basement faults underlying these reservoirs may have provided pathways for migrating low-temperature hydrothermal fluids that thermally altered source rocks in the Mississippian and Pennsylvanian strata. Some of the channel trends are barren of hydrocarbons even though reservoirs are present. Helium accumulations that are south of the main Morrow valley system defined by Bowen and Weimer (2003) generally do not contain economic He.

Figs. 3.60–3.64 are maps of all the He data currently available in the public domain for the LAAHR area and show that the higher He gas concentrations are associated with the LAAHR and relatively low He percentages are associated with the Hugoton Field. Fig. 3.60 is a map of all of the He data in the LAAHR and the western side of the Hugoton Embayment. Note the high He values associated with the Keyes Dome and the northern part of the Las Animas Arch area. Fig. 3.61 is a map of the He data for the Morrow Sandstones and the Mississippian carbonates in the LAAHR. There is a close correlation between basement faulting and the Morrow channel system on the northside of the Las Animas Arch that has numerous producing fields along its trace. There is He-rich area in the Morrow Sandstones on the west side of the Hugoton Embayment and to the northeast of the Keyes Dome.

Fig. 3.62 is a detail map of the main area of the LAAHR and only the He data from the Morrow Sandstone and Mississippian carbonates. Many of the high He values are associated with Precambrian lithologic boundaries. Helium is at its greatest intensity on the northeast side of the Las Animas Arch. This could just represent a sampling bias or actual focus of

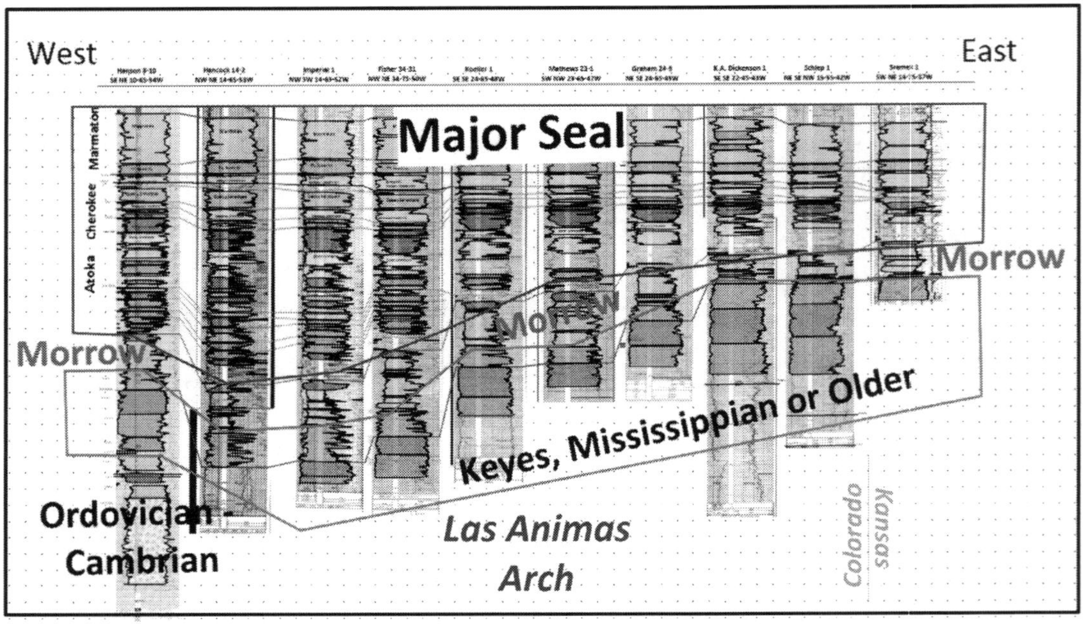

FIGURE 3.58 General west to east cross-section through the LAAHR. Note the thick carbonate-shale section that can act as a seal for He accumulations in the Morrow reservoirs. Colors denote individual formations in the Mississippian (orange and pinks), shale, sandstone, coals and limestones in the Morrow (green, light blues and yellows), carbonate units in the Atoka, Cherokee, and Marmaton and green for carbonaceous mudstones in these same units. The Pennsylvanian section can be divided into allostratigraphic (cyclothem) units and mapped across the basin.

He migration and accumulation. Fig. 3.63 is a detail area of the He data from the Morrow Sandstone and Mississippian carbonates in association with the Precambrian basement in the Keyes Dome area. Helium data are very anomalous across the Keyes Dome but in addition there is anomalous area to the northeast that borders the west side of the Hugoton Field. For the most part, the Morrow and Mississippian reservoirs seem to be barren of significant He under the Hugoton Field. This anomalous under the Hugoton Field may be related to the Keyes Dome or a separate area of He accumulation that is related to a different migration and accumulation history. There are numerous Morrow reservoirs under the Hugoton Field proper but they are devoid of anomalous He. Helium throughout the LAAHR region tends to be elevated suggesting there was an overall regional degassing event in the area.

The Permian strata in the LAAHR (Fig. 3.64) area that produces prolifically from the adjacent PHF lack any significant He in the area. The sources of He for the LAAHR area and the Hugoton area are distinctly different. However, the approach to finding new He production in either area is quite different.

The LAAHR area has successfully produced He as an economic resource. The difference between the LAAHR and the fields in the states of Arizona and New Mexico part of the Colorado Plateau is that the LAAHR producers are putting their gas into the local gathering system and processed at a central facility. This highlights the disadvantage of areas where there are no large natural gas gathering and pipeline systems that allow economics of scale for He extraction.

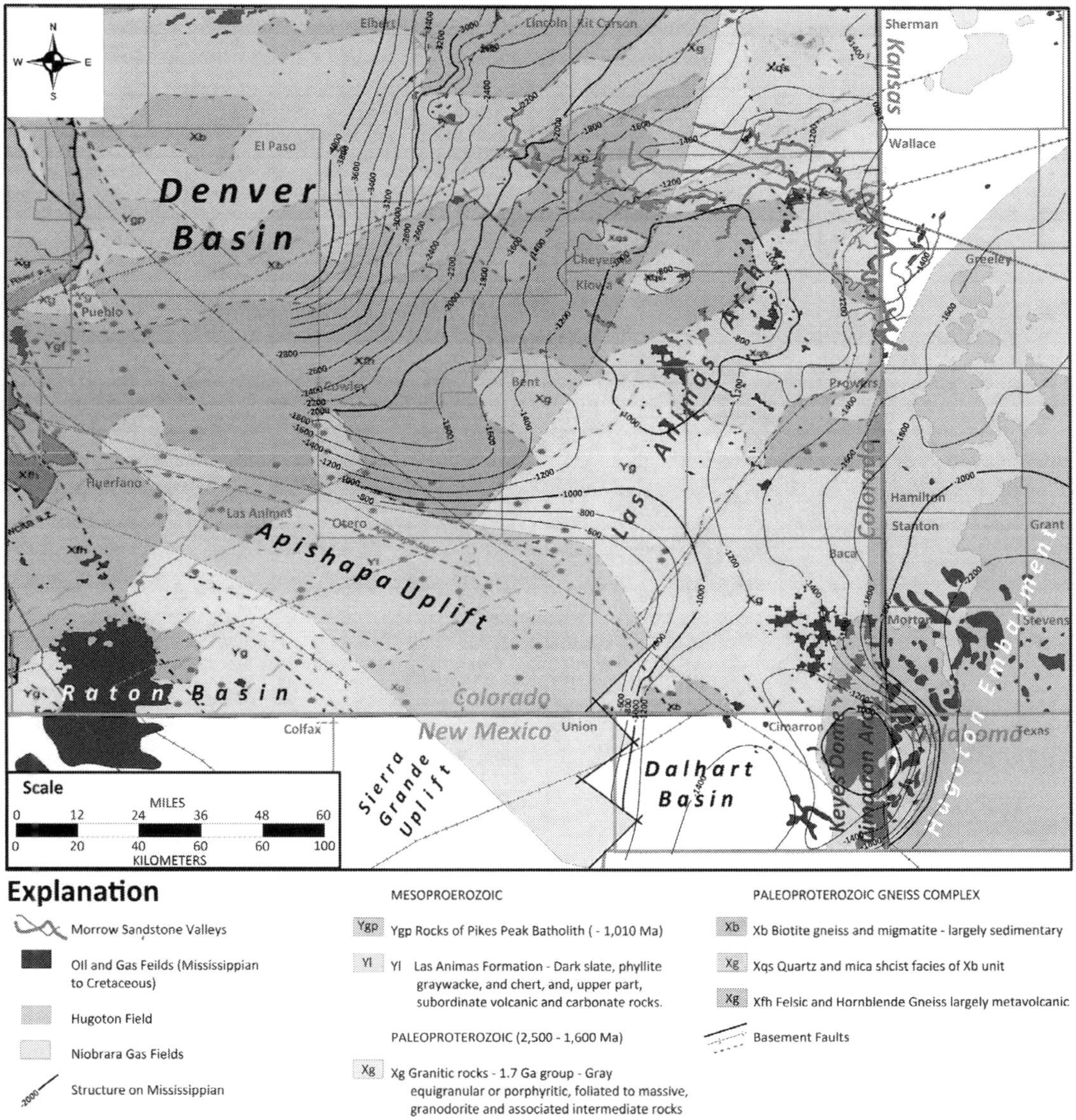

FIGURE 3.59 Basement map of the LAAHR area. The Precambrian rocks are as defined by Sims et al. (2001). There is no similar map for New Mexico, Oklahoma, and Kansas. The oil and gas fields are from the Colorado Oil and Gas Commission (COGCC), 2021, and Kansas Geologic Survey (KGS), 2021. Helium data are from Rogers (1921), Anderson and Hines (1951), Moore (1978, 1981), Moore and Sigler (1985), Wescott (1989), Hamal and Driskill (1996, 2003, 2005).

3.7.7 Arikaree Creek Field

The Arikaree Creek Field is located in sections 10, 11, 14, and 15, Township 6 South, Range 54 West, Lincoln County, Colorado, USA, and does not produce commercial

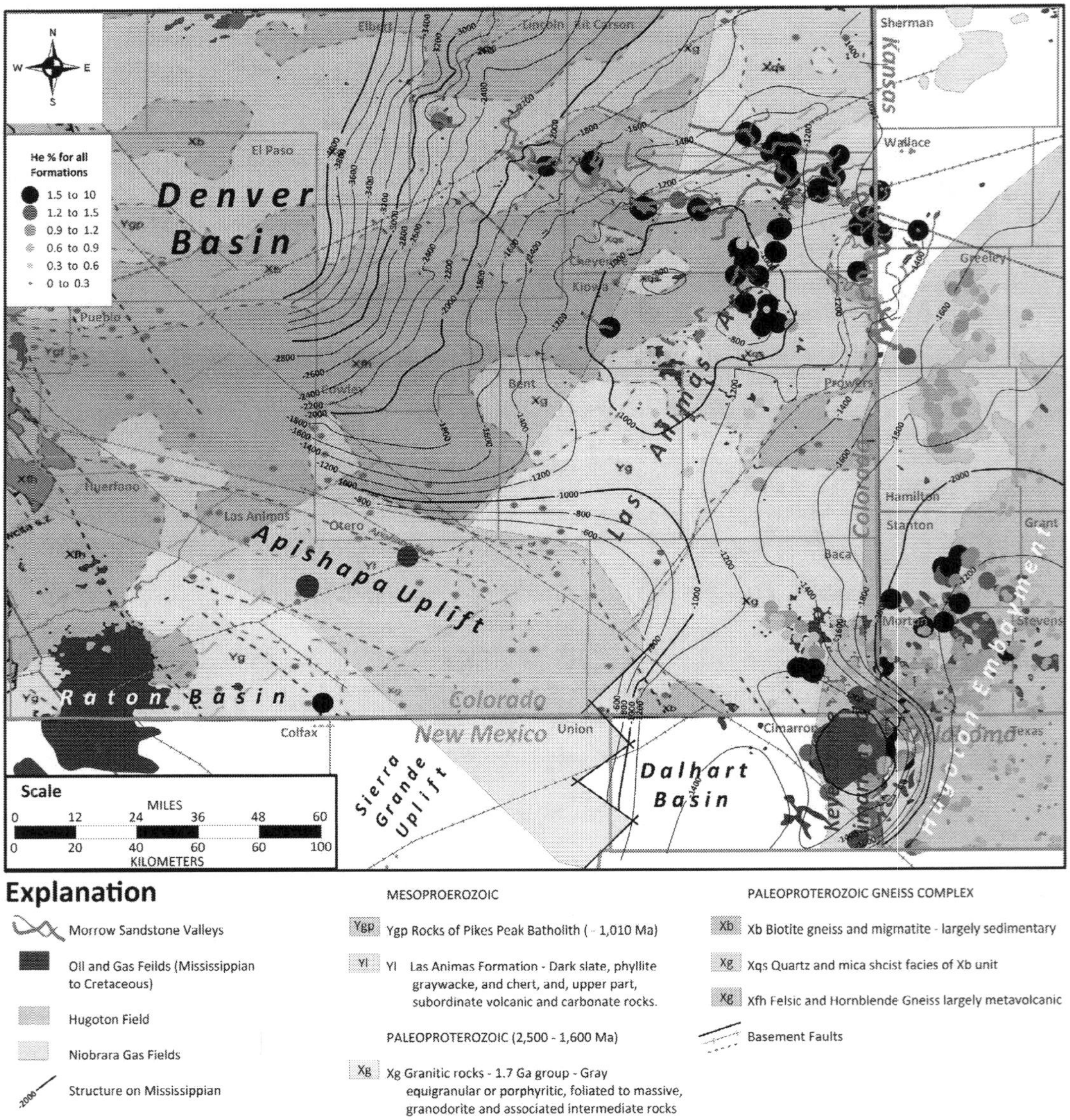

FIGURE 3.60 Basement map of the LAAHR area with all the He data from the USGS, USBM, and COGCC. Note the association of anomalously high He data with basement faulting in the area of LAAHR. Helium data are from Rogers (1921), Anderson and Hines (1951), Moore (1978, 1981), Moore and Sigler (1985), Wescott (1989), Hamal and Driskill (1996, 2003,2005).

He but has produced oil from the Spergen (Mississippian) carbonate at approximately 8000 ft (2070 m). The field consists of a series of structural closures that are controlled by wrench faulting. The field has produced over two MMBO (Tedesco, 2021). There are a number of gas samples analyses that are publicly available from the Colorado Oil and Gas

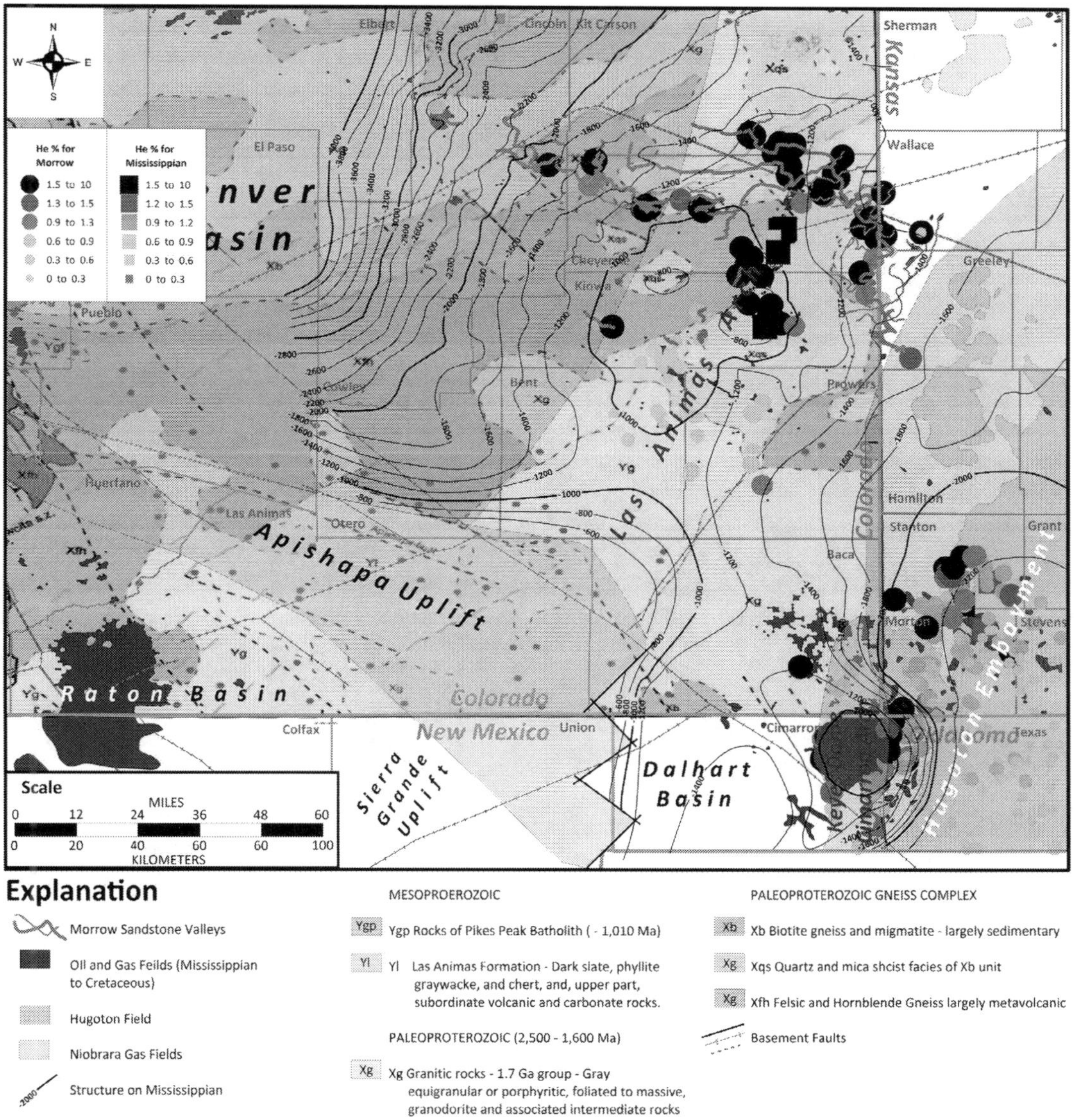

FIGURE 3.61 Basement map of the LAAHR area with He data from the Morrow and Mississippian reservoirs from the USGS, USBM, and COGCC. Note the association of anomalous He data with basement faulting and the associated Morrow channel system as defined by Bowman and Weimer (2003). The anomalous He data are closely associated with the basement faulting but not the Morrow channel system. Helium data are from Rogers (1921), Anderson and Hines (1951), Moore (1978, 1981), Moore and Sigler (1985), Wescott (1989), Hamal and Driskill (1996, 2003, 2005).

Conservation Commission websites that have been taken over the life of the field. Forty-one samples from 14 wells over four distinct time periods were available. Helium, N_2, CO_2, and the ratio of CH_4 to C_2H_6 from the gas analysis over four time periods are presented in

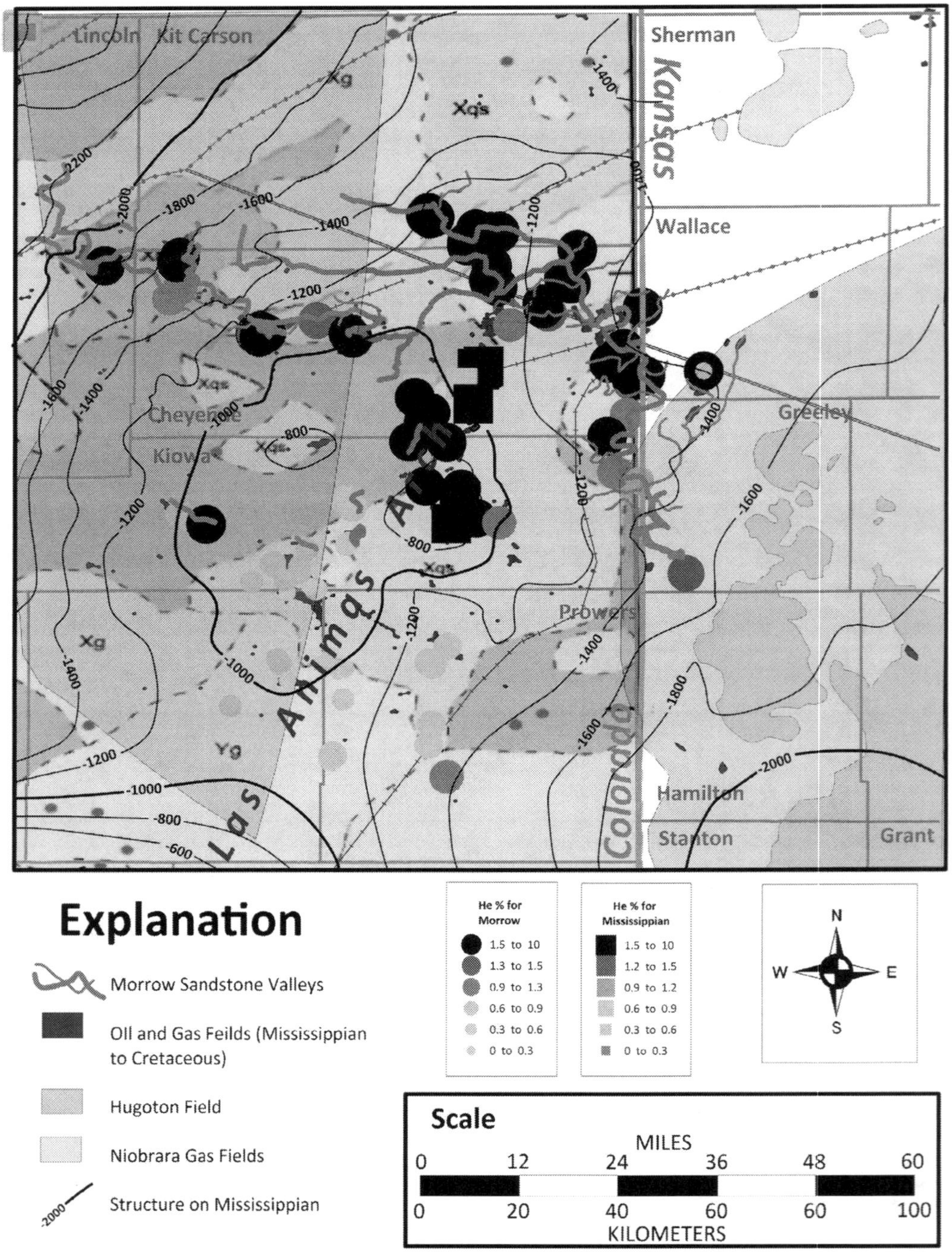

FIGURE 3.62 Detail of Fig. 3.61 for the main Morrow and Mississippian reservoirs He, oil and gas productive area in the Las Animas Arch area. Morrow (circles) and Mississippian (squares) indicate He% and their association with the basement configurations, basement faults, Mississippian structure, Morrow valley system and existing oil fields. For a more detailed explanation of the basement labels see Fig. 3.59.

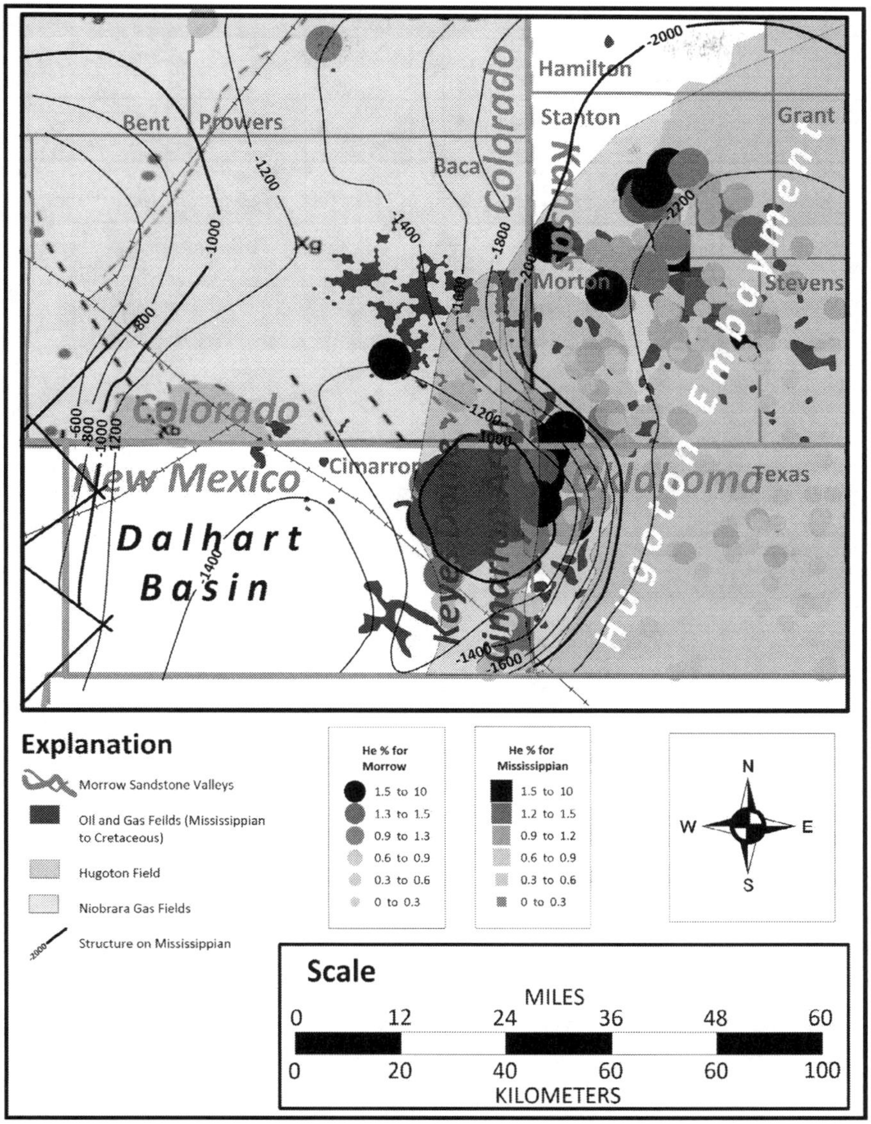

FIGURE 3.63 Detail of Fig. 3.61 for the Keyes Dome area. Morrow and Mississippian reservoirs He% and their association with the basement configurations, basement faults, Mississippian structure, Morrow valley system, and existing oil fields. For a more detailed explanation of the basement labels, see Fig. 3.59.

Figs. 3.66–3.69. Also indicated on Figs. 3.66–3.69 is the wrench fault system defined by the 3D seismic and three distinct horsts (Tedesco, 2021). The majority of oil production from the Arikaree Creek Oil Field was produced from the upper horst; the middle horst has oil and significant water production; and the lower horst does not contain producible hydrocarbons. The upper horst did not start producing water until several months into

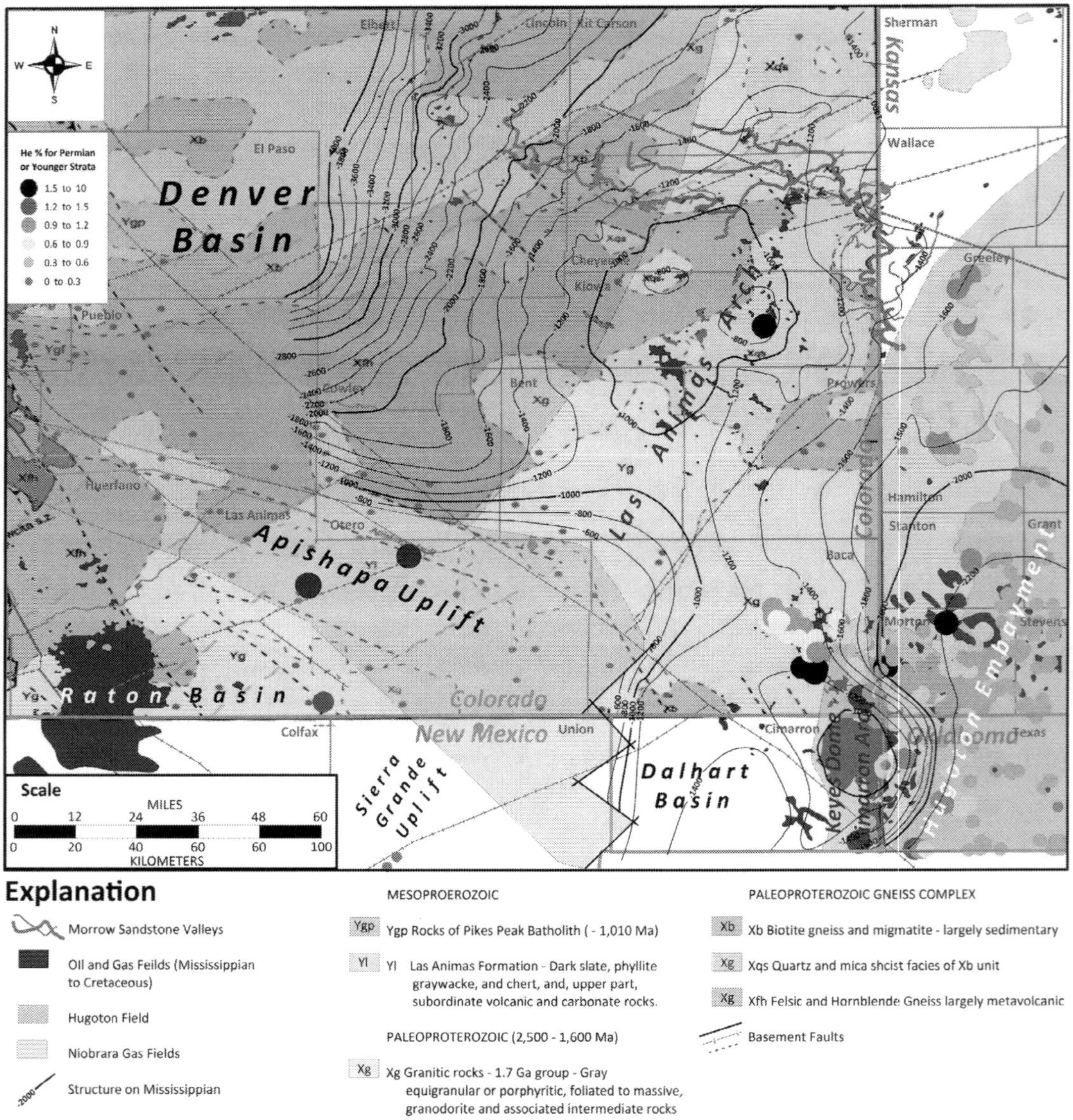

FIGURE 3.64 Basement map of the LAAHR area with He data from the Permian sediments from the USGS, USBM, and COGCC. Note the association of anomalous He data with basement faulting. Helium data are from Rogers (1921), Anderson and Hines (1951), Moore (1978, 1981), Moore and Sigler (1985), Wescott (1989), Hamal and Driskill (1996, 2003, 2005).

production. The operator from 2013 to 2018 noted in public documents that over time an increasing amount of water being produced is from the Precambrian rocks below (Tedesco, 2021). The seal below the reservoir is nonconstraining and fractured allowing migration of fluids from the Cambrian and Precambrian. The Precambrian fluids with higher total dissolved solids mixed with Spergen reservoir fluids. In the area of the wellbore and perforation, changes in salinity, temperature, and pressure caused deposition of various compounds that eventually plugged up perforations reducing fluid flow (Tedesco, 2021).

Gas analysis over the four periods indicated change in He content (Fig. 3.65). Two wells were sampled in all four time periods and three wells sampled in three of the four time periods. Counterintuitively, He data indicate that the element is increasing through time in the natural gas production. In many fields that produce He, He percentage decreases over time with decrease in natural gas production. The Arikaree Creek Oil Field, based on the present data in Fig. 3.66, is a N_2-rich gas system, but N_2 is not increasing through time.

Fig. 3.67 is the CO_2 production data for the Arikaree Creek Oil Field as time increases, specifically in the middle horst of the Arikaree Creek Oil Field but not in the upper horst. Fig. 3.68 is the ratio of CH_4 to C_2H_6 through time and indicates the ratio remains relatively constant. The upper horst has higher CH_4 content than C_2H_6 supporting the concept that the upper and middle horsts are two distinct reservoirs (Tedesco, 2021). The lack of variability in hydrocarbon composition indicates the He and CO_2 are acting independently of hydrocarbon migration and potentially the N_2 content.

If the data are plotted graphically through time as in Fig. 3.69A through D, it can be noted that He in Fig. 3.69A and CO_2 in Fig. 3.69B are increasing through time. The CO_2 could be migrating from the basement or derived from carbonates below the Mississippian reservoir due to chemistry changes caused by upward migrating gases and fluids. Nitrogen in Fig. 3.69C remains relatively constant. The ratio of CH_4 to C_2H_6 in Fig. 3.70D indicates that CH_4 is increasing relative to ethane. This could be due to CH_4 venting from the crustal rocks along with CO_2 and He underlying the Paleozoic strata, or release from the carbonates below the Mississippian reservoir due to chemical reactions caused by the upward migration fluids reacting with rock.

The gas data from the wells of the Arikaree Oil Field, combined with its tectonic, geologic, and production history, suggest that the basal seal under the reservoir is ineffective. This lack of sealing capability in the underlying rock allows migration of Precambrian fluids and gases containing He and CO_2 which, due to depressurization of the reservoir, causes removal of fluids and gases. The gas production from the field is not significant and is typically vented or used to run equipment. The source of the He and CO_2 in the reservoir is unknown but the increasing amounts of He and CO_2 through time are potentially from a crustal source rather than groundwater migration. Halford (2018) suggested that a number of He reservoirs may be going through recharge, such as the DBK Field in Arizona which produced significant amounts of hydrocarbons from an igneous sill reservoir for over 50 years and at that time measured He percentages of over 5%. The data at the Arikaree Creek Field imply that presently He may be still migrating from basement rocks into reservoirs in the LAAHR area.

Table 3.4 is a summary of the general characteristics of some of the fields that produce in LAAHR.

3.8 Raton Basin, Colorado, and new Mexico

The Raton Basin lies in the south-central part of the State of Colorado and north-central New Mexico and is an elongated north-south basin with two significant CO_2 deposits on

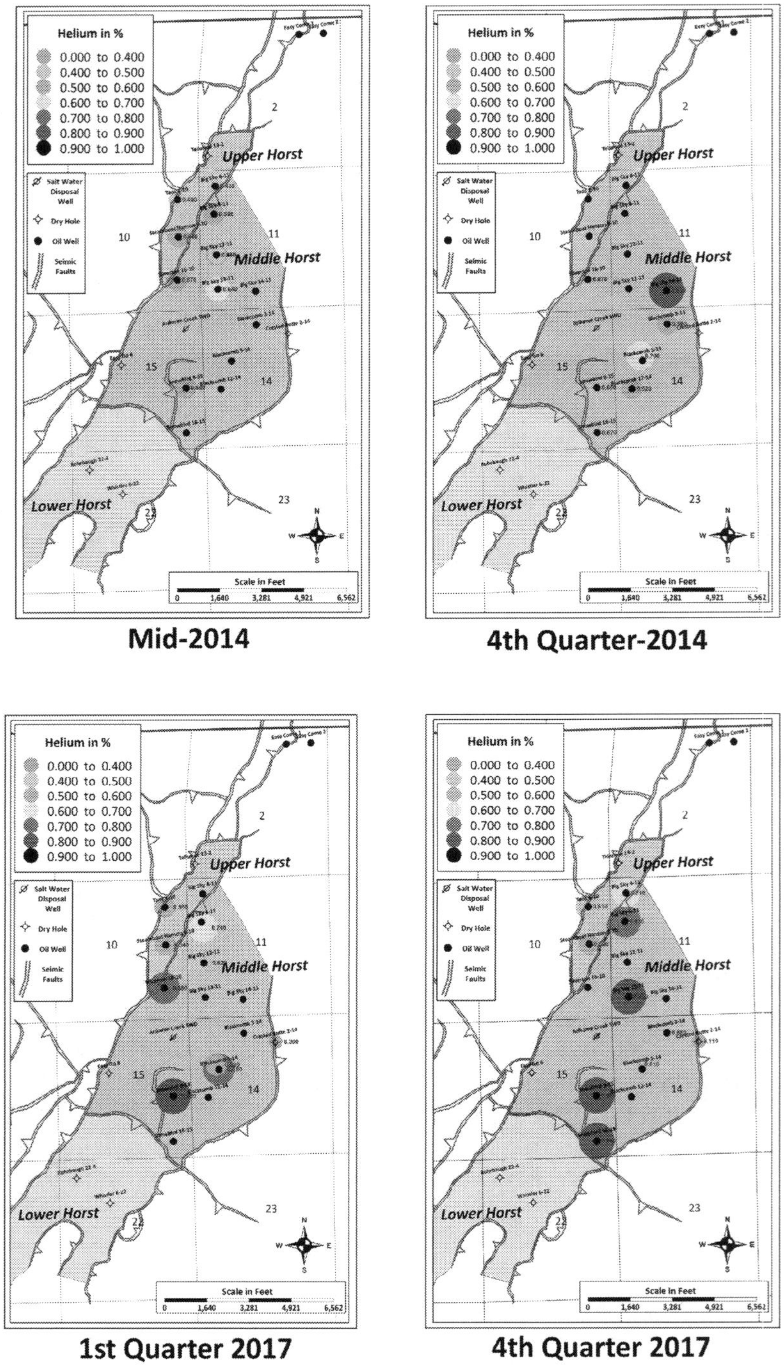

FIGURE 3.65 Helium data through time for the Arikaree Creek Field, sections 10, 11, 14, and 15, Township 6 South, Range 54 West, Lincoln County, Colorado–Denver Basin. The data are presented as a classed post.

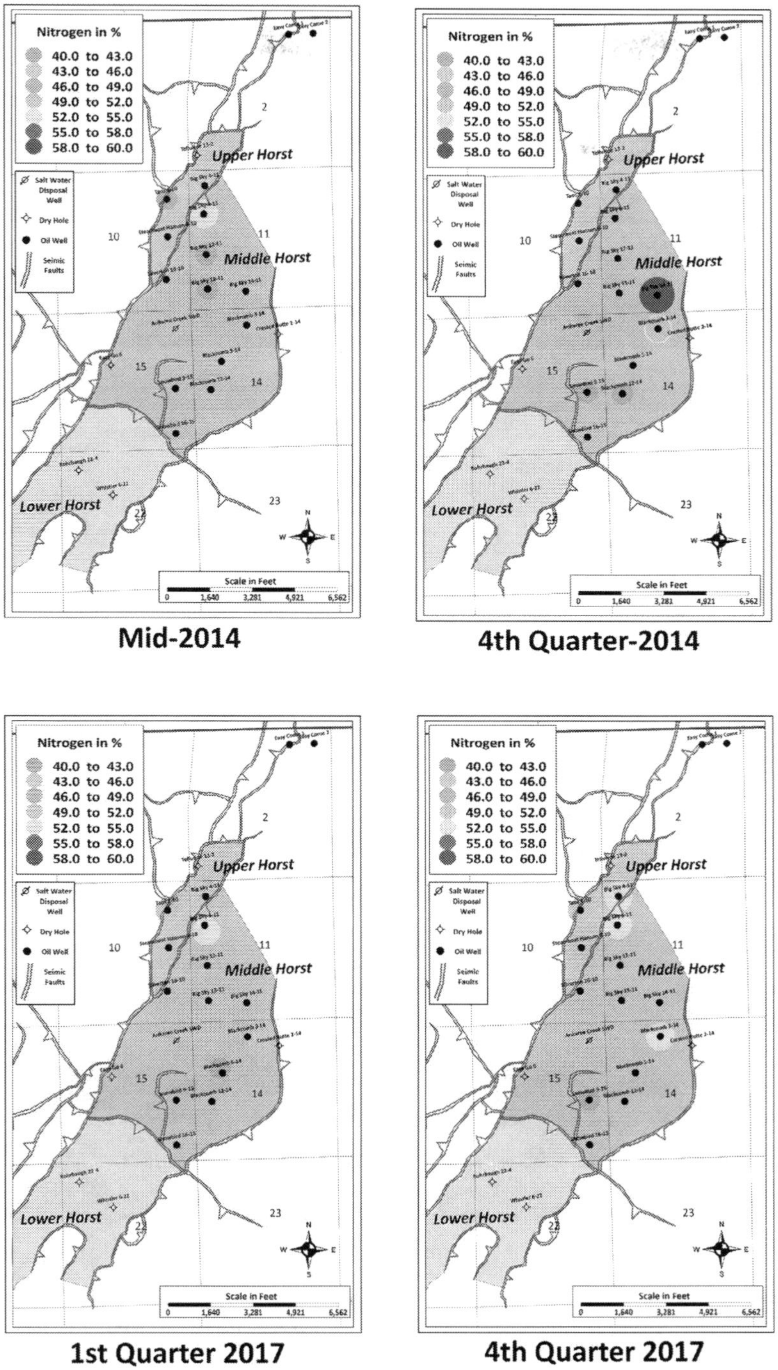

FIGURE 3.66 Nitrogen data through time for the Arikaree Creek Field, sections 10, 11, 14, and 15, Township 6 South, Range 54 West, Lincoln County, Colorado–Denver Basin. The data are presented as a classed post.

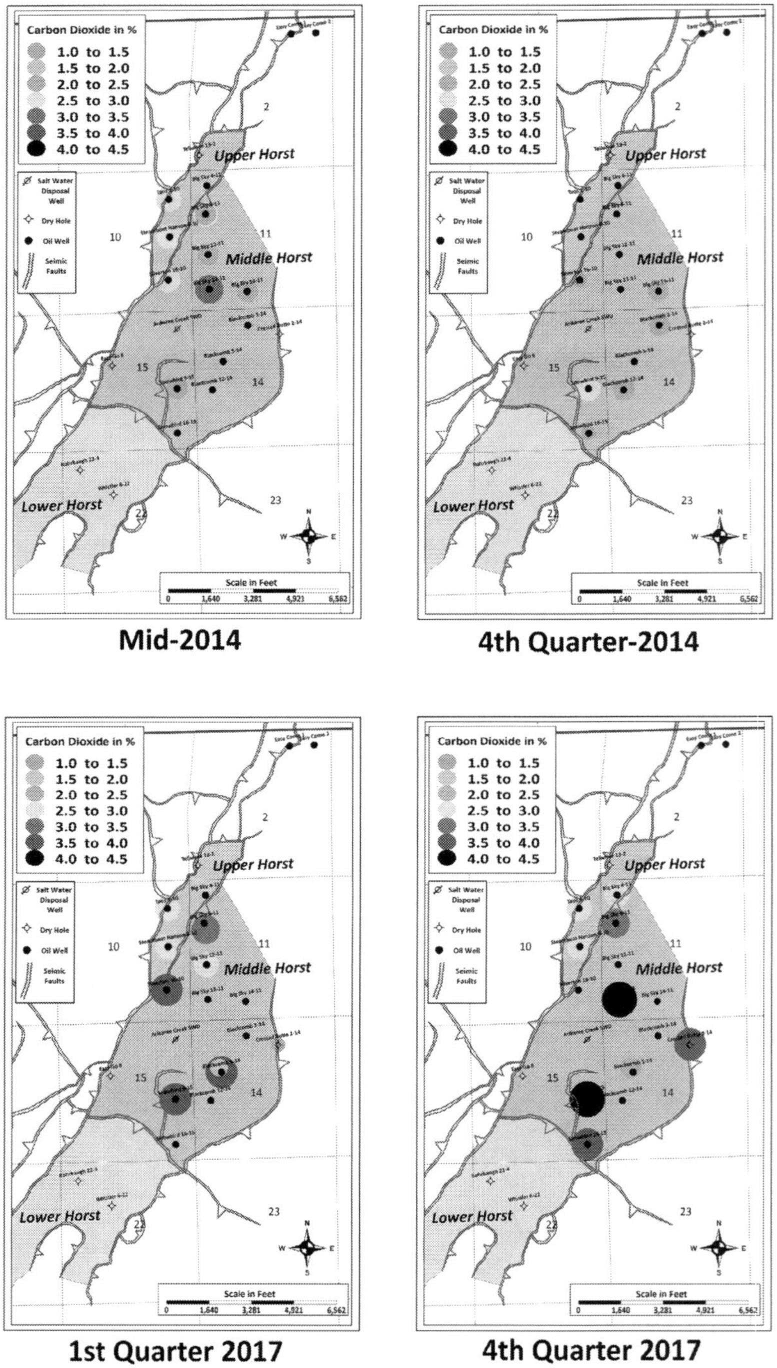

FIGURE 3.67 Carbon dioxide data through time for the Arikaree Creek Field, sections 10, 11, 14, and 15, Township 6 South, Range 54 West, Lincoln County, Colorado–Denver Basin. The data are presented as a classed post.

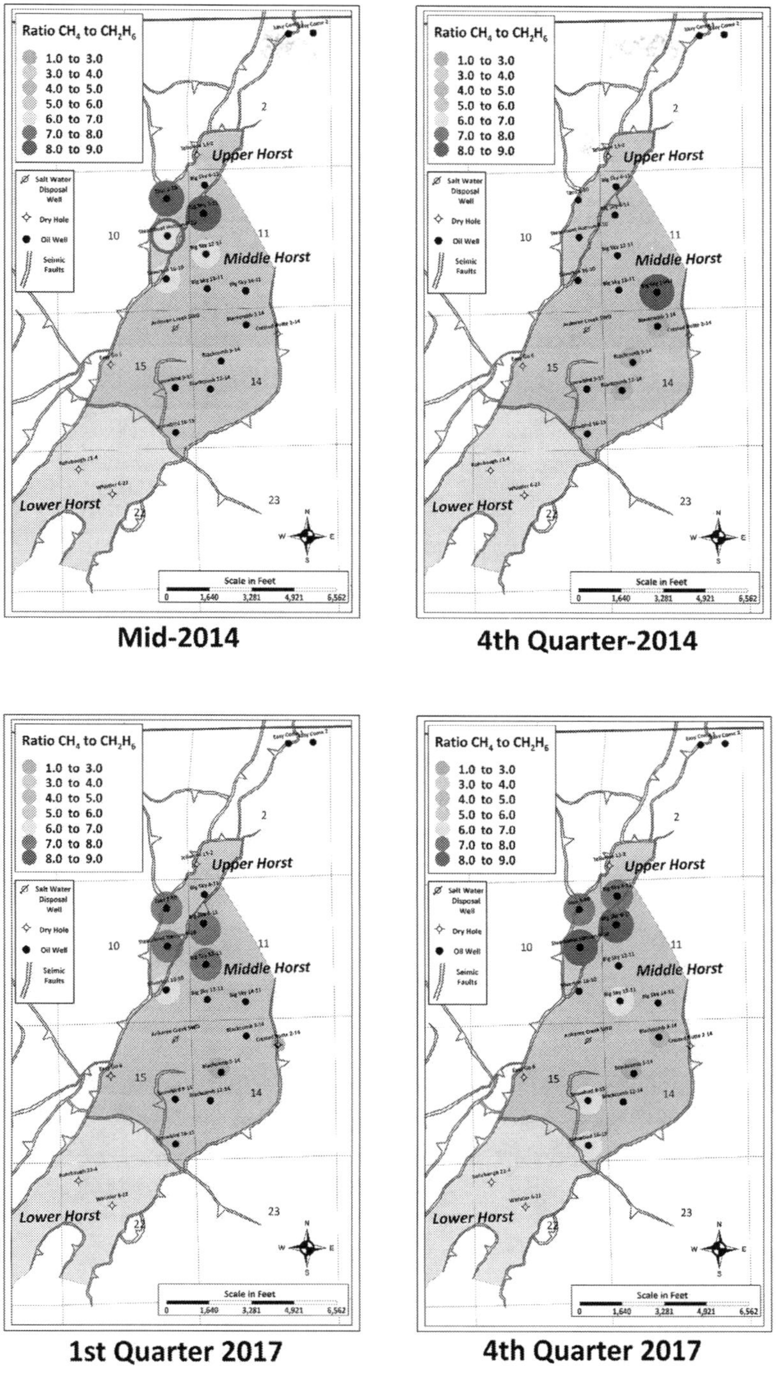

FIGURE 3.68 Methane to C_2H_6 data through time for the Arikaree Creek Field, sections 10, 11, 14, and 15, Township 6 South, Range 54 West, Lincoln County, Colorado–Denver Basin. The data are presented as a classed post.

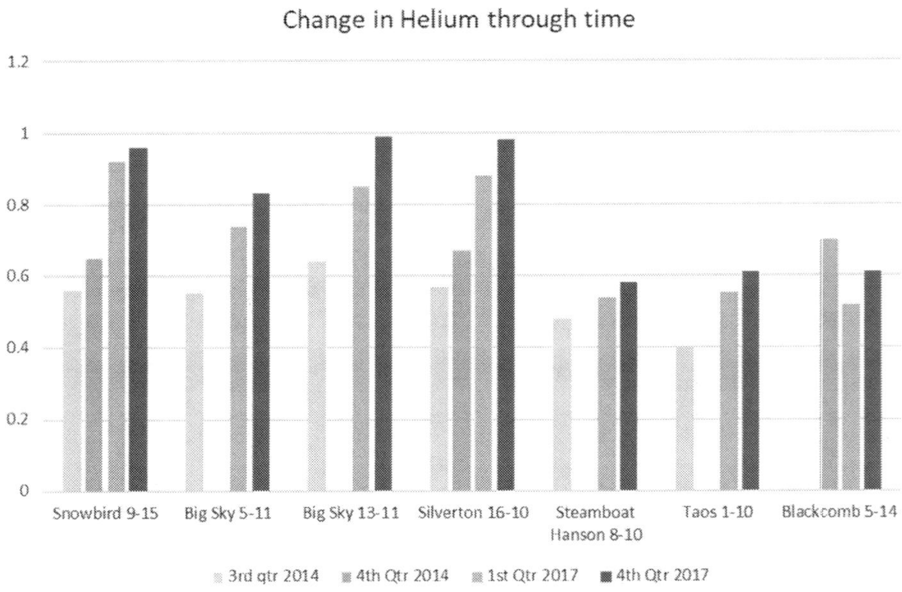

FIGURE 3.69A Helium data in percentage through time for the Arikaree Creek Field, sections 10, 11, 14, and 15, Township 6 South, Range 54 West, Lincoln County, Colorado–Denver Basin.

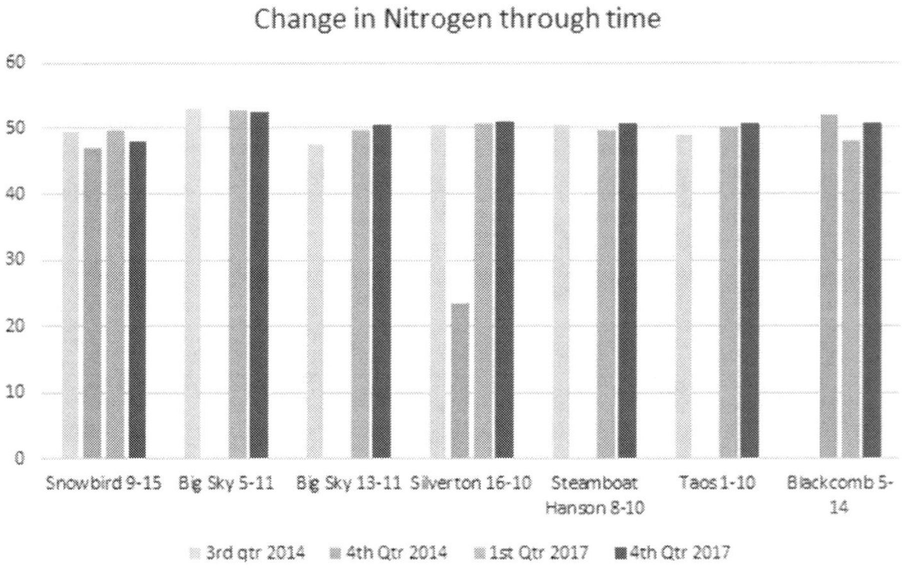

FIGURE 3.69B Carbon dioxide data in percentage through time for the Arikaree Creek Field, sections 10, 11, 14, and 15, Township 6 South, Range 54 West, Lincoln County, Colorado–Denver Basin.

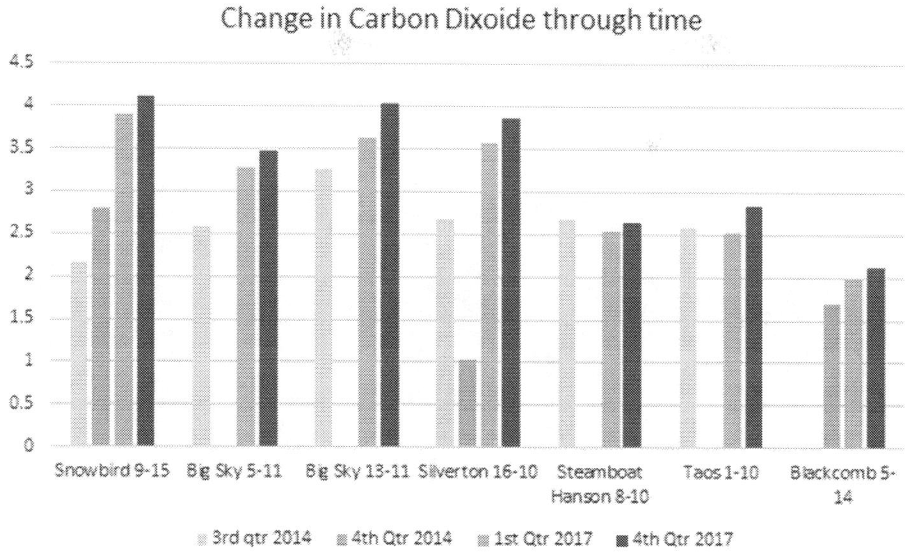

FIGURE 3.69C Nitrogen data in percentage through time for the Arikaree Creek Field, sections 10, 11, 14, and 15, Township 6 South, Range 54 West, Lincoln County, Colorado—Denver Basin.

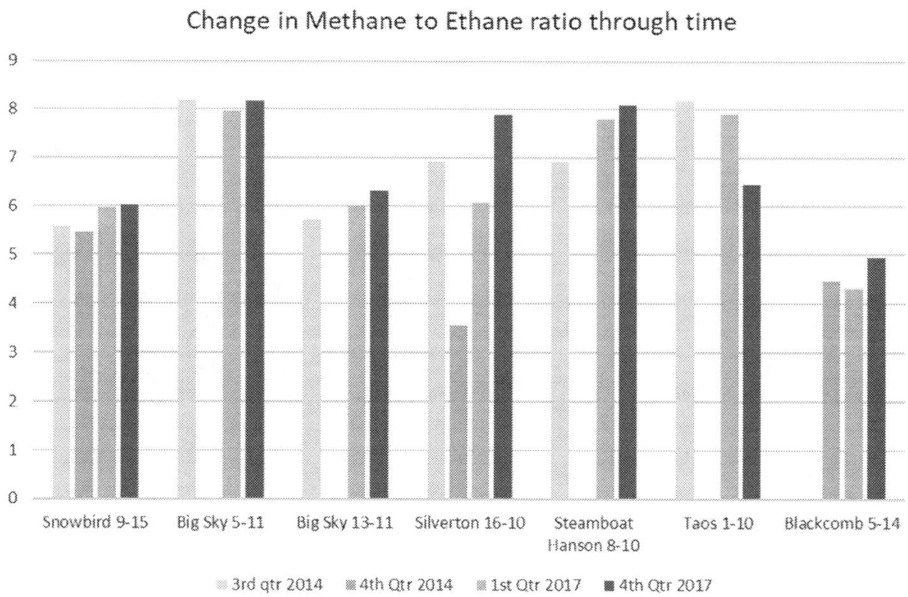

FIGURE 3.69D Methane to C_2H_6 ratio through time for the Arikaree Creek Field, sections 10, 11, 14 and 15, Township 6 South, Range 54 West, Lincoln County, Colorado—Denver Basin.

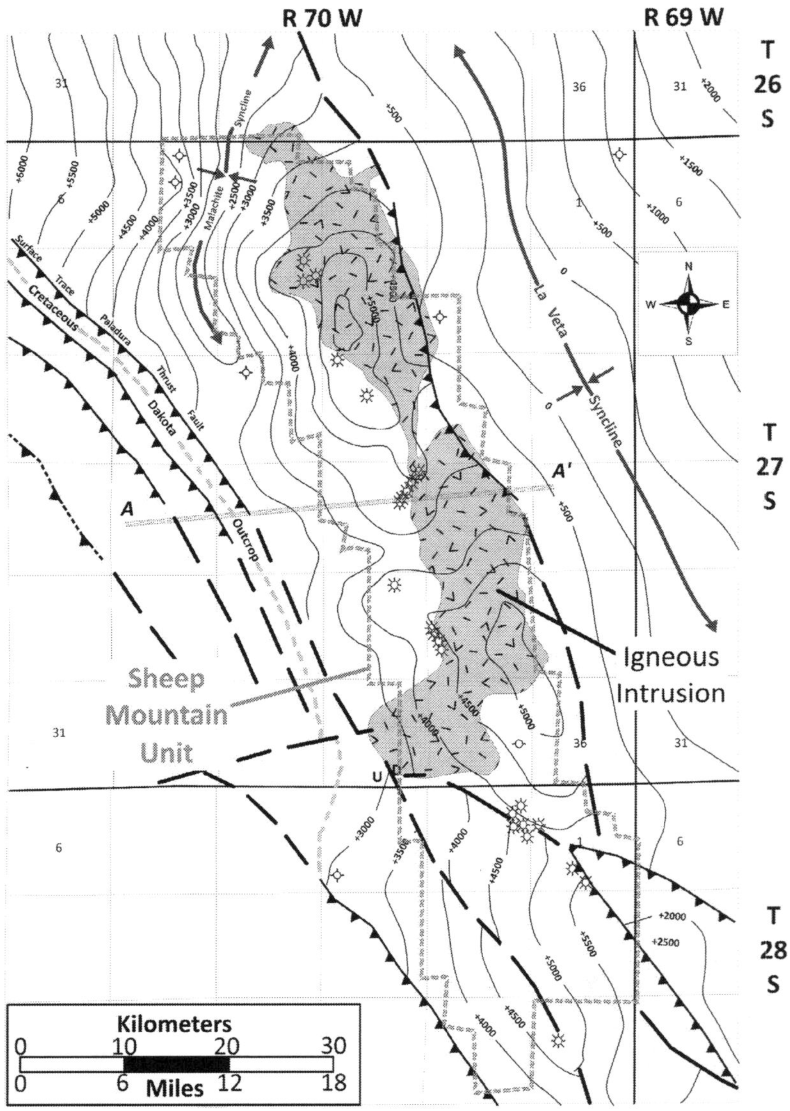

FIGURE 3.70 Dakota structure on top of Sheep Mountain Field located in Huerfano County, Colorado, northern Raton Basin, modified from Roth (1983). The gas wells are blue and dry holes and injectors are black. The original by Roth (1983) has been updated with wells that have been drilled post publication. The cross section for A to A′ is in Fig. 3.72.

its north end related to major tectonic events and a He accumulation on its northeast edge that is related to both igneous activity and proximity to Apishapa Uplift. Historically, the Raton Basin is mostly known for its extensive coal deposits and coal-bed methane production. The basin also has areas of extensive dikes and intrusions that could provide sources for He and/or CO_2.

TABLE 3.4 Summary of He, CO_2, N_2, and CH_4 content of fields in the Las Animas Region, southeast Colorado, western Kansas, and western Oklahoma (Moore, 1981; Avis, 1982; Stoddard, 1982; Moore and Sigler, 1985; Adams, 1990; Kozarek, 1990; Sonnenberg and van Drehle, 1990; Wallace et al., 1990; von Drehle, 1990; Bolyard, 1990; Lange and Wellborn, 1985). The data in some cases may be averages of several samples in a field or a well.

Field	County	Producing zone	Depth ft (m)	BTU of Gas	He%	N_2%	CO_2%	CH_4%
Arapahoe	Cheyenne	Morrow	5196 (1574)	NR	2.235	NR	NR	NR
Arikaree Creek	Lincoln	Spergen	8000 (2070)	839	0.629	47.55	2.56	26.82
Arrowhead	Cheyenne	Morrow	4800 (1450)	NR	2.55	33.985	0.747	47.781
Barrel Springs	Prowers	Morrow	4079 (1236)	950	0.91	30.6	NR	52.2
Bledsoe Ranch	Cheyenne	Morrow	6180 (1873)	720	1.9	37.7	0.72	59.3
Brandon	Kiowa	Lansing–Kansas City/Morrow	4690 (1412)	1124.9	1.02	12.13	0.304	65.487
Channing	Prowers	Morrow	4855 (1471)	1105	0.33	9.54	0.04	73.50
Clyde	Prowers	Clyde	4600 (1400)	NR	0.625	NR	NR	NR
Flank	Baca	Morrow	4691 (1421)	999	0.82	15.49	0.17	70.23
Frontera	Cheyenne	Spergen	5465 (1656)	401	2.05	59.58	1.34	31.32
Great Plains	Lincoln	Morrow	7056 (2138)	683	1.36	49.08	0.14	33.51
Harker Ranch	Cheyenne	Morrow	5425 (1644)	1000	2.603	NR	NR	NR
Haswell	Kiowa	Morrow	5420 (1642)	NR	0.54	NR	NR	NR
Keyes	Cimarron	Morrow/Keyes	4500 (1360)	1014/820	2	24.82	0.3	67.15
Ladder Creek	Cheyenne	Spergen	5270 (1597)	NR	4.2251	50.775	1.62	12.2
Left Hand	Kiowa	Morrow/Keyes	4964 (1504)	1117	0.78	12.4	0.6	68.7
Lookout	Kit Carson	Morrow	6370 (1930)	444	3.874	69.4	0.65	29.9
Mayfield	Cheyenne	Morrow	5120 (1551)	NR	2.6	NR	NR	NR
Jace-Moore Johnson	Kiowa/Greeley	Morrow	5111 (1549)	484	1.69	57.08	0.81	34.63
Nee Noshe	Kiowa	Morrow	4842 (1467)	1.103	0.37	10.87	0.09	72.79

(Continued)

TABLE 3.4 Summary of He, CO_2, N_2, and CH_4 content of fields in the Las Animas Region, southeast Colorado, western Kansas, and western Oklahoma (Moore, 1981; Avis, 1982; Stoddard, 1982; Moore and Sigler, 1985; Adams, 1990; Kozarek, 1990; Sonnenberg and van Drehle, 1990; Wallace et al., 1990; von Drehle, 1990; Bolyard, 1990; Lange and Wellborn, 1985). The data in some cases may be averages of several samples in a field or a well.—cont'd

Field	County	Producing zone	Depth ft (m)	BTU of Gas	He%	N_2%	CO_2%	CH_4%
Penny Packer	Kit Carson	Morrow	5430 (1645)	NR	3.01	NR	NR	NR
Salt Lake	Kiowa	Morrow	5131 (1554)	NR	1.60	NR	NR	NR
Smoky Hill	Kit Carson	Morrow/Marmaton	5755 (1744)	333	4.31	69.31	0.88	25.68
Sorrento–Mt. Pearl	Cheyenne	Morrow	5496 (1665)	NR	1.381	33.78	0.87	36.92
Spelunker	Baca	Red Cave	1684 (510)	440	1.6	59	NR	NR
Wagon Trail	Bent	Morrow	4620 (1400)	1005	0.3	17.6	0.2	81.8

3.8.1 Sheep Mountain Field

The Sheep Mountain CO_2 Field is located in Townships 27 and 28 South, Range 70 West, Huerfano County, Colorado, USA (Fig. 3.70), discovered in 1971 by Arco Oil and Gas and Exxon USA. The field lies underneath Sheep Mountain and is a double plunging foreland fold aligned with imbricate thrust folds of Laramide age (Worrall, 2004; Eppink et al., 2014). The field produces from an Oligocene felsite dike, Dakota (Cretaceous), and Entrada (Jurassic) sandstone reservoirs at 3304 to 6670 ft (919–2021 m). The igneous activity is related to normal faulting in the region. Fig. 3.70 is a structure on top of the Dakota Sandstone from Roth (1983), and the producing wells (blue) are clustered as they are deviated from a small group of locations due to topographic issues for drilling. Fig. 3.71 is a cross section adapted from Roth (1983) and clearly shows the field lies underneath the igneous intrusion. The Dakota reservoir consists of two beds of well-sorted, cross-stratified, fine- to medium-grained sandstones separated by a thin carbonaceous shale (Eppink et al., 2014). The Entrada reservoir is thinner but is similar in characteristics to the Dakota and lies unconformably on the Sangre de Cristo Formation. The felsite dike is 300 ft (90 m) thick and has primary and secondary porosity and permeability (Eppnick et al., 2014). The field has produced over 1.4 TCF of CO_2 at maximum rates of 300 MMCFGPD in 1988 and presently is producing above 100 MMCFGPD from 24 wells, predominantly from the Dakota Sandstone. The CO_2 is used for secondary recovery in several oil fields in Texas. Low recovery is due to the tight nature of the reservoir and access limited by the San Isabel National Forest and Wilderness (Eppink et al., 2014).

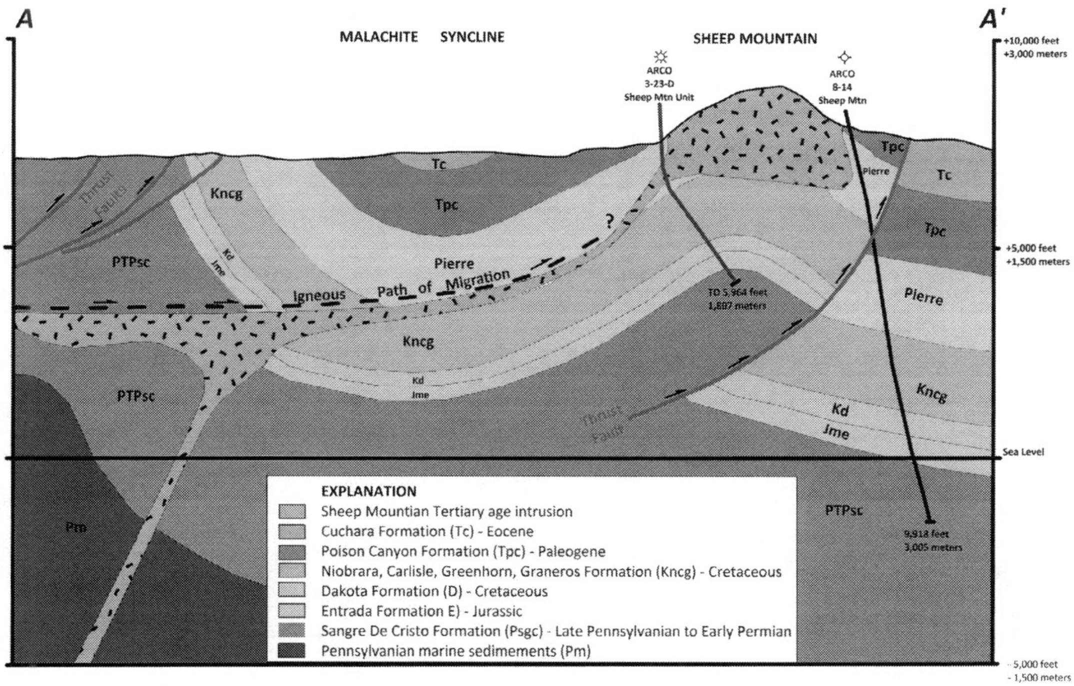

FIGURE 3.71 Cross section of A-A′ through Sheep Mountain modified from Roth (1983).

The structure for the Sheep Mountain and Oakdale fields (see below) was the result of several periods of movement along the Sangre de Cristo Mountains from Paleocene to Late Eocene to Early Oligocene time (Roth, 1983). The sedimentary section at Sheep Mountain and Oakdale fields is represented by Pennsylvanian to Tertiary strata that was folded and overthrusted to form the Sheep Mountain Anticline (Roth, 1983). The emplacement of the laccolith, dikes, and sills in relation to the deformation is not well understood.

Reservoir Characteristics of Sheep Mountain Field (Roth, 1983; Worrall, 2004; Eppink et al., 2014):

Productive area:	43,000 acres (est.)
Pay zone:	Entrada Sandstone (Jurassic), Dakota Sandstone (Cretaceous)
	Felsic dike (Oligocene)
Depth to pay:	3304 to 6670 ft (919–2021 m)
Net pay:	239 ft pay (72 m)
Porosity:	1.6%–25.3% (core, Sheep Mountain Unit 9–7 well)
Permeability:	117 to 742 millidarcies (from core, Sheep Mountain Unit 9–7 well)
Water saturation:	Unknown
Initial field pressure:	2165 psi
Formation temperature:	157°F
Type of drive:	Gas expansion
Type of traps:	Folded thrust sheet
Gas characteristics and analysis:	0.0185% ^{3}He, 97.94% CO_2, 0.41 N_2, 1.14% CH_4 and C_2H_6+ 0.37%, 24.4 BTU, 1.5058 SG (drill stem test from the Dakota, Ferris 1). ^{3}He varies from 0.08% to 0.54% in the Dakota and 0.03% in the Entrada.

3.8.2 Oakdale (Dike Mountain) Field

The Oakland Field, originally called Dike Mountain, is located in section 33, Township 28 South, Range 69 West, and sections 4, 9, and 10, Township 29 South, Range 69 West, Huerfano County, Colorado, USA. The field was discovered in 1997; it is producing primarily CO_2 from a felsic igneous body, the Dakota (Cretaceous) and Entrada (Jurassic) sandstones at 5248 to 5804 ft (1590–1770 m) (Fig. 3.72). The Oakdale is a double-thrusted anticline located in the footwall of the Sangre de Cristo main thrust and the hanging wall of the Oakdale thrust (Eppink et al., 2014). The Dakota reservoir consists of two beds of well-sorted, cross-stratified, fine- to medium-grained quartzite sandstones separated by a thin carbonaceous shale (Roth, 1983; Eppink et al., 2014). The Entrada reservoir is thinner but has similar characteristics to the Dakota and lies unconformably on the Sangre de Cristo Formation. Fig. 3.72 is a structure map on the Dakota Sandstone as modified from Roth (1983). The felsite dike is 300 ft (90 m) thick and has primary and secondary porosity and permeability.

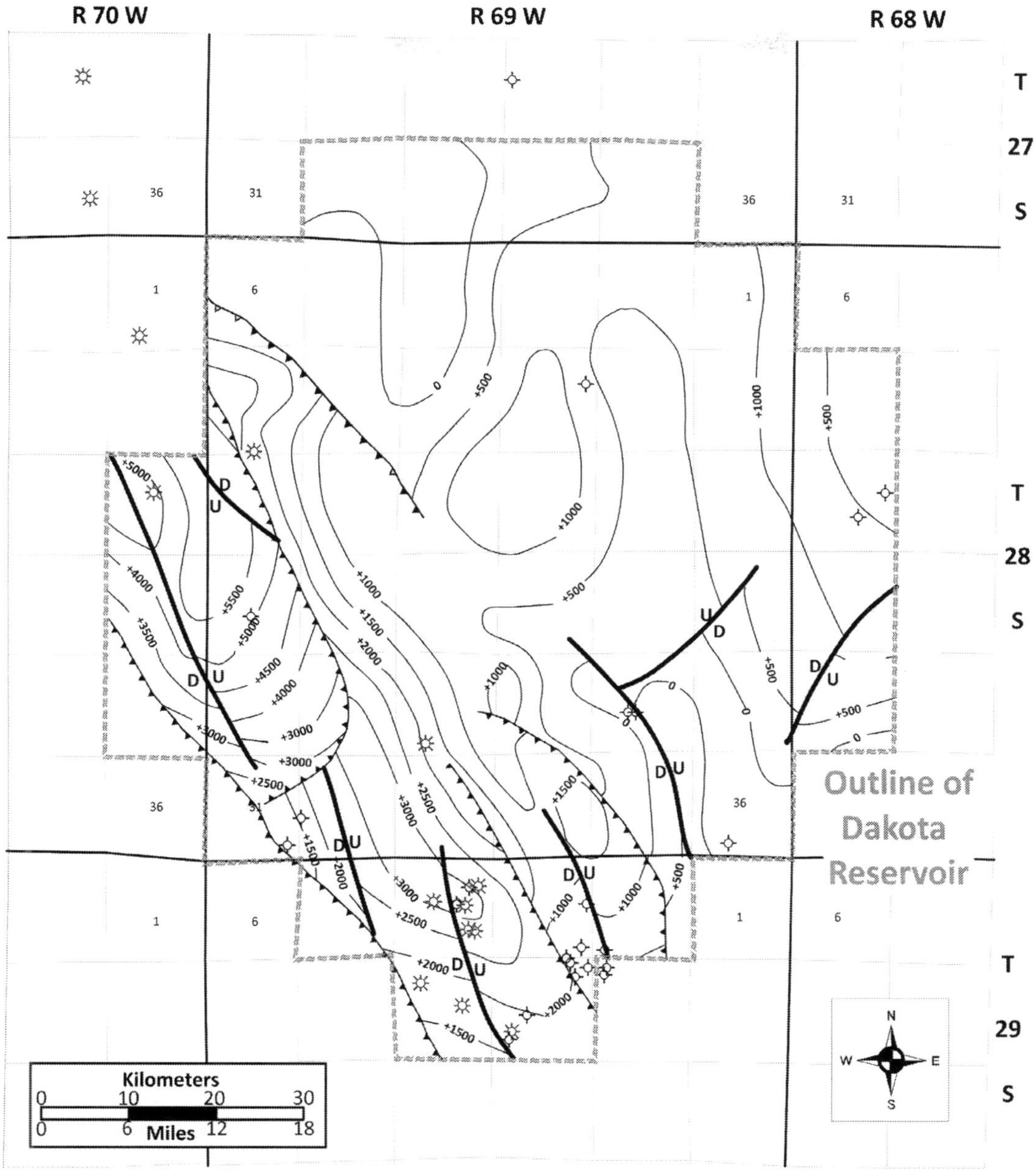

FIGURE 3.72 Structure on top of the Dakota Sandstone at the Oakdale (Dike Mountain) Field, Huerfano County, Colorado, USA, northern Raton Basin, modified from Roth (1983).

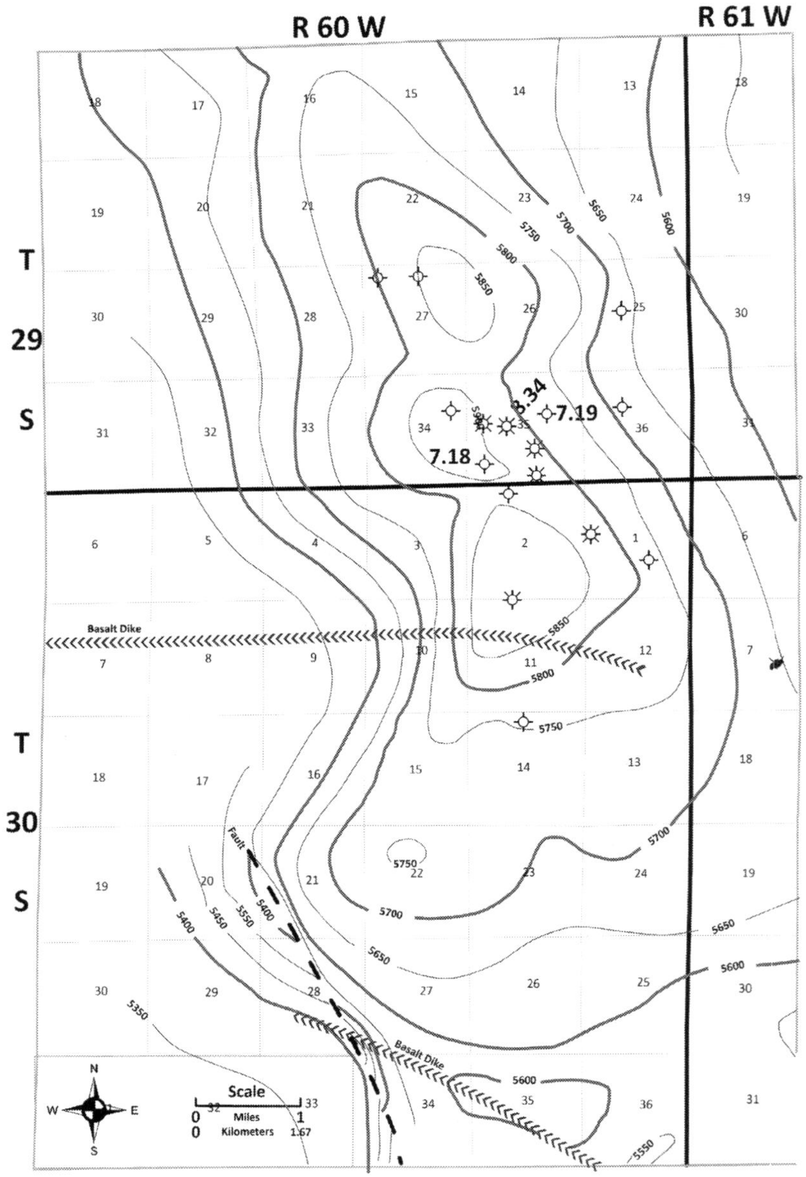

FIGURE 3.73 Structure on top of the Glorieta sandstone of Permian age Model Dome, Las Animas County, Colorado, Apishapa Uplift. The basalt dikes are represented by a chevron pattern. The He data are in percentage.

The He present is ^3He, which is probably sourced along with the CO_2 by the microgranite Maestas stock (370 acres in areal extent) and associated dikes at the Oakdale Field. The field's low recovery from the volcanic reservoir and its location of the field next to the San Isabel National Forest/Wilderness minimizes full exploitation of the accumulation (Eppink et al., 2014). The reservoirs in the Dakota and Entrada sandstones are typically twice as porous as similar reservoirs elsewhere in the Raton Basin but similar to Sheep Mountain reservoirs (Worrall, 2004). The source of the CO_2 is believed to be the thermal alteration of local carbonates (Caffee et al., 1999; Gilfillan et al., 2008).

Reservoir Characteristics of Oakland Field (Worrall, 2004; Eppink et al., 2014):

Productive area:	3400 or 6000 acres depending upon source
Pay zone:	Entrada (Jurassic) and Dakota (Cretaceous)
Depth to pay:	5248 to 5804 ft (1590–1770 m)
Net pay:	Felsic igneous sill: 143 ft (43 m)
Dakota Sandstone:	54–140 ft (16–42 m)
Entrada Sandstone:	19 ft (6.2 m)
Porosity:	16%–20% (Dakota and Entrada sandstones)
Permeability:	Unknown
Water saturation:	Unknown
Initial field pressure:	2790 psi
Formation temperature:	179°F
Type of traps:	Folded thrust sheet, double-thrusted anticlines
Type of drive:	Gas expansion
Gas characteristics and analysis:	24%–97% CO_2, 3%–75% CH_4+, and ^3He 0.40%–0.44%
Reserves:	450 BCF

3.8.3 Model Dome

The Model Dome is located in sections 27 and 35, Township 29 South, Range 60 West, sections 1 and 11, Township 30 South, Range 60 West, Las Animas County, Colorado, USA (Fig. 3.73). The field was discovered in 1926 and produces from the Glorieta or Lyons sandstone (Permian) from approximately 900 ft (270 m). USGS analysis indicated an average of He 8%, CO_2 15%, and N_2 77% (Clair and Bradish, 1956). The productive sandstone lies 140 ft (42 m) above water bearing strata. The field has gone through three periods of exploration and production, 1926 to 1930, in the 1960s to the 1980s and present day (Fig. 3.73). No official production data are available and CO_2 and N_2 was vented to the atmosphere. The exploration period from the 1960s to the 1980s resulted in additional dry holes.

The Model Dome was found by surface mapping (Clair and Bradish, 1956). Two basalt dikes at the surface were identified along with the surface expression of a northwest trending fault. Structure at depth was implied which resulted in its discovery (Fig. 3.73). The presence of the basalt dikes may and probably do provide pathways for He migration into the reservoir rocks. Reserve estimates for He based on the surface structure were put at 261,559,000 MCFHe (Clair and Bradish, 1956). The presence of dry holes in the first two rounds of exploration and exploitation suggests the reservoir is not continuous or the gas saturation is limited to a smaller area than the 4380 acres assigned by the USBM originally (Clair and Bradish, 1956). In addition, there are dry holes on top of the Model Dome structure which further supports compartmentalization or rapid depletion of the reservoir across a large area. Production was indicated at 500,000 to 1,000,000 cubic ft a day with an average of 100–200 MCFGPD with a bottom hole pressure of 11 psi from the Glorieta sandstone of Permian age (Clair and Bradish, 1956). Unfortunately, the original drilling occurred before modern geophysical techniques. No data is publicly available on any of the wells. The He data were analyzed by the USGS (Clair and Bradish, 1956). Recently a public company has placed 3 BFCHe reserves without any public technical data to support the conclusion. The Model Dome may be one of many features of similar type in the area or not. Las Animas County, Colorado, for the most part is a true exploration play and could be comparable to areas found in parts of Arizona and New Mexico that are He rich.

3.8.4 McCallum and South McCallum Fields

The McCallum (also known as North McCallum) and South McCallum fields, discovered in 1927 (Biggs, 1957), are located in Townships 9 to 10 North, Ranges 78 and 79 West, Jackson County, Colorado, USA, in the North Park Basin (Fig. 3.74). Both fields were found by surface

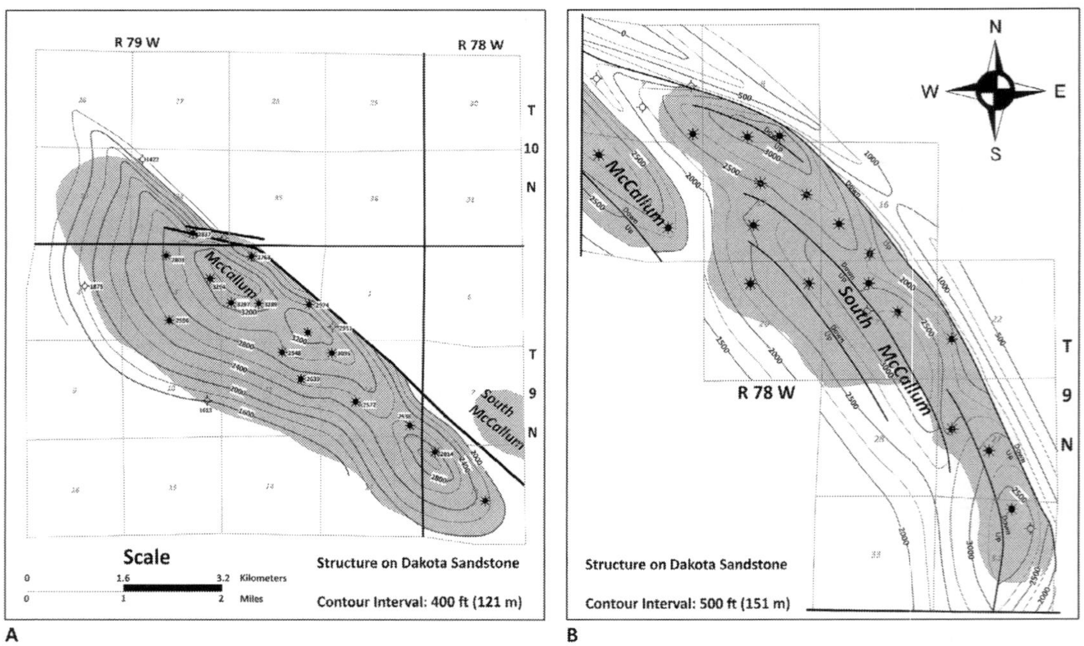

FIGURE 3.74 The McCallum (A) and South McCallum (B) fields, North Park Basin, Jackson County, Colorado, USA, modified from Carpen (1957) and Mark (1961). Anticlinal structures are indicated and the oil and gas fields are indicated as outlined by the Colorado Oil and Gas Commission by a gray patter. The structural contours are on top of the Dakota Sandstone (Cretaceous). The well locations indicated are from Carpen (1957) and Mark (1961) and are not always consistent with those presently identified by the Colorado Oil and Gas Commission. There are numerous wells that have been drilled in both fields post articles Carpen (1957) and Mark (1961) but for were not added.

mapping. The two fields produce CO_2 from the Cretaceous Dakota and Lakota Formations but are also capable of production from the Muddy (Cretaceous) sandstone and felsic dikes (Fig. 3.74). The North Park Basin was created by compressional forces during the Laramide orogeny and has a north-south orientation (Lange and Wellborn, 1985). The McCallum and South McCallum fields consist of two multiple anticlines that have a northwest-southeast orientation. The anticlines are bounded to the northeast by a steeply dipping fault. The McCallum Field has gentle dip to the southwest toward the basin center (Fig. 3.74A). The South McCallum is a complex series of anticlines and faults (Fig. 3.74B).

Reservoir Characteristics of McCallum and South McCallum fields (Carpen, 1957; Mark, 1961; Eppink et al., 2014):

Productive area:	10,000 acres
Pay zone:	Dakota, Lakota, Muddy sandstones (Cretaceous), and felsic dikes
Depth to pay:	South McCallum: 4980 to 5350 ft (1510 to 1620 m)
McCallum:	5300 to 5800 feet (1606 to 1760 m)
Net pay:	South McCallum Field—80 ft (24 m) for the Muddy, 48 ft (14.5 m) for the Dakota, and 86 ft (26 m) for the Lakota.
	McCallum Field—12 ft (3.6 m) for the Muddy, 39 ft (12 m) for the Dakota, and no data for the Lakota. The Morrison Sandstone is also productive.
Porosity:	South McCallum—Muddy 14%, Dakota 16%, and Lakota 14%
	McCallum—Muddy 17.7%, Dakota 16.2%, and Lakota 31%
Permeability:	South McCallum—65 millidarcies (Muddy), 122 millidarcies (Dakota), and 130 millidarcies for Lakota
	McCallum—120 millidarcies (Muddy), 70 millidarcies (Dakota), and 31 millidarcies for Lakota
Water saturation:	Unknown
Initial field pressure:	2316 psi
Formation temperature:	120 degrees is 150°F
Type of traps:	Laramide aged anticline, sealed by the Pierre shale and a laccolith.
Type of drive:	Unknown
Gas characteristics and analysis:	
South McCallum:	CO_2 92%; N_2 3.4%; CH_4+ 4.5%
McCallum:	Lakota–Dakota reservoir: CO_2 92.12%; N_2 4.88%; CH_4+ 3.00%
	Both fields have reported 0.11%–0.32% He
Production:	Approximately 870 BCF

3.9 The rest of Texas

3.9.1 Petrolia Field

The Petrolia Field was first known as the Henrietta Field and is located 16 miles northeast of Wichita Falls, Texas, in Clay County, USA (Fig. 3.75) (Kendrick and McLaughlin, 1929). The field lies along the Electra–Red River Arch that is a basement ridge (Reutinger, 1968). The Petrolia Field was the first field to be developed for He extraction in the United States. The field was initially discovered in 1902 when a local farmer was looking for water but encountered oil at 150 ft (45 m). Several other farmers and ranchers in the area also were having difficulty with oil seeping into their water wells. In 1907, the first gas production was discovered in the field. Data are lacking from the field due to its early development in the oil and gas industry and lack of information from the wells. Production in the field is reported to be from lenticular sandstones of the Pennsylvanian age Cisco strata draped over a Precambrian structural high (Reutinger, 1968) (Fig. 3.75). Many of the sandstones follow a shoreline trend whereas others are related to granite wash type reservoirs. The Cisco gas sands are found from 1500 to 1750 ft (455–530 m) and Precambrian rocks are at 4000 ft (1212 m). The field covers approximately 30,000 acres (Fig. 3.75). Helium analyses from five wells average 1% (Reutinger, 1968).

3.9.2 Val Verde Basin

The Val Verde Basin is located in Terrell, Pecos, Crockett, and Val Verdes counties, Texas, USA, and is a foreland subbasin that is bounded by the Ozona Arch and Central Platform to the north and the Marathon Orogenic Belt (Ouachita thrust belt) to the south (Fig. 3.76) (McMurty, 1982; Montgomery, 1996). The basin produces CO_2 from the Woodford (Lower Mississippian), Hunton Group, Simpson (Middle Ordovician), and Ellenburger (Lower Ordovician) from 12,000 to 15,000 ft (3640 to 4550 m). The Val Verde Basin is a complex structural province characterized by several major tectonic styles that resulted in several major northwest-trending faults that seem to be transform or transform-related wrench faults (Montgomery, 1996).

The Brown Bassett Field located in Terrell and Crocket counties (Fig. 3.76) produces CO_2 from the Ellenburger Limestone reservoir located 250 ft (75 m) from the top of the strata. The reservoir is fractured dolomite with vugs along fractures producing from approximately 14,700 ft (4455 m). The reservoir pressure is 6648 psi, gas column of 500 ft (150 m), 54% CO_2 and 46% CH_4+, 6.7% porosity, and 16,000 acres in areal extent (McMurty, 1982).

The Gray Ranch Field (Fig. 3.76) is located in Pecos County and produces from the Ellenburger Formation at 14,700 ft (4455 m). The reservoir was completed open hole, encompasses 6000 acres, and contains 1700 to 1800 ft (515–545 m) of pay. It is an NW-SE–trending anticline with a fault on the west flank. It has a 1900 ft (575 m) gas column, with 2%–3% porosity enhanced by fracturing, permeability is via fractures, 6674 psi reservoir pressure and 40% CO_2 and 60% CH_4 (Superior Oil Company, 1982). The overlying Devonian and Wolfcamp reservoirs have less than 1.5% CO_2.

Isopach maps of the zones within the Ellenburger (Fig. 3.77A–D) clearly indicate penecontemporaneous and postdepositional thinning along the Val Verde Fault Zone as well as later

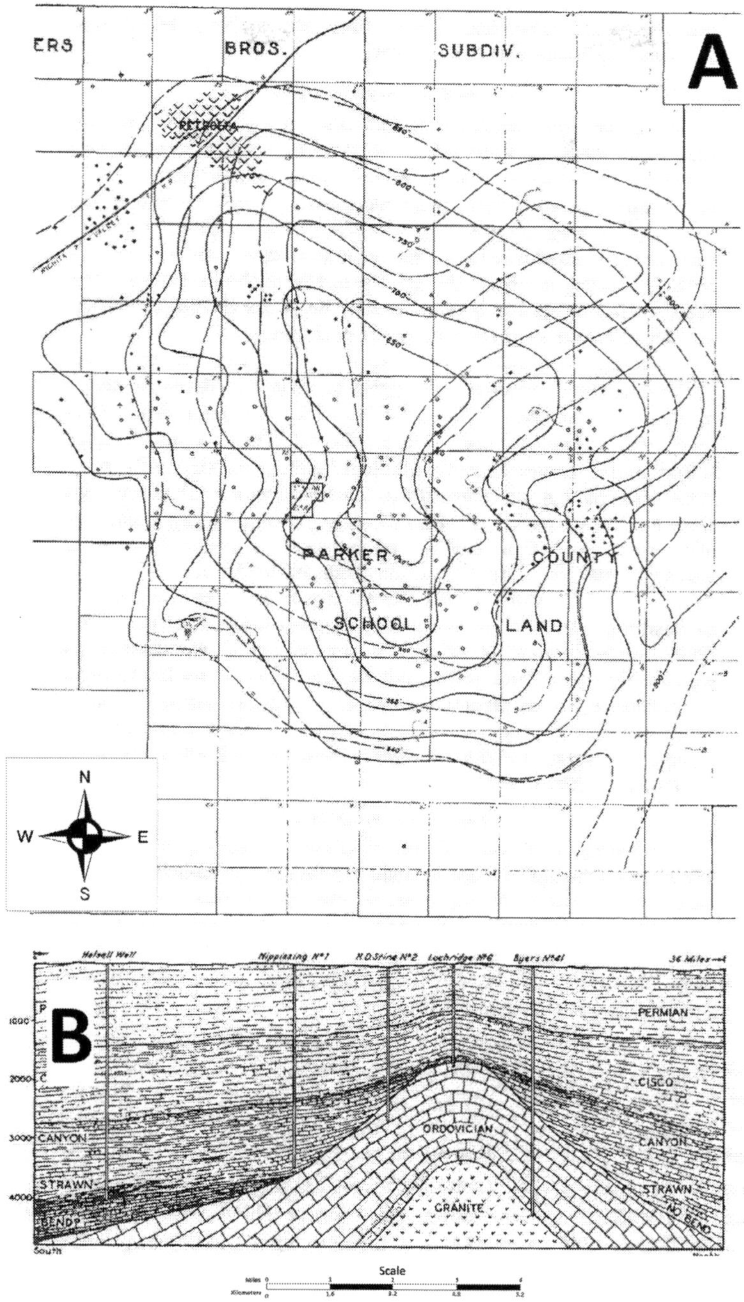

FIGURE 3.75 Petrolia Field in Clay County, Texas, modified from Kendrick and McLaughlin (1929) (*reprinted by permission of the AAPG whose permission is required for future use*). (A) Structure map is on the red beds (broken contours), 20-ft contour interval (6 m) of the gas-producing zone which is of Permian age and referred to as the Wichita Formation. Isopach of the total gas sands is a solid line with a contour interval of 100 ft (30 m) and represents multiple pays. (B) Cross section through the field which represents a paleo high whereby Pre-Pennsylvanian rocks were eroded off and then the Cisco (Pennsylvanian) and the Wichita (Permian) rocks were deposited.

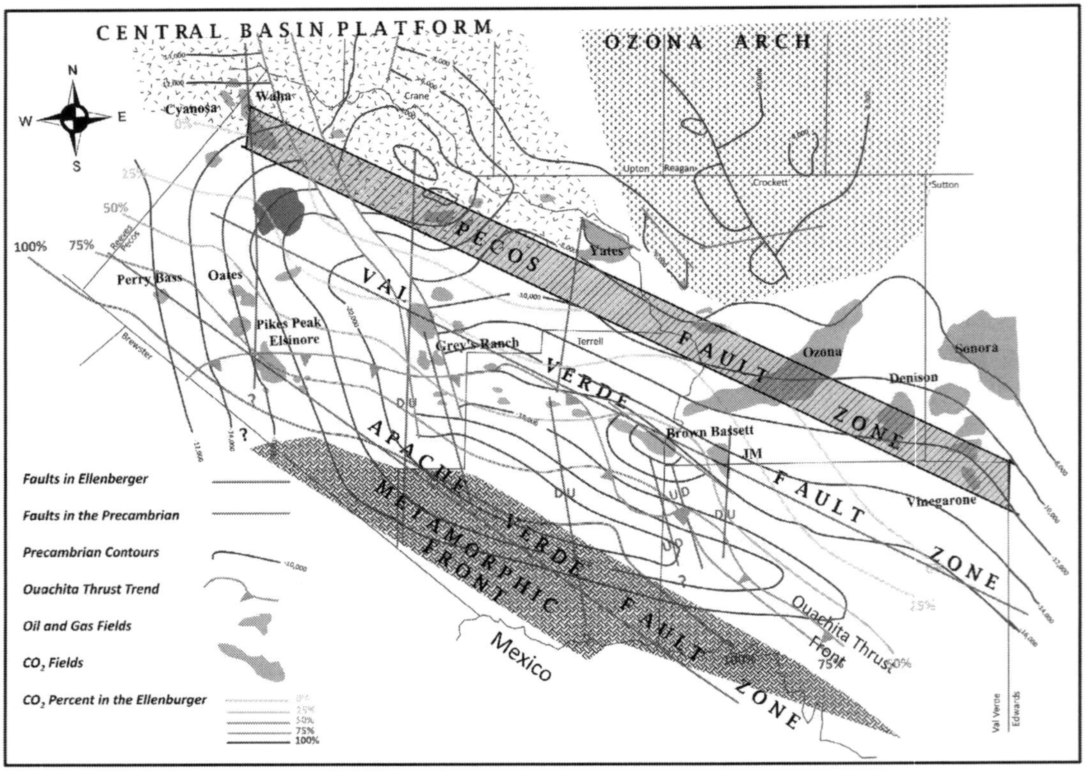

FIGURE 3.76 Val Verde Basin modified from Holmquist (1965), Hills (1968), Montgomery (1996). The map indicates the significant tectonic features adjacent to the basin, Precambrian structure contour, existing oil, gas, and CO_2 fields, and CO_2 percentages in the Ellenburger.

development of karst features. Fig. 3.77A is an isopach of the E4 to E2 zone within the Ellenburger indicating over 100 ft (30 m) thinning of strata along the Brown Bassett Field.

Several other fields are present along the Val Verde Fault system that produce from the Ellenburger with similar reservoir characteristics. Fig. 3.77B indicates mapping of the E-2 Wilberns zone only and thinning is more definitive across several fields along the Val Verde Fault Zone. Fig. 3.77C is an isopach from the top of the Ellenburger to the E-5 zone and clearly indicates thinning related and defined by the northwest to southeast strike of the Val Verde Fault Zone. There is also a strong secondary fault strike to the northeast-southwest. Fig. 3.77D is an example of the cave or karst system in the Ellenburger as defined by Cantor et al. (1993). The presence of CO_2 in the reservoir probably enhanced the porosity and permeability. The gas column within the Ellenburger encompasses the entire Ellenburger section indicating that there was more than sufficient CO_2 to expel the hydrocarbons out of the reservoir and mobilize them from south to north along the Marathon–Ozark orogeny (Ouachita Thrust Front).

The source of the CO_2 in the Val Verde Basin was first believed to be produced by hydrodynamic processes from meteoric waters flushing out reservoirs from the Devonian and deeper strata of hydrocarbons during the Pennsylvanian and later times (Holmquest,

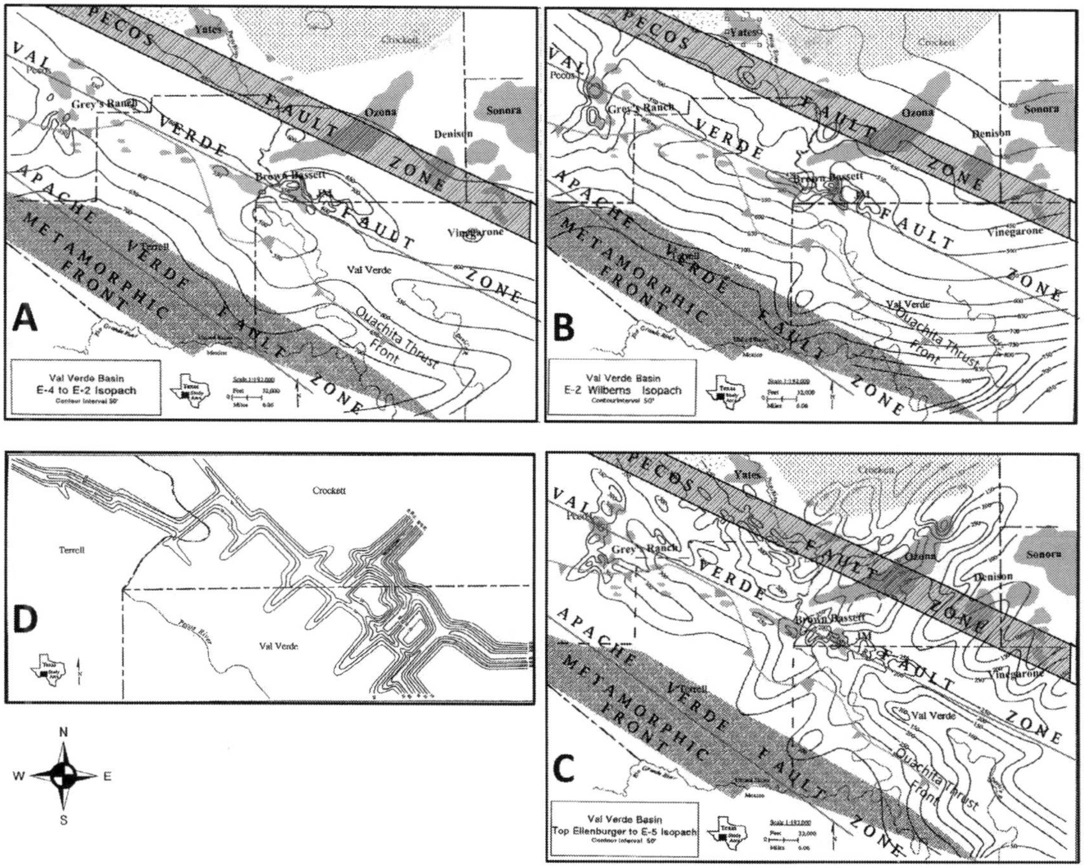

FIGURE 3.77 Val Verde Basin isopachs for the Ellenburger modified from Holmquist (1965); Hills (1968); Cantor et al. (1993); Montgomery (1996). (A) Isopach of the Ellenburger E-4 to E-2 zones notes the thinning associated with the CO_2 fields and Val Verde Fault Zone. (B) Isopach of the Ellenburger E-2 zones notes the thinning associated with the CO_2 fields and Val Verde Fault Zone. (C) Isopach of the top of the Ellenburger to the E-5 zone notes the thinning associated with the CO_2 fields and Val Verde Fault Zone. (D) The Ellenburger reservoir is dominated by a karst system that was controlled by faulting.

1963). More recent works by Weismann (1971) and Eppink et al. (2014) indicate that the CO_2 in the Devonian and older strata were magmatic in origin associated with the thrusting related to the Marathon Orogenic Belt during Pennsylvanian and Permian times.

Reservoir Characteristics of Val Verde Basin (McMurty, 1982; Superior Oil Company, 1982; Eppink et al., 2014):

Productive area:	90,000 acres
Depth to pay:	12,000 to 15,000 ft (3640 to 4550 m)
Pay zone:	Ellenburger (Ordovician), Simpson (Ordovician), Woodford Lower Mississippian

(Continued)

Net pay:	640 ft (195 m)
Porosity:	4.0%, enhanced by fracturing, maybe as high as 8%
Permeability:	1 to 100 millidarcies
Water saturation:	35%
Initial field pressure:	5872 psi average
Reservoir temperature:	218°F
Type of drive:	Gas expansion
Type of traps:	Structures related to thrusting and Marathon Orogenic Belt
Gas characteristics and analysis:	20%–60% CO_2; the rest is hydrocarbons

3.10 The rest of New Mexico

3.10.1 Bravo Dome

The Bravo Dome is located in Townships 18, 19, 20, and 21 North, Ranges 30, 31, 32, 33, 34, and 35 East, Union, Quay, and Harding County, New Mexico, USA, that was accidently discovered in 1916 (Fig. 3.78). The field lies across an anticlinal nose upon a spur of the Sierra Grande Arch (Broadhead, 1998; Baines and Worden, 2004; Broadhead et al., 2009) in which the Tubb sandstone member of the Yeso Formation (Cretaceous) is draped and sealed by the Chinle mudstone and Cimarron anhydrite (Gilfillan et al., 2008; Broadhead et al., 2009). The Tubb member unconformably lies on top of the Precambrian granite. The Tubb sandstone consists of sand-dominated eolian, fluvial, and alluvial reservoirs that were not deposited north of the structure creating a combination structural–stratigraphic trap. The Tubb sandstone grades into impermeable muddy sediments to the northeast (Broadhead et al., 2009). Low recovery is due to the poor quality of the reservoirs that are interbedded with the tight sandstones (Eppink et al., 2014). The CO_2 has been used primarily for EOR recovery in west Texas (Broadhead et al., 2009; Eppink et al., 2014).

The CO_2 at Bravo Dome is believed to be sourced from the mantle based on the small percentages of noble gases present (less than 0.1%), with localized magmas transporting CO_2 from the upper part of the crust (Gilfillan et al., 2008; Broadhead et al., 2009). Isotopic analyses of the CO_2 and associated noble gases at Bravo Dome are based on ^{20}Ne reservoir filled from west to east (Gilfillan et al., 2008). This scenario is supported by distribution of pressure regimes and pressure compartments within the Tubb sandstone reservoir whereby the western side of the field is normally pressured and the eastern side becomes progressively underpressured (Broadhead, 1998). The majority of the CO_2 in the region is penecontemporaneous with the volcanism in the area which is 1.2–3.4 Ma, with a minor volume of CO_2, associated with the recent local Capulin eruptions or temporally related intrusive events (Broadhead et al., 2009). Broadhead (1998) and Cassidy (2006) indicated that the CO_2 may have been generated 10,000 to 8000 years ago; migration and charging may be ongoing based on increasing pressure in at least one well (Baines and Worden, 2004).

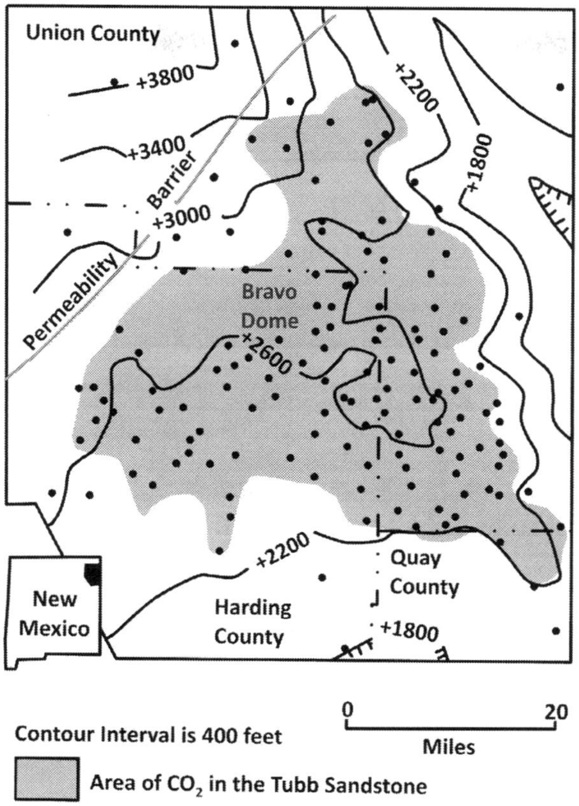

FIGURE 3.78 Structure on top of Tubb sandstone and northwestern, updip permeability barrier within Tubb sandstone, Bravo Dome CO_2 field (Broadhead, 2009).

Reservoir Characteristics of Bravo Dome (Gilfillan et al., 2008; Broadhead, 2009; Eppink et al., 2014):

Productive area:	633,600 acres
Pay zone:	Tubb Sandstone member of the Yeso Formation
Depth to pay:	2000 to 2600 ft (600–790 m)
Net pay:	100 ft (33 m)
Porosity:	20%
Permeability:	42 millidarcies
Water saturation:	0%–45%
Initial field pressure:	1082 psi on the western edge, 336 psi minimum on the eastern edge

(*Continued*)

Temperature of pay:	80°F
Type of drive:	Gas expansion
Type of trap:	Stratigraphic–structural
Gas characteristics and analysis:	96%–99.8% CO_2 and 0.2%–4.0% noble gases
Reserves:	3.5 TCF

3.10.2 Des Moines Field

The Des Moines Field is located in Township 29 North, Range 29 East, Union County, New Mexico, USA, and produces from the Permian Abo Formation lenticular arkosic conglomerates and sandstones lenses sealed by interbedded red shales along the axial crest of the Sierra Grande Uplift (Anderson, 1959; Foster and Jensen, 1972; Broadhead, 2009; Eppink et al., 2014). The field was discovered in 1935 and abandoned in 1966 due to processing issues with the CO_2 (Broadhead et al., 2009). Annual and cumulative production volumes are unknown (Broadhead et al., 2009). The reservoir lies unconformably on top of the Precambrian rocks and interbedded and overlying red shales act as a seal. Low recovery in the field is due to poor quality of reservoirs that are interbedded with the sandstones (Eppink et al., 2014).

Reservoir Characteristics of Des Moines Field (Broadhead et al., 2009; Eppink et al., 2014):

Productive area:	Unknown
Pay zone:	Abo Formation (Permian)
Depth to pay:	2060 to 2600 ft (625–790 m)
Lithology:	Lenticular sandstones and conglomerates
Net pay:	50 ft (15 m)
Porosity:	20%
Permeability:	Unknown
Water saturation:	20%
Initial field pressure:	137 psi, range 64–210 psi
Temperature of pay:	78°F
Type of drive:	Unknown
Type of trap:	Stratigraphic - structural
Gas characteristics and analysis:	CO_2 99%
Production:	Estimated to be 20 BCF

3.10.3 Permian Basin

The Permian Basin is limited to He production from Chaves County, New Mexico, from Abo fluvial-deltaic reds beds of Permian age that rest unconformably in most areas on Precambrian granites (Broadhead and Gillard, 2004). The Pecos Slope and Pecos Slope West gas production varies from 0.09% to 0.974%, with an average of 0.48% He. Helium seems to have migrated from Precambrian granitic rocks based on isotopic ratios and migrated up along basement faults that were sealed by anhydrites of the overlying Yeso Formation. Helium percentages decrease away from the fault systems and the reservoir rocks do lie directly on the Precambrian granites (Broadhead and Gillard, 2004). There are igneous dikes in the area and the traps seem to be stratigraphic. The hydrocarbons were generated from elsewhere downdip in the basin (Broadhead and Gillard, 2004).

3.11 Montana—North Dakota—Idaho

3.11.1 Idaho

The northern Rocky Mountain region of the United States covers three states: Idaho, Montana, and North Dakota. Idaho has developed recent natural gas production in the Weiser—Payette Basin in southwestern part of the state. The two fields, the Willow and Hamilton fields, produce from the Idaho Groups are lacustrine and fluvial deposits interbedded with tuffaceous and bentonitic intervals, rhyolitic and basaltic dikes and flows, and large deposits of evaporite rocks (Breedlovestrout, 2017). The gas from the Willow Field is derived from thermogenic maturation of kerogen (Barton and Ratchford, 2020). Gas from the Hamilton Field is biogenic processes (Barton and Ratchford, 2020). No, He was detected in either field, CO_2 varied from not detectable to 0.069% and N_2 varied from 0.14% to 2.48%. However, one well, not in either field, the Island Capitol 1-19, located as few miles away contained 0.0551% He, 61.32% N_2, and 0% CO_2 (Barton and Ratchford, 2020). The methane is of biogenic origin. There are no indications at this time of the presence of significant He and CO_2 assets in Idaho.

3.11.2 North Dakota

The majority of North Dakota geologically is part of the Williston Basin which is the largest intracratonic basin in North America. Hydrocarbon production from the Williston Basin has been substantial because of both conventional production from several Paleozoic reservoirs and unconventional oil extraction from the Bakken Shale Upper Devonian—Lower Mississippian. But there has been no He production in the state. Saskatchewan to the north has seen economic He production along the flank of the northern part of the basin and will be discussed below. The best He percentages in North Dakota are from Cambrian and Ordovician rocks along the Nesson-Antelope structural trend located in the center. The Newporte structure on the northeastern edge of Williston Basin in North Dakota (Nesheim and Kruger, 2019) also has identified He. The two different areas have distinct N_2 to He ratios suggesting different sources of gas (Nesheim and Kruger, 2019). The highest He value reported by Nesheim and Kruger (2019) so far is 0.46%, but of 65 samples that have been analyzed the average is below 0.04%. The USGS and USBM have data with similar values with the high values being found in the Cambrian–Ordovician strata. There is no indication of economic or anomalous deposits of CO_2 in North Dakota.

3.11.3 Montana

The eastern half of Montana is also part of the Williston Basin but has similar low He percentages based on USGS and USBM data. The Kevin-Sunburst Dome along the Sweetgrass Arch in Glacier, Toole, and Liberty Counties, has significant CO_2 and marginally economic He and the Rudyard field in Hill County has significant He that has not been developed (Fig. 3.79).

3.11.3.1 Kevin-Sunburst Dome

The Kevin-Sunburst Dome (Figs. 3.79 and 3.80) covers a large area of Glacier and Toole counties, Montana, USA, and into southern Alberta, Canada, along the Sweetgrass–

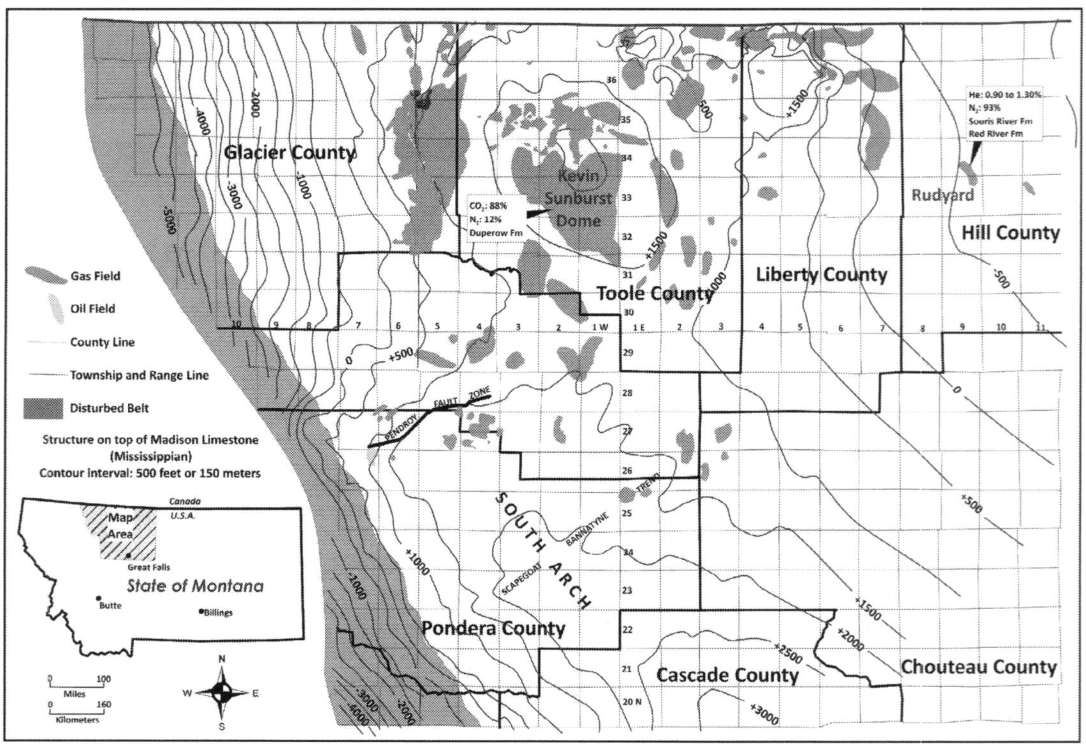

FIGURE 3.79 Location of the Kevin-Sunburst Dome and Rudyard fields in north central Montana. Structure is on top of the Madison Limestone with a contour interval of 500 ft (150 m). Modified from Nordquist and Leskela (1968).

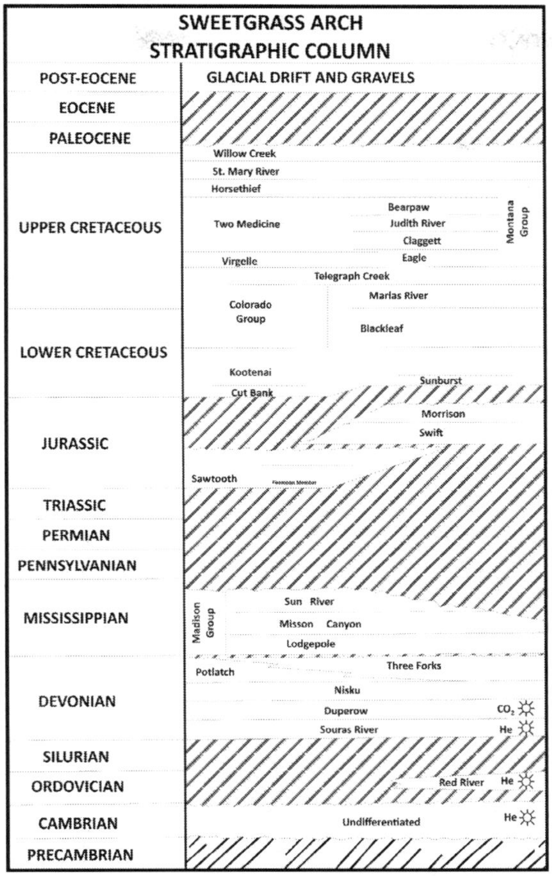

FIGURE 3.80 Stratigraphic column for the Kevin-Sunburst and Rudyard area in north central Montana modified from Nordquist and Leskela (1968). The Red River strata is only present in the Rudyard area.

Battleford Arch (Eppink et al., 2014). The top of the closure of the Kevin Dome is the Kevin-Sunburst Field located in Townships 32 to 36 North, Ranges 1 through 4 West and Range 1 East, Toole County, Montana. The structure has 750 closures and was formed by crustal shortening during the Antler orogeny in Upper Devonian time. During Tertiary time, additional crustal shortening occurred, further enhancing structural closure. The CO_2 is found in the Duperow Formation of Devonian age which is composed of shallowing upward carbonate cycles that typically have anhydrites at the top of the sequences that act as seals (Fig. 3.80). Variation in lithology is both lateral and vertical.

The Kevin Dome is not presently producing CO_2 but there is a recent drilling that is attempting to produce CO_2 for enhanced recovery. Recent events have also redirected industry to focus on sequestration of CO_2 in the dome (Bowen et al., 2012).

Reservoir Characteristics of Kevin-Sunburst Dome (Eppink et al., 2014):

Productive area:	261,000 acres
Reservoir:	Duperow Formation
Lithology:	Carbonates with anhydrite seals
Depth to pay:	3600 ft (1090 m)
Average net pay:	67 ft (20 m)
Average porosity:	9%
Average permeability:	Unknown
Average water saturation:	Unknown
Formation temperature:	91°
Initial field pressure:	1251 psi
Type of drive:	Unknown
Type of traps:	Combination structural–stratigraphic
Gas characteristics and analysis:	CO_2 88% and N_2 12%

3.11.3.2 *Rudyard Field*

The Rudyard Field is located in sections 27, 28, 33, and 34, Township 34 North, Range 9 East; and sections 2, 3, 10, 11, 14, and 15, Township 33 North, Range 9 East, Hill County in Montana, USA (Fig. 3.79), and has approximately 1% He in the Red River (Ordovician) and Souris River (Devonian) reservoirs at 5050 to 5200 ft (1530 to 1575 m) (Fig. 3.81) (Henderson, 1985). The Rudyard Field lies along the northeastern trending Great Falls Tectonic Zone, which lies on the eastern flank of the Sweetgrass–Battleford Arch and trends perpendicular to the fault zone representing a minor structural undulation (Fig. 3.79). The field lies over a significant Tertiary intrusion that is highly magnetized and has been penetrated in at least one well (Henderson, 1985). The Cambrian sediments are completely removed across on the western part of the anticline and then tilted to the east. Gas zones have been encountered in the Cambrian sediments. The overlying Ordovician sediments are represented by a member of the Red River strata that contains gas and is capped by an overlying anhydrite that acts as an effective seal. The zone on a drill stem test (DST) flowed 3.0 MMCFGPD of inert gas. Unconformably overlying the Ordovician strata are Devonian rocks, the oldest being the Souris River which is composed of porous to dense dolomites with interbedded dense anhydrites (Henderson, 1985). The Souris River was tested in at least two wells in the Rudyard Field at 1.15 MMCFGPD that was relatively inert. It also should be noted that an increase in CO_2 was detected through Souris River.

Recent drilling confirmed the high He readings in the Texaco well. However, a second well drilled on the southern part of the anticline (Fig. 3.81) indicated the Souris River and Red River strata were tight. No DST or mudlog was made through the zones and the well was plugged and abandoned (Henderson, 1985). The remaining overlying rock strata has yielded

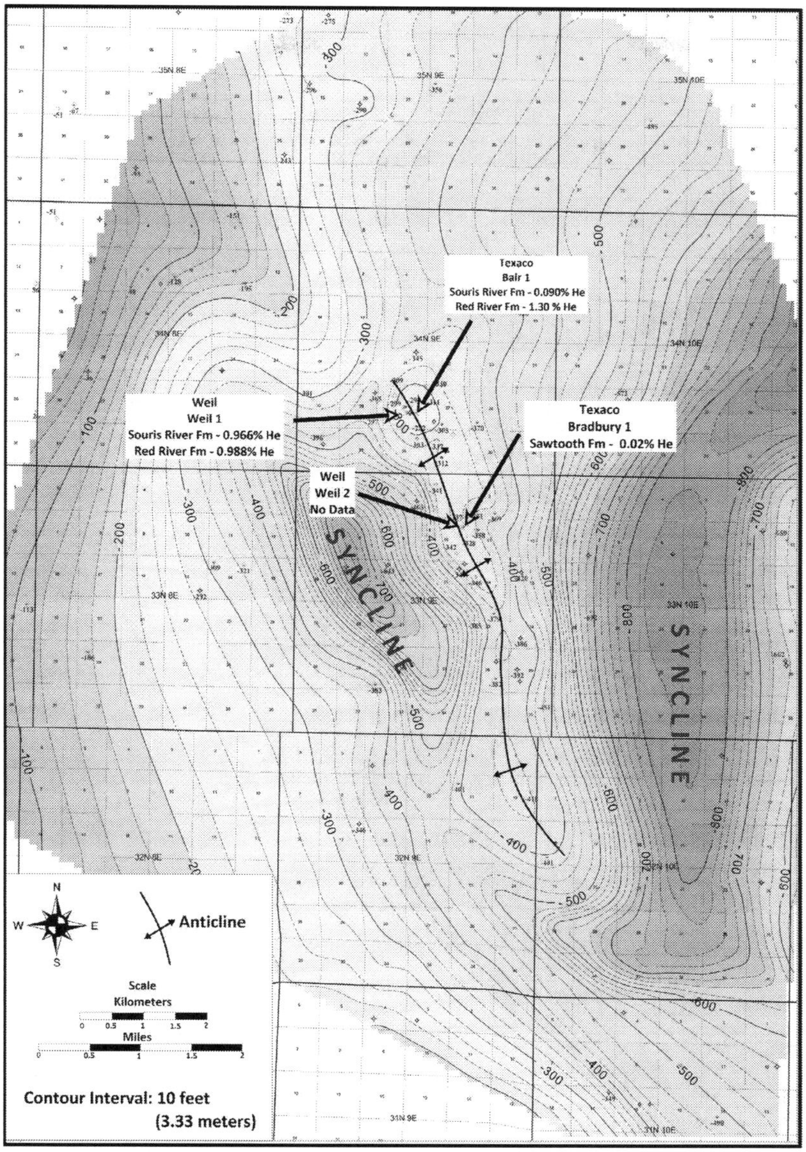

FIGURE 3.81 Structure on top of the Sawtooth Formation (Jurassic) with wells noted with He gas content at the Rudyard Field, Hill County, Montana, USA. Colors on contours represent structural relief, light green is the highest point, light blue is the lowest, contour interval is 10 ft (3 m).

oil and gas shows with very low He. Present oil and gas production in the field is from the shallow Bowes and Firemoon members of the Jurassic Sawtooth and the Eagle formations. Production, based on public records, which are missing several years of data, is 170 MBO and 3.8 BCF. The oil is approximately 15° API gravity. Based on the USGS and USBM database, the Sawtooth has 0.02% of He (Henderson, 1985).

Reservoir Characteristics of Rudyard Field (Henderson, 1985):

Productive area:	Approximately 10,600 acres
Reservoir:	Cambrian strata, Red River (Ordovician), Souris River (Devonian)
Lithology:	Carbonates
Average net pay:	30–100 ft (9–30m)
Depth to pay:	5050 to 5200 ft (1530 to 1575 m)
Pay type:	Structure
Average porosity:	8.0% Souris River; 8.2% Red River
Average permeability:	Unknown
Average water saturation:	Unknown
Réservoir pressure:	2181 psi Souris River; 2238 psi Red River
Formation temperature:	120°
Type of drive:	Gas
Gas characteristics and analysis:	0.90%–0.966% He, 93.95% N_2, CH_4+ 3.4%+, 47 to 72 BTU—Souris River; 0.988%–1.30% He, 93.92% N_2, CH_4+ 3.4%+, 12–47 BTU—Red River

3.12 Mississippi

The Jackson Dome is located in Madison, Hinds, and Rankin cos., Mississippi, USA, and produces CO_2 from the Jurassic age sandstones and carbonates from the Norphlet, Haynesville, and Smackover formations, from an average of 16,000 ft (4850 m) (Bowman, 2016). The dome was formed by piercement of the area by an igneous intrusion during the Cretaceous (Bowman, 2016). The igneous intrusion is the source of the CO_2 and the dome is a series of anticlines that are fault bounded (Fig. 3.82). The sandstones are interpreted as aeolian and sabkha deposits (Studlick et al., 1987).

Reservoir Characteristics of Jackson Dome (Eppink et al., 2014; Bowman, 2016):

Productive area:	90,000 acres
Reservoir:	Norphlet, Haynesville, Smackover Formations
Lithology:	Sandstones and carbonates
Depth to pay:	16,000 ft (4850 m)
Average net pay:	185 ft (56 m)
Average porosity:	13.0%

Average permeability:	Unknown
Average water saturation:	Unknown
Depth to pay:	Average of 16,000 ft (4850 m)
Initial field pressure:	7000 psi (overpressured)
Reservoir temperature:	339°F
Type of drive:	Gas expansion
Type of traps:	Anticlines and salt structures related to salt movement.
Gas characteristics and analysis:	CO_2 65%–99.6%, H_2S 5%–35%, N_2 0%, He 0% and CH_4+ that averages 5%.
Production:	3 TCF produced, with another 9 to 13 TCF potential.

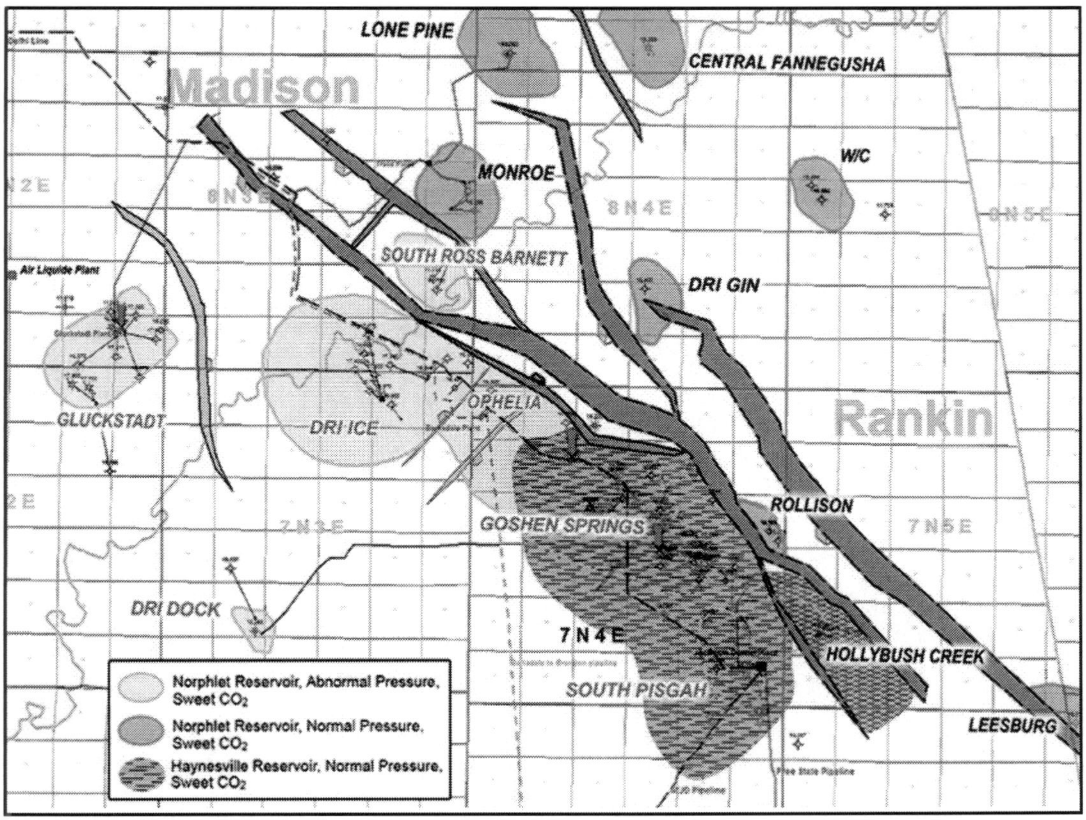

FIGURE 3.82 Jackson Dome area with subfields with identified faulting (Bowman, 2016). Fault systems are the gray linear trends (*reprinted by permission of the AAPG whose permission is required for future use*).

3.13 Eastern USA

3.13.1 West Virginia

The Indian Creek Field is located in Kanawha County, West Virginia, USA, and produces CO_2 from the Lower Silurian Tuscarora sandstones draped over the Warwick Anticline located in the Appalachian Basin (Fig. 3.83) (Gao and Shumaker, 1996). The reservoir is

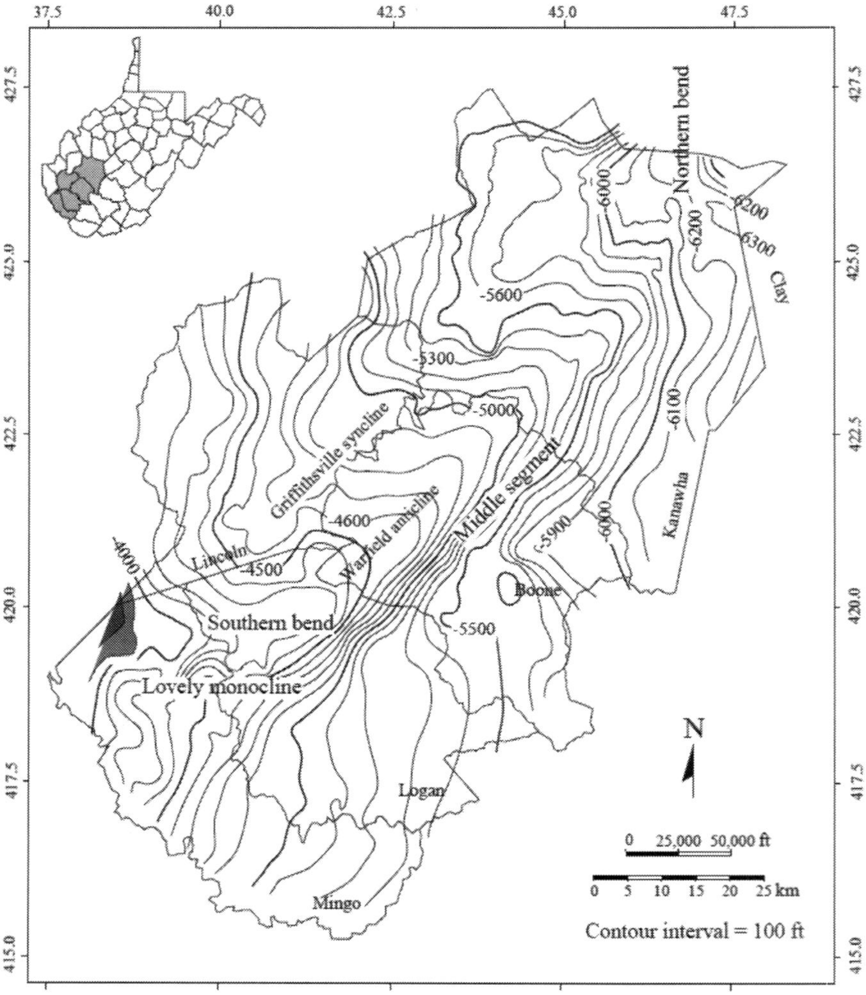

FIGURE 3.83 Structure on the Warwick Anticline, Indian Creek Field, Kanawha County, West Virginia (Gao and Shumaker, 1996). Structure is on top of the Tuscarora Sandstone (Silurian) and contour interval is 100 feet (*reprinted by permission of the AAPG whose permission is required for future use*).

fracture enhanced and the overlying Rose Hill Formation of Middle Silurian age is the seal. The Warfield structure is characterized by an asymmetric half graben within a continental system known as the Rome Trough (Gao and Shumaker, 1996). The structure has been divided into three segments, a north and southern bend and a middle segment (Gao and Shumaker, 1996). The middle segment parallels the New York—Alabama lineament and the structure of the Warfield Anticline because of its formation history is not conducive to hydrocarbon entrapment (Gao and Shumaker, 1996). The CO_2 is derived either from the linings of calcite cement that was deposited within the pores and then partially dissolved by later fluids or as suggested by the presence of framboidal pyrite typically associated with bacterial degradation of organic matter from hydrocarbons (Moore et al., 2018).

Reservoir Characteristics of the Indian Creek Field (Eppink et al., 2014):

Productive area:	18,000+ acres
Reservoir:	Fractured Lower Silurian Tuscarora
Lithology:	Sandstone
Depth to pay:	6700 ft (2030 m)
Pay thickness:	10 ft (3.3 m)
Average porosity:	10%
Average permeability:	Unknown
Average water saturation:	Unknown
Initial field pressure:	3000 psi
Reservoir temperature:	126°F
Type of drive:	Gas expansion
Type of traps:	Fractured anticline
Gas characteristics and analysis:	CO_2 44%—83%; N_2 4%; and the remainder is hydrocarbons.

3.13.2 Kentucky

The State of Kentucky has some relatively high He values between 1.32% and 1.92% in Garrard and Clark Counties from Cambrian reservoirs adjacent to the Lexington—Kentucky Fault System, the bounding feature of the Rome Trough (Bowersox, 2019). The Rough Creek Fault System is an east-west feature that underlies the Illinois Basin and the Cincinnati Arch and is the northern boundary of the Rough Creek Graben. Garrard and Clark counties lie along the Cincinnati Arch which is a north-south feature that extends from Ohio through Kentucky and into Tennessee. These areas have some indication of higher He levels and some remote potential for future He exploration. The rest of the State of Kentucky has generally had He values below 0.05%, and data are limited and scattered, so there is no overall conclusion on the potential for region (Bowersox, 2019).

3.13.3 Michigan

The Michigan Basin covers all of the State of Michigan and extends into the adjacent Province of Ontario (discussed below) and states of Ohio, Indiana, and Wisconsin. The basin is a prolific oil and gas producer but commonly has background levels of He as that seen throughout most of the world of less than 0.04% based on USGS and BLM database. The basin has a north-south Precambrian Rift system running underneath. It has significant oil, gas, and CO_2 production from Silurian reefs capped with thick salts and anhydrites that would be capable traps for He. There are numerous Paleozoic traps from Ordovician to Devonian strata which have significant hydrocarbon accumulations with efficient and effective traps but with no He production directly from these fields. If He is being extracted from production facilities, it is not being reported. Both the USGS and BLM database (Rogers, 1921; Anderson and Hines, 1951; Moore, 1978, 1982; Moore and Sigler, 1985; Wescott, 1989; Hamal and Driskill, 1996; Gage and Driskill, 2003; Gage and Driskill, 2005; Ma, 2009) have reported elevated He% in the Berea Sandstone (Early Mississippian) and Michigan Stay Sandstone (Mississippian). The Berea Sandstone has a total of 12 samples, average of 0.603%, median of 0.665%, minimum of 0.31%, maximum of 0.75%, and a standard deviation of 0.158% He. The Michigan Stray has a total of 53 samples, average of 0.461%, median of 0.40%, minimum of 0.22%, maximum of 1.11%, and a standard deviation of 0.237% He. In summary, the He exploration potential is mostly unknown but not nonexistent.

3.13.4 Triassic basins eastern USA

There is a series of north-northeast to south-southwest discrete trending Triassic age rift basins that extend from South Carolina to Massachusetts and New Hampshire in the United States that contain indications of petroleum, thin, gassy coal seams, and interesting He percentages. The sediments in these basins are predominantly lacustrine. Reid et al. (2015) reported crustal derived He with 0.22%–0.24% volume with 30%–45.49% N_2 with significant gas flow from the Cumnock Formation in the Sanford subbasin of the Deep River Basin (Reid et al., 2015). The wells were drilled for oil, coal-bed methane, and shale gas from the Triassic rocks. The basins are surrounded and underlain by granitic and metamorphic rocks which can be a major source of the He. The difficulty with these basins is the likely limit of any significant oil, gas, and He accumulations.

3.14 California

California is a major producer of oil and gas in the United States. It produces from numerous basins both onshore and offshore. Some of these basins are bounded by granite and metamorphic rocks and contain both normal and reverse faults that penetrate from these basement rocks into the sedimentary section. Yet, very little He has been detected in these deposits. California production is predominantly from Mesozoic and younger sediments in areas of active subduction, rifting, and mountain building which tends to be destructive to He accumulations. Based on UGS and BLM database, data from 81 samples include an average of 0.02%, a median of 0.010%, a minimum of 0.009%, a maximum of 0.15%, and a standard deviation of 0.025% He. The Sacramento Basin and Northern San Joaquin is noted for very prolific dry gas producing reservoirs with some that are up to 87.5% N_2

(Jenden et al., 1988). The N_2 concentrations increase toward the basement rocks along the shallow eastern margin of the valley suggesting that the N_2-rich component is derived from metasedimentary rocks deep within the crust (Jenden et al., 1988). Based on $^3He/^4He$ and N_2/Ar isotopic ratios and CH_4 $\delta^{13}C$ less negative $-25‰$, the most likely source of the majority of the N_2 is the Coast Ranges Franciscan thrust complex, beneath the Sacramento Basin and northern San Joaquin during Late Mesozoic and Early Cenozoic subduction. Some of the N_2 is derived from sedimentary, intrusive metasedimentary sources. The source of N_2 is not related to and independent of hydrocarbon generation (Jenden et al., 1988). The low He reflects the lack of "old" pore water being preserved.

3.14.1 Imperial Field

The Imperial Field is located in Niland County, California (Fig. 3.84), adjacent to the south side of the Salton Basin and on the southeastern shore of the Salton Sea (Eppink et al., 2014). The field produced from 1934 to 1954 (Muffler and White, 1968). The reservoirs are Cenozoic age sandstones with five small rhyolite domes in the area (Eppink et al., 2014). The reservoirs produce CO_2 and no He and the gas is of magmatic origin. The field is part of the Salton Sea geothermal system of the Salton Trough which is the northernmost of a series of active hydrothermal systems found along East Pacific Rise–San Andreas fault zone plate boundary (Williams et al., 1991). The Colorado River Delta formed over a fragmented rift system with discontinuous lacustrine and fluvial sandstones being deposited (Muffler and White, 1968; Tarbet, 1971). The reservoirs have been subject to significant alteration by the active hydrothermal system to a greenschist facies with carbonates being destroyed (Muffler and White, 1968; Williams et al., 1991). They have limited access to the field and development due to its location in relation to the Imperial National Wildlife Refuge. The area is now being utilized for geothermal energy.

Reservoir Characteristics of the Imperial Field (Muffler and White, 1968; Eppink et al., 2014):

Productive area:	1700+ acres
Pay zone:	Cenozoic sandstone reservoirs
Depth to pay:	590 ft (180 m)
Average net pay:	230 ft (70 m)
Average porosity:	12%
Average permeability:	Heterogenous
Average water saturation:	Unknown
Initial field pressure:	339 psi
Reservoir temperature:	245°F
Type of drive:	Gas expansion
Type of traps:	Fractured anticline
Gas characteristics and analysis:	CO_2 44%–83%, N_2 4%, and the remainder is hydrocarbons.
Production:	650 $MMCFCO_2$ was produced for dry ice production. 158 BCF remaining.

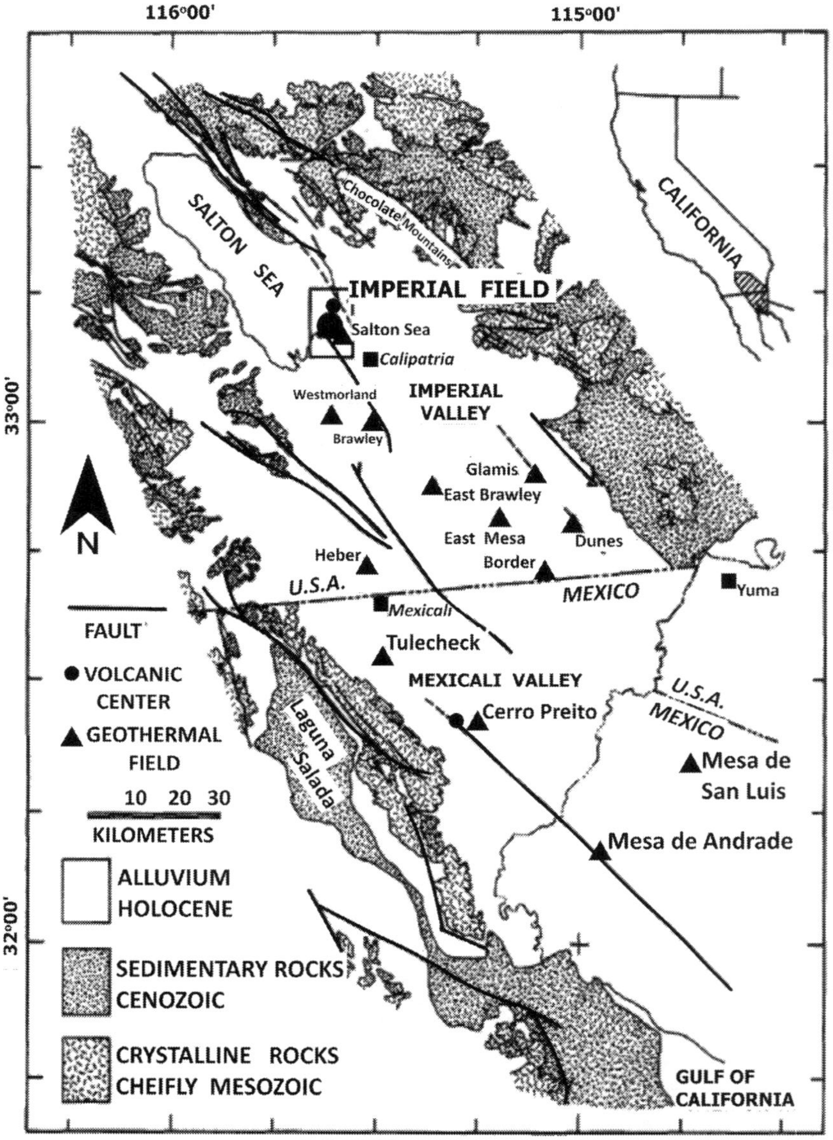

FIGURE 3.84 The imperial Field, Niland County, California, USA (Williams et al., 1991) (*reprinted by permission of the AAPG whose permission is required for future use*).

There are other CO_2 reservoirs in California such as Mammoth Mountain that are actively degassing from basaltic intrusions that are forming 2.4 miles (4 km) above a rising hydrothermal fluid front (Pfeiffer et al., 2018). Isotopic analysis of the He, CO_2, and N_2 indicates that the gases are derived from a common source (Sorey et al., 1998). The CO_2 is being released through faulting either caused by the magmatic activity or the degassing of the CO_2 fluids. Active systems like these can be used to better understand the source and emplacement of CO_2 reservoirs.

3.15 Canada

Canada is similar to the United States in that it has significant oil and gas reserves, a small to medium size number of energy companies, and, in association with that, significant He and CO_2 accumulations. Both the He industry and the petroleum industry started in Ontario but eventually moved to Alberta, British Columbia, Manitoba, and Saskatchewan. As of 2010, the industry in Canada has seen growth in He exploration especially in southeastern Alberta and southwestern Saskatchewan. A distinct advantage to working in Canada over the United States is the requirement by provinces to submit significant amounts of data derived from exploration, production and development, and specifically gas analysis.

3.15.1 Ontario

Southwestern Ontario is part of the Michigan Basin and produces oil and gas from reservoirs of Cambrian to Devonian age. Helium had been produced during and after World War I from some of these fields. The productive gas fields were in the Woodhouse Township in Norfolk County, Ontario (Dobbin, 1935; Lee, 1962), and produced gas containing 0.22% to 0.52% He with as much as 20.0% N_2 from the Clinton and Medina formations (Silurian) at depths ranging from 850 to 1500 ft (260—455 m). Lambton—Kent, Haldimand, and Welland (now Regional Municipality of Niagara) counties produced natural gas containing less than 0.50% of He from the Guelph, Clinton, Medina, and Trenton reservoirs from less than 1400 ft (425 m). The gases found in reservoirs in the Trenton-Black River (Ordovician) are at depths ranging from 350 to 950 ft (105—290 m) in Caledon Township, Regional Municipality of Peel (Dobbin, 1935). Helium was found to contain up to 1.0% from scattered wells between Georgian Bay and Lake Ontario (Dobbin, 1935). By 1935, He extraction had ceased in Ontario as the productive fields were in significant decline.

Lollar et al. (1994) analyzed a number of wells from the Trenton, Black River, and Cambrian reservoirs and determined that the highest He content is found in the Cambrian reservoirs (Table 3.5). Three of the fields, Innerkip, Innerkip East, and Gobles, are stratigraphic traps of sandstone reservoirs (Figs. 3.85—3.87). The Cambrian sandstone reservoirs tend to form isolated traps. Under current economic and technological conditions, these fields would now be considered potential opportunities for extraction of He. These types of Cambrian reservoirs may also be present deeper on the US side of the Michigan Basin.

3.15.2 Saskatchewan

Helium in Saskatchewan, Canada, is found in the northwestern part of the Williston Basin, the north edge of the Bowdoin Dome, and the southeastern part of the Alberta Basin

TABLE 3.5 Helium, N_2, CO_2, and CH_4 data from Lollar et al. (1994) from various fields in Southwestern Ontario, Michigan Basin, Canada.

Well	Field	He	N_2	CO_2	CH_4
Ordovician—Trenton					
1	Gosfield South	0.04	2.14	0.04	77.91
2	Colchester South	0.15	8.21	0.04	60.95
3	Malden 3-41-IV	0.11	14.27	0.18	57.09
4	Dover	0.09	6.02	0.09	82.90
5	Dover	0.09	2.41	0.20	85.63
6	Dover	0.13	2.93	0.22	85.62
7	Hillman	0.03	1.69	0.05	65.17
8	Hillman	0.10	3.54	0.03	86.43
9	Hillman	0.08	2.81	0.02	85.47
10	Hillman	0.03	1.41	0.05	72.21
11	Hillman	0.09	3.22	0.31	84.70
12	Hillman	0.05	2.69	0.31	80.85
13	Goldsmith	0.01	0.99	0.11	70.23
14	Goldsmith	0.02	0.55	0.05	66.05
15	Wheatley	0.04	1.82	0.24	70.94
16	Wheatley	0.08	3.38	0.02	92.42
17	Wheatley	0.13	3.71	0.02	88.25
18	Wheatley	0.09	4.16	0.03	86.95
19	Wheatley	0.03	1.80	0.10	73.16
20	Wheatley	0.08	2.31	0.05	84.93
21	Wheatley	0.06	3.20	0.02	89.54
Ordovician—Black River					
22	Fergus	0.38	4.79	0.03	86.02
23	Acton	0.39	7.96	0.03	82.71
24	Sombra	0.13	6.21	0.03	80.77
25	Hornby	0.39	7.12	0.02	84.20
26	Dover	0.11	2.71	0.18	85.62
27	Dover	0.13	2.87	0.25	87.61
28	Egremont	0.56	7.59	0.08	83.46

TABLE 3.5 Helium, N_2, CO_2, and CH_4 data from Lollar et al. (1994) from various fields in Southwestern Ontario, Michigan Basin, Canada.—cont'd

Well	Field	He	N_2	CO_2	CH_4
Cambrian					
29	Willey	1.26	2.55	0.04	86.57
30	Innerkip	0.74	5.10	0.02	84.05
31	Innerkip	0.63	6.07	0.01	85.21
32	Innerkip	0.75	6.16	0.03	84.36
33	Innerkip	0.51	5.50	0.11	86.06
34	Clearville	2.70	3.83	0.22	80.56
35	Clearville	2.31	4.46	0.38	79.81
36	Gobles	0.73	6.06	1.00	82.82
37	Raleigh	1.36	5.21	0.41	80.92
38	Raleigh	1.07	15.76	0.08	72.50
39	Arthur	0.45	7.55	0.03	83.09
40	North Gosfield	0.55	8.78	0.06	82.57

(Fig. 3.88) (Dobbin, 1935; Lee, 1962). There have been numerous shows of He in strata spanning from the Cambrian to Cretaceous age rocks (Fig. 3.88) (Yurkowski, 2016). Recent activity seems to be focused on the Cambrian Deadwood formation. The Deadwood sand is not only He productive but also oil productive in some fields (Potter, 2010), which complicates exploring specifically for He (Fig. 3.89). Other zones, such as the Devonian Duperow and Dawson Bay zones, can also contain significant He gas (Fig. 3.89). Carbon dioxide–rich reservoirs are also present and may contain high contents of He but may have a different source, migration, and trapping history (Lake and Whittaker, 2006).

United Canso-Consumers Co-op, in 1952, drilled the Battle Creek No.4-3 well (101/04-31-003-26W3M) (Fig. 3.88, number 3 on the map, and Fig. 3.90, number 2 on the map), near Swift Current area north of Bowdoin Dome in Saskatchewan (Yurkowski, 2016). The well yielded nonflammable gas from the Devonian Duperow Formation with 81.7% CO_2, 13.5% N_2, 0.14% He, and 5.66% other gases and the Dawson Bay Formation with 95.16% N_2, 0.47% He, and 4.47% other gases (Thompson, 1964; Burwash and Cumming, 1974).

In 1958, British American, present-day BP Petroleum, found anomalous He content in lenticular zones of porosity in the Upper Cambrian Deadwood Sandstone well at the Wilhelm 101/01-09-017-14W3M; 58H045 at 6480 ft (1964 m) (Fig. 3.90, number 7 on the map) (Thompson, 1964). The reservoir contains silicified siltstones and mudstones draping over a Precambrian topographic high about 8.7 mi (14.5 km) north of Swift Current (Thompson, 1964; Yurkowski, 2016). British American used seismic data for structural control in an effort to find Upper Jurassic sandstones which are productive in the area (Thompson, 1964).

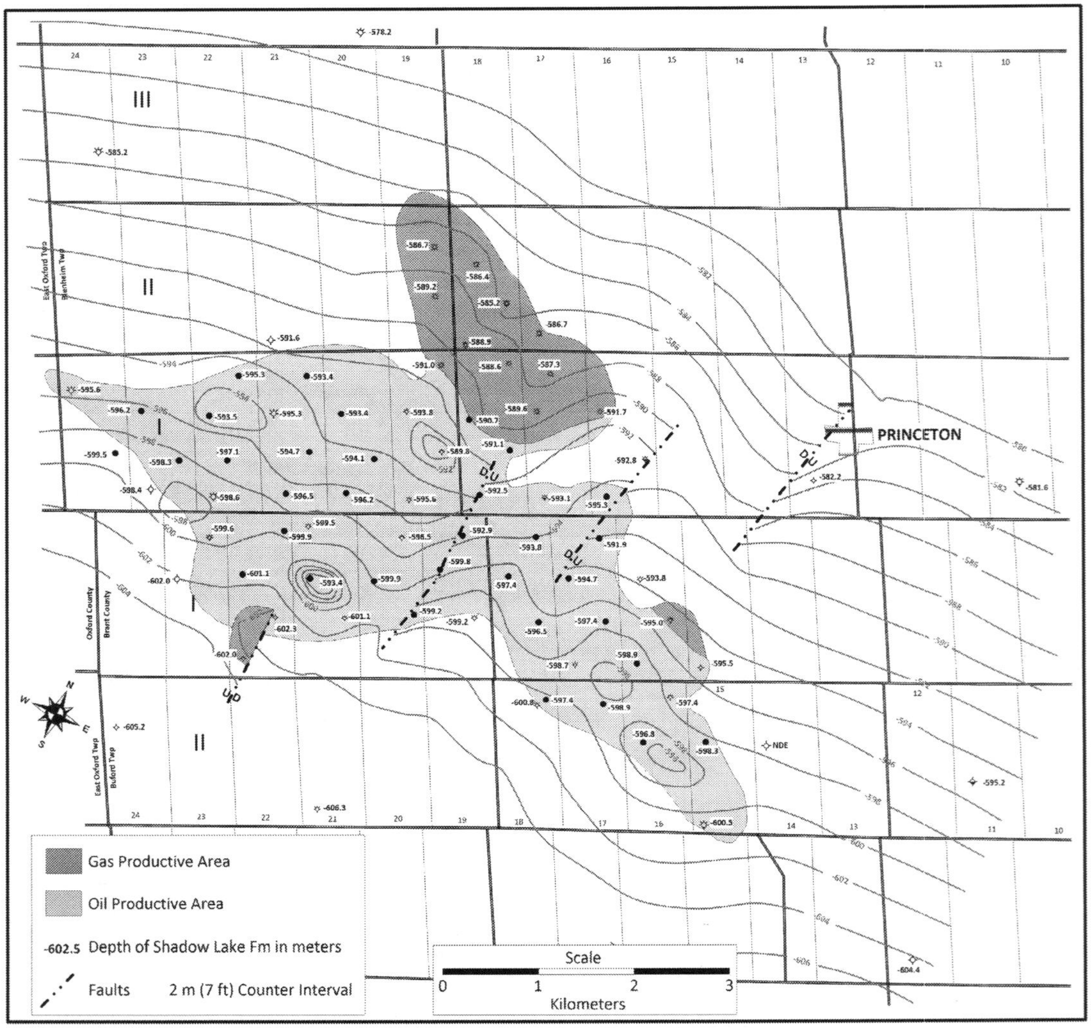

FIGURE 3.85 Structure map on top of the Shadow Lake Formation at the Gobles Field, Blandford and Blenheim Township, Oxford County, Ontario, Canada, modified from the original Ministry of Natural Resources, 1983a map.

Seismic maps indicated a Precambrian closure covering several square miles that diminished moving up the stratigraphic section. Inert gas flowed from the well at a rate of 1–5 MCF (0.02–0.14 Mcm^3) per day and was composed of 97% N_2, 2% He, and 1% CO_2 (Yurkowski, 2016). The Wilhelm Field basement feature well appears to be a Precambrian volcanic plug, covers several square miles, and rises several hundred feet above the level of the surrounding Precambrian surface. The Precambrian high affected Cambrian sedimentation in the area (Thompson, 1964). A second well, British American Wilhelm No. 1-9A (Fig. 3.90, number 7 on the map as well) (Lsd. 1 Section 9, T17, R14 W3M), was drilled in May, 1961, approximately 150 feet southwest of the original well in Section 9, T17, R14 W3M, and equipped for He production. A third well, British America in Saskatchewan Landing Wilhelm No. 3,

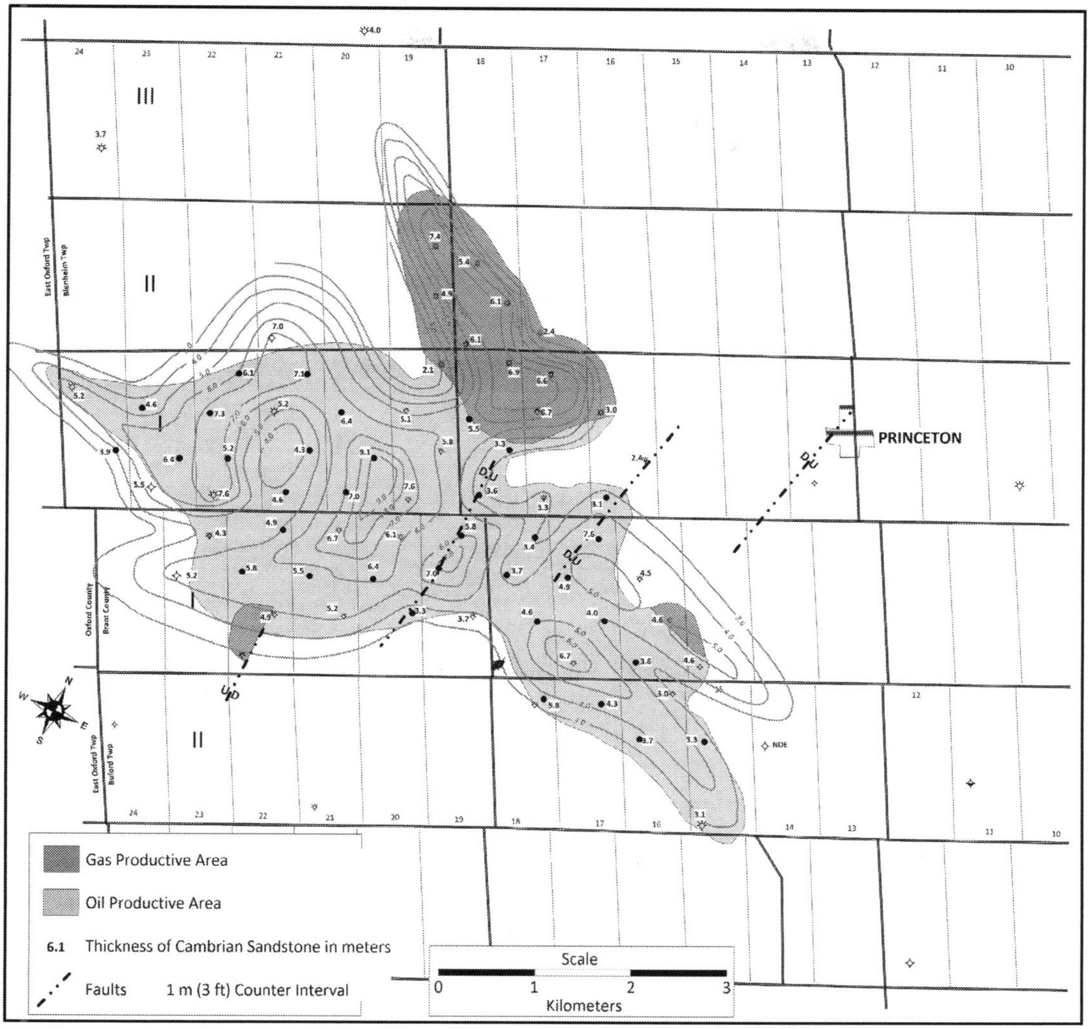

FIGURE 3.86 Isopach map of the Cambrian Sandstone at the Gobles Field, Blandford and Blenheim Township, Oxford County, Ontario, Canada, modified from the original Ministry of Natural Resources, 1983b map.

10-17-1 4, was drilled in Section 10, T17, R14 W3M, one-half mile east of the discovery well, in August 1962 (Thompson, 1964; Yurkowski, 2016). This well is lower structurally with a thinner pay section. Helium production began in 1963 and continued until 1977 from four wells in Ranges 14–17 W3M, producing a combined total of 2 BCF (57.2 × 106 MCm3). It is estimated that 1.4% of this production was He (Yurkowski, 2016). At the Tidewater Johnston Lake Crown No. 1 Well Section 20, T 12, R 2 W3M, some 6.5 mi (8.7 km) east southeast of the Wilhelm wells, the Precambrian surface was several hundred feet higher than expected. A rhyolite plug forms the accumulation and published gas composition data are available (Thompson, 1964).

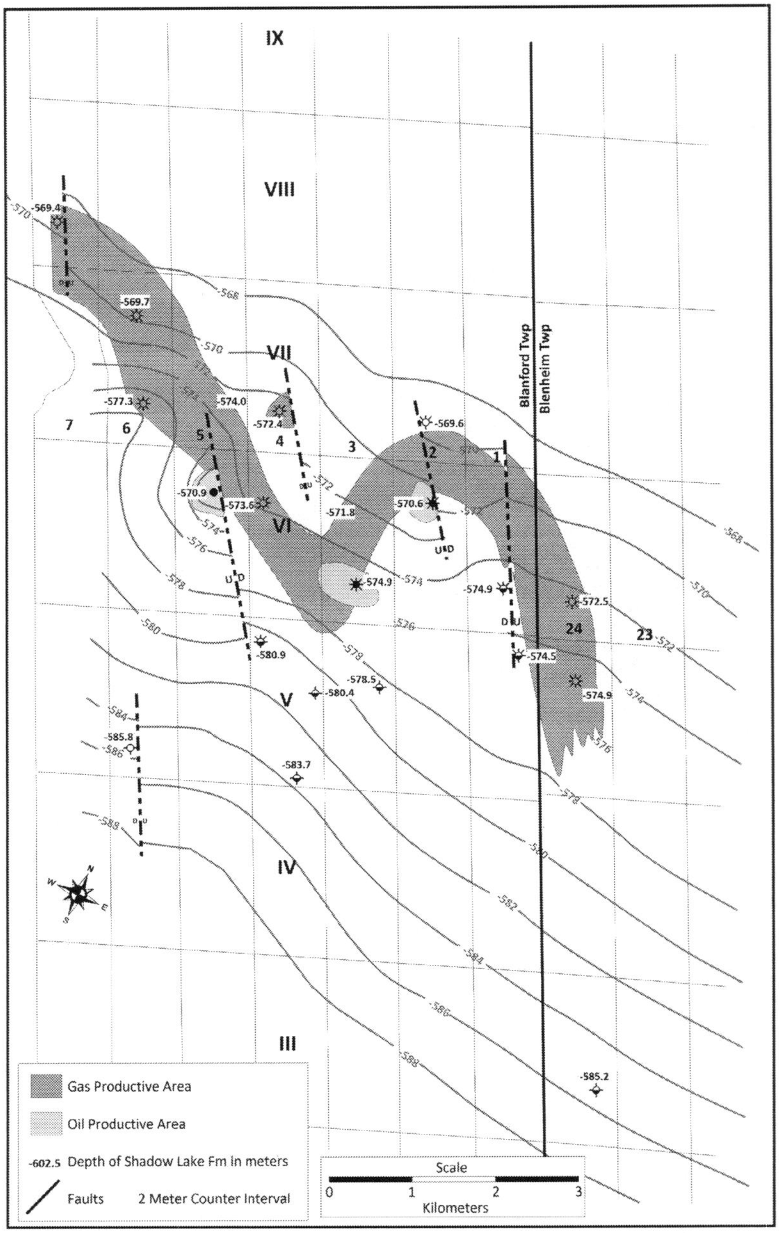

FIGURE 3.87 Isopach map of the Cambrian Sandstone at the Innerkip and Innerkip East fields, Blandford and Blenheim Township, Oxford County, Ontario, Canada, modified from the original Ministry of Natural Resources, 1983c map.

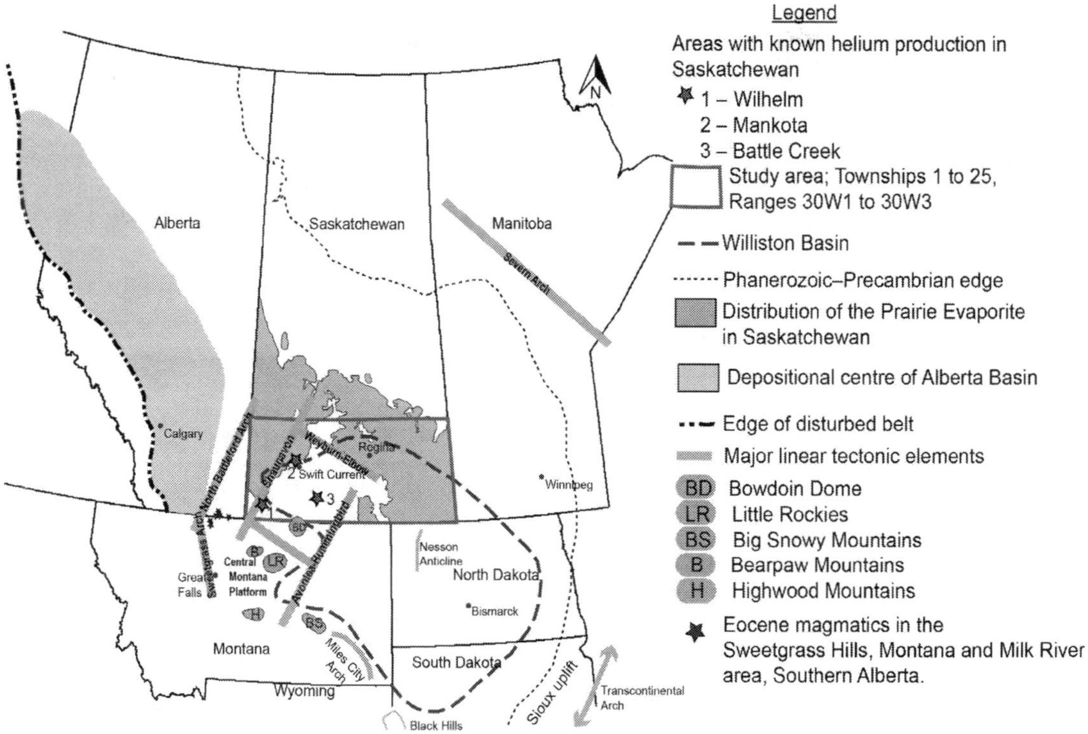

FIGURE 3.88 Map showing the location of the present-day structural Williston Basin, the approximate center of the Alberta Basin, and the location of wells that have produced or are producing He (past-producing wells 02-09-017-14W3M and 12-10-005-08W3M as of August 2016). Also included are the structural elements within the Williston Basin and surrounding area (Yurkowski, 2016).

In 1960, He was discovered in a second well at Texaco Wood Mountain 101/12-10-005-08W3M; 59L036 now CVE Mankota (well 2 on Fig. 3.90); southeast of the town of Swift Current on the east flank of the Bowdoin dome, 80 mi (133 km) southeast of Swift Current in the Cambrian sands at a depth of 8200 ft (2485 m) (Thompson, 1964). A DST over a 20 ft (6 m) sandstone interval flowed at the rate of 16.3–20.0 mcf (0.46–0.57 Mcm^3) of inert gas per day, composed of 96.35% N_2, 1.08% He, and 2.5% other gases (Yurkowski, 2016). International He Wood Mountain took over the well in T5, R8 W3M; the He percent is less than at the Wilhelm deposit and the pay zone much thinner. The well was plugged and abandoned before a gas analysis was obtained (Thompson, 1964).

In 2004, Industrial Air Corporation re-entered the B.A. Wilhelm 102/01-09-017-14W3M; 61E006 well (IAC Wilhelm 102/01-09-017-14W3M; 03L306) in 2004 and in 2008 drilled a second well at IACX Wilhelm 141/02-09-017-14W3M; 08F296 (Yurkowski, 2016). In June 2014, Canadian Helium Inc. began to produce from the 141/02-09-017-14W3M; 08F296 well and as of the end of September 2015, 8.8 × 106 m^3 of gas (310,600 mcf; He is not broken out) has been produced. The well, B.A. Wilhelm 102/01-09-017-14W3M; 61E006, has since been abandoned (Yurkowski, 2016).

North American Helium Inc. began drilling for He at NA Helium 102/10-25-003-27W3M; 52127 (yellow star no. 2 on Fig. 3.90) in 2016 and in August 2016, Weil Group Resources, LLC opened a He processing plant near Mankota, approximately 90 miles (150 km) southeast of

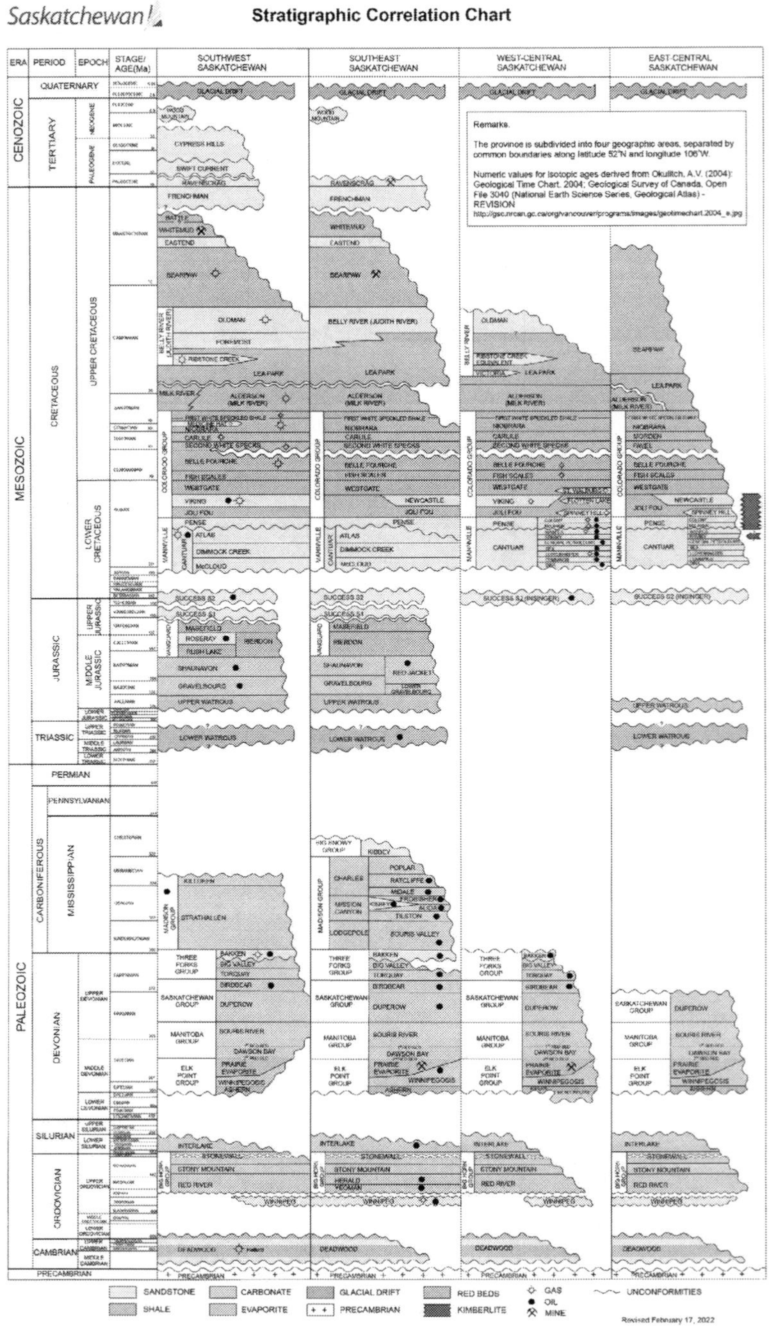

FIGURE 3.89 Saskatchewan's stratigraphic correlation chart (from Saskatchewan Ministry of the Economy, 2014). Helium has been reported in gas analyses from stratigraphic intervals spanning the Cambrian to Cretaceous in southern Saskatchewan. Sequences identified from Sloss, 1963. The Deadwood and Souris River formations in southwestern Saskatchewan are the only helium producing horizons at the time of publication and are outlined in red (Saskatchewan Ministry of the Economy, 2014; Yurkowski, 2021).

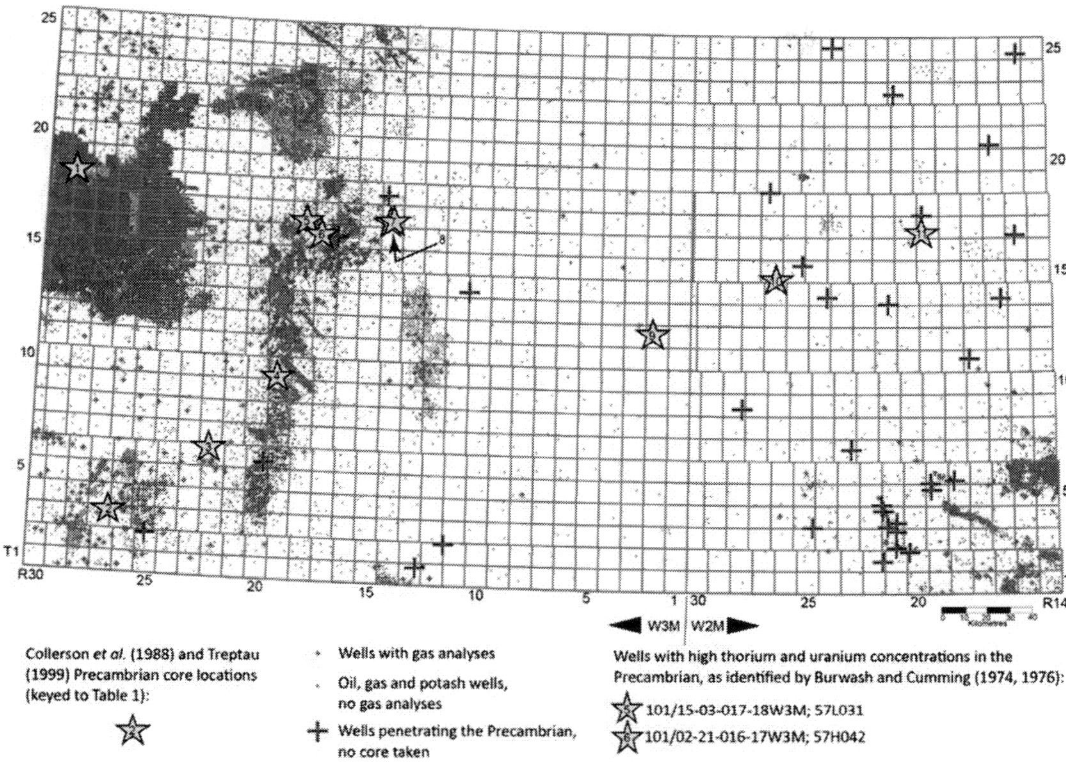

FIGURE 3.90 Map outlining the distribution of oil, gas and potash wells (small grey crosses) in the study area and the wells with gas analyses (small red crosses). Also included is the distribution of Precambrian well penetrations, Precambrian drill cores and rock types are shown as numbered yellow stars and correspond to the well locations in Table 1 in Yurkowski, 2021. Wells that were drilled into the Precambrian that do not have cores are identified by the larger blue crosses. Wells identified as having high thorium and uranium concentrations contained >46 ppm Th and >15 ppm U. Abbreviations: T (Yurkowski, 2016, 2021).

Swift Current (Yurkowski, 2016). At the end of August 2016, there were two wells reporting He production in southwest Saskatchewan: well 12-10-005-08W3M and well 02-09-017-14W3M.

Lake and Whittaker (2006) discussed CO_2 accumulations on the western flank of the Williston Basin in southwestern Saskatchewan that occur within Devonian carbonates and Cambrian sandstones (Yurkowski, 2016). Inert gases from some carbonate reservoirs contain upwards of 80% CO_2 and have sustained flow rates up to 15 MMCFPD (425,000 m^3/day). Lake and Whittaker (2006) found that the Duperow Formation that contains the CO_2 is a succession of thin cycles of carbonates sealed by evaporite units. The shallow cyclic sedimentation in periodically restricted waters in a regional carbonate platform developed the reservoirs (Yurkowski, 2016). The strata are dolomitized limestones to limestones that within reservoir intervals porosities are generally from 6% to 8% (Lake and Whittaker, 2006). The porous intervals are sealed by dense anhydrite layers 6.5 ft (2 M) to over 30 ft (10m) thick. The southernmost area in Saskatchewan has an anomalous high geothermal gradient that is likely related to Tertiary age tectonic and igneous processes in the Bearpaw and Little Rocky Mountains of Montana that occurred 50 MY ago (Lake and Whittaker, 2006). Carbon dioxide found in Devonian and Cambrian rocks of southwestern Saskatchewan was generated during intrusion of hot alkaline magma into Lower Paleozoic carbonates in Montana. The CO_2 migrated north about 60 mi (100 km) to its present location where dissolution of salt layers below the reservoirs had formed local structures that trapped the gas (Fig. 3.91) (Lake and Whittaker, 2006). Traps are both structural and stratigraphic from 5530 6765 ft (1675 to 2050 m) in the Devonian age rocks.

3.15.3 Manitoba

Southern Manitoba, Canada, is in the northeastern part of the Williston Basin (Fig. 3.93). The number of He analysis is limited (Nicolas, 2018). Helium analysis from the Lower Amaranth member, of Triassic age (Fig. 3.92), is typically below 0.2% He and the reservoirs are typically moderate to rich in N_2. The stratigraphic column is indicated in Fig. 3.93 and the

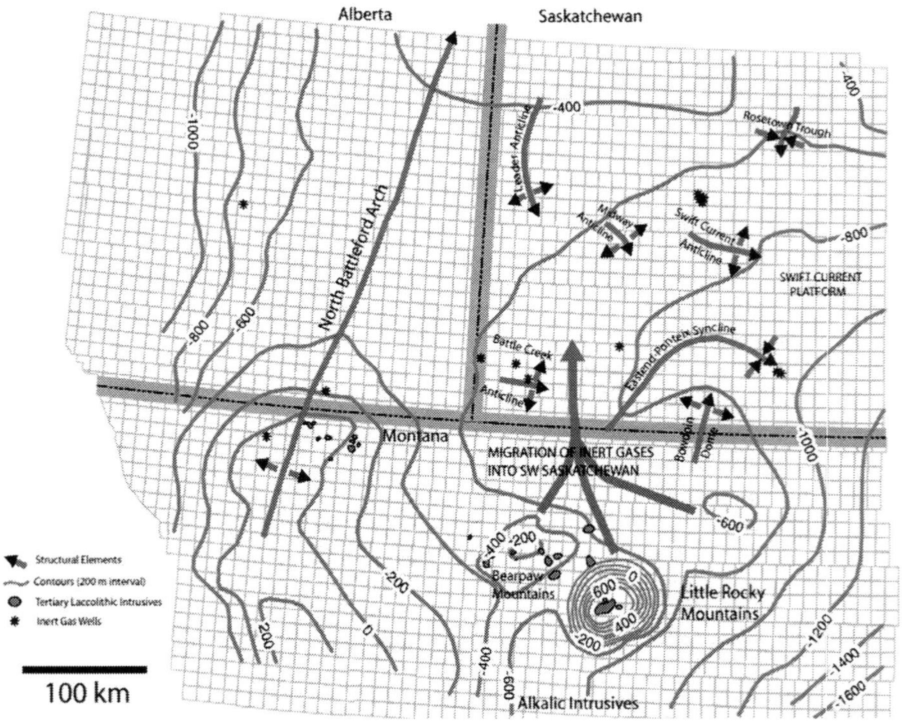

FIGURE 3.91 Map indicating the proposed migration pathway of CO_2 generated in Montana through the contact of alkalic magma with limestones approximately 50 million years (MY) ago. The locations of known natural inert gas occurrences in the area are shown along with several basement and intrasedimentary structures. Structure contours represent Duperow and equivalent strata in region (Lake and Whitaker, 2006).

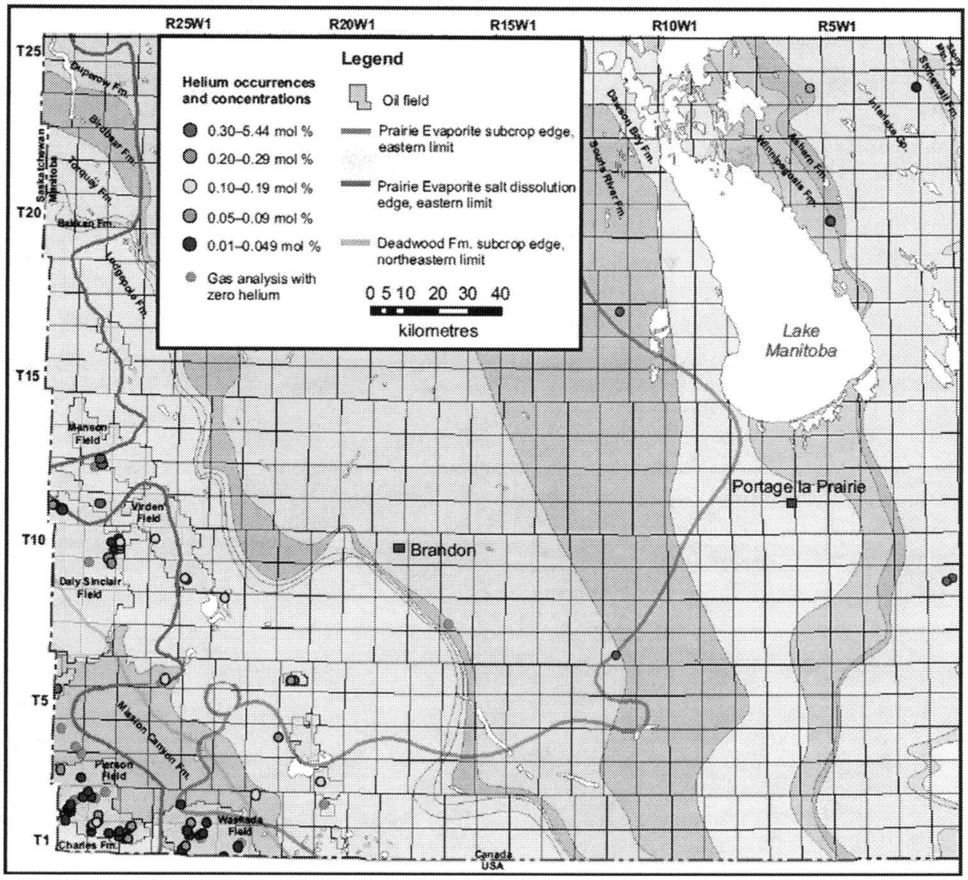

FIGURE 3.92 Paleozoic stratigraphic map showing the distribution of gas analyses and He occurrences in southwestern Manitoba. Select oil fields are labeled. *Fm.*, formation; *Grp.*, group; *Mtn.*, mountain (Nicolas, 2018).

location of He samples in relation to existing production and outline of the present geologic landscape in the Williston Basin is shown in Fig. 3.92 (Nicolas, 2018). The Cambrian He-productive fields in Ontario are located in a similar intercratonic basin setting and can be used as a model for the Saskatchewan and Manitoba side of the Williston Basin.

3.15.4 Alberta

The Alberta Basin dominates the province of Alberta and is part of the Western Canada Sedimentary Basin (WCSB) and the very southern part of the province is along the flank of the Kevin-Sunburst Dome–Sweetwater Arch and Williston Basin. The WCSB is an uplifted basin with oil and gas reservoir rocks ranging in age from Cambrian to Tertiary (Hutcheon, 1999). The Alberta Basin can be broadly divided into a Paleozoic passive margin section, comprised largely of limestone and dolomite, and a predominantly clastic Mesozoic section that formed as a foreland basin during the Laramide orogeny (Hutcheon, 1999). The Alberta

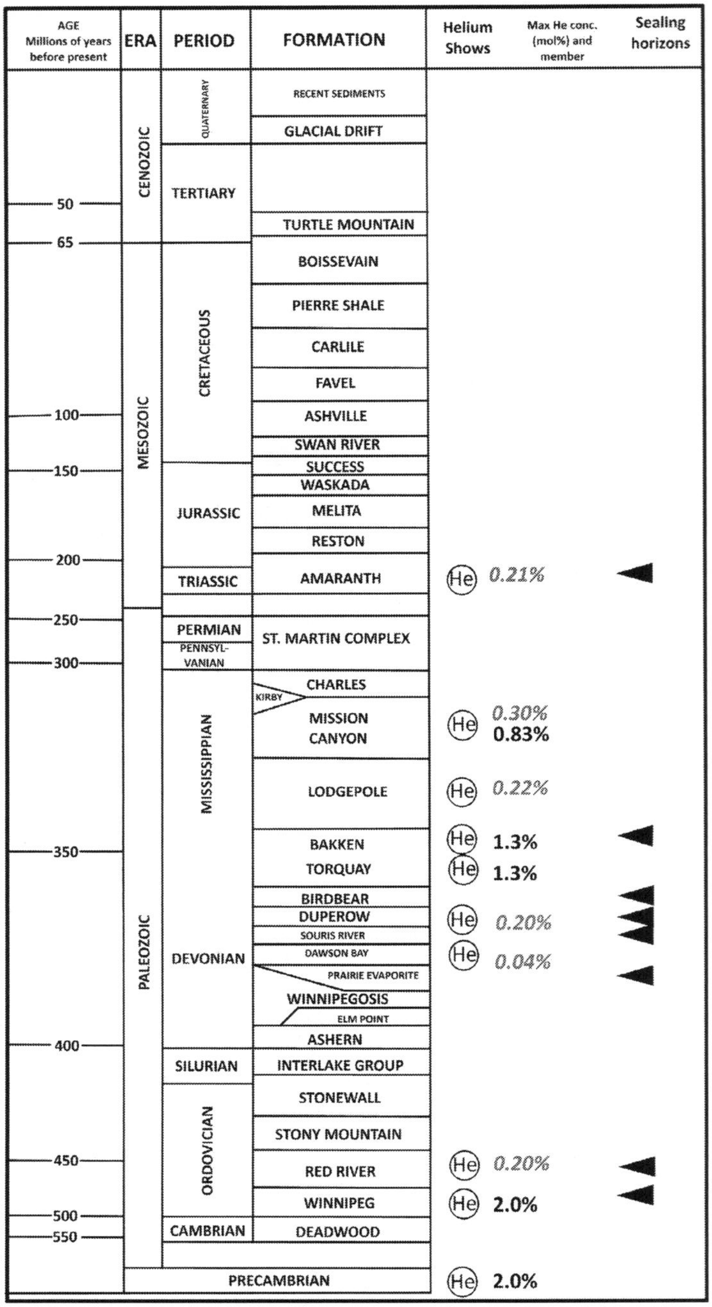

FIGURE 3.93 Stratigraphic column of southwestern Manitoba showing He occurrences: maximum He concentrations reported for that interval; the abbreviation in parentheses indicates the member in which the occurrence was measured, if known; and known potential sealing horizons. Red text indicates values equal to or above 0.30 mol % He. *LA*, Lower Amaranth Formation; *LW*, Lower Winnipeg Formation; *MC1*, MC-1 member; *MC3*, MC-3 member; *WL*, Whitewater Lake Member (Nicolas, 2018).

Basin contains numerous oil and gas fields and some contain nonhydrocarbon gases in varying amounts. The southern boundary of the Alberta Basin is the Sweetgrass Arch that extends to the northeast and merges with the Battleford Arch that extends from the Canadian Shield (Herbaly, 1973). The present He exploration activity in Alberta is focused along the Sweetgrass Arch, as the majority of the Alberta Basin, except in the Peace River Arch area, is generally devoid of economic (in terms of present prices) He.

The Bow Island Field was the first field to produce He in Alberta, and in 1914, pre-He era supplied Calgary and Lethbridge with 75 MMCFGPD (Irwin, 1923). Some wells were capable of producing between 5 and 29 MMCFGPD from 1980 feet (600 m). The Bow Island sandstone reservoir is present in two benches of 10 ft (3m) separated by 34 ft (11m) of shale, pinches out updip, and has porous and tight streaks (Irwin, 1923). The field was nearly depleted by 1935. Dobbin (1935) reported that the He content of the gas varies from 0.29% to 0.34% and that N_2 averages about 8.5%. The Foremost field also produces gas from the Bow Island sandstone at a depth of about 2150 ft (650m) with He of 0.22% and 10% N_2. Dobbin (1935) reported the Barnwell or Chin Coulee fields contained at that time the richest known He of 0.38% in Alberta from a sandstone in the lower Colorado shale at 2200 ft (667 m) but was uneconomic.

Hutcheon (1999) reviewed the existing database available publicly and described N_2 and CO_2 in several geologic units and his results for the Devonian and Mississippian are presented in Figs. 3.94 and 3.95. Carbon dioxide and N_2 in Devonian rocks indicate an anomalous increase and trend toward the Montana, US border with Alberta that reflects Kevin-Sunburst Dome—Sweetgrass Arch area (Hutcheon, 1999).

In 2020, the Alberta Geological Survey released He data which is shown in Fig. 3.96 as a class post or bubble map. The Peace River Arch area indicates the only significant anomalous area of He in Alberta (Fig. 3.96). The other two major structural trends have a few scatted high He percentages and it should be noted there are significantly less data in both those areas. The southeastern part of Alberta has scattered anomalous He percentages but is generally on the low end of what may be economic (Fig. 3.96). The data were not segregated to reflect geologic unit or reservoir by the Alberta Geological Survey. The lack of data to the northeast represents thinning of and removal of geologic units onlapping to the Canadian Shield. Alberta has good potential for He production in the Peace River Arch area. The present focus of exploration is on the southern and southeastern edge of the province that is related to activity in southwestern Saskatchewan and northern Montana. The majority of the reported results are generally less than 1% He.

3.15.5 British Columbia

The petroleum producing sedimentary basin in British Columbia is the western half of the Western Canadian Sedimentary Basin and is subdivided into the Horn River, Montney, and Peace River Arch areas (BMEMPR, 2006). The three areas are predominantly gas productive with secondary oil production. Johnson (2012) evaluated the amount of He in northeastern British Columbia and reported an He range of 0.04%–0.36% with the higher numbers in the Peace River Arch area. The Peace River Arch is a northeast-southwest basement feature that extends into Alberta and has anomalous He percentages associated with it (see above).

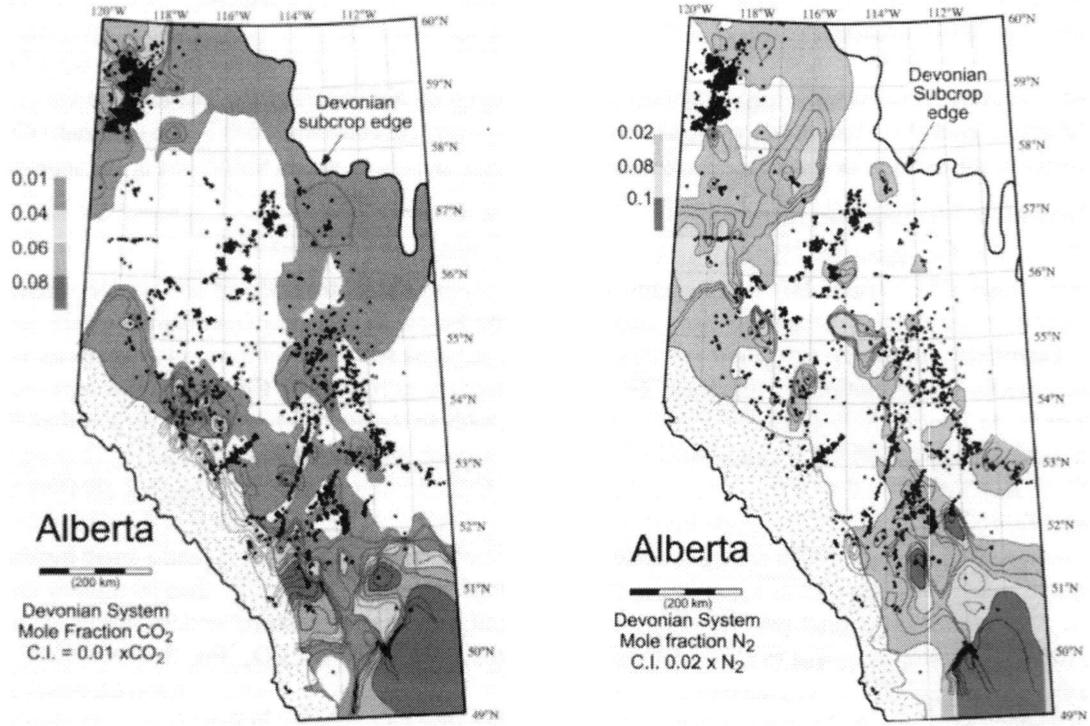

FIGURE 3.94 Carbon dioxide and N_2 percentage in Devonian strata in the Alberta Basin, Alberta (Hutcheon, 1999).

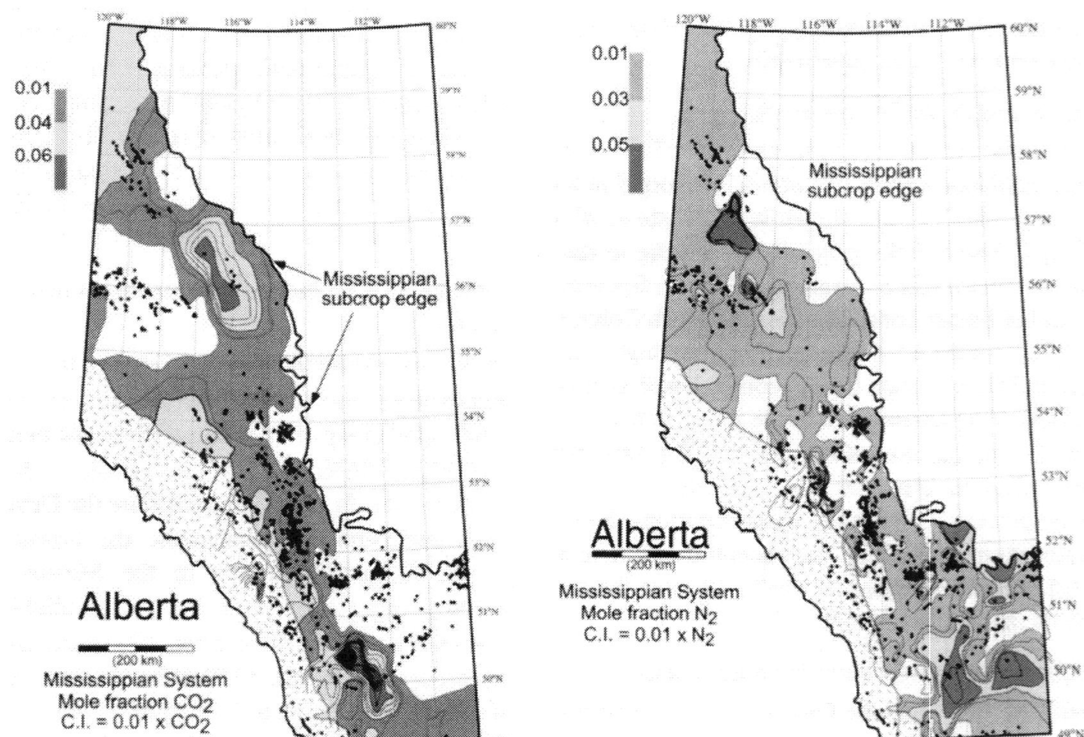

FIGURE 3.95 Carbon dioxide and N_2 percentage in Mississippian strata in the Alberta Basin, Alberta (Hutcheon, 1999).

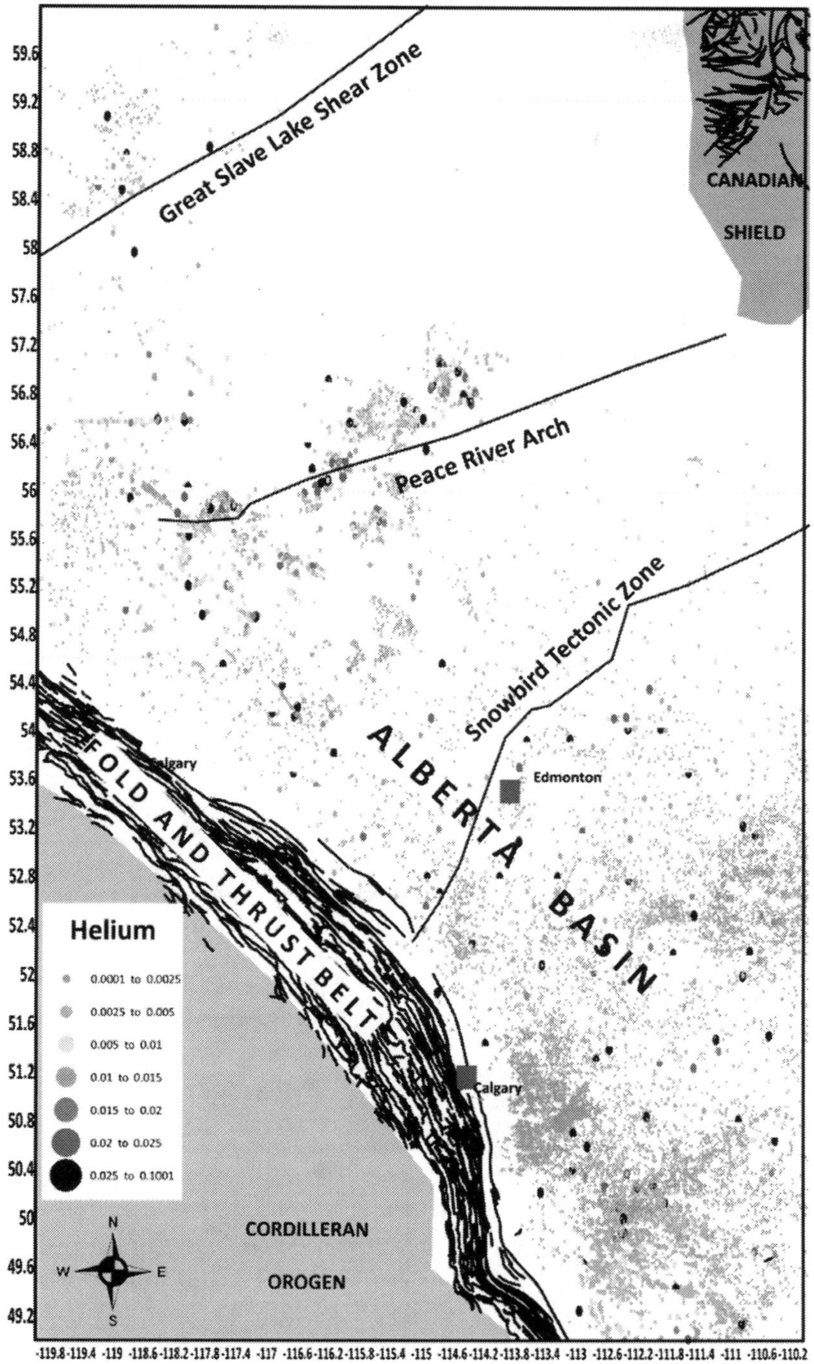

FIGURE 3.96 Helium data in a classed post presentation from the Alberta Geological Survey, 2020.

Many of the reservoirs have been subject to secondary low-temperature hydrothermal fluids that dolomitize limestone reservoirs, increasing porosity and permeability. These fluids typically migrate upward from the Precambrian and crustal rocks along faults and fracture systems (Johnston, 2012). The fluids can also carry noble gas as well as petroleum gases during migration. Because there are large volumes of gas production from these fields, there may be opportunities for economic He extraction.

3.16 Middle East

The Middle East is well endowed with petroleum from a variety of tectonic setting and stratigraphic units; therefore, it would make sense that He is probably present. So far, the amount of He that has been reported is below 0.30% and to date it is only economic as an additional extraction from natural gas processing facilities.

3.16.1 Qatar

The North and South Pars Field in Qatar and Iran (Fig. 3.97) is the largest gas field in the world, containing original estimated reserves of 1800 TCF (5.097^{E12} m^3) of gas in primarily the Dalan (Permian) and Kangan (Triassic) formations equivalent to the Khuff (Permian–Triassic) Formation reservoir and 50 BBO from overlying and satellite fields producing from younger reservoirs (Fig. 3.98) (Al-Siddiqi, and Dawe, 1999; Tavakoli et al., 2011, van Buchem et al., 2014). The North and South Pars Field lies along the Qatar Arch which is a Precambrian ridge that is a major northwest trending structure that is bounded by two salt basins to the north and south (Fig. 3.99). The north and northwest arch is bounded by the Zagros fold and thrust belt (Fig. 3.99) (Tavakoli et al., 2011).

The main gas reservoirs are the dolomites and limestones of the K2 and K4 units with interbedded anhydrites of the Dalan and Kangan strata (Fig. 3.100) at a depth of 9075 to 10,560 ft (2750 to 3200 m) with a reservoir thickness in the South Pars Field of 1480 ft (450 m) thinning to the North Field to 1263 ft (385 m) (Fig. 3.99) (Rahimpour-Bonab et al., 2010; Tavakoli et al., 2011). Lithological variations from limestone to dolomite and anhydrite control the reservoir properties that cause immense lateral and vertical changes in the porosity types and percentages (Tavakoli et al., 2011). The reservoir is dissected by north northwest–south southeast trending faults and the reservoir is modified extensively by diagenesis that has a significant effect on production.

The source for the hydrocarbons in the field is the Early Silurian shale of the Sarchahan Formation. The amount of He present in the field is approximately 0.04% (Tavakoli et al., 2011). The viability of the field is based on the politically stability of the Persian Gulf region. In 2016, Iran indicated that they would not be extracting He from the South Pars Field even though in 2012 the government said they were pursing He extraction. Lack of exploiting He from the field may be related to nuclear treaties signed with several other countries including the United States, Russia, Germany, France, England, etc. Recently, Iran has announced a restart of their nuclear weapons program and they will need a source of He to carry it out.

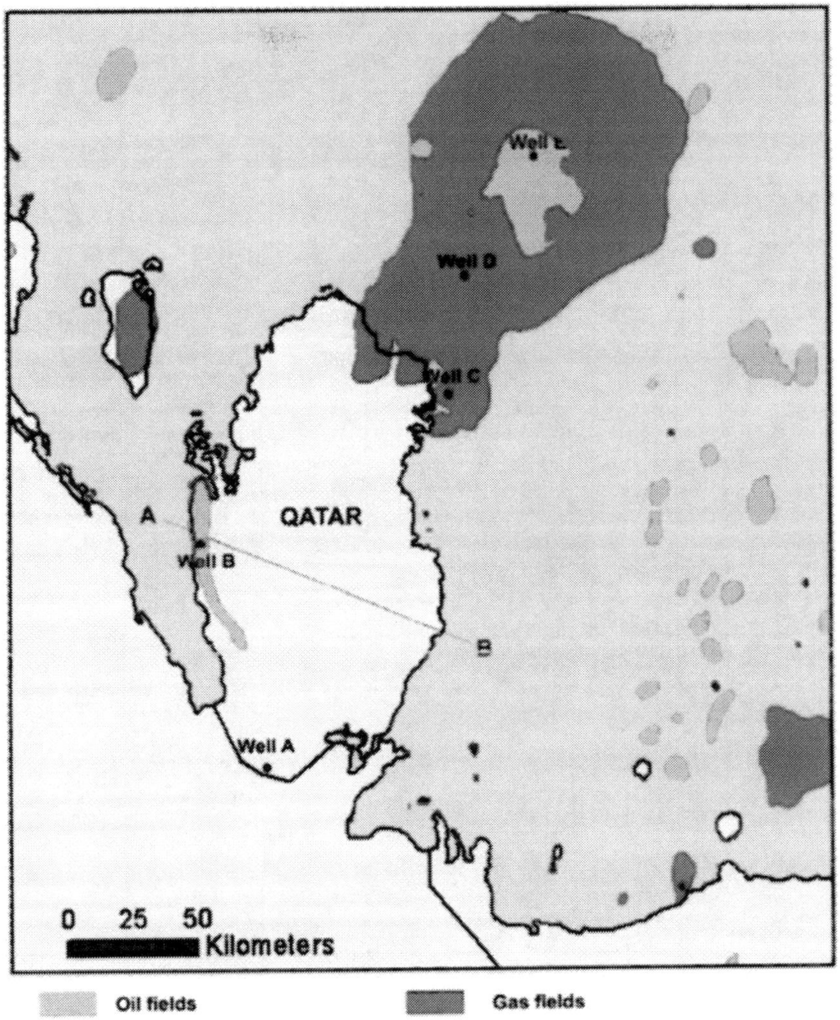

FIGURE 3.97 Geographical map of Qatar with onshore and offshore locations of oil and gas fields specifically to the North Dome Field (van Buchem et al., 2014 and *reprinted by permission of the AAPG whose permission is required for future use*).

Iran has represented that they have the largest He reserves in the world. Consequently, it seems likely that Iran would begin He extraction at some time and may even eventually have a significant supply to sell on the worldwide market. Qatar is presently capable of producing approximately 35% of the world's He. Qatar also has political issues with its neighbors which historically have inflicted blockades on the country. Therefore, presently the largest field with He reserves in the world is in a politically and militarily unstable setting presenting uncertainty in the worldwide supply chain.

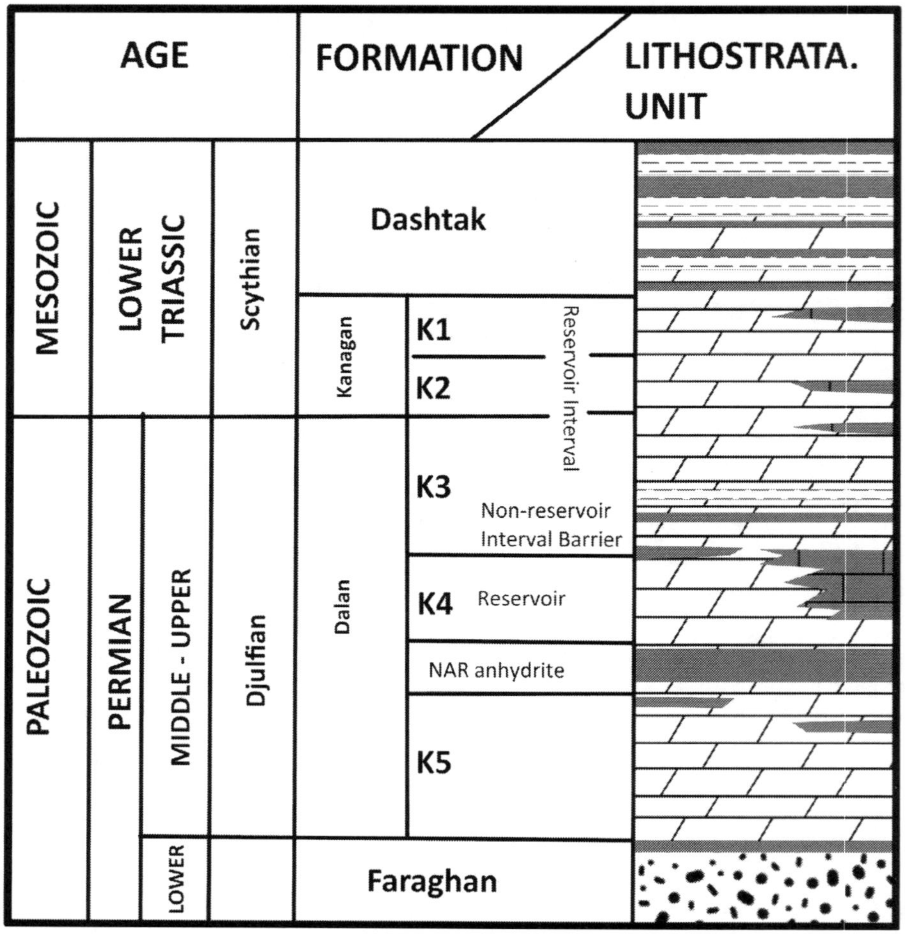

FIGURE 3.98 Generalized stratigraphic chart of the South Pars gas field (not to scale). Formations and main lithology from Lower Silurian to Cretaceous have been shown (Rahimpour-Bonab, 2007).

Reservoir Characteristics of the North and South Pars Field (Rahimpour-Bonab et al., 2010; Tavakoli et al., 2011):

Productive area:	3700 sq. miles or 9700 sq. kilometers or 2,368,000 acres—literally three times the size of the State of Rhode Island (USA)
Reservoir:	Dalan Member (Permian) and the Kangan Formation (Triassic) of the Upper Khuff Formation. They are highly heterogenous
Depth to pays:	9075 to 10,560 ft (2750 to 3200 m)

Lithology:	Dolomites and limestones of the K2 and K4 units of the Dalan and Kangan strata
Gross pay:	South Pars Field: 1480 ft (450 m)
	North Field: 1263 ft (385 m)
Net pay:	South Pars Field—K2 Unit: 51 ft (15.5 m) of limy dolomite and 96 ft (29 m) of dolomitic limestone
	South Pars Field—K4 Unit: 561 ft (170 m) of limestone at the top and dolomite at the base
Porosity:	South Pars Field—K2 Unit: 9.71%
	South Pars Field—K4 Unit: 15.61%
Permeability:	South Pars Field—K2 Unit: 38.1 md
+	South Pars Field—K4 Unit: 36.27 md (K4 unit)
Water saturation:	14%
Type of drive:	Gas expansion

(Continued)

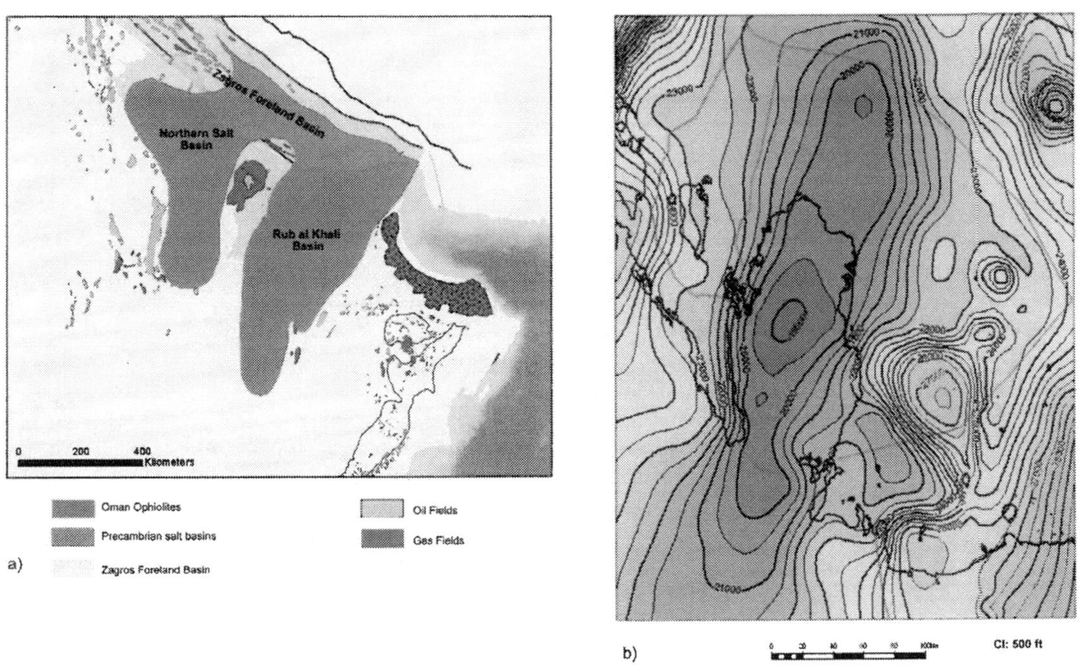

FIGURE 3.99 (A) Structural element map of Qatar and surrounding area comprising the fault-bounded Qatar High, the Precambrian salt basin consisting of the Northern Salt Basin, the Rub' al Khali Basin, and the Zagros fold belt (Konert et al., 2001; Sharland et al., 2001; van Buchem et al., 2014), (B) Basement structure map based on gravity and magnetic data (GeTech, 2002; van Buchem et al., 2014 and *reprinted by permission of the AAPG whose permission is required for future use*).

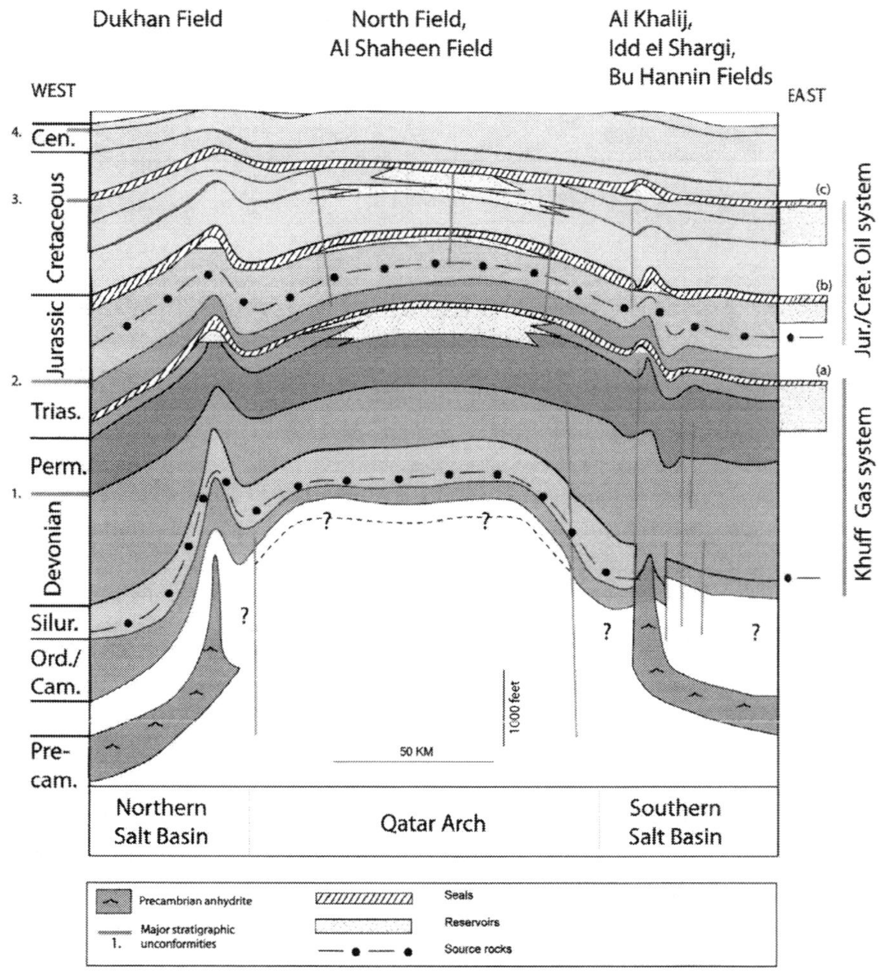

FIGURE 3.100 Schematic E–W cross section of the Qatar High illustrating the main stratigraphic and structural elements of the petroleum systems (Konert et al., 2001; Alsharhan and Nairn, 1997; van Buchem et al., 2014 and *reprinted by permission of the AAPG whose permission is required for future use*). The map clearly displays the importance of the Qatar High as the dominant tectonic feature and how it was the focus of oil migration. The presence of salts was the main factor for the creation of the satellite fields in the west and the east. Also note the importance of the Late Cenomanian/Turonian uplift and erosion that cut out the Mishrif Formation over the crest of the structure. Unconformities: (1) Early/Middle Carboniferous unconformity; (2) Late Triassic–Early Jurassic unconformity; (3) Late Cenomanian/Turonian unconformity; and (4) (Early) Oligocene unconformity. Petroleum systems: (A) Qusaiba source rock–Khuff reservoir–Sudair seal; (B) Hanifa source rock–Arab reservoirs–Hith seal; (C) Hanifa source rock–Middle Cretaceous reservoirs–Laffan seal.

Type of traps:	Fractured and highly porous strata on an anticline
Gas characteristics and analysis:	He is 0.04%, and the remainder is methane and condensate.
Production:	Typical well produces 100 MMCFGPD (2.8 × $10^6 m^3$).

3.16.2 Algeria

The Hassi R'Mel Gas Field is located 340 miles (550 km) south of the city Algiers located in the Triassic Basin, along the M'zad Arch that separates the Western Org Paleozoic Basin to the west and Oued Mya Basin to the east (Magloire, 1970). The field was discovered in 1956. The depth to top of the pay is 6890 to 7550 ft (2085 to 2285 m). The arch has a strike of north northeast to south southwest and has been stable since Cambrian time. The arch is composed of Cambrian rhyolite that is overlain by Tassili Cambro–Ordovician sandstone group (Fig. 3.101) (Magloire, 1970). The Ordovician rocks are overlain by Silurian–Devonian (Hunton Group) carbonaceous mudstone that is probably the source rock for the field (Magloire, 1970). The Paleozoic rocks have been partly eroded down to the Ordovician strata across the structure but the topographic lows have been filled in by sandstones interbedded with andesite flows which are probably of Permian to Triassic age (Fig. 3.101). The reservoirs are fluvio-lacustrine of Triassic age. The lowest reservoir rock of this Lower Permian sequence is the C sandstone reservoir which is Triassic age and varies of from zero to 200 ft thick (0–60 m). Overlying is the middle unit or B sandstone reservoir that is 0–70 ft (0–20 m) thick (Magloire, 1970). The upper unit or A sandstone is 50–120 ft (15–36 m) thick. All three reservoirs lie across the arch as mapped on the Hassi R'Mel Sandstone as indicated in Fig. 3.102A. The A reservoir is the most contiguous of the three reservoirs and is present across the entire field and thins in the area of the arch (Fig. 3.102B). Overlying the sandstone reservoirs is salty shale, salt, and anhydrite that acts as an effective seal to trap the gas and is 1000 to 1400 ft (300–420 m) thick (Magloire, 1970).

The field is presently in significant decline. The estimated gas in place was 110 TCF (3000×10^9 m^3) with a recovery of 85.3 TCF (2415×10^9 m^3) with 209 BCF of He in place and with a potential recoverable of 161.5 BCF (Magloire, 1970).

Reservoir Characteristics of Hassi R'Mel Gas Field (Magloire, 1970):

Productive area:	1333 sq. miles or 3500 sq. kilometers or 832,000 acres—literally the size of the State of Rhode Island (USA)
Reservoir:	A, B, and C sandstone reservoirs (Triassic)
Depth to pay:	6890 to 7550 ft (2085 to 2285 m)
Lithology:	Sandstones
Net pay:	C sandstone reservoir: 0–200 ft (0–60 m)
	B sandstone reservoir: 0–70 ft (0–20 m)
	A sandstone reservoir: 50–120 ft (15–36 m)
Porosity:	Grading from 5% on the northwest part of the field to 20%–22% in the center of the field.
Permeability:	Up to several Darcy
Water saturation:	5%–20%
Type of drive:	Gas expansion

(Continued)

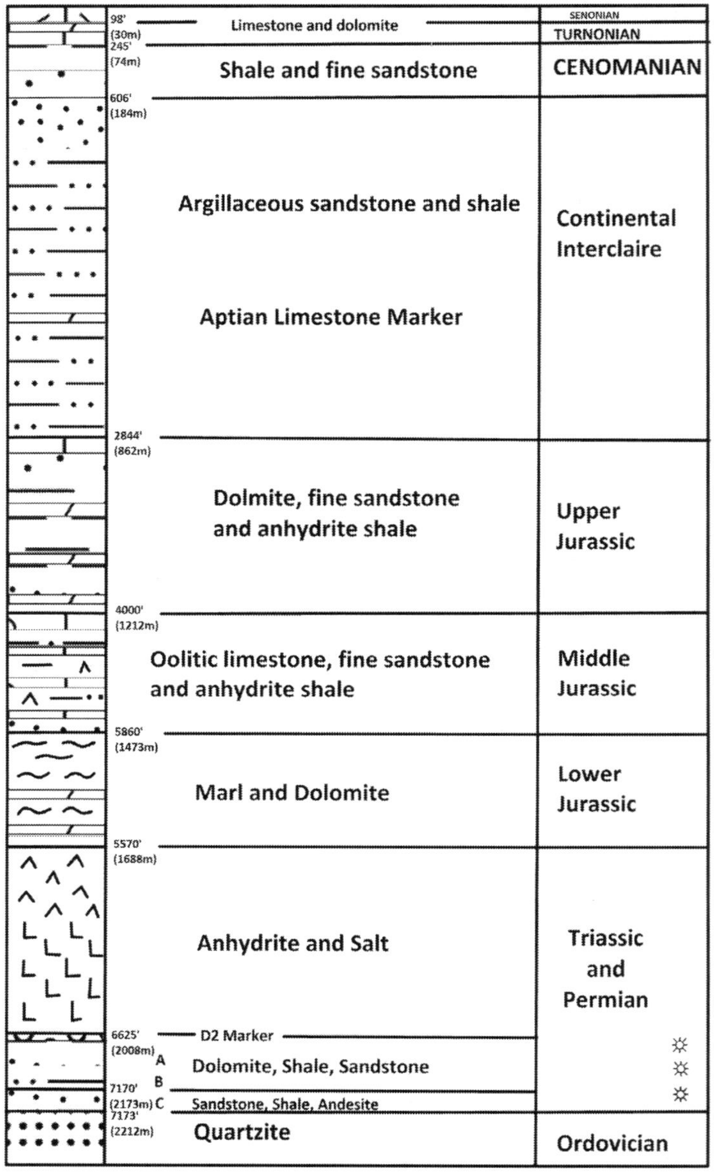

FIGURE 3.101 Stratigraphic column for the Hassi R'Mel Field in Algiers (Magloire, 1970, *reprinted by permission of the AAPG whose permission is required for future use*).

Type of traps:	Anticline
Gas characteristics and analysis:	CH_4 78.5%, C_2H_6 7.5%, C_3H_8 3.0%, $C_4H_{12}+$ 5.0%, CO_2 and N_2 6.0%, He 0.19%
Production:	As of 2019, the field was producing 2.4 BCFGPD (70 × 10^9 m^3).

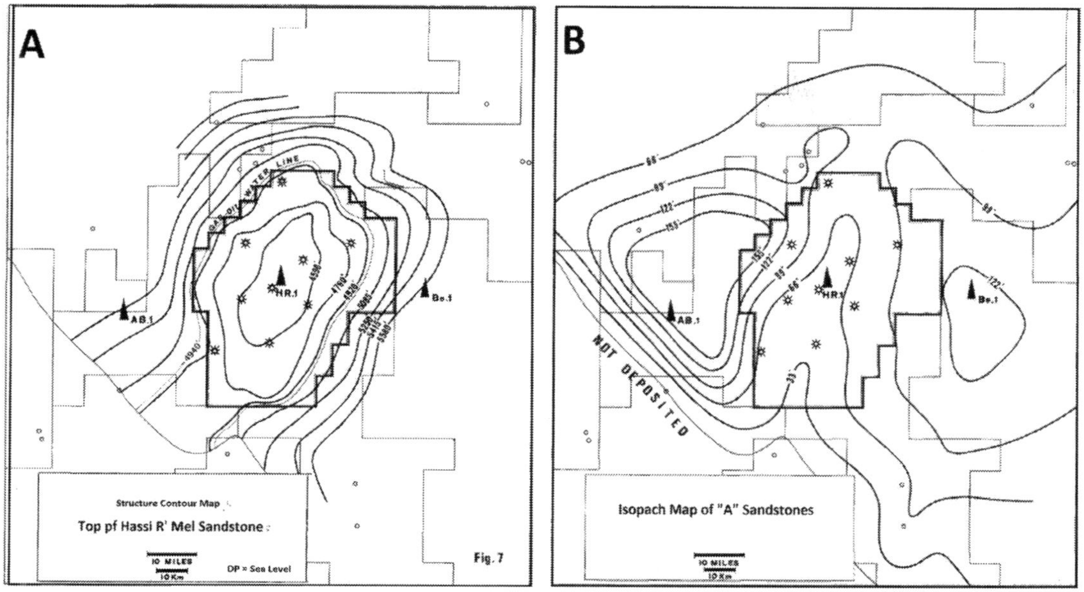

FIGURE 3.102 Structural contour map on top of the Hassi R'Mel Sandstone (A) and isopach map of the "A" Sandstone (B) after Magloire (1970), *reprinted by permission of the AAPG whose permission is required for future use.*

3.17 Russia

According to Nuttall et al. (2012), Russia contains approximately 13% of the world's He resources. A tremendous problem with understanding the actual He resources of Russia is their lack of systematic data reporting and the ability to verify data from that country. This limits evaluation from a geological and engineering perspective and means that the conclusions must be used with caution as it is generally not clear how these reservoirs are defined in terms of He in place, porosity, permeability, deliverability, etc. These data are generally not publicly available. Russian reserves are concentrated in three regions: North Caspian, Yakutia, and Eastern Siberia with 0.03%–0.06%, 0.3%–0.6%, and 0.2%–0.3%, respectively (Nuttall et al., 2012). The Northern Caspian area of He production is estimated to have a resources of three BCF in 2012. The main field for supplying He for the Northern Caspian facility is Karachaganak. Chayanda, Yakutia, as of 2012 has been identified as the area with the greatest potential for Russia's He reserves with 90 BCF recoverable He from 31 TCF of gas. As of 2020, 31 fields have been discovered in the Siberian Platform, Russia, that have potential He production (Rapatskaya et al., 2020).

Production of He in Russia is generally from processing large volumes of natural gas at facilities such as the Amur plant (Fig. 3.103). The Amur facility derives its gas from the Chayandinskoye and Kovyktinskoye fields in the Irkutsk Region of Russia with an annual capacity of 1.3 BCF of He gas per year. The Amur plant also recovers N_2 to be sold to a fertilizer facility. The Amur plant at the time of this publication has suffered a series of fires and equipment failure, thereby reducing its ability to extract He for potentially a year or more (Table 3.6).

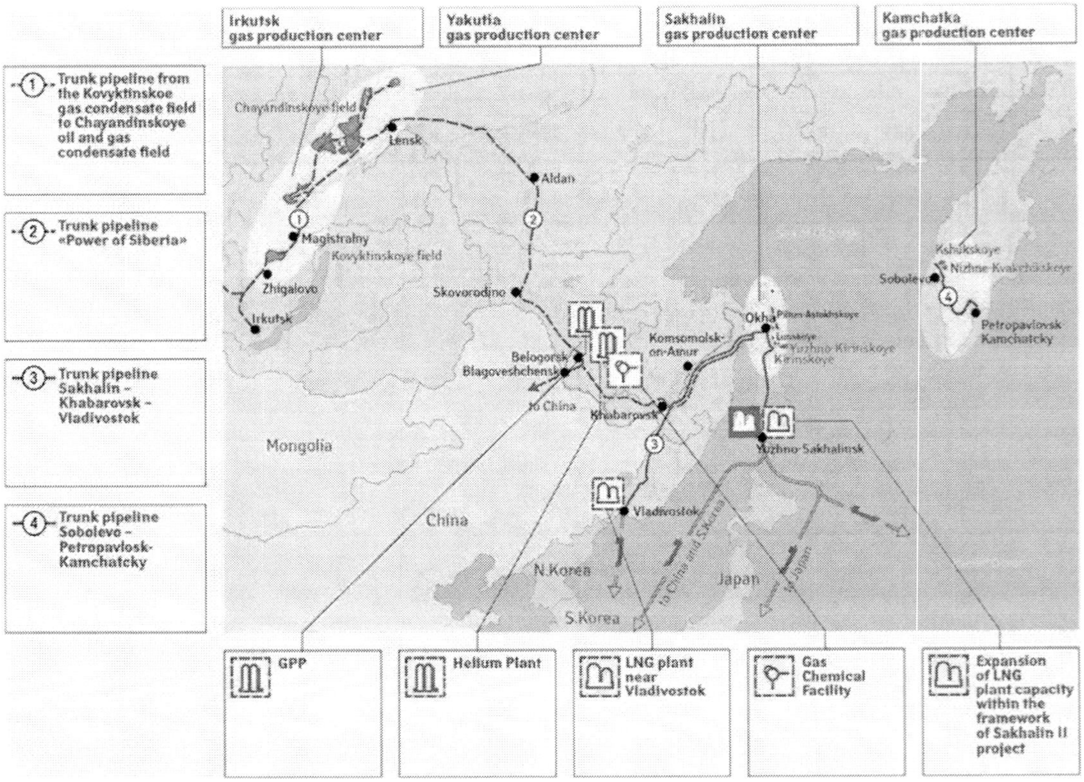

FIGURE 3.103 Amur Plant facility along with the Kovyktinskoye and Chayandinskoye (Lang, 2017).

The Kovyktinskoye, gas and condensate field, discovered in 1987, has the largest gas reserves within the Zhigalovsky and Kazachinsko-Lensky Districts of the Irkutsk Region, eastern Siberia, Russia. It serves as the foundation for the Irkutsk gas production center and, together with the Chayandinskoye Field, discovered in 1989, in the Lensky District of Sakha (Yakutia) Republic in the Far East region of Russia, form the major natural gas source for the Amur processing facility. The reserves for the field are 49 BCF of He. There are probably numerous other small fields that also send their gas to Amur. The Kovyktinskoye gas field has potential of 48.9 TCF (Everett, 2012).

3.18 Africa

Africa is a geologically diverse continent with numerous metalliferous and petroleum deposits. A wide range of religious, political, and cultural beliefs makes exploration, exploitation, and development of natural resources of any sort a complicated process. There are two areas in Africa that are currently the focus of He exploration, South Africa and Tanzania. Both countries host unique He deposits that do not seem to be present anywhere else in the world.

TABLE 3.6 Important He deposits in Eastern Siberia as of 2009 (Rapatskaya et al., 2020).

Region of gas recovery, deposit, its type	Helium reserves billions MCF (M^3)		Basic chemical composition (%) (vol)			
	ABC_1	C_2	He	CH_4	T.U	N_2
Krasnoyarsk						
Sobinsk, OGC	10.6 (0.3)	3.53 (0.1)	0.576	67.5	6.4	25.26
Yurubchen-Tokhom, OGC	28.25 (0.8)	17.66 (0.5)	0.183	83.0	8.4	7.8
Irkutsk region						
Dulisma, OGC	6.35 (0.18)	1.41 (0.04)	0.260	84.1	6.8	6.8
Kovykta, GC	137 (3.88)	42.37 (1.2)	0.276	92.3	5.7	1.5
Sakha (Yakutia)						
Verkhnevilyuchan, OC	6.35 (0.18)	3.53 0.1)	0.13	84.5	7.5	7.46
Srednebotuobia, OGC	26.5 (0.75)	1.41 (0.04)	0.67	83.8	6.9	8.0
Chayanda, OGC	65.33 (1.85)	187 (5.3)	0.63	85.6	6.4	8.2

GC, gas and condensate; *OC*, oil and condensate; *OGC*, oil, gas, and condensate. ABC are primary reserves and C is additional reserves.

3.18.1 South Africa

South Africa has significant offshore potential in oil and gas but onshore at present it is generally limited to microbial and coal-bed methane gas (Petroleum Agency, SA, 2012). The Virginia Gas Field is a microbial gas accumulation that overlies the Welkom Goldfield (Fig. 3.104) which extracts Au from Witwatersrand Precambrian age Supergroup of metasediments (Truswell, 1977; MHS, 2019, 2020). Fig. 3.104 is the surface geology for the Witwatersrand Gold District which has produced almost 40% of gold consumed in the world. There are 19 wells in the field (Fig. 3.105) that have been evaluated by the operator of the project and have estimated proved, developed, and nonproducing reserves and resource of 14,7765.3 MMCFG and 346 MMCF He (MHS, 2019).

This is a unique type of natural gas and He deposit but one that may be applicable elsewhere in similar geologic settings. Gas wells in the area are known as "blowers." The basement strata had been tectonically folded into a large north-to-south trending anticline that is in turn bisected by a large extensional graben. There are numerous deep-seated faults that allow pathways of degassing of He from the earth's crust. The volcanic Ventersdorp Supergroup unconformably overlies the Witwatersrand Supergroup (Truswell, 1977; MHS, 2019). The Permian Karoo Supergroup overlies the Ventersdorp Supergroup and is composed of sandstone, coal, and carbonaceous mudstones. Locally there is the Dwyka Tillite between Karoo and Ventersdorp Supergroup that is a Cambrian glacial deposit (Truswell, 1977; MHS, 2019). The basement faults generally do not extend beyond the upper Ventersdorp Supergroup.

The Virginia Gas Field Project is a microbial methane gas reservoir that is sourced from within the Witwatersrand Supergroup as the result of groundwater circulating

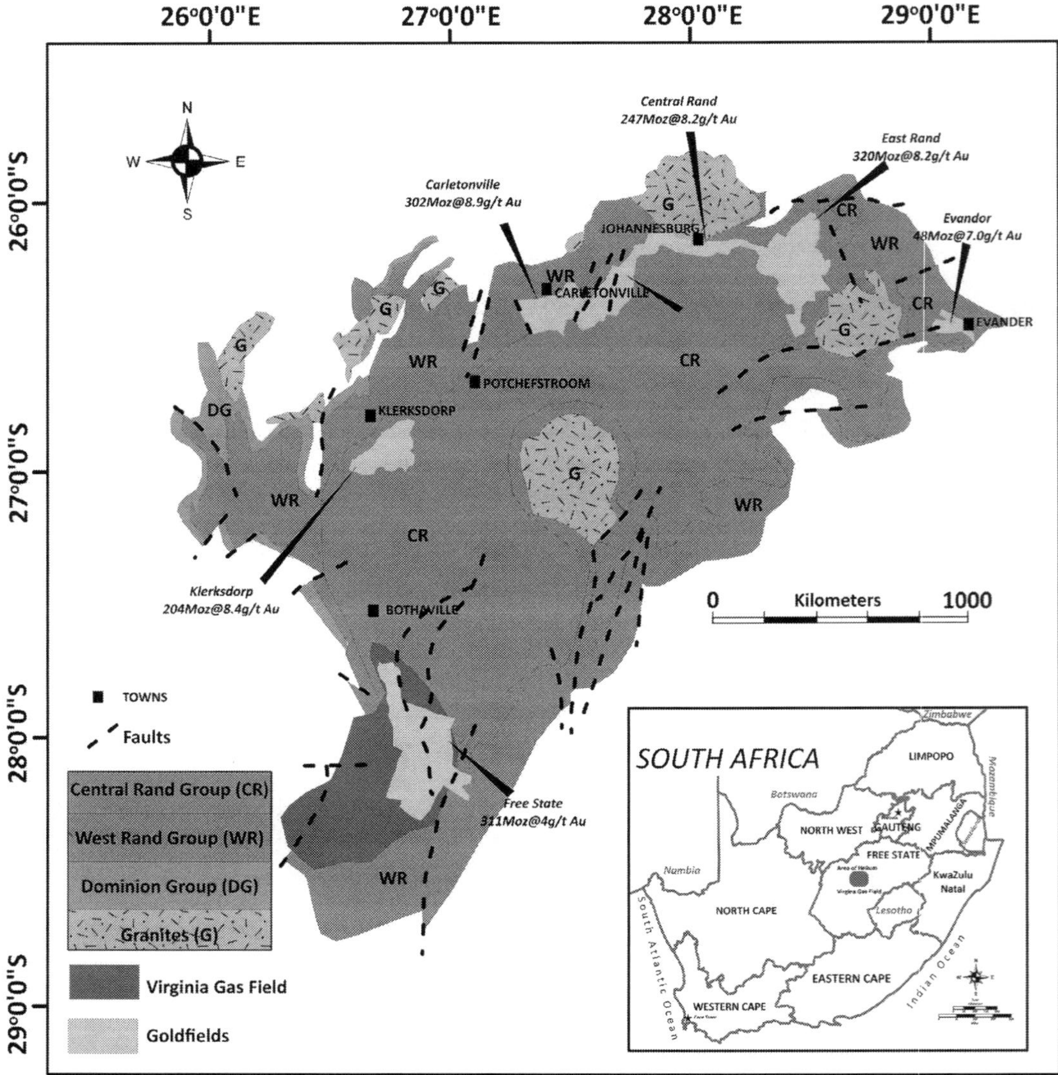

FIGURE 3.104 Location of Virginia Gas Field Project in the Free State, South Africa, and surface geology of the area. Also indicated is He labeled Evander Project. *Modified from MHS, 2019, https://www.renergen.co.za/mha-final-report/.*

downward into the basement faults contacting with bacteria living within the crustal rocks (MHS, 2019). Methane isotope studies demonstrate that very little, if any, of the CH_4 can be attributed to the Karoo coal-beds or the carbonaceous mudstones. The CH_4 because it is biogenic in origin will continually be generated by microbes at a relative slow rate. The natural gas and He reservoir have been identified as the "White Sand" of the Permian Karoo Supergroup (MHS, 2019). Helium migrates up with the CH_4 along the basement faults that tie into minor fault and fracture systems. The rate of recharge for He from the crustal rocks and

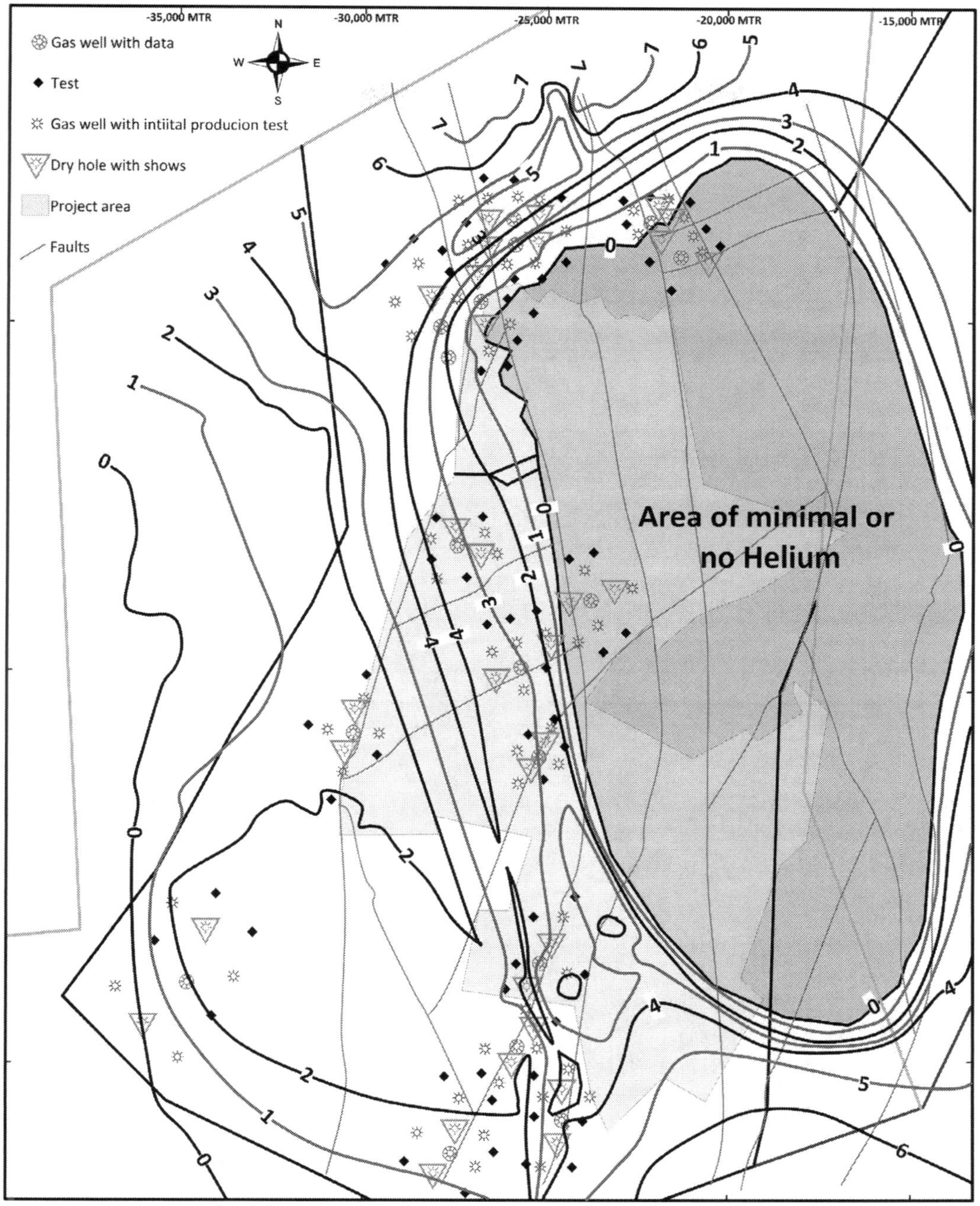

FIGURE 3.105 Well location of the Virginia Gas Field in the Free State land block (yellow), contour interval for He in %; faults are in orange, are of no He is in indicated in gray, gas wells and various test wells indicated. *Modified from MHS, 2019, https://www.renergen.co.za/mha-final-report/.*

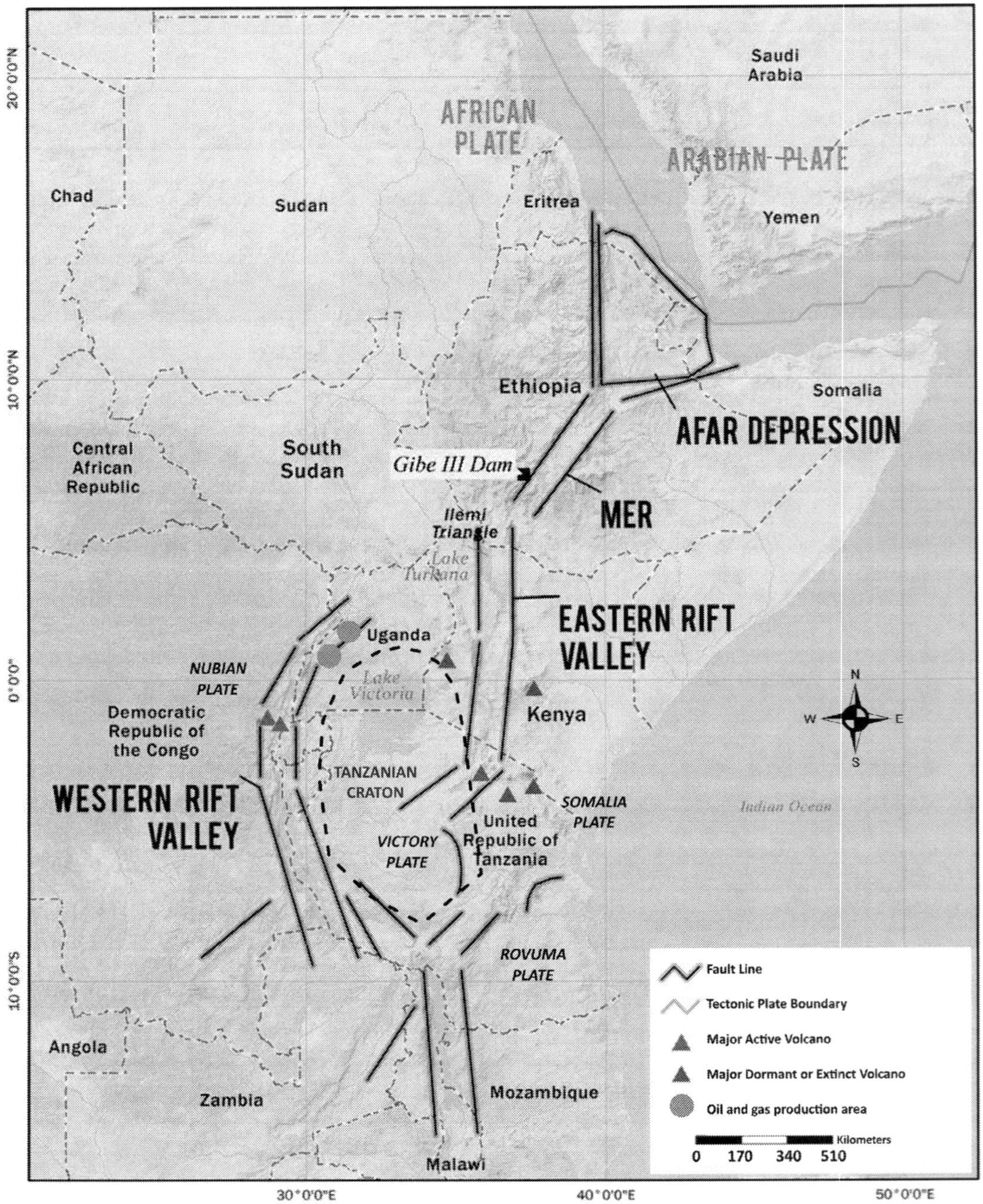

FIGURE 3.106 Location of the East African Rift System as modified from Carr (2017).

Witwatersrand Supergroup gas is not known but probably very slow. Fig. 3.106 is the percentage of He in the Virginia Gas Field Project. There is a large area of minimal to zero He which correlates with a structural high at the base of the Karoo (MHS, 2020). The faults (orange color) do not correlate with the location of the higher He values suggesting there is not a direct relationship between the two variables.

The Virginia Gas Field is unique in that the generation of microbial or biogenic gas in the shallow strata provides a host gas migrating He to concentrate. Like many He deposits, the economic viability of He extraction is linked to the production of CH_4+ from the same reservoir. This is a unique deposit and at least locally there seems to be some similar leads around the Witwatersrand Gold complex.

3.18.2 Tanzania

Tanzania is located in central East Africa just south of the equator (Fig. 3.107) and lies on the Tanzania Craton which is approximately 2.5 BY (Archean) old and may be divided into a number of unique areas. It consists of quartzites, aplite, schist, pegmatite, and ironstone in the Dodoman Belt in the center of the country and basalt, dolerite, trachyte, rhyolite, and tuff as the Nyanzian Belt in the north (Morley et al., 1999). During the Proterozoic the Ubendian Belt, a highly metamorphosed folded metamorphic fold belt, was emplaced with intrusive granites that formed parallel to present-day Lake Rukwa and Tanganyika. The Karagwe-Ankole Belt lies in northwestern Tanzania and is composed of moderately metamorphosed argillite, phyllite, quartzites, and schists (Morley et al., 1999). The Mozambique Belt is a complex structural terrane that lies adjacent to the Tanzania craton and contains pyroxene-gneiss, charnockite, and hornblende with minor amounts of schist, quartzites, and limestones. During the Paleozoic, there was extensive erosion across the entire country (Morley et al., 1999).

In the Mesozoic to Cenozoic time, the East African Rift System (EARS) was emplaced on the west, east, and north central areas of the country with the deposition of the Karoo Supergroup during the Late Carboniferous to Early Jurassic time (Morley et al., 1999; Guangya, 2019). The Karoo is composed of sandstone, conglomerates, tillite, shale, and a local limestone. The Late Jurassic was dominated by stable shelf limestone deposition and the Cretaceous through Quaternary period saw a long period of marine transgression with deposition of shale, limestone, and sandstone intervals (Morley et al., 1999; Guangya, 2019). From the Jurassic to present day, EARS began opening up with the movement of Madagascar plate to the south and east along the Davies Fracture zone. This extensional phase is still active today causing volcanism in Tanzania (Mt. Kilimanjaro) and numerous areas of surface volcanic rocks. The extension formed several grabens (Fig. 3.106) that are responsible for the present-day Lakes Tanganyika and Nyasa with secondary rift basins represented by Selous and Lake Rukwa grabens (Morley et al., 1999; Guangya, 2019).

The focus of the Tanzania Geological Survey in the 1940s was to locate sources of salt (James, 1967). It was discovered in 1947 that gas venting from the Nyanza cold brine spring at Uvinza contained 2.5% He with 94.6% N_2 (James, 1967). Fig. 3.108 is a modification of the map done by James (1967) indicating the location of He and brine seeps. In 1955, three groups of springs on the northern border of Tanzania were found to emit similar rich N_2 gases with 13.2%–17.9% He (Fig. 3.107) (James, 1967). The United Kingdom Atomic Agency then took

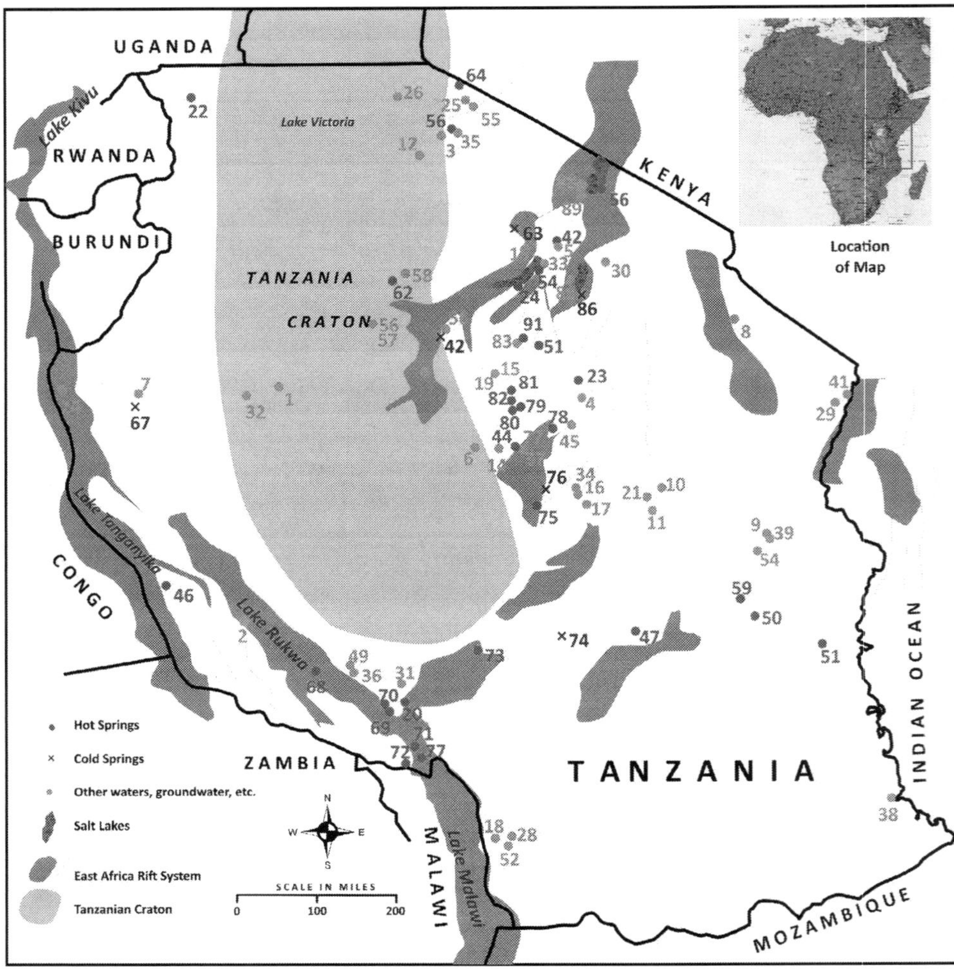

FIGURE 3.107 Location and map of Tanzania with the East Africa Rift System, Tanzania craton, and location of samples taken from seeps and groundwater locations as outlined in James (1967) for analysis of brines for He.

an active role and several more springs were discovered with similar high He percentages throughout the country coming from cold and hot brine springs. Two areas, the Maji Moto and Manyeghi Springs, underwent drilling in order to determine if the rate of He production could be enhanced (James, 1967). Several wells were drilled in both areas reaching a maximum depth of 800 ft (242 m), produced brines, and the flow from the adjacent natural springs was not affected. The amount of He only slightly increased from the wellbore production compared to the He venting from the springs at the surface (James, 1967). Seismic was not used to determine if there were underlying traps or accumulations. The interest in He production was abandoned but brine production for salt recovery did continue from several springs.

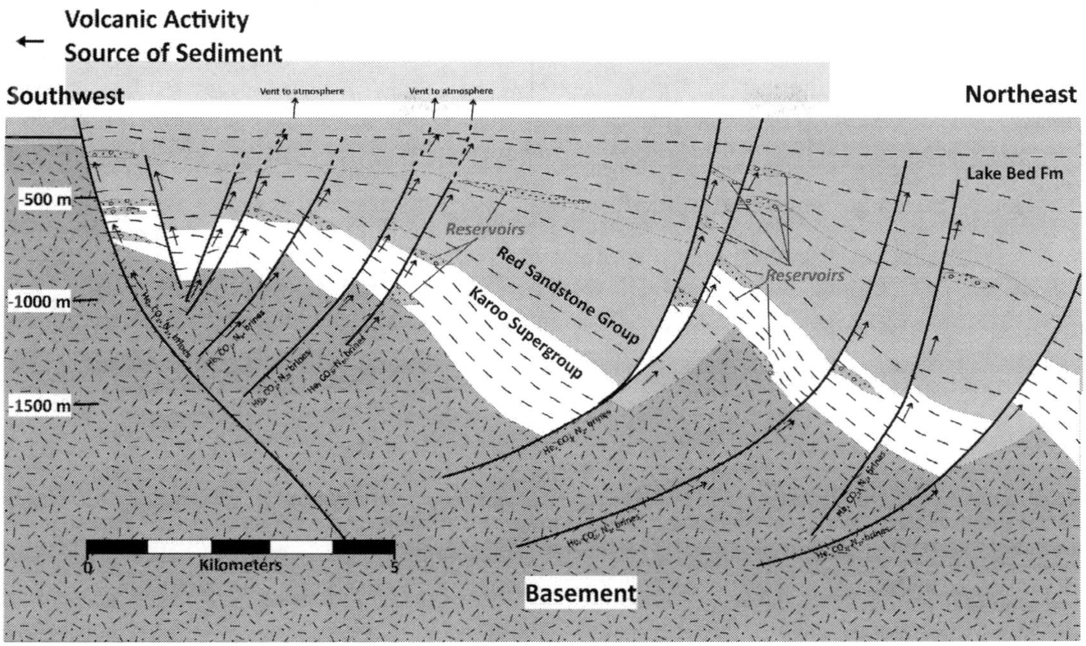

FIGURE 3.108 Cartoon model modified from Helium One's presentation, 2021, indicating the complex nature of the East African Rift System. The yellow sandstones represent proposed traps of accumulation of He, CO_2, and N_2. The fault system is active and probably in many areas open providing pathways to the surface for He, CO_2, and N_2.

In 2002, the Turaco 1 discovery by Heritage Oil in the northern Kenyan and southern Ethiopian portion of the East Africa Rift system has led to the discovery of over 4 billion barrels of oil reserves from at least 20 oil fields (Davison et al., 2014). The rift system is distinctly segmented into subbasins with each having its own structural style and tectonic history. Both compressed anticlinal footwall and hanging wall traps are productive, but there is risk of ongoing tectonic activity and lateral variation in a trap's seal (Davison et al., 2014). The volcanic deposits at the surface make acquiring seismic data problematic. The western rifts are filled with fluvial deltaic and lacustrine clastic deposits, and the eastern rifts are filled with significant lava flows and thick sandstone beds. The source rocks for petroleum are lacustrine shale beds with up to 20% TOC and are highly variable in thermal maturation, from immature to over mature (Davison et al., 2014). There are numerous macroseeps at the surface and some have been determined to be from source rocks that are 25,000 years old or less (Davison et al., 2014). Macroseeps are a useful but not definitive guide to the presence of a petroleum system being present. There are shallow CO_2 deposits in some of the present-day lake sediments and present a natural hazard if released by natural or human causes.

The increase in the price of He in the 2010s caused a resurgence in interest in the element because of the high percentage found in many of Tanzania's active springs. A company called Helium One based out of London, England, has been exploring in the southern part of the EARS based on previous work done by the British in the 1950s and Danabalan (2017).

In Fig. 3.108 is a cartoon of the rift system Lake Rukwa. As of this publication drilling of the first well by the operator has been carried out without success. In the Rukwa Rift Basin, work by Mtili et al. (2019) identified three different groups of gas seeps, Ivuna–Itumbula–Tete, Ibaya–Ngwilo–Kajunda, and Songe–Nanyara, which suggest different hydrothermal systems that are active. Helium One's has had an independent resource estimate of 138 BCF of He in Lake Rukwa area and another area in Central Tanzania but this is not based on actual drilling. As with all macro oil and gas seeps around the world, the He seeps in Tanzania along the EARS so may represent just seepage and venting from a basin or rift area. The Rufiji Basin is a Karoo aged rift running east west through central Tanzania and may also have potential that has not been explored for He. The He is likely derived from crustal and sedimentary rocks undergoing intense pressure and temperature variations and migrating along open fracture and fault systems (Fig. 3.109).

3.19 Poland

The only country presently producing He in Europe is Poland. Europe depends upon Russia and the Middle East for its He because of the European Union's present focus on moving away from fossil fuel extraction. Poland produces small amounts of He from a series of fields that lie along a production trend that runs east northwest to southeast east on the border of the Fore Sudetic Monocline and the Fore Sudetic Block in the Polish Lowlands of western Poland (Fig. 3.109) (Szamalelek et al., 2017). These fields produce from the Permian Rotliegend and Zechstein limestone and main dolomite formations (Szamalelek et al., 2017). Helium production, except for one field, is constrained to the boundary between Fore Sudetic Monocline and the Fore Sudetic Block that seems to be heavily faulted. The future for new He and N_2 production in Poland is unknown.

Helium contents vary from 0.02% to 0.45% in many producing gas fields but the 16 producing natural reservoirs vary from 0.22% to 0.42%. Fig. 3.110 is the production history of He production in Poland which in decline.

3.20 Asia-pacific region

Southeast Asia consumes a tremendous amount of He that is used in the medical and electronics industries. Unfortunately, there are only a few small areas of He production in the area and there are little signs of encouragement for additional significant He discoveries. Australia like the United States and Canada allows small and mid-size companies to prosper that have the ability to find small- to medium-sized deposits that can be profitable. Unfortunately, to date there has been little He exploration success either onshore or offshore, except through extraction from a processing facility near Darwin. The conventional wisdom is that most of the basins throughout the area are young and have not had sufficient time to accumulate He in traps. In addition, in many areas of the Asia-Pacific region, the tectonic activity as with the Pacific Rim of Fire is young geologically lack so there would be a lack of "old pore water." The very active tectonic and volcanic activity would inhibit or destroy any potential traps or accumulations that would form.

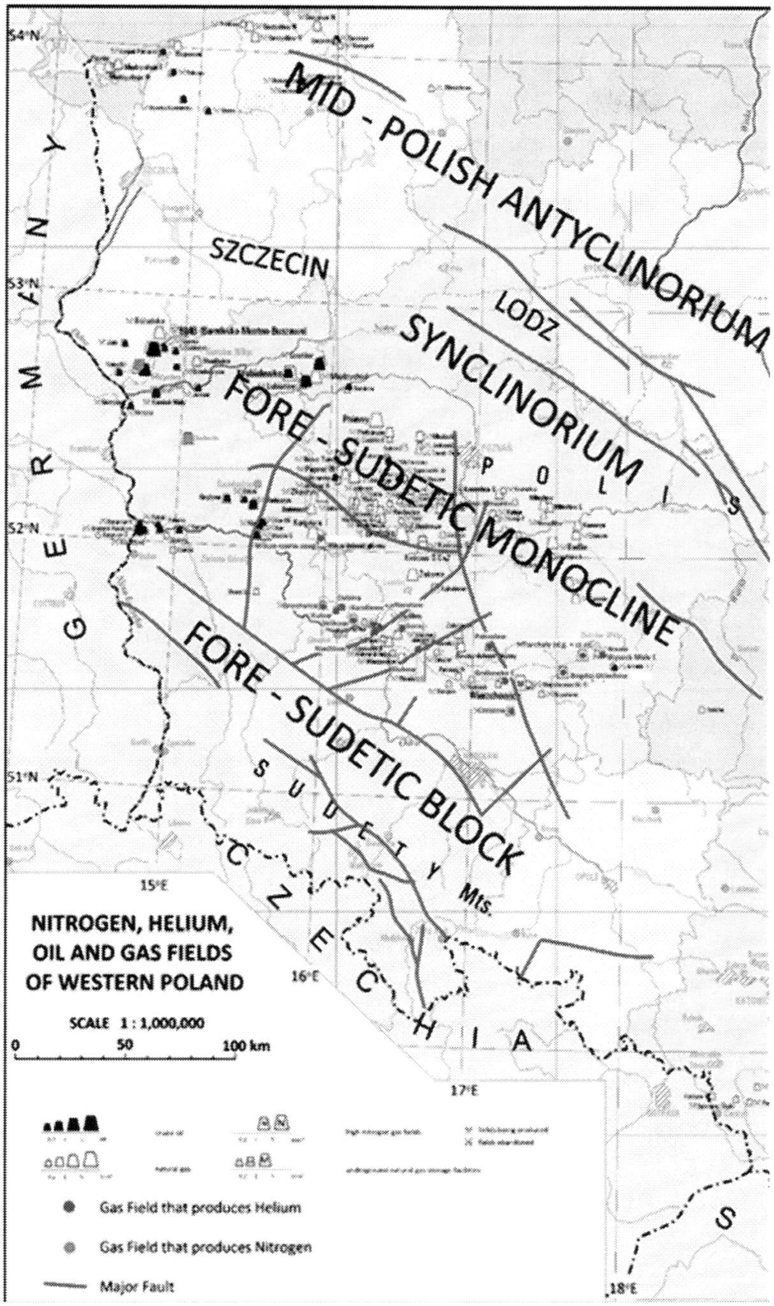

FIGURE 3.109 Nitrogen, He, oil, and gas fields of western Poland modified from Polish Geological Institute National Research Institute (2017). Helium-producing fields are indicated in purple and N_2 is indicated in green.

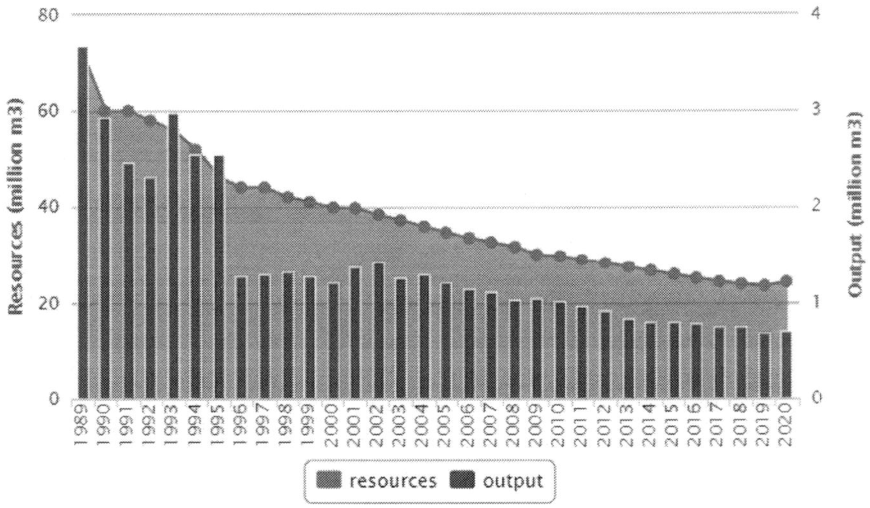

FIGURE 3.110 Helium production for Poland from 1989 to 2015. (Czapigo-Czapia and Breziznski, 2020).

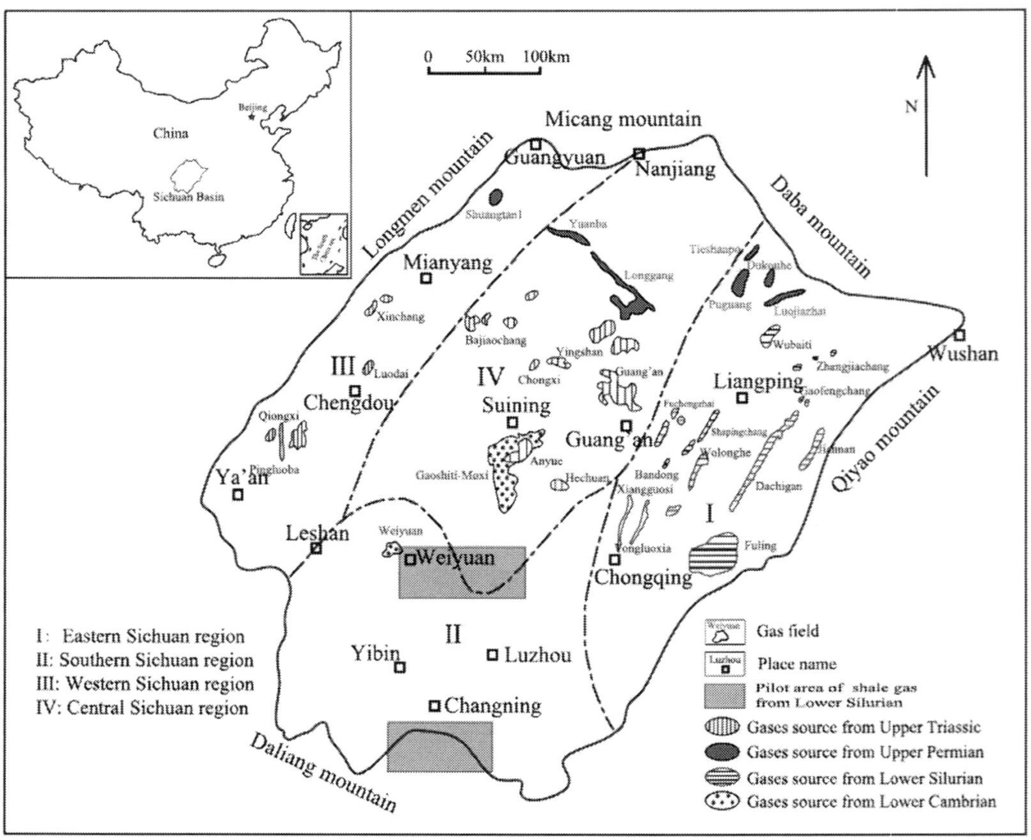

FIGURE 3.111 Location of the only He-producing field in China, Weiyuan Gas Field (Wang et al., 2020).

3.20.1 People's Republic of China

China has one field producing He commercially that is located in the southern Sichuan Basin, Weiyuan Gas Field (Fig. 3.111 (Wang et al., 2020). The Sichuan Basin is a Mesozoic–Cenozoic foreland basin overlying a passive margin and surrounded by several mountain ranges (Wang et al., 2020). The basement is composed of metamorphic and granitic rocks that are overlain by Paleozoic through Cenozoic strata (Fig. 3.111). The Weiyuan Field, discovered in 1964, has proved reserves of 400×10^8 m^3 He from the Sinian age Dengying carbonate reservoir (Fig. 3.112) (Zhang et al., 2015; Wang et al., 2020). The amount of He present ranges 0.15%–1.88% with an average of 0.27% in the Dengying carbonate, decreases up section until in the Permian it averages 0.11% and the N_2 content is typically 7.84% in the producing reservoir (Wang et al., 2020). Wang et al. (2020) indicated that the source of the He was from the Presnian granite based on $^3He/^4He$ ratios (Fig. 3.113). The accumulation was formed during Late Triassic to Cretaceous times when the strata above the Presinian granite was depressurized causing fracture and fault systems to open up and release the He into the overlying strata.

There are two fields to the north of Weiyuan called the Ziyang and Gaoshiti-Moxi that have a similar structural configuration but have significantly lower He (Fig. 3.111) (Zhang et al., 2015; Wang et al., 2020). Both fields are located along the Leshan-Longnusi uplift. The trap that formed at the same time as the Weiyuan trap. This was followed by a separate tectonic event that caused the disappearance of that trap and possibly whatever He was present was released. The Gaoshiti-Moxi Field did not undergo the same overlying uplift and depressurization and He gas did not migrate from the Presinian granites into the lower Paleozoic reservoirs (Wang et al., 2020).

Zhang et al. (2019) analyzed 12 wells subdivided into three groups from the Weihe Basin with depths 6270 to 12,045 ft (1900 to 3650 m) with ranges of 58.57%–91.66% for N_2, 0.03%–31.69% for CO_2, 2.23%–71.7% for CH_4, and 0.006%–0.294% for He. The basin as described by Zhang et al., (2019) has geothermal reservoirs and the He is accumulating in Pliocene aged sediments.

In the eastern part of the People's Republic of China, there are numerous basins with CO_2 resources that have become a detriment in pursuing hydrocarbons. The middle part of China has low amounts of CO_2 due to the lack of tectonic activity (Dai et al., 1996). The eastern basins are rift related or part of or adjacent to major subduction zones or plate boundaries and many of these CO_2 reservoirs are devoid of He and N_2. Prolific areas such as the Bohai Bay Basin have numerous CO_2 accumulations that has inhibited exploration for hydrocarbons (Zhang et al., 2008). An example of CO_2 accumulations includes wells in the Ordos (Cratonic) and Bohai Bay (rift) basins have greater than 0.05% He and many fields have CO_2 (Dai et al., 2017). The Bohai Bay Basin has seven CO_2 gas fields and a He percentage overall greater than the Ordos Basin. The CO_2 source in the Ordos Basin is both associated with hydrocarbon generation and thermal decomposition of carbonates. The Bohai Bay Basin has similar sources of CO_2 but in addition, a significant amount of the gas is derived from volcanic activity and degassing of magmas (Dai et al., 2017).

The South China Sea contains oil, gas, and CO_2 in numerous productive basins including Beibuwan, Yinggehai, Qiongdongnan, and Pearl River Mouth (Fig. 3.114). According to Huang et al. (2015), the CO_2 reservoirs were derived from thermal decomposition of Miocene and Paleozoic carbonates with mantle CO_2 only providing a minor contribution. In the

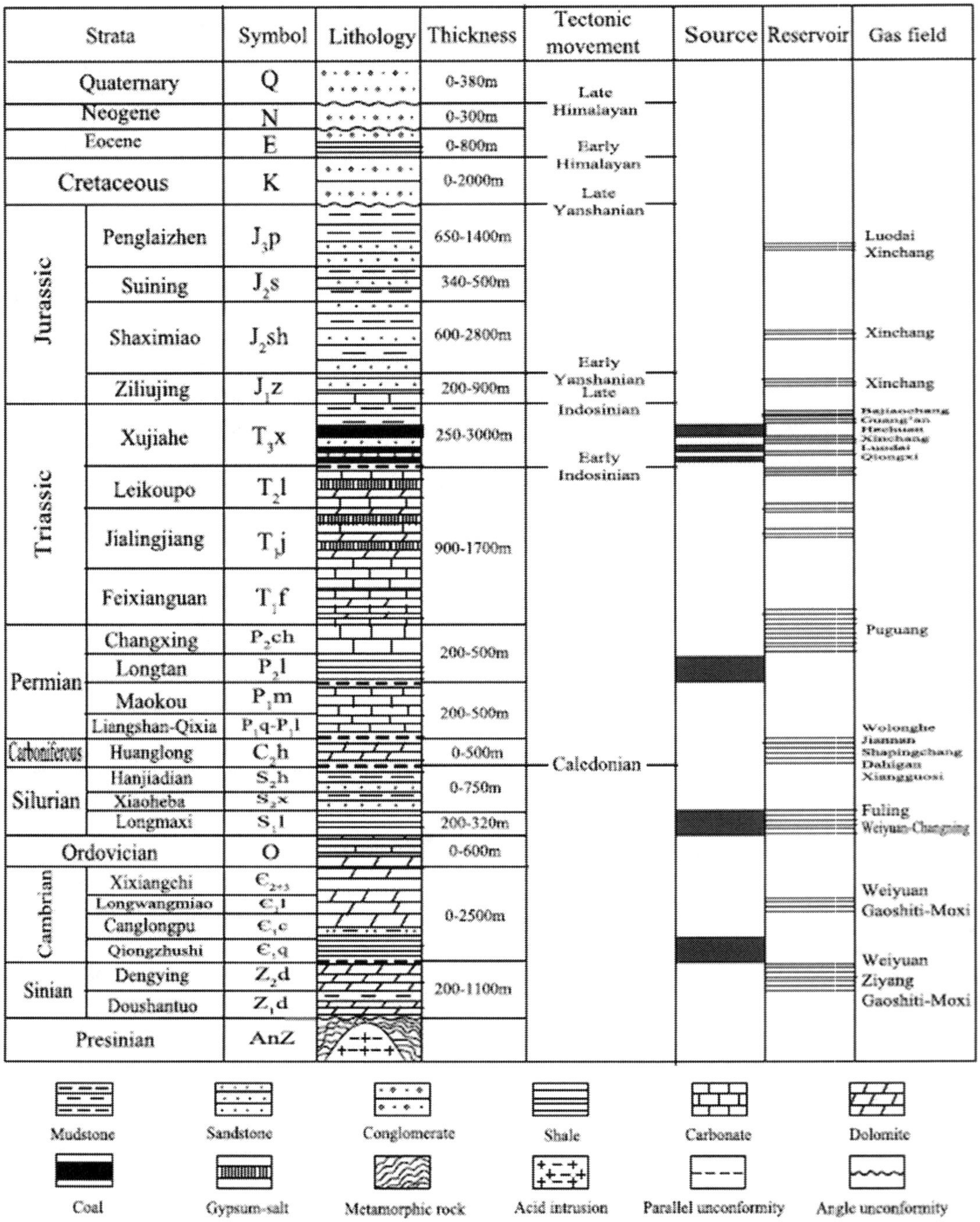

FIGURE 3.112 Stratigraphic column for the Sichuan Basin (Wang et al., 2020).

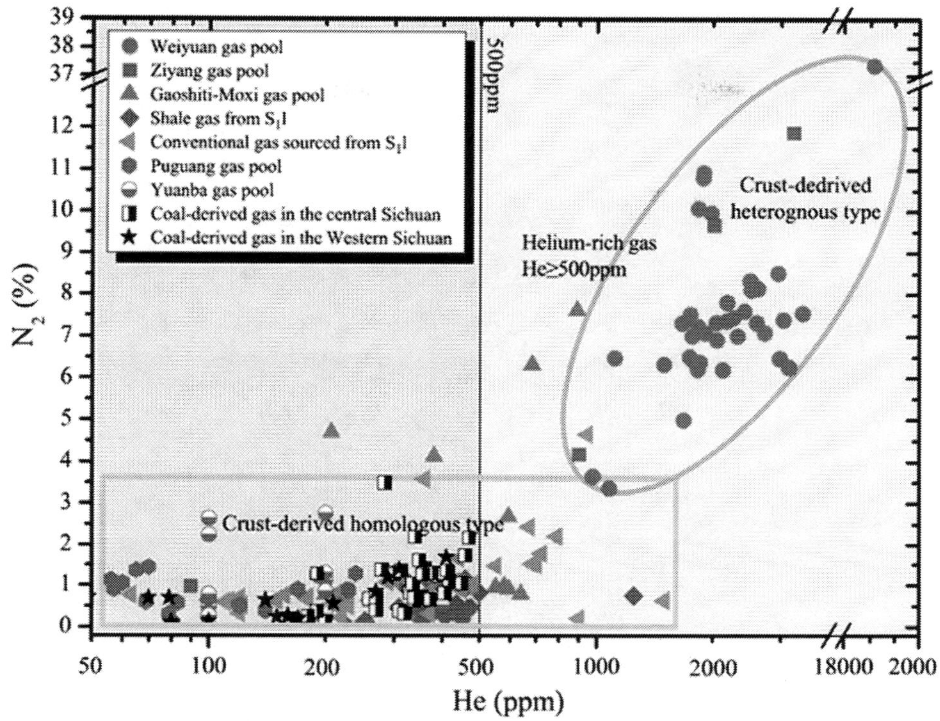

FIGURE 3.113 Relationship of He (ppm) and N_2 (%) in natural gas of the Sichuan Basin. Helium concentrations in crust-derived homologous type reservoirs are relatively low, and the concentration increases with increasing age of natural gas source rocks. Helium is relatively abundant in crust-derived heterogenous type reservoirs, and these reservoirs often exhibit high N_2 levels (Wang et al., 2020).

Qiongdongnan and Pearl River Mouth basins, the CO_2 was derived from mantle degassing. Fig. 3.115 is a cross section through the BD5-13 Field and looks similar to structures in the adjacent Yinggehai Basin but the source of the CO_2 is different between the two areas (Li et al., 2008). The Beibuwan Basin has numerous hydrocarbon and CO_2 fields where the primary source of CO_2 is mantle derived with only a small organic and crustal component (Li et al., 2008).

China in July 2020 announced the opening of a He processing facility in Yanchi County, Ningxia, to process waste gas from one or more of its processing facilities (Chen, 2020). China uses 4300 tons of He a year and does not expect that its processing plant will provide a competitive advantage over imports (Chen, 2020).

3.20.2 Australia

Australia has a number of oil and gas productive basins both onshore and offshore (Fig. 3.116), but He production has been limited historically to He extraction from a production facility located in Darwin (Boreham et al., 2018; Haines and Allen, 2019). The Darwin facility processes natural gas from the Bayu-Undan Field stream that is located in the offshore

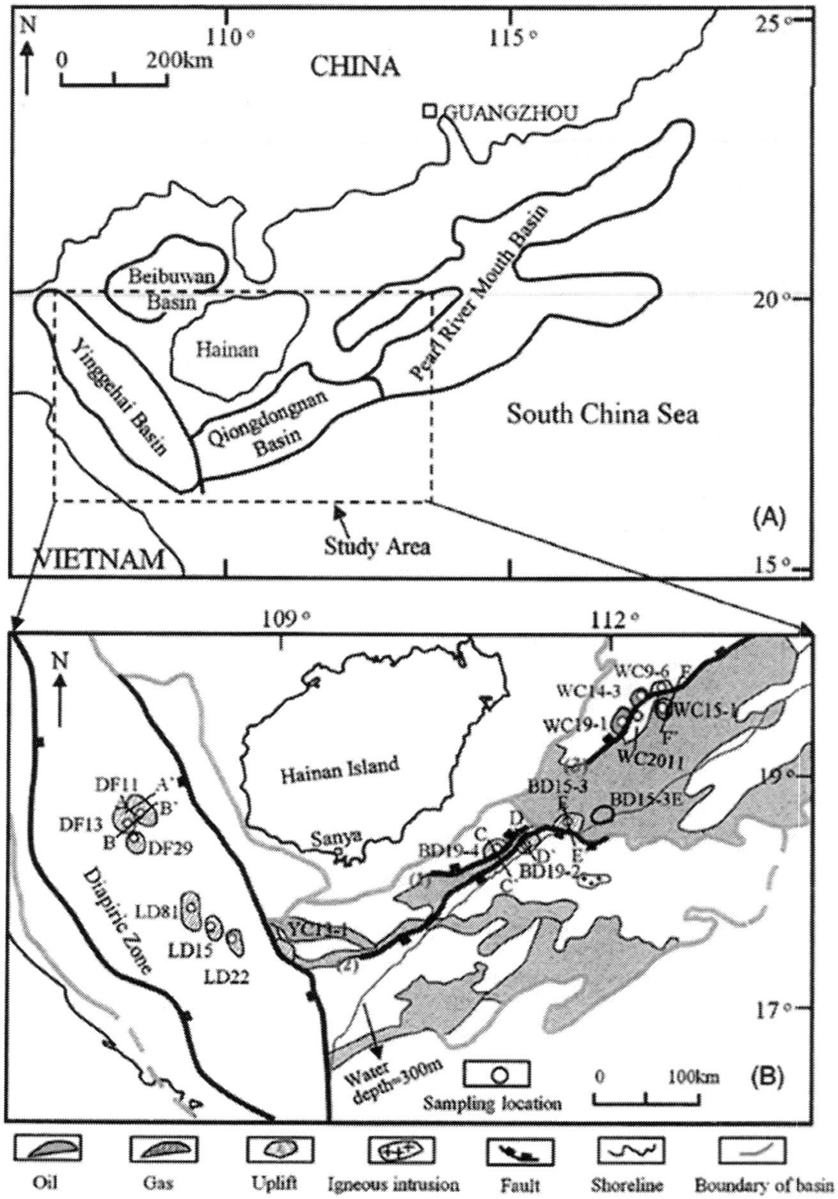

FIGURE 3.114 Location of continental margin basins of North China Sea (Huang et al., 2015 *reprinted by permission of the AAPG whose permission is required for future use*).

Bonaparte Basin in the Timor Sea (Boreham et al., 2018; Haines and Allen, 2019). The plant produces 4.25 MScm^3 a year by refining the waste stream at the processing plant facility (Palmer and Ambrose, 2012; Haines and Allen, 2019). Fig. 3.117 is from Boreham et al. (2018), which indicates that He% in Australia is typically low.

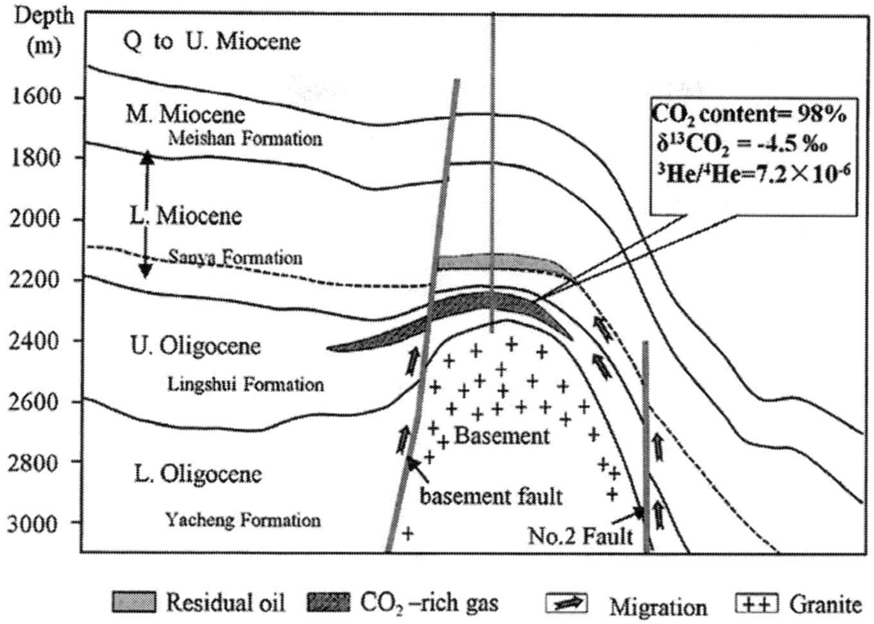

FIGURE 3.115 Geological cross section through the BD 15-3 Field in the Qiongdongnan that hosts both hydrocarbons and CO_2 gases (Huang et al., 2015 *reprinted by permission of the AAPG whose permission is required for future use*). The migration pathways and displacement of oil by late charged mantle-derived CO_2 in original reservoirs are indicated.

Helium in Australian natural gases is dominated by crustal sources with the older (Late Paleozoic–Proterozoic) reservoirs commonly having the highest crustal ^4He input. High ^3He percentages are concentrated in eastern Australia and are related to more recent younger tectonic/magmatic events (Fig. 3.117).

The Amadeus Basin has had two significant He results from the Neoproterozoic Pioneer sandstone Ooraminna-2 Heavitree Quartzite in Magee-1 well and granitic basement from the Mt. Kitty 1 of 6.2% and 12%, respectively (Ambrose et al., 2012). The reservoirs are overlain by the Proterozoic Gillen Salt but the gas flow was insufficient for economic natural gas production (Fig. 3.118). The area has extensive basement fractures/thrusts common in this compressive tectonic regime with a competent seal being provided by the regionally extensive Gillen Salt. Helium is found the Neoproterozoic rocks which provides a significant timespan for the accumulation of this resource (Palmer and Ambrose, 2012). In addition, Ooraminna-2 in 2010 recorded a gas flow of 150 MSCFD with a He content of 0.22% from the Pioneer Sandstone and the Palm Valley Field has produced gas with a He content of 0.15% from the Pacoota and Stairway sandstones (Palmer and Ambrose, 2012; Haines and Allen, 2019).

Ozimic (2016) reviewed the Otway, Bass, and Gippsland basins for CO_2, N_2, and He and determined that they were all CO_2 prone and lacking in N_2 and He (Table 3.7). The Otway Basin trends northwest-southeast is of late Jurassic–Cenozoic origin on the southern

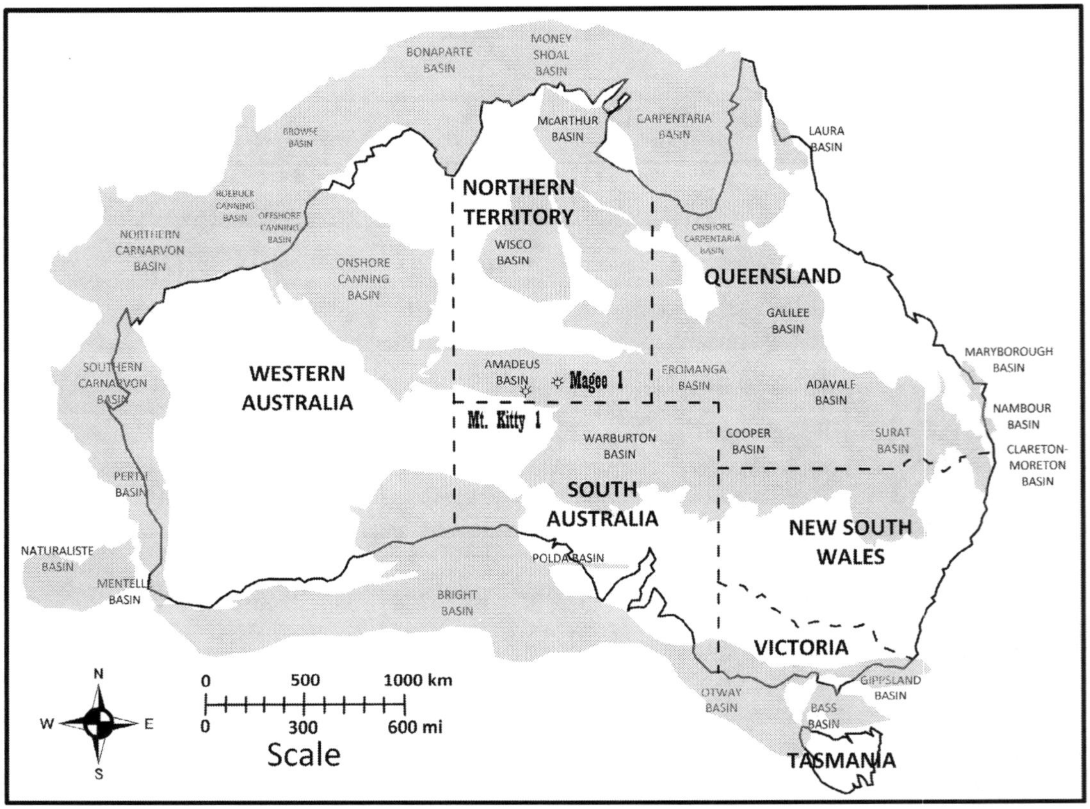

FIGURE 3.116 Sedimentary basins in Australia with two recent tests that yielded significant He, the Mt Kitty 1 and the Magee 1.

Australian passive margin (Geoscience Australia, 2021a). The basin is formed by multistage rifting and inversion phases. The first round of rifting in the Late Jurassic to Early Cretaceous resulted in the east-west trending Inner Otway Basin. Late Cretaceous rifting, culminating in continental breakup in the Maastrichtian produced northwest-southeast trending depocenters beneath the outer shelf and slope. Multiple phases of compression in the Cretaceous to recent times resulted in inversion and wrenching of preexisting structures (Geoscience Australia, 2021a). Exploration has resulted in over 200 wells in South Australia, Victoria, and Tasmania that have discovered natural gas and industrial grade CO_2 fields but no commercial oil discoveries.

The Gippsland Basin contains in excess of 6 mi (10 km) of Cretaceous to Neogene sediments composed of in ascending order nonmarine clastic, to marginal marine clastic, marine clastic, and uppermost marine carbonates (Geoscience Australia, 2021b). The Gippsland Basin is bounded to the north by Paleozoic basement of the Eastern Uplands, to the west by uplifted Lower Cretaceous fault blocks, and to the southwest by the Bassian Rise, which separates it from the Bass Basin to the west. More than 300 exploration wells have been

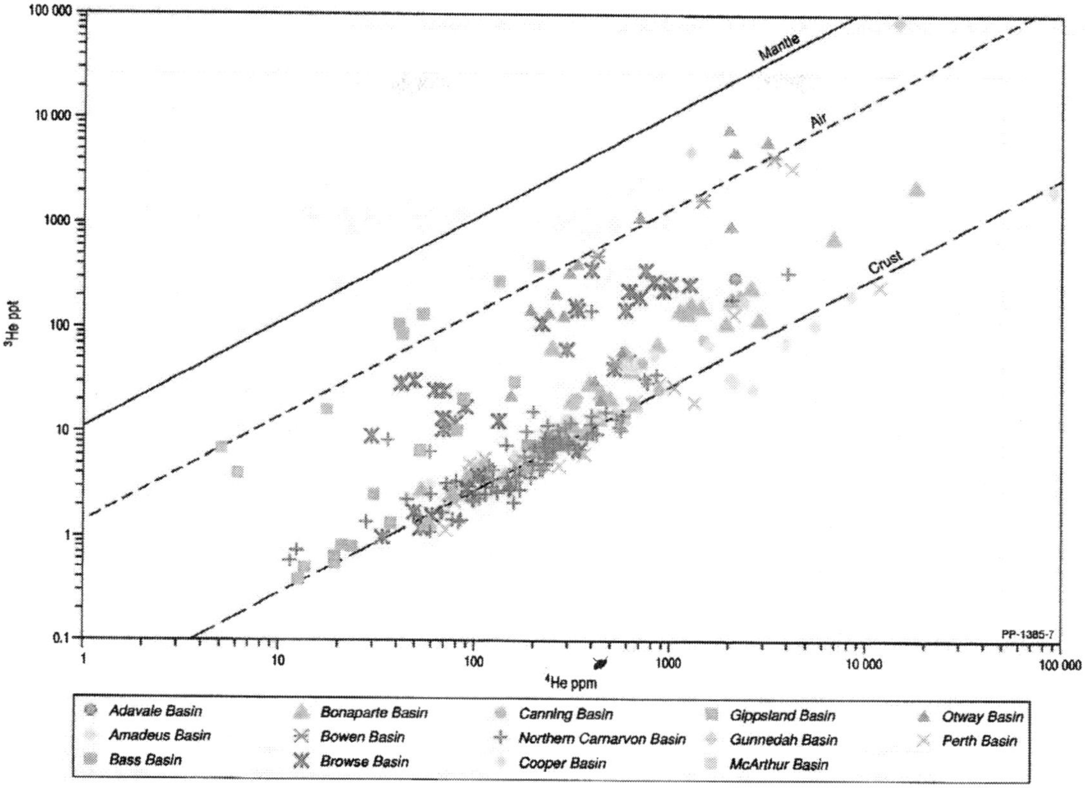

FIGURE 3.117 Helium concentration in several fields and wells in Australia (Boreham et al., 2018).

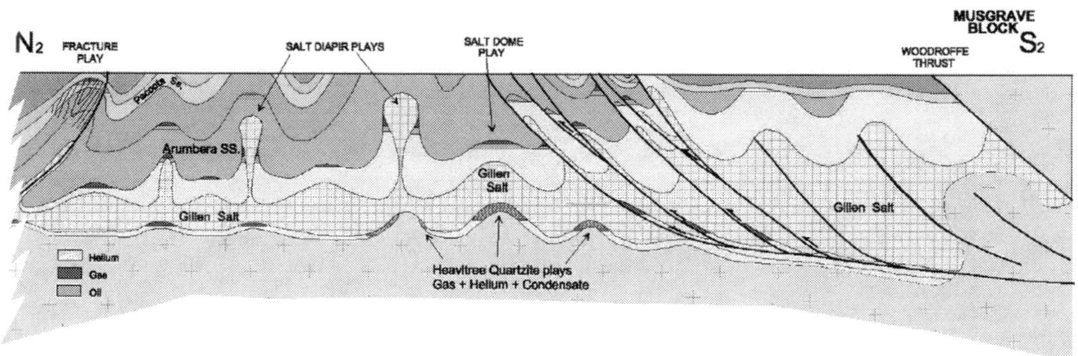

FIGURE 3.118 Structural styles and hydrocarbon play types in the southern Amadeus Basin, highlighting structures of the Petermann Ranges Orogeny (Palmer and Ambrose, 2012).

drilled leading predominantly to gas discoveries. The Gippsland Basin is part of the Early Cretaceous rift system between Antarctica and Australia that continued into the Late Cretaceous and generated a broader depocenter (Geoscience Australia, 2021a). The Rosedale and Lake Wellington Fault systems marked the northern margin of the Central Deep and

TABLE 3.7 Percentages of CO_2, N_2, and He in the Gippsland, Bass, and Otway basins (Ozimic, 2016).

Field	Type	CO_2	N_2	He
Gippsland Basin				
Barracouta	Gas	0,83	2.80	
Seahorse	Oil			
Bream	Gas	1.80	0.92	
Golden Beach-1A Marlin (Eocene)	Gas	0.01	6.30	
Marlin (Eocene)	Gas	2.00	0.55	
Marlin (Paleocene)	Gas	17.4	1.62	
Snapper	Gas	1.22		
West Seahorse	Oil			
Sperm Whale	Gas	2.59	0.69	0.002
Turrum	Gas	8.2	0.16	
Wirrah	Gas	5.00		
Basker	Gas	19.00		
Volador	Gas	36.00		
Bass Basin				
Bass-3	Gas	3.71		
Pelican	Gas/condensate	7.00	0.6	
Otway Basin				
Wangoom	CO_2	95.00		
Port Campbell (Paaratte Formation)	Oil	0.5		
Caroline	CO_2	98.9	0.60	0.08
North Paaratte	Gas	0.28	1.45	
Port Campbell (Waare Formation)	Oil	23.00	6.80	0.05
Pecten	Gas		1.6	
Tullich (Eumeralla Formation)	Gas		18.5	
Tullich (Pretty Hill Sandstone)	Gas		32.62	
Ferguson Hill	Gas		12.0	
Geelong Bore	Gas		13.5	
Garvoc	CO_2	96.7		
Kalangadoo	CO_2	88.0	12.60	0.029
Amadeus Basin				
Ooraminna-2				6.2%
Mt. Kitty 1				12%

Northern Terrace, respectively, with the Darriman and Foster Fault systems defining the southern margin of the Central Deep and the northern boundary of the Southern Platform (Geoscience Australia, 2021a). To the east, the Central Deep is characterized by rapidly increasing water depths, which exceed 3000 m in the Bass Canyon. The eastern boundary of the basin is defined by the Cape Everard Fault System, a prominent north-northeast–striking basement high. During the Late Eocene, a compressional period began to affect the Gippsland Basin, initiating the formation of a series of northeast to east-northeast–trending anticlines (Geoscience Australia, 2021a). Compression and structural growth climaxed in Middle Miocene and resulted in partial basin inversion. All the major fold structures at the top of the Latrobe Group became the hosts for the large oil and gas accumulations in the basin, such as Barracouta, Tuna, Kingfish, Snapper, and Halibut fields that are related to this tectonic episode (Geoscience Australia, 2021a).

Australia's lack of He deposits similar to the United States makes its ability to establish a viable He industry unlikely. However, Australia has a tremendous amount of natural gas production with significant reserves, so it may be able to establish a He industry that is similar to Russia's or to the PHF in the United States. The large volumes of natural gas in relatively young Cretaceous formations and younger strata have probably not had enough time for He to accumulate in traps and "old pore water" to develop. Australia has large volumes of natural gas that are processed with variable amounts of low percentage He (Boreham et al., 2018). Boreham et al. (2018) in Table 3.8 identified several present or future LNG facilities that could be used for He extraction and they estimated natural gas resources for all potential basins. The most prospective areas are the Greater Sunrise (Bonaparte Basin), Ichthys (Browse Basin), and Goodwyn–North Rankin North West Shelf Venture (Northern Carnarvon Basin) gas fields (Table 3.8). Boreham et al. (2018) indicated that 14 of the 18 current and proposed/potential LNG projects would have a waste gas of >0.5% He that could be optimal for He extraction. The authors then indicated an estimated potential resource of 56.15 BCF and a tail resource of 54.15 BCF of potentially recoverable He. Unless a yet undiscovered high concentration source of He is found, Australia's growth potential is in extracting He from waste gas such as presently utilized at extraction facilities at the PHF in the United States, North/South Pars Field in Qatar and Iran, and AMUR in Russia and China.

3.20.3 Papua New Guinea

The difficulty in finding He resources worldwide is the lack of the necessary information on the percent present of He in many gas accumulations. Bird and Seggie (1990) reported on He in the Barikewa and Iehi Gas Fields having reserve estimates ranging from 251 BCF (proven) to up to 2.0 TCF (Fig. 3.119). The Barikewa and Iehi fields are doubling plunging anticlines. The Barikewa Field lies along a broad basement arch of Pliocene–Pleistocene origin, whereas the Iehi Field is an asymmetric fault propagation fold which appears to have formed during the Pliocene by reverse movement on a normal fault of Mesozoic origin. Gas analysis indicates He 0.1%–0.2%, N_2 17%–20% CO_2 0.1%, CH_4 80.6%–82.0%, C_2H_6+ 0.7%–1.3%, and BTU ranging between 839 and 841 (Bird and Seggie 1990). Both fields produce gas from the Basal Cretaceous Toro formation which is composed of shallow marine

TABLE 3.8 List of LNG facilities by Boreham et al. (2018) that could potentially extract He.

Field or project (reservoir)	Basin	No. of samples	N_2%[a]	CO_2%[a]	He%[a]	N_2/He	He resource BCF[b]
LNG development east coast							
Gladstone LNG	Bowen/Surat	3	2.72(0.85)	0.10(0.04)	0.045(0.05)	60.00	2.94
Queensland Curtis LNG	Bowen/Surat	7	2.79(1.94)	0.56(0.41)	0.025(0.015)	112.20	6.62
Australian Pacific LNG	Bowen/Surat		2.79[c]		0.025[c]		3.90
LNG development west coast							
Bayu-Undan LNG	Bonaparte	3	3.66(0.26)	5.81(0.15)	0.128(0.07)	4.35[d]	3.28
Barossa, Caldita (Plover)	Bonaparte	4	0.13(0.10)	15.48(3.98)	0.017(0.003)	0.73	11.22
Ichthys	Browse		0.55		0.051	6.26	8.23
Ichthys (Brewster Mbr)	Browse	13	0.56(0.10)	9.66	0.058(0.42)		
Ichthys (Plover)	Browse	6	0.62(0.08)	17.12(0.81)	0.057(0.057)		
Prelude (Brewster and Plover Mbr)	Browse	3	0.30(0.18)	10.22(0.94)	0.028(0.013)	10.40	0.99
Northwest Shelf Venture GWF (Mungaroo)	Northern Carnarvon	26	1.10(0.55)	2.90(1.12)	0.038(0.027)	29.30	4.45
Wheatstone	Northern Carnarvon		9.91		0.02	471.90	1.43
Wheatstone (Iago, Mungaroo)	Northern Carnarvon	16	6.69(0.85)	2.10(0.69)	0.021(0.0006)	323.60	
Julimar (Mungaroo)	Northern Carnarvon	8	16.34(0.43)	3.71(1.55)	0.022(0.0006)	752.00	
Pluto (Mungaroo)	Northern Carnarvon	21	7.80(1.09)	1.92(0.06)	0.023(0.0004)	339.00	1.04
Greater Gorgon	Northern Carnarvon		2.56		0.13	196.90	3.25
Gorgon (Mungaroo)	Northern Carnarvon	27	2.52(1.62)	13.05(2.0)	0.16(0.013)	152.50	
Io (Jansz Sandstone)	Northern Carnarvon	1	2.82	0.31	0.015	189.00	
Janz (Jansz Sandstone)	Northern Carnarvon	5	2.74(0.36)	0.14(0.02)	0.013	210.70	

LNG Potential development							
Greater Sunrise LNG (Plover)	Bonaparte/Sahul	5	3.43(1.18)	4.96(0.55)	0.21(0.01)	16.50	16.01
PTTEP FLNG (Plover)	Bonaparte/Vulcan	1	0.93	12.65	0.03	32.80	0.99
Bonaparte FLNG (Kinmore Group)	Bonaparte/Petrel	7	2.88(1.19)	1.60(1.1)	1.60(1.1)	53.00	1.46
Greater Poseidon (Plover)	Browse	8	0.48(0.28)	12.51(2.68)	12.51(3.48)	49.30	0.39
Greater Browse (Plover)	Browse	20	0.53(0.32)	9.44(2.68)	9.44(2.68)	42.60	1.98
Greater Scarborough Gas (Barrow Group)	Northern Carnarvon/Exmouth Plateau	9	4.35(0.85)	0.05(0.05)	0.05(0.05)	173.70	2.18
Equus	Northern Carnarvon/Exmouth Plateau		2.28		0.011	207.30	0.22
Mentorc (Lower Barrow Group)	Northern Carnarvon/Exmouth Plateau	3	6.46(0.07)	0.34(0.01)	0.007(0.0002)	900.60	
Chester, Glencoe (Murat, Siltst, Jansz Mungaroo)	Northern Carnarvon/Exmouth Plateau	18	1.59(0.97)	2.23(1.87)	0.012(0.0003)	136.60	

[a]*Average (standard deviation).*

[b]*Remaining resources × average He%/100; 35.515 BCF = 1 Bm^3 (1000 MMm^3).*

[c]*Assumed same as Queensland Curtis LNG.*

[d]*Based on ultimate recoverable natural of 3.4 TCF.*

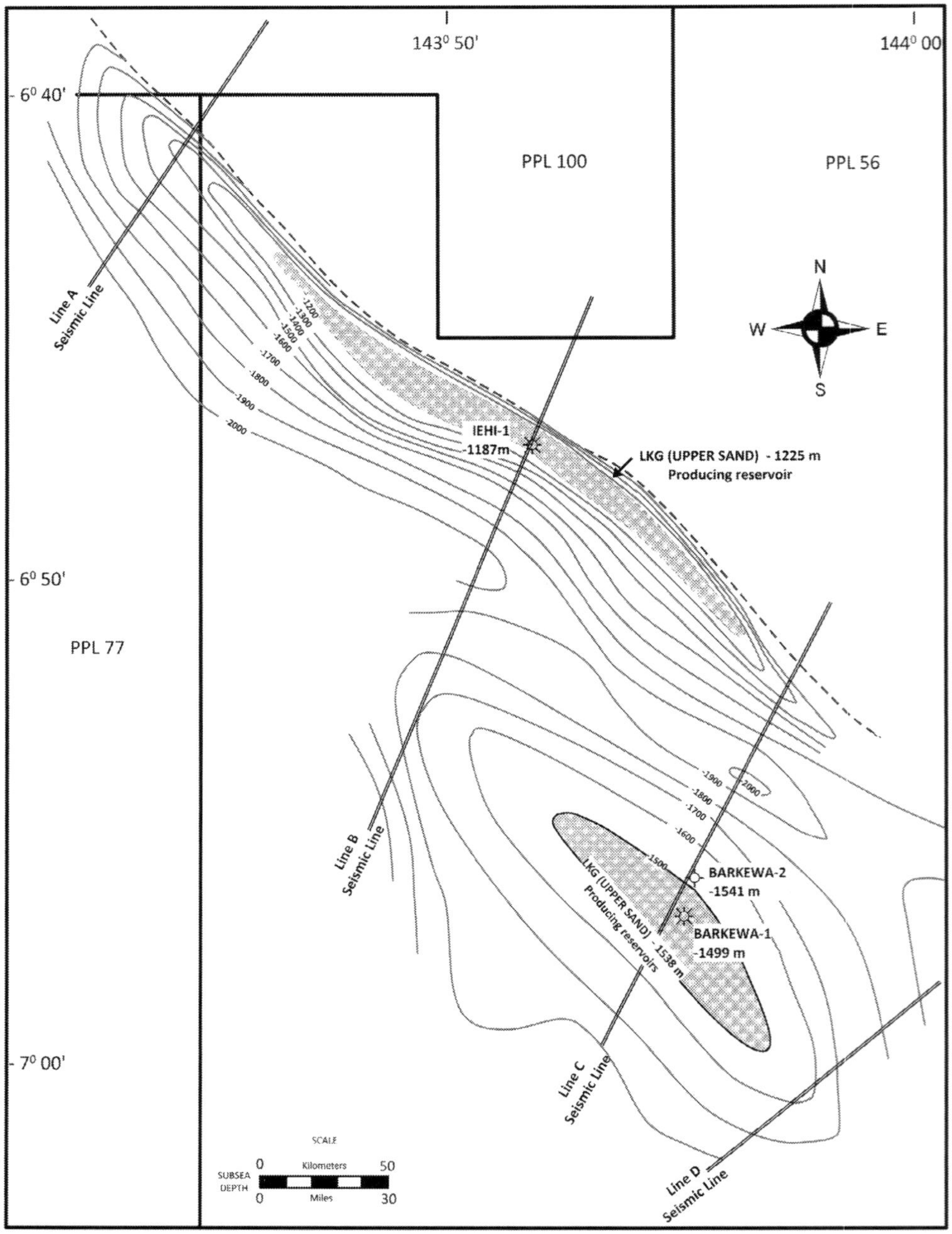

FIGURE 3.119 Structure on top of the Barikewa and Iehi fields in Papua New Guinea (Bird and Seggie, 1990).

sands and shales. The net pay is 49–86.5 ft (15–26.2 m) and has 9%–17.6% porosity (Bird and Seggie, 1990). Whether the He will eventually be extracted is unknown but it does suggest that this and other fields in the area may be potential sources of the molecule.

3.21 South and Central America

There is very little information on the exploration for He and CO_2 resources in South and Central America. Many of the oil and gas productive basins are young and not associated with Precambrian cratons. Argentina is making an effort to find and extract He and in 2018 put out a report in an attempt to identify potential areas to look for the resource (BID, 2018). According to a study of analogous plays, the Claromecó and Salado basins are the only intracratonic basins that are associated with Precambrian granite basement and the most likely to be associated with crustal generated He. Neither of these basins has, to date, had commercial hydrocarbons production, minor gas shows and no indication of significant He (BID, 2018). He analysis that has been done indicates very low He percent except at La dorsal (Neuquén Basin) and in southern Mendoza (BID, 2018). Within La dorsal, the fields known as Octógono Fisca have 0.2%–0.3% of He and are mature in terms of production (BID, 2018).

3.22 The Earth's moon

Virtually all ^3He that is commercially manufactured on Earth is from the decay of tritium that is sourced from the US and Russian military. ^3He is projected to be a major component of fission reactors because as discussed in Chapter 1, it is the most viable and stable component for fusion reactions. A major potential source of ^3He for fusion is the moon (Dilmare, 2004; Slyuta et al., 2007; Williams, 2007; Beike, 2011; Schmitt, 2013, 2014; Ambrose and Cutright, 2017). ^{3}Helium is generated by solar processes and is ejected from the sun with several other light elements as the solar wind. The solar wind is typically composed of He, Ne, C, and N_2. The solar wind moves away from the sun in all directions and any objects it intersects will deposit He, Ne, C, and N_2 onto the object's surface. The lunar surface is perfect for collecting the ^3He from the solar wind as it has no atmosphere and no ongoing erosional, tectonic, or weathering processes other than meteoric impacts. The moon is considered to be the major source of ^3He that can be extracted for use in fusion reactors (Williams, 2007). ^{3}Helium is concentrated between 1.4 and 15 ppb in sunlit areas, and it is postulated the dark side of the moon and permanently shadowed regions contain concentrations as much as 50 ppb (Hedman, 2006; Slyuta et al., 2007). Kulcinski and Schmitt (1992) estimated there is a million metric tons of ^3He resource on the moon.

Moon geology, also known as selenology, is simplistic in comparison to the Earth (Fig. 3.120). Lunar geological studies of the moon are based on a combination of Earth-based telescope observations, measurements from orbiting spacecraft, moon samples, and geophysical data (Wilhelms, 1987). The NASA Apollo landings from 1969 to 1972 collected

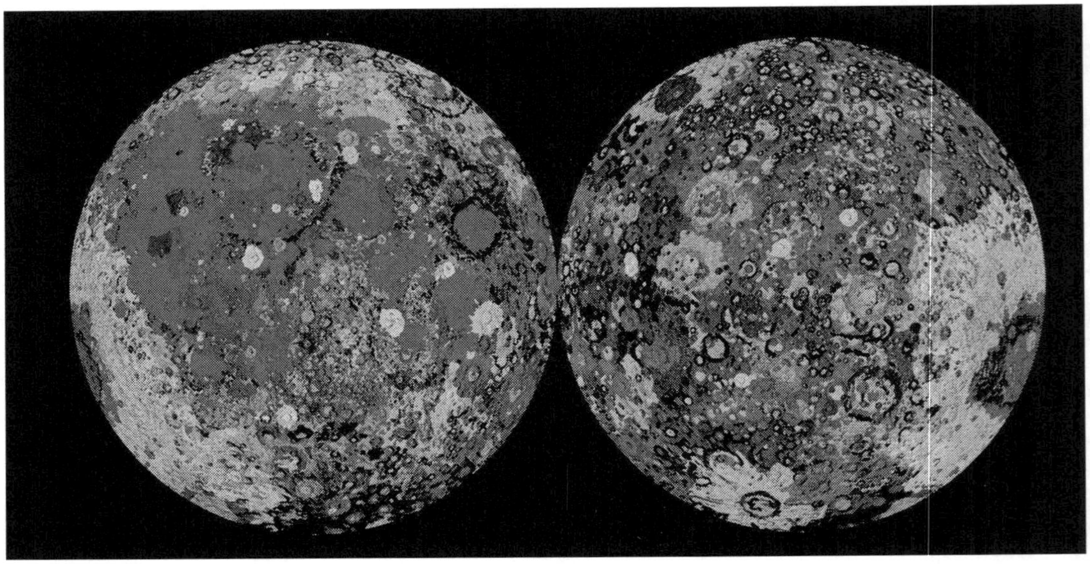

FIGURE 3.120 USGS geologic map of the moon (Fortezzoi et al., 2020).

samples from six locations and returned 380.96 kg (839.9 lb) of lunar and lunar soil to Earth. Three robotic Soviet Luna spacecrafts returned another 326 g (11.5 oz) from 1970 to 1976. The estimation of ^{3}He reserves and resources on the moon is based on the lunar rock samples collected during these expeditions (Wilhelms, 1987).

The lunar landscape is divided into distinct geomorphic units, highlands and lowlands (mare). The mare is basaltic in nature representing volcanic flood flow eruptions that are silica poor and magnesium and iron rich (Wilhelms, 1987). The highlands are anorthosites, rich in aluminum and silica, and older than the lowlands as defined by the greater number of craters. The lunar surface has undergone over 3–3.5+ billion years of erosion from meteorite impacts that injected material across the landscape forming what is called the regolith (Wilhelms, 1987).

The regolith varies from 2 to 20 m (6.6–66 ft) in thickness, is very fine grained, and is composed of minerals from the original bedrock fused with glassy particles that were formed during the impacts. The approach to ^{3}He extraction has been discussed by Kulinski and Schmitt (1992) and Schmitt (2013) and that He^3 from the moon would supply the Earth with the element for decades if not centuries.

There obviously are a number of hurdles to mining ^{3}He from the moon and transporting it back to Earth (Fig. 3.121). First and foremost are the political issues of who can legally mine the moon. Japan, China, India, Europe Union, and other countries are developing space programs to establish bases with the sole purpose of mining the moon (Burke, 2019; Jamasmie, 2019; Kundaliya, 2019). The Chinese are purposefully landing on the dark side of the moon which reduces but not eliminates direct observation from Earth. The technical problems of mining may be decades away from solution. The type of extraction process could result in the largest strip mine(s) in the solar system. However, on a positive note there is no waste or radioactivity associated with the mining of ^{3}He. In 2009, a movie by Sony Pictures called

FIGURE 3.121 Artists depiction of moon mining (Nave, 2016).

"The Moon" uses mining of ^3He on the moon as backdrop for the film's plot. There is potential, based on how the moon collects ^3He, that a number of large asteroids may yield similar recoverable grade deposits. However, getting pure He back to Earth is a whole different issue and problem.

References

Adams, C.W., 1990. Jace and Moore-Jonson fields. In: Sonnenberg, S.A., Shannon, L.T., Rader, K., von Drehle, W.F., Martin, G.W. (Eds.), Morrow Sandstones. Rocky Mountain Association of Geologist, pp. 157–161.

Adams, J.G., Gonzales, D., Darrah, T., 2015. Application of Noble Gas Isotopic Signatures at McElmo Dome-DOE Canyon Field to Investigate CO_2 Source and System Characterization. American Association of Petroleum Geologists. Search and Discovery 41711.

Alberta Geological Survey, 2020. Helium Concentrations from Natural Gas Analysis Tests in Alberta. Info Series 153. https://ags.aer.ca/publication/inf-153.

Al-Siddiqi, A., Dawe, R.A., 1999. Qatar's oil and gas fields: a review. J. Petrol. Geol. 22 (4), 417436.

Allen, W.E., 1978a. Pinta dome springs. In: Fassett, J.E., Thomaidis, N.D., Matheny, M.L. (Eds.), Oil and Gas Fields of the Four Corners Area, vol. I. Four Corners Geological Society, pp. 85–89.

Allen, W.E., 1978b. Navajo springs. In: Fassett, J.E., Thomaidis, N.D., Matheny, M.L. (Eds.), Oil and Gas Fields of the Four Corners Area, vol. 1. Four Corners Geological Society, pp. 80–84.

Alsharhan, A.S., Nairn, A.E.M., 1997. Sedimentary Basins and Petroleum Geology of the Middle East, first ed. Elsevier, Amsterdam, p. 843.

Ambrose, W.A., Cutright, B.L., 2017. Living on the Moon: Lessons Ofr Mars. American Association of Petroleum Geologists. Search and Discovery Article #70279, 24 p.

Ambrose, G.J., Heough, J.P., Shortt, T., July 16–17, 2012. Ooraminna-2, a gas discovery in the neoproterozoic pioneer sandstone, eastern Amadeus Basin. In: Central Australian Basins Symposium III. Petroleum Exploration Society of Australia, Alice Springs, NT, 9 p.

Anderson, C.C., Hinson, H.H., 1951. Helium-bearing natural gases of the United States. USGS, Bull 486, 146 p.

Anderson, E.C., 1959. Carbon Dioxide in New Mexico, 1959, New Mexico Bureau of Mines and Mineral Resources, Circular 43, 13 p.

Avis, L.E., 1982. Left handed field. In: Crouch III, M.A., et al. (Eds.), Oil and Gas Fields of Colorado-Nebraska and Adjacent Areas. Rocky Mountain Association of Geologist, pp. 286–299.

Baars, D.L., 1983. Rattlesnake Mississippian. In: Fassett, J.E., Thomaidis, N.D., Matheny, M.L. (Eds.), Oil and Gas Fields of the Four Corners Area, vol. III. Four Corners Geological Society, pp. 1006–1007.

Baars, D.L., Stevenson, G.M., 1981. Tectonic evolution of the Paradox Basin Utah & Colorado. In: Geology of Paradox Basin. RMAG, Rocky Mountain Association of Geologist, pp. 23–31.

Bagley, M.E., Ouckette, J., Boardman, D., Watney, W.L., 2011. Lithofacies and Reservoir Assessment for the Thirteen Finger Limestone Hugoton Embayment, Kansas. American Association of Petroleum Geologists. Search and Discovery #50513.

Baines, S.J., Worden, R.H., 2004. The long-term fate of CO_2 in the subsurface: natural analogues for CO_2 storage. In: Baines, S.J., Worden, R.H. (Eds.), Geological Storage of Carbon Dioxide. Geological Society, pp. 59–85.

Ballentine, C.J., Burnard, P.G., 2002. Production, release, and transport of noble gases in the continental crust. Rev. Mineral. Geochem. 47, 481–538.

Ballentine, C.J., Lollar, B.S., 2002. Regional groundwater focusing of nitrogen and noble gases into the Hugoton-Panhandle giant gas field, USA. Geochem. Cosmochim. Acta 66, 2483–2497.

Barnard, P.C., Cooper, B.S., 1983. A Review of Geochemical Data Related to the Northwest European Gas Province, vol. 12(1), pp. 19–33. Geological Society, London, Special Publications.

Barrett, W.J., Hubley, M.D., M.D., 1969. The Madden Field Fremont County, Wyoming Geological Association, Wyoming, 21st Symposium on Tertiary Rocks of Wyoming, pp. 115–120.

Barry, P.H., Hilton, D.R., Fischer, T.P., De Moor, J.M., Mangasini, F., Ramirez, C., 2013. Helium and carbon isotope systematics of cold "mazuku" CO_2 vents and hydrothermal gases and fluids from Rungwe Volcanic Province, southern Tanzania. Chem. Geol. 339, 141–156.

Barton, M., Ratchford, M.E., 2020. Executive Summary-Petroleum Geochemistry: Insights into the Origin of Natural Gases and Fluid Hydrocarbons from the Western Snake River Plain, Idaho, Idaho Geological Survey, USGS, Grant G15AC00284: Reservoir Characterization & Petroleum Assessment of Miocene Sedimentary Rocks, Southwestern Idaho, 17 p.

Beams, R.J., 1982. Greenwood Topeka trend. In: Crouch III, M.A., et al. (Eds.), Oil and Gas Fields of Colorado-Nebraska and Adjacent Areas. Rocky Mountain Association of Geologist, pp. 217–220.

Beck, H.V., 1959. Geology and Ground-Water Resources of Kansas River Valley Between Wamego and Topeka Vicinity: State Geological Survey of Kansas Bulletin 135, 88 p.

Becker, T.P., Lynds, R., 2012. A geologic deconstruction of one of the world's largest natural accumulations of CO_2, Moxa Arch, southwestern Wyoming. Am. Assoc. Pet. Geol. Bull. 96 (9), 1643–1664.

Beike, D., 2011. Lunar Helium-3 Value Chain: Investment and Funding. American Association of Petroleum Geologists. Search and Discovery Article #80154, 14 p.

BID, 2018. Prospeccion de helioenreservporios de gas, FDG EnergiaEficiente. www.fdgsa.com.

Biggs, P., 1957. CO_2 at North and South McCallum. In: Munoz, R.R. (Ed.), Guidebook to the Geology of the North and Middle Parks Basin. Rocky Mountain Association of Geologist, pp. 115–118.

Bolyard, D.W., 1990. Upper Morrow "B" sandstone reservoir, flank field, Baca County, Colorado. In: Sonnenberg, S.A., Shannon, L.T., Rader, K., von Drehle, W.F., Martin, G.W. (Eds.), Morrow Sandstones. Rocky Mountain Association of Geologist, pp. 191–205.

Bon, R.L., 1999. Petroleum Geology of the Harley Dome Field, Grand County, Utah, Utah Geological Survey, OG-21.

Bird, K.J., Seggie, R., February 12–14, 1990. Barikewa and Iehi fields revisited. In: Petroleum Exploration in Papua New Guinea: Proceedings of the First PNG Petroleum Convention, Port Moresby, p. 565.

Boreham, C.J., Edwards, D.S., Poreda, R.J., Darrah, T.H., Zhu, R., Grosjean, E., Main, P., Waltenberg, K., Henson, P.A., 2018. Helium in Australian liquefied natural gas economy. AAPEA J. 58, 209–237.

Bowen, D.W., Weimer, P., 2003. Regional sequence stratigraphic setting and reservoir geology of Morrow incised-valley sandstones (lower Pennsylvanian), eastern Colorado and eastern Kansas. Am. Assoc. Petrol. Geol. 87 (5), 781–815.

Bowen, D.W., Talbott, J., Lageson, D.R., Jackson, F.D., Eby, D.E., DeVault, B., 2012. Montana Characterization Study - Comprehensive Report (Task 15). Montana State University, Bozeman, MT, p. 63.

Bowman, K., 2016. The History of Central Mississippi's Naturally Occurring CO_2 Fields. American Association of Petroleum Geologists, Explorer March.

Bowersox, J.R., 2019. Assessing the Potential Helium Resources in Central Kentucky. American Association of Petroleum Geologists. Search and Discovery #51573, 22 p.

Breedlovestrout, R.L., 2017. Hydrocarbon Potential of Arkosic Sands in the Weiser-Payette Basin, Southwest Idaho: A Discussion on Basin Fill, Subsurface Structure, and Syndepositional Maturation and Migration History. American Association of Petroleum Geologists. Search and Discovery Article #10904, 35 p.

British Columbia Ministry of Energy, Mines and Petroleum Resources (BMEMPR), 2006. Conventional Gas Play Atlas Northeast British Columbia. Petroleum Geology Publication 2006-01, 50 p. https://www2.gov.bc.ca/assets/gov/farming-natural-resources-and-industry/natural-gas-oil/petroleum-geoscience/oil-gas-reports/og_report_2006-1_nebc_atlas_part1.pdf.

Broadhead, R.F., Gillard, L., 2004. Helium in New Mexico: Geologic Distribution and Exploration Possibilities. New Mexico Bureau of Geology and Mineral resources. OFR No. 483, 62 p.

Broadhead, R.F., 1998. Natural accumulations of carbon dioxide in the New Mexico region – where are they, how do they occur and what are the uses for CO_2. Geology 20, 2–6.

Broadhead, R.F., Mansell, M., Jones, G., 2009. Carbon Dioxide in New Mexico- Geologic Distribution of Natural Occurrences. NMBGMR, OFR 514, 131 p.

Brown, A.A., 2010. Formation of High Helium Gases: A Guide for Explorationist. American Association of Petroleum Geologists. New Orleans Convention, Poster session.

Brown, A.A., 2017. Possible Origins for Low Thermal Maturity, High-Nitrogen Natural Gases. American Association of Petroleum Geologists. Search and Discovery Article #42065.

Brown, A.A., 2019. Origin of helium and nitrogen in the Panhandle-Hugoton field of Texas, Oklahoma, and Kansas, United States. Am. Assoc. Pet. Geol. Bull. 103 (2), 369–403.

Brown, H.H., 1978. Beautiful mountain mississippian. In: Fassett, J.E. (Ed.), Oil and Gas Fields of the Four Corners Area, vol. III. Four Corners Geological Society, pp. 207–208.

Burke, J.H., June 13, 2019. China's New Wealth Creation Scheme: Mining the Moon, Nation Review. https://www.nationalreview.com/2019/06/china-moon-mining-ambitious-space-plans/.

Burwash, R.A., Cummings, G.L., 1974. Helium source-rock in southwestern Saskatchewan. Can. Petrol. Geol. Bull. 22 (4), 405–412.

Burwash, R.A., Cumming, G.L., 1976. Uranium and thorium in the Precambrian basement of western Canada. I. Abundance and distribution. Can. J. Earth Sci. 13 (2), 284–293.

Cantor, K.L., Stearns, D.B., Geesaman, R.C., Wilson, J.L., 1993. Paleostructural and related paleokarst controls on reservoir development in the lower Ordovician Ellenburger group, val Verde Basin, Texas, in Paleokarst Related Hydrocarbon Reservoirs. Society of Sedimentary Geology, pp. 61–100. Core Workshop No. 18, C.

Caffee, M.W., Hudson, G.U., Velsko, C., Huss, G.R., Alexander, E.C., Chivas, A.R., 1999. Primordial noble cases from Earth's mantle: identification of a primitive volatile component. Science 285, 2115–2118.

Cappa, J., Rice, D., 1995. Carbon Dioxide in Mississippian Rocks of the Paradox Basin and Adjacent Areas, Colorado, Utah, New Mexico and Arizona. USGS, Bulletin No. 2000-H.

Carr, C.J., 2017. The seismic threat to the Gibe III Dam: a disaster in the waiting. In: River Basin Development and Human Rights in Eastern Africa, - a Policy Crossroads, pp. 43–52. Creative Commons, Chap. 3.

Carver Jr., H.S., 1956. Keyes field. In: Guidebook Panhandle of Oklahoma, Northeastern New Mexico, South-Central Colorado. Oklahoma City Geological Society, pp. 130–135.

Carpen, T.R., 1957. North McCallum field, Jackson county, Colorado. In: Guidebook to the Geology of the North and Middle Parks Basin. Rocky Mountain Association of Geologist, pp. 109–112.

Cassidy, M., 2006. Occurrence and Origin of Free Carbon Dioxide Gas Deposits in the Earth's Continental Crust (Ph.D.). University of Houston, Texas.

Chen, S., 2020. China Opens First Large-Scale Helium Plants as it Tries to Reduce Reliance on US Imports, South China Morning Post, 2/7/2020, China Opens First Large-Scale Helium Plant as it Tries to Reduce Reliance on US Imports (msn.com).

Clair, J.R., Bradish, B.B., 1956. Model Dome Gas Field. Las Animas Arch Co., CO, Rocky Mountain Association of Geologist, Guidebook to the Geology of the Raton Basin Guidebook, p. 80.

Clark, C.R., 1978. Lisbon field. In: Fassett, J.E. (Ed.), Oil and Gas Fields of the Four Corners Area, vol. II. Four Corners Geological Society, pp. 662–665.

Craddock, W.H., Blondes, M.S., DeVera, C.A., Hunt, A.G., 2017. Mantle and crustal gases of the Colorado Plateau: geochemistry, sources, and migration pathways. Geochem. Cosmochim. Acta 213, 364–374.

Czapigo-Czapia, M., Brzezinski, D., 2020. The balance of mineral resources deposits in Poland as of 31 XII 2020. Polish Geological Institute National Research Institute. http://geoportal.pgi.gov.pl/surowce/energetyczne/hel/2020.

Dai, J., Song, Y., Dai, C., Wang, D., 1996. Geochemistry and accumulation of carbon dioxide gases in China. Am. Assoc. Pet. Geol. Bull. 80 (10), 1615–1626.

Dai, J., Ni, Y., Qin, S., Huang, S., Gong, D., Liu, D., Feng, Z., Peng, W., Han, W., Fang, C., 2017. Geochemical characteristics of He and CO_2 from the Ordos (cratonic) and Bohaibay (rift basins. Chem. Geol. 468, 192–213.

Danabalan, D., 2017. Helium: Exploration Methodology for a Strategic Resource (Ph.D.). Durham University, United Kingdom, 296 p.

Danie, T.C., 1978. Dinah-Bi-Keyah. In: Fassett, J.E., Thomaidis, N.D., Matheny, M.L. (Eds.), Oil and Gas Fields of the Four Corners Area, Four Corners Geological Society, vol. 1, pp. 73–76.

Davison, I., Steel, I., Taylor, M., Beirne, E.O., Faull, T., 2014. Exploration in the east African Rift system. GeoExPro 11 (5), 8 p.

De Bruin, R.H., 1991. Wyoming's Carbon Dioxide Resources. Wyoming Geological Survey, OFR 91-6, 23 p.

De Bruin, R.H., 1995. Helium resources of Wyoming. In: Resources of Southwestern Wyoming: Field Conference Guidebook. Wyoming Geological Association, pp. 191–201.

De Voto, R.H., 1980a. Mississippian stratigraphy and history of Colorado. In: Kent, H.C., Porter, K.W. (Eds.), Colorado Geology. Rocky Mountain Association of Geologist, pp. 57–70.

De Voto, R.H., 1980b. Pennsylvanian stratigraphy and history of Colorado. In: Kent, H.C., Porter, K.W. (Eds.), Colorado Geology. Rocky Mountain Association of Geologist, pp. 71–102.

Dilmare, D., October 2004. Mining the Moon. Popular Mechanics.

Dobbin, C.E., 1935. Geology of Natural Gas, Rich in Helium, Nitrogen, Carbon Dioxide and Hydrogen Sulfide. American Association of Petroleum Geologists, pp. 1053–1072. Special Volume 7.

Dockrill, B., Shipton, Z.K., 2010. Structural controls from a natural gas CO_2 geologic storage site, Central Utah, U.S.A. J. Struct. Geol. 32 (11), 1768–1782.

Dubois, M.K., Byrnes, A.P., Bhattacharya, S., Bohling, G.C., Doveton, J.H., Barba, R.E., 2007. Hugoton Geomodel Final Report, Kansas Geological Survey 2007-06-1.

Dutton, S.P., Garrett Jr., C.M., 1989. PN-13, Pennsylvanian fan-delta sandstone, Anadarko basin. In: Kosters, E.C., Bebout, D.G., Seni, S.J., Garrett Jr., C.M., Brown Jr., L.F., Hamlin, H.S., Dutton, S.P., Ruppel, S.C., Finley, R.J., Tyler, N. (Eds.), Atlas of Major Texas Gas Reservoirs. Gas Research Institute, pp. 146–147.

Eppink, J., Heidrick, T.L., Alvarado, R., Marquis, M., 2014. Subsurface Sources of CO_2 in the Contiguous United States. National Energy Technology Laboratory. DOE/NETL-2014/1637, 90 p.

Everett, M.A., 2012. The Precambrian Petroleum Systems of Eastern Siberia: Evidence from Oil Geochemistry and Basin Modeling. American Association of Petroleum Geologists. Search and Discovery Article 10433, 19 p.

Foos, A., 1999. Geology of the Colorado Plateau. Geology Department, University of Akron, National Park Service.

Fortezzoi, C.M., Spudis, P.D., Harrel, S.L., 2020. Release of the Digital Unified Global Geologic Map of the Moon at 1:5,000,000- SCALE. USGS.

Foster, R.W., Jensen, J.G., 1972. Carbon Dioxide in Northeastern New Mexico: New Mexico Geological Society, Guidebook to 23rd Field Conference, pp. 192–200.

Frye, J.C., Byron Leonard, A., 1952. Pleistocene Geology of Kansas. Bulletin 99. State Geological Survey of Kansas, Lawrence, 230 p.

Gage, B.D., Driskill, D.L., 2003. Analysis of Natural Gas 1998–2001. USBLM Technical Note 412, 179 p.

Gage, B.D., Driskill, D.L., 2005. Analysis of Natural Gases 2002–2004. USBLM Technical Note 418, 252 p.

Gao, D., Shumaker, R.C., 1996. Subsurface geology of the Warfield structures in southwestern West Virginia: Implications for tectonic deformation and hydrocarbon exploration in the Central Appalachian Basin. Am. Assoc. Petrol. Geol. Bull. 80 (8), 1242–1261.

Garvin, R.F., 1978. Slick Rock Dakota. In: Fassett, J.E., Thomaidis, N.D., Matheny, M.L. (Eds.), Oil and Gas Fields of the Four Corners Area, Four Corners Geological Society, vol. III, pp. 497–499.

Gerhard, L.C., 2004. A new look at an old petroleum province: Kansas Geological Survey, Current Research in Earth Sciences. Bulletin 250 (1), 27.

Geoscience Australia, 2021a, www.ga.gov.au/scientific-topics/energy/province-sedimentary-basin-geology/petroleum/offshore-southern-australia/otway.

Geoscience Australia, 2021b, www.ga.gov.au/scientific-topics/energy/province-sedimentary-basin-geology/petroleum/offshore-southern-australia/gippsland#heading-4.

Gerling, C.R., 1983. McElmo Dome Leadville carbon dioxide field, Colorado. In: Fassett, J.E. (Ed.), Oil and Gas Fields of the Four Corners Area, vol. III. Four Corners Geological Society, pp. 735–739.

GeTech, 2002. The Basin Development of Arabia, Non-exclusive Study.

Gilfillan, S.M., 2006. Deep Magmatic Degassing and the Colorado Plateau Uplift. University of Manchester (Ph.D. thesis), 141 p.

Gilfillan, S.M.V., Ballentine, C.J., Holland, G., Blagburn, D., Lollar, B.S., Stevens, S., Schoell, M., Cassidy, M., 2008. The noble gas geochemistry of natural CO_2 gas reservoirs from the Colorado Plateau and Rocky Mountain provinces, USA. Geochim. Cosmochim. Acta 72, 1174–1198.

Goff, W.T., Hundley-Goff, E., 2014. Spelunker Field. In: Rogers, J.P., Milne, J.J., Cumella, S.P., Dubois, D.P., Lillis, P.G. (Eds.), Oil and Gas Fields of Colorado. Rocky Mountain Association of Geologists, pp. 335–338.

Gorham, F.D., 1975. Tectogenesis of the central Colorado Plateau Aulacogen. In: Four Corners Geological Society – Canyonlands Country, Eighth Field Conference, 1975, Four Corners Geological Society, 8th Field Conference, pp. 211–216.

Guangya, J., 2019. Tectonic Evolution and Hydrocarbon Distribution of Basins in Africa. American Association of Petroleum Geologists. Search and Discovery Article #30602.

Haines, P.W., Allen, H.J., 2019. Hydrocarbon and helium prospectivity of the Amadeus and Murraba basins in Western Australia. In: Keep, M., Moss, S.J. (Eds.), The Sedimentary Basins of Western Australia V: Proceedings of the Petroleum Exploration Society of Australia Symposium, Perth, WA, 2019, 17 p.

Halford, D.T., 2018. Isotopic Analysis of Helium from Wells Located in the Four Corners Area, Southwestern US, Colorado School of Mines PHD Dissertation, 376 p.

Hamal, J.E., Driskill, D.L., 1996. Analysis of Natural Gases - 1994–1995. USBLM Technical Note 399, 76 p.

Han, G., Han, W.S., Kim, K.-Y., Park, J.G., Piao, J., Yun, T.K., 2019. Role of fault structures and regional formations on CO_2 migration and distribution in shallow saline aquifer in Green River, Utah. J. Hydrol. 570, 786–801.

Hawkins, C.M., Baugh, R.A., 2006. Madden Field: a growing giant in the Wind River Basin, Wyoming, Rocky Mountain Association of Geologist. Mt. Geol. 43 (3), 177–180.

Heath, J.E., Lachmar, T.E., Evans, J.P., Kolesar, P.T., Williams, A.P., 2009. Hydrogeochemical characterization of leaking, carbon dioxide-charged fault zones in East-Central Utah, with implications for geologic carbon storage. Geophys. Monogr. 183, 147–158.

Hedman, E.R., January 16, 2006. A Fascinating Hour with Gerald Kulcinski, the Space Review.

Henderson, L.B., 1985. Rudyard field. In: Montana Geological Society Oil and Gas Fields Symposium, vol. 2, pp. 995–998.

Herbaly, E., 1973. Petroleum Geology Sweetgrass Arch, Alberta, Wyoming Geological Association, Earth Science Bulletin March, 21 p.

Hills, J.M., 1968. Gas in Delaware and Val Verde Basin, West Texas and Southeastern New Mexico, Association of Petroleum Geologists, Natural Gases of North America, Volume Two, M9, pp. 1394–1432.

Holmquest, H.J., 1965. Deep Pays in Delaware-Val Verde Basins, AAPG, Memoir, Fluids in Subsurface Environments, M4, pp. 257–279.

Hoppe, W.F., 1978. Table Mesa Mississippian (Gas), Table Mesa Pennsylvanian "C". In: Fassett, J.E. (Ed.), Four Corners Geological Society, Oil and Gas Fields of the Four Corners Area, vol. III, pp. 1,046–1049.

Huang, B., Tian, H., Huang, H., Yang, J., Xiao, X., Li, L., 2015. Origin and accumulations of CO_2 and its natural displacement oils in the continental margin basins, northern South China Sea. Am. Assoc. Petrol. Geol. 99 (7), 1349–1369.

Hutcheon, I., 1999. Controls on the distribution of non-hydrocarbon gases in the Alberta basin. Can. Petrol. Soc. Bull. 47 (4), 573–593.

Irwin, J.S., 1923. Oil and Gas Possibilities of Eastern Alberta, American Association of Petroleum Geologists. Bulletin 7 (2), 147–158.

Jacobs, E.K., 1998. Non-hydrocarbon and Hydrocarbon Gas Association with Hydrocarbon Associations with Hydrocarbon Reservoirs along the Las Animas Arch, Southeastern Colorado. Colorado School of Mines (MS thesis), 206 p.

Jamasmie, C., June 11, 2019. India Unveils Spacecraft that Will Look for Energy Sources on the Moon, Education Energy Asia Canada Europe Gold Rare Earth. https://www.mining.com/india-unveils-spacecraft-that-will-look-for-water-energy-sources-in-the-moon/.

James, T.C., 1967. Thermal springs in Tanzania. Trans. Appl. Earth Sci. 76 (section B), B1–B18.

Jenden, P.D., Newell, K.D., Kaplan, I.R., Watney, W.L., 1988. Composition and stable-isotope geochemistry of natural gases from Kansas, Midcontinent, U.S.A. Chem. Geol. 71, 117–147.

Johnson, E.G., 2012. Helium in northeastern British Columbia. In: Geoscience Reports 2013, BC Ministry of Natural Gas Development, pp. 45–52.

Johnson, K.S., 1989. Geologic evolution of the Anadarko Basin. In: Johnson, K.S. (Ed.), Anadarko Basin Symposium. Circular 90. Oklahoma Geological Survey, Norman, pp. 3–12.

Kendrick, F.E., McLaughlin, H.C., 1929. Relation of petroleum accumulation to structure, Petrolia Field, Clay County, Texas. In: American Association of Petroleum Geologists, Structure of Typical American Oil Fields, vol. 2. Special Publication 4, pp. 543–555.

Konert, G., Afifi, A.M., Al-Hajri, S.A., Droste, H.J., 2001. Paleozoic stratigraphy and hydrocarbon habitat of the Arabian plate. GeoArabia 6, 407–442.

Kozarek, R.J., 1990. Frontera Field, Cheyenne County, Colorado. In: Sonnenberg, S.A., Shannon, L.T., Rader, K., von Drehle, W.F., Martin, G.W. (Eds.), Morrow Sandstones, Rocky Mountain Association of Geologist, pp. 153–156.

Krivanek, C.M., 1978. Elmo Field. In: Oil and Gas Fields of the Four Corners Are: Four Corners Geological Society, pp. 148–1951.

Krivanek, C.M., 1981. New fields and exploration drilling Paradox Basin Utah and Colorado. In: Geology of the Paradox Basin – Field Conference. RMAG, pp. 77–81.

Kulinski, G.L., Schmidt, H.H., 1992. Fusion power from lunar resources. Fusion Technol. 21, 2221–2229.

Kundaliya, D., January 22, 2019. European Union Considers Mining Mission to Launch in 2025. https://www.computing.co.uk/news/3069703/european-union-considers-moon-mining-mission-to-launch-in-2025.

Lake, J., Whittaker, S., 2006. Occurrences of CO_2 within southwest Saskatchewan - Natural analogues to Weyburn CO_2 injection site. In: Summary of Investigations 2006, Saskatchewan Geological Survey, Miscellaneous Reports 2006-4.1, 14 p.

Lang, M., Afanasyev, I., Slutskiy, B., Schmid, F., 2017. Monetizing Gas of a Giant High Helium and Nitrogen Gas Reservoir – Amur Gas Processing Plant, F13-Sustainable Value Chains for Unconventional Gas, 22nd World Petroleum Congress, Istanbul, 10 p.

Lange, J.K., Wellborn, R.E., 1985. Seismic profile: North Park Basin, RMAG. In: Greis, R., Dyer, R.C. (Eds.), Seismic Exploration of the Rocky Mountain Region. Rocky Mountain Association of Geologist, pp. 239–246.

Lauth, R.E., 1978a. Slick Rock Dakota. In: Fassett, J.E. (Ed.), Oil and Gas Fields of the Four Corners Area, Four Corners Geological Society, vol. III, pp. 497–499.

Lauth, R.E., 1978b. Table Mesa Dakota. In: Fassett, J.E. (Ed.), Oil and Gas Fields of the Four Corners Area, Four Corners Geological Society, vol. III, pp. 507–509.

Lee, H., 1962. Technical and Economic Aspects of Helium Production in Saskatchewan, Saskatchewan Department of Mineral Resources. 39041-Report-72, 41 p.

Li, M., Wang, T., Liu, J., Lu, H., Wu, W., Gao, L., 2008. Occurrence and origin of carbon dioxide in the Fushan Depression, Beibuwan Basin, South China Sea. Mar. Petrol. Geol. 25 (6), 500–513.

Lollar, B.S., Weise, S.M., Frape, S.K., Barker, J.F., 1994. Isotopic constraints on the migration of hydrocarbon and helium gases of southwestern Ontario. Can. Petrol. Geol. Bull. 42 (3), 283–295.

Ma, L., 2009. Noble Gases Dissolved in Groundwaters of the Michigan Basin: Implications for Paleoclimatology, Hydrogeology, Tectonics and Mantle Geochemistry (Ph.D. Dissertation). University of Michigan, 229 p.

Magloire, P.R., 1970. Triassic gas field of Hassi er R'Mel, Algeria, American Association of Petroleum Geologists. In: Geology of Giant Petroleum Fields, Memoir, vol. 14, pp. 489–502.

Mark, A., 1961. South McCallum Field. In: Oil and Gas Field Volume, Colorado-Nebraska. Rocky Mountain Association of Geologist, p. 186.

Mason, J.W., 1968. Hugoton Panhandle Field, Kansas, Oklahoma and Texas. American Association of Petroleum Geologists, pp. 1539–1547. Natural Gases of North America, Memoir, Chapter 9.

Maynard, J., 1978a. Hogback Mississippian. In: Fassett, J.E., Thomaidis, N.D., Matheny, M.L. (Eds.), Oil and Gas Fields of the Four Corners Area, Four Corners Geological Society, vol. 1, p. 333.

Maynard, J., 1978b. Hogback Pennsylvanian. In: Fassett, J.E., Thomaidis, N.D., Matheny, M.L. (Eds.), Oil and Gas Fields of the Four Corners Area, Four Corners Geological Society, vol. 1, p. 334.

McKenny, J.W., Masters, J.A., 1968. Dineh-Bi-Keyah Field, Apache County, Arizona, American Association of Petroleum Geologists. Bulletin 52 (10), 2045–2057.

McMurty, R.P., 1982. Brown-Bassett. In: Selected Oil and Gas Fields in West Texas Symposium, West Texas Geological Society, Symposium V. 1, II and III, pp. 111–124.

Merewether, E.A., 1987. Oil and Gas Plays of the Las Animas Arch, Southeastern Colorado: U.S. Geological Survey Open-File Report 87-450D, 22 p.

Merrill, M.D., Hunt, A.G., Lohr, C.D., 2014. Noble gas geochemistry investigation of high CO_2 natural gas at the La Barge Platform, Wyoming, USA. Energy Proc. 63, 4186–4190.

Merrill, M.D., Hunt, A., 2017. Updated Regional and Field Scale He Accumulation Geochemistry, La Barge Platform, WY. American Association of Petroleum Geologists. Search and Discovery 90301.

MHS, 2019, https://www.renergen.co.za/mha-final-report/.

MHS, 2020, https://www.renergen.co.za/sproule-evaluation-of-certain-helium-prospective-resources-on-the-tetra4-virgina-gas-project-july-2020/.

Miller, R.R., 1968. Geology of Otis-Albert Field, Rush and Barton Counties., Kansas, AAPG. In: Memoir 9 Natural Gases of North America, Volume Two, Chapter G, pp. 1588–1615.

Ministry of Natural Resources, 1983a. Gobles Pool, Structure on Top of the Shadow Lake, Fig D-5, Petroleum Section, Map.

Ministry of Natural Resources, 1983b. Gobles Pool, Cambrian Isopach, Fig D-6, Petroleum Section, Map.

Ministry of Natural Resources, 1983c. Innerkip and Innerkip East Pools, Cambrian Isopach, Fig D-9, Petroleum Section, Map.

Montgomery, S.L., 1996. Val Verde Basin: Thrusted Strawn (Pennsylvanian) carbonate reservoirs, Pakenham area. Am. Assoc. Pet. Geol. Bull. 80 (7), 987–998.

Moore, B.J., 1978. Analysis of Natural Gas. USBM Info Circular 8810, 136 p.

Moore, C.A., 1981. Geologic factors which may affect the occurrence of natural gas in the Oklahoma Panhandle. In: Reckamp, J.U. (Ed.), Shale Shaker Digest Volumes XVIII – XX, 1967-1970. Oklahoma City Geological Society, pp. 203–210.

Moore, B.J., 1982. Analysis of Natural Gas, 1917-80. USBM Info Circular 8870, 1074 p.

Moore, B.J., Sigler, S., 1985. Analysis of Natural Gas. USBM, Info Cir, 9096, 200 p.

Moore, J.P., Dinterman, P.A., McDowell, R.R., Sminchak, J.R., Daft, G.W., 2018. From the Tuscarora Sandstone to the Bubbles in Your Beverage: Naturally – Occurring CO2 in the Indian Creek Field, Kanawha Country, West Virginia. American Association of Petroleum Geologists. Eastern Section Joint Meeting, Abstract, October 7 to 11.

Morgan, C.D., 1994. Exploring for New Oil in Old Field, Salt Wash Field: A Case Study. Utah Geological Survey, OFR 307, 128 p.

Morgan, C.D., 2007. Structure, reservoir and characterization, and carbon dioxide resources of Farnham Dome Field. Carbon County, Utah, In Central Utah – Diverse geology of dynamic landscape, Utah Geological Association, p. 295–310.

Morgan, C.D., Chidsey Jr., T.C., 1991. Gordon Creek, Farnham Dome, and Woodside fields, Carbon and Emery counties, Utah. In: Chidsey Jr., T.C. (Ed.), Geology of East-Central Utah, UGS, pp. 301–309.

Morley, C.K., Ngenoh, D.K., Ego, J.K., 1999. In: Morley, C.K. (Ed.), Geoscience of Rift Systems – Evolution of East Africa, American Association of Petroleum Geologists Studies in Geology, No. 44, pp. 1–18.

Mtili, K.M., Kazimoto, E.O., Barry, P., 2019. In: Helium Gas Seeps and Processes Involved in the Formation of the Helium Gas Field within the Rukwa Rift Basin Tanzania, Abs. Tanzania Geological Society, 2019 Annual Meeting and Workshop, Dar Es Salaam, September 25th to 27th, p. 18.

Muffler, P., White, D., 1968. Origin of CO_2 in the Salton Sea Geothermal System, pp. 185–194. Southeastern California, U.S.A., XXIII International Geologic Congress, 1968, V. 17.

Nave, K., 2016. Space Mining Will Take a Giant Leap. https://www.wired.co.uk/article/space-mining-a-reality-in-2016.

Nesheim, T.O., Kruger, N.W., 2019. Helium Trends in North Dakota, GI 223. North Dakota Geological Survey, 12 p.

Newell, K.D., Corsair, S., Chafin, S., Pennybaker, K., 2003. Low-BTU Gas in the Permian Chase Group in the Ryersee Field in Western Kansas: A Case History where Technology Creates a Marketable Commodity. Kansas Geological Survey, OFR 2003-57, Poster.

Newell, K.D., 2021. Stratigraphic and Regional Trends in Chemistry and Quality of Natural Gas in Central and Western Kansas, Mid-continent, USA, Kansas Geological Survey (in press).

Nicolas, M.P.B., 2018. Summary of helium occurrences in southwestern Manitoba. In: Report of Activities 2016, Manitoba Growth, Enterprise and Trade. Manitoba Geological Survey, pp. 110–118.

Nordquist, J.W., Leskela, W., 1968. Natural gas in Sweetgrass arch area, Northwestern Montana, American Association of Petroleum Geologists. In: Natural Gas of North America, vol. 1, pp. 736–759. Chapter M9.

Norton, J.A., 1978. Salt Wash. In: Fassett, J.E. (Ed.), Oil and Gas Fields of the Four Corners Area, Volume II: Four Corners Geological Society, pp. 697–699.

Nuttall, W.J., Clarke, R.C., Glowacki, D.G., 2012. The Future of Helium as a Natural Resource. Routledge Publishing, 376 p.

Oty, J.W., 2014. Clyde field. In: Rogers, J.P., Milne, J.J., Cumella, S.P., Dubois, D.P., Lillis, P.G. (Eds.), Oil and Gas Fields of Colorado, Rocky Mountain Association of Geologist, pp. 68–73.

Ozimic, S., 2016. Non-petroleum gas accumulations, Gippsland, Bass and Otway basins. In: Glenie, R.C. (Ed.), Second South-Eastern Australia Oil Exploration Symposium, November 14–15, Melbourne, pp. 375–382.

Palmer, N., Ambrose, G., 2012. Mount Kitty Prospect, Southern Amadeus Basin. In: Keep, M., Moss, S.J. (Eds.), The Sedimentary Basins of Western Australia V, Proceedings of Petroleum Exploration Society of Australia Symposium, Perth, WA, 10 p.

Pan American, 1964. Well Files for Suniland State Unit A1, State of Utah. https://oilgas.ogm.utah.gov/oilgasweb/live-data-search/lds-files/files-lu.xhtml.

Parker, J.M., 1981. Lisbon Field Area, San Juan County, Utah, Rocky Mountain Association of Geologist. Geology of Paradox Basin Field Conference, pp. 89–100.

Petroleum Agency SA, 2012, http://info.matchdeck.com/hubfs/Petroleum_Potential_of_the_Karoo_Basins5.pdf.

Pfeiffer, L., Wanner, C., Lweicki, J., 2018. Unraveling the dynamics of magmatic CO_2 degassing at Mammoth Mountain, California. Earth Planet Sci. Lett. 484, 318–328.

Pierce, A.P., Gottand, G.B., Mytton, J.W., 1964. Uranium and helium in the Panhandle gas field, Texas, and adjacent areas. In: USGS Prof. Pap., vol. 454-G, 57 p.

Pippin, L., 1968. Panhandle-Hugoton field, Texas-Oklahoma-Kansas—the first fifty years. In: Halbouty, M.T. (Ed.), Geology of Giant Petroleum Fields. American Association of Petroleum Geologists Memoir 14, Tulsa, pp. 204–222.

Polish Geological Institute National Research Institute, 2017. Oil and Gas Fields in Poland as of 2019-12-31, Map.

Potter, D., 2010. Relationships of Cambro-Ordovician Stratigraphy to Paleotopography on the Precambrian Basement, Williston Basin, Saskatchewan Geological Society, Saskatchewan and Northern Plains Oil & Gas Symposium, 2006, pp. 63–73.

Rahimpour-Bonab, H., 2007. A procedure for appraisal of a hydrocarbon reservoir continuity and quantification of its heterogeneity. J. Petrol. Sci. Eng. 58 (1), 1–12.

Rahimpour-Bonab, H., Esrafili-Dizaji, B., Tavakoli, V., 2010. Dolomitization and anhydrite precipitation in Permo-Triassic carbonates at the South Pars Gas Field, offshore Iran: controls on reservoir quality. J. Petrol. Geol. 33 (1), 43–66.

Rapatskaya, L.A., Tonikh, M.E., Ustyuzhanin, A.O., 2020. Natural reservoir as a geological body for storing helium reserves. IOP Conf. Ser. Earth Environ. Sci. 408, 5 p.

Rascoe Jr., B., 1968. Permian System in Western Mid-Continent, vol. 5 (3), pp. 127–138. TRocky Mountain Association of Geologist The Mountain Geologists.

Rascoe Jr., B., 1988. Permian system in western mid-continent. In: Morgan, W.A., Babcock, J.A. (Eds.), Permian Rocks of the Midcontinent, pp. 3–12. Midcontinent SEPM Special Publication 1.

Rascoe Jr., B., Adler, F.J., 1983. Permo-Carboniferous hydrocarbon accumulations, mid-continent, U.S.A. Am. Assoc. Pet. Geol. Bull. 67 (6), 979–1001.

Rauzi, S.L., 1999. Carbon Dioxide in the St. John's – Springerville Area, Apache County, Arizona. Arizona Geological Survey, OFR 99-2.

Rauzi, S.L., 2003. Review of Helium Production and Potential in Arizona. Arizona Geological Survey, OFR 03-05, 30 p.

Reid, J., Hunt, A.G., Ellis, G.S., Taylor, K.B., Patterson III, O.F., 2015. Noble Gas Isotopes, Major Element Isotopes, and Gas Composition from the Cumnock Formation: Sanford Sub-Basin, Deep River Basin, Lee County, North Carolina, U.S.A. American Association of Petroleum Geologists. Search and Discovery Article #80492, 4 p.

Reutinger, C.A., 1968. Natural gas in Permian and Cisco Beds of Red River and Electra Arch, American Association of Petroleum Geologists. In: Memoir 9, Natural Gas of North America, vol. 2, pp. 1455–1458.

Riggs, E.A., 1978. Tocito Dome Mississippian North. In: Fassett, J.E. (Ed.), Oil and Gas Fields of the Four Corners Area, Volume II Four Corners Geological Society, pp. 519–530.

Rogers, G.S., 1921. Helium Bearing Natural Gas, Kansas Geological Survey Professional Paper 121, 107 p.

Rogers, J.P., 2014. Arrowhead gas field. In: Rogers, J.P., Milne, J.J., Cumella, S.P., Dubois, D.P., Lillis, P.G. (Eds.), Oil and Gas Fields of Colorado. Rocky Mountain Association of Geologist, pp. 20–23.

Ross Jr., R.J., Tweto, O., 1980. Lower Paleozoic sediments and tectonics in Colorado, in Geology of Colorado Symposium. RMAG, pp. 47–56.

Roth, G., 1983. Sheep Mountain and Dike Mountain fields, Huerfano County, Colorado. In: Fassett, J.E. (Ed.), Oil and Gas Fields of the Four Corners Area, Volume II: Four Corners Geological Society, pp. 740–744.

Saskatchewan Ministry of the Economy, 2014. Stratigraphic Correlation Chart; Saskatchewan Ministry of Energy and Resources. http://www.economy.gov.sk.ca/stratigraphiccorrelationchart.

Schmitt, H.H., 2013. Lunar Helium-3 energy resources. In: Ambrose, W.A., Reilly, J.F., Peters, D.C. (Eds.), Energy Resources for Human Settlement in the Solar System and Earth's Future in Space, American Association of Petroleum Geologists Memoir 101., pp. 33–51.

Schmitt, H.H., 2014. Synthesis of Current Data on Helium and Hydrogen Concentration in Lunar Regolith. American Association of Petroleum Geologists. Search and Discovery #70162, 26 p.

Sellars, R.T., 2014. Bledsoe Ranch. In: Rogers, J.P., Milne, J.J., Cumella, S.P., Dubois, D.P., Lillis, P.G. (Eds.), Oil and Gas Fields of Colorado. Rocky Mountain Association of Geologist, pp. 44–46.

Shannon, C.W., 1962. Oil and gas developments in Utah and Nevada in 1961. Am. Assoc. Pet. Geol. Bull. 46 (6), 975.

Sharland, P.R., Archer, R., Casey, D.M., Davies, R.B., Hall, S.H., Heward, A.P., Horbury, A.D., Simmons, M.D., 2001. Arabian Plate Sequence Stratigraphy: GeoArabia Special Publication No. 2, Gulf PetroLink, Bahrain, 371 p.

Sims, P.K., Bankey, V., Finn, C.A., 2001. Preliminary Precambrian Basement Map of Colorado – a Geologic Interpretation of an Aeromagnetic Anomaly Map, USGS, OFR 01-0364.

Slyuta, E.N., Abdrakhimov, A.M., Galimov, E.M., 2007. Lunar and Planetary Science XXXVIII (1338), 2175.

Smouse, 1993. Salt Wash. In: Hill, B.G., Bereskin, S.R. (Eds.), Oil and Gas Fields of Utah, Utah Geological Association Publication 22, Unpaginated.

Sorenson, R.P., 2005. A dynamic model for the Permian Panhandle and Hugoton fields, western Anadarko Basin. Am. Assoc. Petrol. Geol. 89 (7), 921–938.

Sorey, M.L., Evans, W.C., Kennedy, B.M., Farrar, C.D., Hainsworth, L.J., Hausback, B., 1998. Carbon dioxide and helium emissions from a reservoir of magmatic gas beneath Mammoth Mountain, California. J. Geophys. Res. Solid Earth 103 (87), 15 p.

Spencer, C.W., 1978a. Tohache Wash Area. In: Fassett, J.E., Thomaidis, N.D., Matheny, M.L. (Eds.), Oil and Gas Fields of the Four Corners Area, Four Corners Geological Society, vol. 1, pp. 92–93.

Spencer, C.W., 1978b. Tocito Dome Pennsylvanian "D". In: Fassett, J.E., Thomaidis, N.D., Matheny, M.L. (Eds.), Oil and Gas Fields of the Four Corners Area, Four Corners Geological Society, vol. 1, pp. 522–527.

Stilwell, D.P., 1989. CO_2 Resources of the Moxa Arch and the Madison Reservoir: Gas Resources of Wyoming 40th Annual Field Conference Guidebook. Wyoming Geological Association (1989), pp. 105–115.

Sonnenberg, S.A., van Drehle, W.F., 1990. Morrow gas composition in southeast Colorado. In: Sonnenberg, S.A., Shannon, L.T., Rader, K., von Drehle, W.F., Martin, G.W. (Eds.), Morrow Sandstones, Rocky Mountain Association of Geologist, pp. 233–235.

Stoddard, J.N., 1982. Brandon Field – Pennsylvanian, Morrow Formation and Lansing-Kansas City Group. In: Crouch III, M.C., et al. (Eds.), Oil and Gas Fields of Colorado and Nebraska and Adjacent Areas, Rocky Mountain Association of Geologist, p. 7881.

Studlick, J.R.J., Shaw, R.D., Basye, G.L., Ray, J.R., 1987. Carbon Dioxide Source Development, Northeast Jackson Dome, Mississippi. American Association of Petroleum Geologists. Search and Discovery Article #91038.

Superior Oil Company, 1982. Grey Ranch, in Selected Oil and Gas Fields in West Texas Symposium, West Texas Geological Society, Symposium V. 1, II and III, pp. 276–287.

Szamalelek, K., Szuflicki, M., Malon, A., Tyminski, 2017. Mineral Resources of Poland. Polish Geological Institute National Research Institute, 191 p.

Tarbet, L.A., 1971. Petroleum Potential of Imperial Valley, California, American Association of Petroleum Geologists, Memoir: Future Petroleum Provinces of the United States – Their Geology and Potential, pp. 384–391.

Tavakoli, V., Rahimpour-Bonab, H., Esrafili-Dizaji, B., 2011. Diagenetic controlled reservoir quality of South Pars gas field, an integrated approach. Comptes Rendes Geosci. 343, 55–71.

Tedesco, S.A., 2020. The Pennsylvanian Desmoinesian Mudstone and Carbonate Reservoirs in Southern Denver Basin. American Association of Petroleum Geologists. Search and Discovery Article 11340, 44 p.

Tedesco, S.A., 2021. Evaluation of the pilot waterflood at Arikaree Creek Field Case Study for enhancing oil recovery from a low temperature hydrothermal unconventional carbonate reservoir. In: Gaci, S., Olga, H., Nicolis, O. (Eds.), Methods and Applications in Petroleum and Mineral Exploration and Engineering Geology.

Thompson, J.A., 1964. The occurrence of helium in the Cambrian near Swift Current, Saskatchewan. In: North Dakota Geological Survey and Saskatchewan Geological Survey, Third Williston Basin Symposium, pp. 179–185.

Truswell, J.F., 1977. The Geological Evolution of South Africa, Cape Town, Purnell.

Urquhart, A.S.M., 2011. Structural Controls on CO_2 and Diagenesis in a Natural Long-Term Carbon Sequestration Analogue: Little Grand Wash Fault. University of Texas A & M, Utah (MS thesis), 461 p.

van Buchem, F.S.P., Svandsen, N., Hoch, E., Pedersen-Tatalovic, R., Habib, K., 2014. Depositional history and petroleum habitat of Qatar. In: Petroleum Systems of the Tethyan Region, American Association of Petroleum Geologists, Memoir 106, pp. 641–677.

von Drehle, W.F., 1990. Barrel Springs Field, Prowers County, Colorado. In: Sonnenberg, S.A., Shannon, L.T., Rader, K., von Drehle, W.F., Martin, G.W. (Eds.), Morrow Sandstones, Rocky Mountain Association of Geologist, pp. 183–189.

Wagner, C.R., 1961. Keyes Field. In: Oil and Gas Fields of Texas and Oklahoma Panhandles, Panhandle, Texas Geological Society, pp. 239–243.

Walker, R.,N., 1978. Big Gap Pennsylvanian. In: Fassett, J.E., Thomaidis, N.D., Matheny, M.L. (Eds.), Oil and Gas Fields of the Four Corners Area, Four Corners Geological Society, vol. III, pp. 936–937.

Wallace, K., Robinson, S., Heinz, J., 1990. Nee Noshe Field, Kiowa County, Colorado. In: Sonnenberg, S.A., Shannon, L.T., Rader, K., von Drehle, W.F., Martin, G.W. (Eds.), Morrow Sandstones, Rocky Mountain Association of Geologist, pp. 175–182.

Wang, X., Liu, W., Li, X., Li, Q., Tao, C., Xu, Y., 2020. Radiogenic helium concentration and isotopic variations in crustal pools from Sichuan Basin, China. Appl. Geochem. 117, 9 p.

Weaver, B.D., 2021. Keyes Helium Extraction Facility," the Encyclopedia of Oklahoma History and Culture. https://www.okhistory.org/publications/enc/entry.php?entry=KE018.

Weinlich, F.H., Bräuer, K., Kämpf, H., Strauch, G., Tesař, J., Weise, S.M., 1999. An active subcontinental mantle volatile system in the western Eger rift, Central Europe: Gas flux, isotopic (He, C, and N) and compositional fingerprints. Geochem. Cosmochim. Acta 63 (21), 3653–3671.

Weismann, T.J., 1971. Stable carbon isotope investigation of natural gases from Sacramento and Delaware-Val Verde Basins – Possible igneous origin. Am. Assoc. Pet. Geol. Bull. 55 (2), 369.

Wescott, P., 1989. Analysis of Gases in Earth's Crust, January 1982 - May 31, 1988, Final Report. Gas Research Institute, 88-262, 571 p.

Wilhelms, D.E., 1987. The Geologic History of the Moon. USGS. Professional Paper 1348, 309 p.

Williams, A.E., McKibben, M.A., Eldridge, C.S., 1991. An Overview of Fluid Geochemistry and Ore Genesis in the Salton Sea Geothermal System and Comparison with the Gulf of California Hydrothermal Systems, AAPG, Memoir 47, The Gulf and Peninsular Province of California, Chapter 37, pp. 781–792.

Williams, M., August 23, 2007. Mining the Moon: Lab Experiments Suggest that Future Fusion Reactors Could Use Helium-3 Gathered from the Moon. MIT Technology Review.

Wilson, M.S., Grove, K.W., Wilson, A.B., 2003. Exploration of the Mississippian Leadville Limestone in the Piceance Basin, western Colorado, Chapter 15. In: Piceance Basin Guidebook, Rocky Mountain Association of Geologist, pp. 252–270.

Wingerter, H.R., 1968. Greenwood Gas Field, Kansas, Colorado and Oklahoma, American Association of Petroleum Geologists Memoir 14, Geology of Giant Petroleum Fields, pp. 1557–1566.

Wiseman, T., 2019. Personal Communication, Helium Production at Harley Dome. SITLA, Grand County, Utah.

Wiseman, T., Eckels, M.T., 2020. Proven and hypothetical helium resources in Utah, 174. SITLA and Utah Department of Natural Resources, Misc Pub, 43 p.

Worrall, J., 2004. Preliminary Geology of Oakdale Field, Northwest Raton Basin, Huerfano County, Colorado, American Association of Petroleum Geologists, Search and Discovery Article 20017.

Yurkowski, M.M., 2016. Helium in Southwestern Saskatchewan - Accumulation and Setting, Saskatchewan Ministry of Economy. OFR 2016-1, 26 p.

Yurkowski, M.M., 2021. Helium in Southern Saskatchewan: Geological Setting and Prospectivity; Saskatchewan Ministry of Energy and Resources, Saskatchewan Geological Survey, Open File Report 2021-2, 77p. and two Microsoft® Excel® files. https://publications.saskatchewan.ca/#/products/115248.

Zhang, T., Zhang, M., Bai, B., Wang, X., Li, L., 2008. Origin and accumulation of carbon dioxide in the Huanghua depression, Bohai Bay Basin, China. Am. Assoc. Pet. Geol. Bull. 92 (3), 341–358.

Zhang, J., Wei, Y., Haiyong, Y., Wuren, X., Zengye, X., Fuying, Z., Yongjing, C., 2015. Feasibility of high-helium natural gas exploration in the Presinian strata, Sichuan Basin. Nat. Gas. Ind. B 2, 88–94.

Zhang, W., Li, Y., Zhao, F., Han, W., Li, Y., Wang, Y., Holland, G., Zhou, Z., 2019. Using noble gases to trace groundwater evolution and assess helium accumulation in Weihe Basin, central China. Geochem. Cosmochim. Acta 251, 229–246.

Further reading

Colorado Oil and Gas Conversation Commission, 2021. https://cogcc.state.co.us/#/home.

Helium One, 2021. Corporate Presentation. www.helium-one.com.

Kansas Geological Survey, 2021, www.kgs.ku.edu.

CHAPTER

4

Exploration methods for finding helium deposits

Occam's Razor the simplest explanation is usually the best one. ***English Franciscan friar William of Occham (c. 1287–1347).***

4.1 Introduction

Until recently, He exploration has been an afterthought of and/or part of gas and petroleum development programs. Helium is an elusive element. It is commonly associated with natural gas, but is found in economic quantities only 17.6% of the time (Broadhead and Gillard, 2004). So it is a very difficult target to find. The Appalachian Basin seems to be to date devoid of economic He whereas the Black Mesa and Four Corners Platform region are rich in He. This distribution of He has helped focus exploration to some degree but does not encourage exploration in new areas where no He data presently may yet exist, or areas constrained by nongeologic issues. Solving the mechanisms and processes that cause sourcing and accumulation of He has taken on more urgency in recent years as the element's price has increased. Finding the simplest atom seems to be getting significantly complex and so Occam's razor is an appropriate guide to how to approach exploration for He.

Exploration specific to CO_2 and N_2, like He, has been for the most part nonexistent and discovery of new reserves has been developed as an afterthought of either an unsuccessful petroleum project or a byproduct of a successful one. Worldwide He production overall more commonly been the result of processing large volumes of natural gas from supergiant gas fields. The recent surge in the price of He from 2014 to present caused several companies to specifically target areas known or thought to be rich in He. Nitrogen is in petroleum reservoirs but has rarely been specifically targeted in the industry. The element when used for commercial use for fertilizers is mined from potash beds or extracted from the air. Nitrogen from the natural gas stream has not been considered a viable resource. Many companies extract N_2 from natural gas to be rejected into the atmosphere to make natural gas sellable.

Helium is almost always found with N_2 but not necessarily the reverse is true. The ratio of He to N_2 varies between areas as will be discussed below, but provides no guidance in exploration. The percentage of He production from the supergiant gas fields is typically well below 0.05% by volume. These types of large-scale deposits are comparable to low-grade porphyry

https://doi.org/10.1016/B978-0-323-90988-4.00003-7

copper ore deposits in mining. Porphyry ore deposits extract low-grade Cu from large volumes of rock over a very large area, because of the uniform low grade, large areal extent, and economies of scale, they are very profitable. Examples of these types of low-grade He accumulations include the gas fields in Algeria, Qatar, Amur Processing Facility (Russia), Big Piney-La Barge, and Texas Panhandle-Hugoton Embayment, USA. These supergiant fields can have anywhere from a handful to several thousands of wells over a small to large area. These fields have significant long-lived reserves. Some of these large scales He reservoirs are probably internally a series of small- to mid-size coalesced reservoirs formed into a larger deposit of great areal extent.

The small to medium high-grade accumulations, >0.3% He or greater, can be compared to high-grade gold or base metal deposits that are hosted by quartz veins, skarn, and Mississippi Valley Type (MVT) ore bodies. They are relatively small in areal extent but very (relatively) high (concentration) such as the Harley Dome (Utah), Ryersee (Kansas), Salt Wash Field (Utah), Pinon Field (Arizona), and Battle Creek, Swift River, and Cypress Fields (Saskatchewan). These types of accumulations can initially have high flow rates and excellent He concentrations but have a relatively limited reservoir compared to supergiant fields. Most of these deposits are shallow from depths of less than 9000 feet (2700 m).

Exploring for He is historically based on its association with natural gas. It has been suggested that exploration for the gas can be based on a similar work and data flow (McDowell et al., 2017). This approach is based on the assumption that He models can be developed similar to petroleum using a source to reservoir and trap model.

There have been millions of wells drilled for natural gas around the world and many in areas that produce high-grade He. Potentially less than 1% of all wells have any potential for the element to be present economically either as a significant component of the natural gas or as an alternative recoverable product.

Helium does not form any compounds, so it has no minerals, oxides, carbonates, and sulfides that can be used as pathfinder tools. Nitrogen, on the other hand, is part of a wide variety of compounds, minerals, soils and found in almost every fluid in the subsurface thus limiting its use as a specific pathfinder combination for identifying He bearing deposits. Carbon dioxide is like N_2 in that it has limited use as a pathfinder tool. It is present in the sedimentary rocks, part of a wide variety of compounds, minerals, and soils; a byproduct of biological activity in the soil section; and found in many fluids or compounds in the subsurface.

We can summarize He deposits' general characteristics as follows:

(1) Always associated with N_2;
(2) Commonly associated with natural gas;
(3) Derived from mantle or crustal rocks rich in U and Th;
(4) Possibly derived from U and Th decay from Li and carbonaceous mudstones;
(5) Economic accumulations related to structural closures either as defined by tectonics or internal stratigraphic controls;
(6) Reservoir underlying the first best seal from the basement is likely where He will be trapped and accumulated;
(7) The older the pore water, generally Paleozoic in age, the more likely the He will be in significant economic quantities (Brown, 2010);

(8) Helium is inert and not reactive with elements or compounds except at extremely high temperatures and pressures;
(9) No associated minerals or compounds to act as pathfinders;
(10) Aeromagnetic and gravity information defines basement faults and associated defined granitic intrusions and plutons correlate with He and N_2 accumulations. However, most aeromagnetic and gravity defined basement faults and granitic intrusions and plutons have no associated overlying He, CO_2, and N_2 accumulations.

Helium and N_2 accumulations are typically found within the USA and Canada on the flanks of cratonic basins such as the Anadarko (Texas Panhandle-Hugoton Embayment), Denver Basin (Las Animas Arch), Central Kansas Uplift, Four Corners area (Arizona-New Mexico), Uncompahgre Uplift (Harley Dome), Williston Basin (southwestern Saskatchewan) and Battle Creek and Cypress Fields (Saskatchewan). Carbon dioxide deposits tend to correlate with volcanic activity and/or significant tectonic activity such as Huerfano County, Colorado (Sheep Mountain and Oakdale), thrust areas such as the Big Piney—La Barge Gas Field, Wyoming, and venting of areas within the Colorado Plateau related to basement activity such as Gordon Creek and Woodside fields.

4.2 Basin analysis and modeling

Basin analysis and modeling is a series of analytical methods for understanding physical and chemical processes that cause oil and gas generation, migration and entrapment, along with associated fluids and gases such as He, CO_2 and N_2 in sedimentary basins (Kleinspehn and Paola, 1988; Lerche, 1989; Miall, 1990; Einsele, 1991; Allen and Allen, 2013). This type of analysis starts from the mega scale overview of a basin and eventually refines analysis to the prospect level where the actual targets are subject to drilling. There are numerous detailed treatise on basin analysis. Writings by Kleinspehn and Paola (1988), Lerche (1989), Miall (1990), Einsele (1991), Allen and Allen (2013) are a few well known ones. Basin analysis is an extensive subject, but many of its elements have limited use in defining reservoir, source, generation and migration characteristics for He, CO_2 and N_2. Therefore an overview of the specific processes for He exploration will be discussed here. The sequence of data flow analysis and workflow can be summarized as follows:

(1) ***Regional-scale analysis.*** Overall defines the extent of the basin, important regional structural features, the tectonic framework, source rocks, geologic history and basin location, and plate movement within the subject area. In terms of He, CO_2, and N_2 geologic history and tectonic framework can be related to where plutonic and intrusive activity has occurred or related to migration of basinal fluids during stages of compression in basin formation. The parameters needed to define this analysis are:
   (a) Identify the main structural features and the basin depocenter. For petroleum exploration this type of analysis revolves around depositional systems and their associated source and reservoir rocks. Large or small structural closures possibly with a strong stratigraphic component, are typically critical for identifying potential areas of He deposits. Areas of CO_2 and N_2 accumulations are typically related to

areas of igneous activity, thrusting related to basinal compression, zones of weakness in the crust, and changes in the chemistry of basinal fluids or stratum that releases them into the gas phase.

(b) For petroleum exploration source rock analysis is critical in determining the origin of hydrocarbons by typing its total organic content, maturation, thickness, areal extent, migration pathways, and volume of hydrocarbons generated. Petroleum source rocks can be identified from existing wells, outcrops, and evaluated. On the other hand, He source rocks, U and Th, are in the crust and mantle out of the reach of direct or indirect observation. Helium within the crust can migrate upward independently or with fluids after being generated from the decay of U and Th. Low percentages of CO_2 and N_2 typically found with many He and associated natural gas deposits are related to processes defined with the chemistry of the reservoir due to change by chemical parameters or during basin brine and associated gas migration from source to trap. He gasses that are related to high volumes of CO_2 and N_2 are related to degassing of mantle and crustal rock and volcanic activity.

(c) If there are no wells of significance, the initial basin analysis generally requires the use of aeromagnetic and gravity surveys, satellite imagery analysis, regional seismic evaluation, and information from outcrops. For He and CO_2 exploration evaluating regional aeromagnetics and gravity data to determine likely areas of granitic intrusions and plutons as potential source rocks is key. In petroleum exploration aeromagnetics and gravity are used to define and augment interpretations of the present-day structural framework and tectonic history. Although rarely used for prospect generation, this type of analysis can define large scale anticlines and synclines, along with fault and fracture systems that are typically critical for petroleum accumulations. Outcrop information is often correlated with seismic data to create a preliminary subsurface basin model which includes major structural features as well as summarizing the likely depth of reservoirs, there thickness, there depositional characteristics and presence of a seal. If there are intrusions or plutons in outcrop, the percentage of U and Th will be useful in predicting He rates of generation and the potential amount being generated deeper in the basin. Unfortunately, the igneous activity on the fringes of a basin may not be present throughout the basin. On the other hand, aeromagnetic and gravity analysis can identify intrusives within the basin. This factor is unlike sandstone or carbonate reservoirs that can outcrop on the margins of a basin extend into the subsurface, allowing the depositional and reservoir characteristics analyzed in outcrop to be extrapolated to potential hydrocarbon producing areas in the basin.

(d) Define the timing of major structural events in a basin in relation to timing of hydrocarbon generation, expulsion and migration through faults, fracture networks, and porous sediments. For He and CO_2 exploration, the process can be independent of structural timing within a basin. He generation and accumulation is a very slow process.

(e) Large-scale analysis identifies sub-basins, sub-basin-level depocenters, areas of deltaic deposition, carbonate ramps, areas of significant organic richness, and local structural movements.

(f) Basins where He is historically associated with natural gas can potentially lead to thousands of wells to evaluate. The He present usually has no relationship with the natural gas in associated reservoirs, but may have dissolved in and migrated with basinal fluids, and accumulated within the same natural gas reservoirs. Underlying plutonic and intrusion activity may have provided a local heat source for petroleum generation from source rocks. Within the productive reservoirs of the Central-Kansas Uplift, USA, the He and CO_2 may have migrated with various brines generated from underlying intrusions and plutons and/or as part of the fluid expulsion from the Anadarko Basin during the Ouachita-Marathon Orogeny. The presence of CO_2 and N_2 could lead to potential economic He accumulations but their relationship, if any, is not well understood.

(2) ***Medium-scale analysis***. Petroleum exploration at this level evaluates deposition trends and processes, which are directly related to ongoing or intermittent episodes of tectonic activity (subsidence, uplift, or hiatus of inactivity), compartmentalization of the basin into subbasins, minibasins, or basin breakup leads to areas to evaluate to oil and gas potential. However, the igneous or metamorphic basement is not necessarily intricately related to later sedimentary basinal processes on a local scale. If the intrusion or plutonic activity occur prior to, during, or postsedimentary deposition, the timing for He and CO_2 generation and accumulation may be affected. Evaluation of the sedimentary cover will hopefully indicate high-grade areas with potential He traps. Some standard studies at the medium-scale analysis level are:

(a) Cross-sections which incorporate all available information and data, are created to define the structural and stratigraphic architecture of the basin. 2D, 3D, and 4D seismic data can be used to identify faults, thicknesses of sediments, and tectonic changes through time. However, the usefulness of this application in He and CO_2 exploration may be limited beyond identifying areas of potential structural closures that are undrilled or not drilled deep enough to test for accumulations of this element or compound.

(b) Medium scale basin analysis includes the evaluation of any existing petroleum fields in the study area. Select the most prospective subbasins and start to map major depositional events. In petroleum an example is the Cretaceous onlap of the Gulf of Mexico. This is a sand-prone sediment system whose mapping helps create age-specific studies that incorporate the time sequence and movement over the deposits. However, for He and CO_2 sandstone timing of sequences is not useful. He and CO_2 will sometimes migrate vertically from below a trap where the reservoir may be a sandstone-dominate fluvial system or a carbonate system. But, the key for the He and CO_2 accumulation is the quality of the seal overlying the trap. The He molecule is so small, only the lowest permeable seals will be effective. The quality of the seal, therefore, is more critical than the characteristics of the reservoir in the trap.

(c) A study of hydrocarbon types and associated reservoir fluids and gases within a structural framework that includes faults and fracture systems from mega to nano scale. Combine this information with thermal flow data where possible in order to discern maturation kitchens, expulsion areas, and migration pathways. Identify areas of low or high temperature hydrothermal fluids flow and diagenesis patterns.

Diagenetic alteration can either enhance and/or destroy porosity. Sparry calcite/chert and anhydrite can clog pore space, whereas dolomitization can result in enhanced porosity. If possible, analyze regionally and field-specific He, CO_2, and N_2 content from natural gas, water, and oil wells.

(d) Specific to He exploration is the usually old age of the pore water. Brown (2010, 2014, 2017) and Zhang et al. (2017, 2019a,b) identified this characteristic, the age of the pore water, as a critical factor in economic He accumulations. In the majority of cases in Chapter 3, the age of the pore water, typically Paleozoic, can be shown to be closely tied to the accumulation of He. Exceptions to this rule of thumb include the Harley Dome in Grand County, Utah, Virginia Gas Field in South Africa, and He accumulations in rift areas.

(e) The above analysis would identify possible seals and traps (structural and stratigraphic), from which the relative likelihood of commercial He and CO_2 accumulations can be assessed. The size of CH_4 is 400 picometer (pm), C_6H_{14} (Hexane) is 500, He is 62 pm, CO_2 is 232 pm, and N_2 is 71 pm. Obviously, to contain or, at least slow down, the dissipation of He requires a very effective seal or a seal whose vertical factures, unconformities, and discontinuities are plugged with heavier compounds and elements, such as CH_4 and heavier hydrocarbons to slow He migration from the reservoir.

(f) Petroleum exploration at this point assumes analysis that creates a time-focused (chronostratigraphic) model of the sediments. Sequence boundaries (unconformities or correlative conformities) correlate to depositional sequences. This will map where the transgressive flooding surfaces occurred, and also the point of maximum flooding. Mapping of the identified generic depositional sequences are extremely helpful in identifying tops, reservoirs, seal, and migration pathways. This type of analysis is difficult to apply to He and CO_2 exploration, as the accumulation of these elements is independent of dynamic petroleum processes of maturation, expulsion, migration, and accumulation. However, as previously mentioned, knowledge of flooding surfaces allows one to understand seals and reservoirs. Knowing where these components of a hydrocarbon system may be important to He exploration as well.

(3) ***Small-scale analysis.*** This stage for petroleum exploration is where drillable prospects are identified by integrating new techniques of data mining, geochemical analysis, geomechanical modeling, and 2D and 3D seismic. This stage identifies specific conventional targets for acquisition or drilling, plan fluids for drilling and completion, and in some cases develop a reservoir model appropriate for a waterflood or CO_2 flood for enhanced oil recovery. Some standard studies at the small-scale analysis level are:

(a) Movement along faults, the opening and closing of fractures and fracture networks, combined with nanoscale structural behavior due to hydrocarbon generation and expulsion can be used to high-grade migration pathways and focus areas of comparative enrichment. Being able to map facies patterns can also identify relative permeability and porosity pathways along with, the ultimate storage capacity in the reservoir. Based on these volumes, initial and very preliminary reserve potentials can be estimated.

(b) Large-scale analysis of surface lineaments as correlated to different subsurface structural features can provide new insights further helping high-grade areas to be explored in detail.

(c) Using a combination of well log interpretation, petrophysical analysis, and regional 2D or 3D seismic mapping can also help develop minibasin models that can pinpoint likely focus areas. There is presently no way to identify He on mechanical logs in a well.

(d) Understanding low-temperature hydrothermal fluid movement, and its interaction with the reservoir, plus an understanding the reservoir rock's chemical composition can help determine where there may be porosity enhancement via dissolution or secondary dolomitization. Likewise, it is possible to project where there may be porosity destruction due to precipitation/mineral filling of pore spaces and fracture networks.

(e) Data mining techniques for finding relationships between vast datasets can be helpful. Small-scale depositional modeling with typing of grains/mineralogy, along with reservoir fluids. Map facies patterns that identify the depositional lithologies and their chemical composition.

(f) The analytical process will assist in selecting drilling and completion fluids to avoid formation damage. It is possible to fine-tune this process with new geochemical tests and fingerprinting using isotopes.

(g) Fine-scale depositional system elements can help one determine chronostratigraphic relationships, and thus determine connected reservoirs, especially in clastic depositional systems. Microfossils can also be very helpful in carbonates, particularly where there are correlations between the paleoecological environment and reservoir quality.

(h) Low level methane seeps are an important indication of subsurface hydrocarbon deposits. Surface geochemical studies combined with large-scale reconnaissance work using hyperspectral sensors from satellites, the new capabilities of drones with methane sensors, can be integrated with surface geochemical and structure maps to further refine the hydrocarbon and He development potential of acreage where there is sparse subsurface control (Tedesco, 1994). Soil gas surveys for He can be very useful in identifying areas of potential leakage. This analysis is the most effective tool for identifying probable He accumulations locally other that analysis of natural gas or other types of gas wells. Nitrogen from the subsurface is difficult to identify in surface geochemical surveys, as it is 78% of the atmosphere by volume. In addition, there is a variety of N_2-based compounds present in the soil section and associated microbial and plant life. Carbon dioxide has similar issues.

(i) 2D and 3D seismic surveys are used to pinpoint structural framework, reservoirs, and stratigraphic architecture. This analysis is very applicable to both petroleum and He, CO_2, and N_2 traps.

In summary, petroleum basin analysis has limited application to identifying He and CO_2 accumulations. The tools that are useful for regional basin analysis include tectonic and geologic history, location of structural features and areas with potential productive natural gas reservoirs with a He component. These are used in conjunction with regional and detailed aeromagnetics and gravity to define basement faulting (migration pathways) and granitic intrusions and plutons (source rock). In conjunction with this regional work, mining of historic gas data or acquiring new He, CO_2 and N_2 gas analysis from existing petroleum wells are also used. Once general target areas are defined, surface geochemical surveys and 2D or 3D seismic surveys can be used to define drill sites.

What is even more interesting is that the He molecule is just over one-sixth the size of the CH_4 molecule. The N_2 molecule is only slightly bigger than the He molecule. This brings on a whole new concept on the effects of permeability in that the He and N_2 molecules are far more mobile and fracture stimulation may not be required if the CH_4^+ molecule is absent.

4.3 Isotopes

Isotopes analyses of He and N_2 are used to determine the sources of these elements. The details of the science of isotopes will not be discussed. The study of the isotopes of He and N_2 have not been discussed in depth in this book, as historically, they have not been used as an exploration tool. Typically their collection and interpretation are after the discovery and production of the elements. However, the inert nature of noble gases is not affected by chemical reactions or interactions and can be used with some confidence to determine their source and analyze details of their migration patterns (Halford, 2018). Tables 4.1–4.4 outline how isotopes are used to determine the sources of various isotopes.

Krypton has six stable isotopes but only ^{84}Kr is used to determine its source. Xenon has nine stable isotopes but only ^{132}Xe is used to determine its source.

TABLE 4.1 Source of He based on $^3He/^4He$ ratios.

Isotope	Range	Source
$^3He/^4He$ (R/R_a)	0.01–0.1	Crustal
	5–50	Mantle
	120	Primordial
	336	Cosmogenic

TABLE 4.2 Source of Ne based on individual isotopes.

Isotope	Source
^{20}Ne	Atmospheric, made in stellar nucleosynthesis
^{21}Ne	Crustal and mantle, more significant in mantle systems
^{22}Ne	Crustal, discharged from the decay of U and Th (Halford, 2018)

TABLE 4.3 Source of Ar based on individual isotopes.

Isotope	Source
^{36}Ar	Atmospheric (very insignificant crustal source)
^{38}Ar	Can be formed in the crust or cosmic ray spallation of terrestrial materials. If present in shallow reservoirs it is atmospheric.
^{40}Ar	Result of electron capture during radioactive decay of potassium (^{40}K) and is related to amount of K present (Halford, 2018)

TABLE 4.4 Table showing a summary of different noble gas signatures from different reservoirs (Halford, 2018).

Sources	^{4}He	^{20}Ne	^{36}Ar	^{84}Kr	^{132}Xe	R/RR_a	$^{20}Ne/^{22}Ne$	$^{21}Ne/^{22}Ne$	$^{40}Ar/^{36}Ar$
Air	5.24	16.5	31.4	0.65	0.023	1	9.8	0.029	295.5
ASW	2.04	7.67	47.8	1.83	0.039	1	9.8	0.029	295.5
Mantle	36,600	1.04	1	0.029	0.004	8	12.5–13.8	0.06–0.063	40,000
Crust						0.02	0.08–0.10	0.4–0.52	3000

ASW is fresh water (μmol/m^3 at 20°C).

4.4 Existing gas wells, production, and the helium–nitrogen relationship

The present primary method of He exploration has been by analyzing natural gas from existing wells for the element. These methods are low cost but rely on clustered results, tend to be focused on known areas that have already been picked over in terms of potential assets. The USGS and US Bureau of Mines have an extensive database for the USA. Canada has a similar database that has provided He information for prospect generation. Most CO_2 fields discoveries were found by accident in the pursuit of hydrocarbon deposits.

Helium's association with N_2, CO_2, and CH_4 is a difficult one to fully assess, as it is not necessarily a chemistry issue. Helium and N_2 have a strong relationship in that they tend to be found together and must therefore either be related by source or by molecular size; it cannot be explained by simple chemistry. Helium does not occur without N_2 but N_2 can be present without He. Helium's direct relationship with CO_2 is nonexistent. Helium typically has a strong negative relationship with CH_4 as well. The relationship between N_2 and CO_2 is a variable relationship. Nitrogen and CO_2's relationship with CH_4 is typically a strong negative association whereby when one is in abundance the other is in small amounts.

Evaluating natural gas compositional data is similar to analyzing any other geochemical data. Geochemical data sampling can vary from the same source(s) over time, subject to collection and laboratory errors. The errors that occur in the collection are (1) the capabilities of the collector to follow procedure; (2) the quality of the sample container; (3) internal variations in the laboratory equipment; and (4) the capabilities of the laboratory technician. The collection point which can be (1) at the wellhead, (2) from casing, (3) from tubing, (4) from a drill stem test (DST), (5) the meter run, (6) tank battery, and (7) a variety of various orifices related to production can create or cause variations in the gas composition from the reservoir. Helium has an additional problem: as a very small molecule, if a collected sample is not tested quickly, the He diffuses through the wall of the sample chamber. Usually if the sample is analyzed within 2 weeks of collection it is considered by the industry to be representative of the gas in the reservoir. There are mobile He analytical systems, but while they can analyze data in real time their shortcomings are accuracy, quality control, and repeatability. They typically are not as accurate as laboratory measurements. Quality control in the chain of custody for a sample is critical. The good news is that most samples are usable if collection and analysis procedures are followed carefully.

The number of samples in any dataset is critical. The smaller the dataset the less likely that the results are representative of the entire population and thus statistically insignificant. Gas produced from a reservoir typically varies with time. A pure CH_4 field with minor amounts of impurities may show minimal variation in gas composition through time but a reservoir with a significant, N_2, CO_2, and/or petroleum concentrations will see measurable changes.

There are a number of datasets have been synthesized over the year that are available for review: Anderson and Hinson (1951), Boone (1958), Wagner (1961), Zartman and Wasserburg (1961), Munnerlyn and Miller (1963), Miller and Norrell (1964a,b, 1965), Moore et al. (1964), Moore and Shrewsbury (1966), Cardwell and Benton (1967, 1968, 1969, 1970, 1971, 1972, 1973), Magloire (1970), Moore (1973, 1975, 1976a,b, 1977, 1978a,b, 1979, 1980, 1981, 1982), Miller and Hertweck (1982, 1983), Hertweck and Fox (1984), Moore and Hamak (1985), Jenden et al. (1988), Hamak and Sigler (1989, 1990, 1991a,b, 1993), Sonnenberg and van Drehle (1990), Hamak and Gage (1992), von Drehle (1990), Hamak and Driskill (1996), Gage and Driskill (1998, 2001, 2003, 2004, 2005), Rauzi (2003), Broadhead and Gillard (2004), Driskill (2008), Brown (2005, 2017), Newell (2017, 2021), Alberta Geological Survey (2020), Singh et al. (2020), Wiseman and Eckels (2020), and COGCC (2020), to name a few.

Table 4.5 is a summary of several fields discussed in Chapter 3 plus some additional fields and wells that have been included for statistical analysis. The gas analysis results presented in Table 4.5 were taken from the literature and are assumed to have been collected and analyzed using standard methods with the proper quality control. Included are also three Canadian datasets: (1) from the Alberta Geological Survey, (2) from the Saskatchewan Geological Survey, and (3) representing Ontario data from Lollar et al. (1994). There is a dataset covering the State of Kansas from Anderson and Hinson (1951), Boone (1958), Wagner (1961), Zartman and Wasserburg (1961), Munnerlyn and Miller (1963), Miller and Norrell (1964a,b, 1965), Moore et al. (1964), Moore and Shrewsbury (1966), Cardwell and Benton (1967, 1968, 1969, 1970, 1971, 1972, 1973), Magloire (1970), Moore (1973, 1975, 1976a,b, 1977, 1978a,b, 1979, 1980, 1981, 1982), Miller and Hertweck (1982, 1983), Hertweck and Fox (1984), Moore and Hamak (1985), Jenden et al. (1988), Hamak and Sigler (1989, 1990, 1991a,b, 1993), Sonnenberg and van Drehle (1990), Hamak and Gage (1992), von Drehle (1990), Hamak and Driskill (1996), Gage and Driskill (1998, 2001, 2003, 2004, 2005), Rauzi (2003), Broadhead and Gillard (2004), Driskill (2008), Brown (2005, 2017), Newell (2017, 2021), Alberta Geological Survey (2020), Singh et al., 2020, Wiseman and Eckels (2020), and COGCC (2020).

A way to evaluate the association of He with N_2, CO_2, and CH_4 is to use Pearson's correlation coefficients statistical analysis. This procedure determines the correlation between two elements. A zero correlation means there is no correlation between the two elements: a−1

TABLE 4.5 Fields used to evaluate the relationship between He, N_2, CO_2, and CH_4 (Anderson and Hinson, 1951; Boone, 1958; Wagner, 1961; Zartman and Wasserburg, 1961; Munnerlyn and Miller, 1963; Miller and Norrell, 1964a,b, 1965; Moore et al., 1964; Moore and Shrewsbury, 1966; Cardwell and Benton, 1967, 1968, 1969, 1970, 1971, 1972, 1973; Magloire, 1970; Moore, 1973, 1975, 1976a,b, 1977, 1978a,b, 1979, 1980, 1981, 1982; Miller and Hertweck, 1982, 1983; Hertweck and Fox, 1984; Moore and Hamak, 1985; Jenden et al., 1988; Hamak and Sigler, 1989, 1990, 1991a,b, 1993; Sonnenberg and van Drehle, 1990; Hamak and Gage, 1992; von Drehle, 1990; Hamak and Driskill, 1996; Gage and Driskill, 1998, 2001, 2003, 2004, 2005; Rauzi, 2003; Broadhead and Gillard, 2004; Driskill, 2008; Brown, 2005, 2017; Newell, 2017, 2021; Alberta Geological Survey, 2020; Singh et al., 2020; Wiseman and Eckels, 2020; and COGCC, 2020).

Field	Age	Zone	Lithology	He%	N_2%	CO_2%	CH_4%
Kevin Dome	D	Duperow	carb	0	0	80	5
Farnham Dome	Jr	Navajo	ss	0.42	0	72.5	27
Indian Creek	S	Tuscarora	ss	0.2	0.1	99.42	0.13
Jackson Dome	Jr	Norphlet, Haynesville, Smackover	ss and carb	0.2	0.1	99.42	0.13
Escalante	Pm/PC	Cedar Mt/White Rim Kaibab	ss/ls	0.0185	0.41	97.94	1.51
Sheep Mountain	Jr – K	Entrada, Dakota, Felsic Dike	ss/felsic dyke	0.1	0.5	0.5	84.2
Gordon Creek	Pm/Tr/K	Coconino/Sinbad/Cretaceous	ss/ls/ss	0.51	0.8	0.8	90.2
Harley Dome Dakota	K	Dakota	ss	0	0.9	98.9	0
Swift River	D	Duperow	Carb	0.03	1.03	98.82	0.14
Iehi	Pl – Pt	Toro Formation	ss	0.7	1.6	98.2	0.2
Hassi R'Mel	Tr	A, B and C sandstone	ss	0.5	3	85	9
St. Johns	Pm/PC	Supai/Granite	ls/granite	0.19	3	3	78.5
St. Johns	Pm/PC	Supai	ss/granite	0.69	3.3	91.7	3
Coalbed Canyon	M	Leadville	ls	0.38	3.9	93.4	1
Escalante	Pm	Kaibab	ss	0.5	4	81	11
Wagon Trail	Pn	Morrow	ss	0	4	64	32
Stateline	Jr	Morrison	ss	0.53	4	0.4	78.9
Gothic Mesa	M	Madison	ls	0.5	4.33	80	11.67
Bar X	Jr	Morrison	ss	0.33	4.65	0.375	86.99
Bluff	M	Leadville	ls	0.3	5.5	93.1	0.4
Oakdale	Jr – K	Entrada, Dakota, Felsic Dike	ss	0.01	5.5	93.1	0.4
San Arroyo	Jr	Entrada	ss	0.5	6	74	15
Clay Basin	Pn	Weber	ss	0.72	6.7	26.2	38.75
Big Piney – La Barge	M	Madison	ss and dolo	0.4	6.75	0.4	86.15
Big Piney – La Barge Basinal	M	Madison	ss and dolo	0.73	7.42	0.64	91.21

(Continued)

TABLE 4.5 Fields used to evaluate the relationship between He, N_2, CO_2, and CH_4 (Anderson and Hinson, 1951; Boone, 1958; Wagner, 1961; Zartman and Wasserburg, 1961; Munnerlyn and Miller, 1963; Miller and Norrell, 1964a,b, 1965; Moore et al., 1964; Moore and Shrewsbury, 1966; Cardwell and Benton, 1967, 1968, 1969, 1970, 1971, 1972, 1973; Magloire, 1970; Moore, 1973, 1975, 1976a,b, 1977, 1978a,b, 1979, 1980, 1981, 1982; Miller and Hertweck, 1982, 1983; Hertweck and Fox, 1984; Moore and Hamak, 1985; Jenden et al., 1988; Hamak and Sigler, 1989, 1990, 1991a,b, 1993; Sonnenberg and van Drehle, 1990; Hamak and Gage, 1992; von Drehle, 1990; Hamak and Driskill, 1996; Gage and Driskill, 1998, 2001, 2003, 2004, 2005; Rauzi, 2003; Broadhead and Gillard, 2004; Driskill, 2008; Brown, 2005, 2017; Newell, 2017, 2021; Alberta Geological Survey, 2020; Singh et al., 2020; Wiseman and Eckels, 2020; and COGCC, 2020).—cont'd

Field	Age	Zone	Lithology	He%	N_2%	CO_2%	CH_4%
Big Piney – La Barge Foreland	M	Madison	ss and dolo	0.7	9.3	0.1	82.4
Big Piney – La Barge Highland	M	Madison	ss and dolo	1	10	5	84
Little Valley	M	Leadville	ls	0.5	10	16.8	63.15
Tocito Dome	Pn	Paradox	carb	0.42	11.3	85.7	1.3
White Mesa	Pn	Hermosa/Paradox	ss	0	12	88	0
Lisbon	D	McCracken	ss	1.02	12.1	0.3	86.6
Greenwood	Pm	Wolfcampian	carb	0.78	12.4	0.6	85.8
Westwater	Jr	Entrada	ss	0.56	12.7	0.21	63
Grassy Trail	Tr	Moenkopi	ss	1.55	12.8	0	76.4
Emery Unit	Pm	Coconino	ss	1.37	13.1	77.5	6.3
BrysonCanyon	Jr	Morrison	ss	0.66	13.1		82.1
McElmo	M	Leadville	dolo ls	0.14	13.5	81.7	5.66
Lisbon	M	Redwall	ls	1.7	13.7	0.3	84.2
Black Rock	Pn	Paradox-Pinkerton	ls	0.82	15.5	0.17	83.5
Bar X	Jr	Entrada	ss	0.8	15.8	15.8	56.7
Chinle Wash	Pn	Hermosa/Paradox	ss	1.9	16	0.2	82
Boundary Butte	Pn	Hermosa	ss	0.46	17	23.9	51.4
Left Hand	Pn	Morrow	ss	0.3	17.6	0.2	81.8
Big Indian	M	Leadville	ls	0.9	18.3	69.2	4.5
Boundary Butte	Pn	Paradox	ss	0.77	18.4	1.15	3.3
Flank	Pn	Morrow	ss	0.15	18.5	0.1	81.3
Gothic Mesa	M	Leadville	ls	0.78	19.7	8.625	61.2
Rolling Mesa	M	Leadville	ls	0.81	19.7	4	64.31
Barrel Springs	Pn	Morrow	ss	1.47	20.4	0	76.12
Rudyard	D	Souris River	carb	1.9	20.9	0.1	77
San Arroyo	Jr	Morrison	ss	0.6	21.09	0.03	57.15
Big Flat	M	Leadville	ls	0.9	21.3	73.2	1.5

TABLE 4.5 Fields used to evaluate the relationship between He, N_2, CO_2, and CH_4 (Anderson and Hinson, 1951; Boone, 1958; Wagner, 1961; Zartman and Wasserburg, 1961; Munnerlyn and Miller, 1963; Miller and Norrell, 1964a,b, 1965; Moore et al., 1964; Moore and Shrewsbury, 1966; Cardwell and Benton, 1967, 1968, 1969, 1970, 1971, 1972, 1973; Magloire, 1970; Moore, 1973, 1975, 1976a,b, 1977, 1978a,b, 1979, 1980, 1981, 1982; Miller and Hertweck, 1982, 1983; Hertweck and Fox, 1984; Moore and Hamak, 1985; Jenden et al., 1988; Hamak and Sigler, 1989, 1990, 1991a,b, 1993; Sonnenberg and van Drehle, 1990; Hamak and Gage, 1992; von Drehle, 1990; Hamak and Driskill, 1996; Gage and Driskill, 1998, 2001, 2003, 2004, 2005; Rauzi, 2003; Broadhead and Gillard, 2004; Driskill, 2008; Brown, 2005, 2017; Newell, 2017, 2021; Alberta Geological Survey, 2020; Singh et al., 2020; Wiseman and Eckels, 2020; and COGCC, 2020).—cont'd

Field	Age	Zone	Lithology	He%	N_2%	CO_2%	CH_4%
Bita Peak	Pn	Ismay	carb	0.61	21.5	1.5	71.65
Brandon	Pn	Morrow	ss	2	22	0.2	75.8
Rudyard	O	Red River	carb	2.1	23.4	0.2	73.4
Salt Wash	M	Leadville	ls	0.3	24.2	62.3	7.4
Woodside	Pm	Kaibab	ss	2	24.82	0.3	67.15
Long Canyon	M	Leadville	ls	0.77	26	37.33	43.51
Bartlett Flat	M	Leadville	ls	0.91	30.6		68.3
Ryersee	Pm	Chase	carb	1.58	31.9	56.8	7.5
Ryersee	Pm	Chase Group	carb	2.554	33.99	0.747	47.78
Bow Knot	M	Leadville	ls	3.26	34.2	1.2	49.6
Otis-Albert	C	Rush	ss and carb	4.4	35.2	11.2	49.2
Boundary Butte	D	Ouray	carb	6.03	36	22	32
Jace-Moore-Johnson	Pn	Morrow	ss	1.9	37.7	0.72	59.3
Reichel	Pn	Pennsylvanian Conglomerate	cglm	2.25	43.8	0.3	50
Beldsoe Ranch/Castle Peak	Pn	Morrow	ss	1.44	48.62	0.04	48.52
Reichel	Pn	Kansas City	carb	1.44	48.62	0.04	48.52
Reichel	C	Granite wash	ss	5.91	54.1	22.1	17.1
Keyes Field	Pn	Keyes	ss	7.17	54.6	2	35.4
Reichel	Pn	Lansing–Kansas City	carb	1.2	56	0.1	43
Wilheim	C	Deadwood	ss	1.69	57.08	0.81	34.63
Reichel	Pn	Topeka	carb	1.6	59		
Harley Dome Morrison	Jr	Morrison	ss	0.65	60.7	4.6	46.6
Ar	Pn	Morrow	ss	1.31	61	31.7	5.7
Doe Canyon	M	Leadville	ls	1	63.4	3.85	13.5
Temple Springs	D	Elbert	ls	5.15	63.5	20	10.9
Tocito Dome North	Pn	Paradox	carb	8.07	70.72	0.02	21.19
Dineh-Bi-Keyah	D	McCracken	ss	5.9	73.3	0.4	20.3

(Continued)

TABLE 4.5 Fields used to evaluate the relationship between He, N_2, CO_2, and CH_4 (Anderson and Hinson, 1951; Boone, 1958; Wagner, 1961; Zartman and Wasserburg, 1961; Munnerlyn and Miller, 1963; Miller and Norrell, 1964a,b, 1965; Moore et al., 1964; Moore and Shrewsbury, 1966; Cardwell and Benton, 1967, 1968, 1969, 1970, 1971, 1972, 1973; Magloire, 1970; Moore, 1973, 1975, 1976a,b, 1977, 1978a,b, 1979, 1980, 1981, 1982; Miller and Hertweck, 1982, 1983; Hertweck and Fox, 1984; Moore and Hamak, 1985; Jenden et al., 1988; Hamak and Sigler, 1989, 1990, 1991a,b, 1993; Sonnenberg and van Drehle, 1990; Hamak and Gage, 1992; von Drehle, 1990; Hamak and Driskill, 1996; Gage and Driskill, 1998, 2001, 2003, 2004, 2005; Rauzi, 2003; Broadhead and Gillard, 2004; Driskill, 2008; Brown, 2005, 2017; Newell, 2017, 2021; Alberta Geological Survey, 2020; Singh et al., 2020; Wiseman and Eckels, 2020; and COGCC, 2020).—cont'd

Field	Age	Zone	Lithology	He%	N_2%	CO_2%	CH_4%
Boundary Butte East	Pn	Ismay	ls	1.4	74.7	7.4	11.9
Dineh-Bi-Keyah	T	Tertiary	syenite sill		75.12		
Hogback - Slick Rock	M	Leadville	carb	5.37	77.83	0.19	16.61
Table Mesa	Pn	Paradox	carb	3.49	78.3	9.9	2.5
Table Mesa	M	Leadville	carb	1.47	79.1	2.7	12.6
Reichel	O	Reagen	ss	0.48	80.1	14.7	3.3
Tohache Wash	M	Leadville	carb	5.7	83.83	1	8.3
Teec Nos Pos	Pn	Ismay	ls	7.06	84.6	0.8	5.5
Harley Dome Entrada	Jr	Entrada	ss	7.41	87.83	1.62	3.41
Hogback - Slick Rock	Pn	Paradox	ss	7.2	89.7	0.1	1.8
Tocito Dome North	M	Leadville	carb	8.2	89.9	0.2	1.7
Beautiful Mountain	M	Leadville	carb	8.2	90	0.6	0.2
Black Rock	M	Leadville	ls	1.39	92.5	0.5	3.65
Navajo Springs and Navajo Springs East	Pm	Coconino	ss	5	93	1	
Pinto Dome	Pm	Coconino	ss	1.09	93.92		3.4
Gosfield South	O	Trenton	dolo ls	0.04	2.14	0.04	77.91
Colchester South	O	Trenton	dolo ls	0.15	8.21	0.04	60.95
Malden 3-41-IV	O	Trenton	dolo ls	0.11	14.27	0.18	57.09
Dover	O	Trenton	dolo ls	0.1033	3.787	0.17	84.72
Hillman	O	Trenton	dolo ls	0.0633	2.56	0.128	79.14
Goldsmith	O	Trenton	dolo ls	0.015	0.77	0.08	68.14
Wheatley	O	Trenton	dolo ls	0.0729	2.911	0.069	83.74
Acton	O	Black River	dolo ls	0.39	7.96	0.03	82.71
Sombra	O	Black River	dolo ls	0.13	6.21	0.03	80.77
Hornby	O	Black River	dolo ls	0.39	7.12	0.02	84.2
Dover	O	Black River	dolo ls	0.12	2.79	0.215	86.62

TABLE 4.5 Fields used to evaluate the relationship between He, N_2, CO_2, and CH_4 (Anderson and Hinson, 1951; Boone, 1958; Wagner, 1961; Zartman and Wasserburg, 1961; Munnerlyn and Miller, 1963; Miller and Norrell, 1964a,b, 1965; Moore et al., 1964; Moore and Shrewsbury, 1966; Cardwell and Benton, 1967, 1968, 1969, 1970, 1971, 1972, 1973; Magloire, 1970; Moore, 1973, 1975, 1976a,b, 1977, 1978a,b, 1979, 1980, 1981, 1982; Miller and Hertweck, 1982, 1983; Hertweck and Fox, 1984; Moore and Hamak, 1985; Jenden et al., 1988; Hamak and Sigler, 1989, 1990, 1991a,b, 1993; Sonnenberg and van Drehle, 1990; Hamak and Gage, 1992; von Drehle, 1990; Hamak and Driskill, 1996; Gage and Driskill, 1998, 2001, 2003, 2004, 2005; Rauzi, 2003; Broadhead and Gillard, 2004; Driskill, 2008; Brown, 2005, 2017; Newell, 2017, 2021; Alberta Geological Survey, 2020; Singh et al., 2020; Wiseman and Eckels, 2020; and COGCC, 2020).—cont'd

Field	Age	Zone	Lithology	He%	N_2%	CO_2%	CH_4%
Egremont	O	Black River	dolo ls	0.56	7.59	0.08	83.46
Willey	O	Black River	dolo ls	1.26	2.55	0.04	86.57
Innerkip	C	Mt. Simon	ss	0.63	5.91	0.05	85.21
Clearville	C	Mt. Simon	ss	2.505	4.145	0.3	80.19
Gobles	C	Mt. Simon	ss	0.73	6.06	1	82.82
Raleigh	C	Mt. Simon	ss	1.215	10.49	0.245	76.71
Arthur	C	Mt. Simon	ss	0.45	7.55	0.03	83.09
North Gosfield	C	Mt. Simon	ss	0.55	8.78	0.06	82.57
Greenwood	Pm/Pn	Wolfcampian/Virgilian	carb	0.6	21.09	0.03	53.37
Battle Creek	C	Deadwood	ss	0.9	96.82	1.76	0

Age abbreviations: *C*, Cambrian; *carb*, carbonates; *cglm*, conglomerate; *D*, Devonian; *dolo*, dolomite; *Jr*, Jurassic; *K*, Cretaceous; *ls*, limestone; *M*, Mississippian; *O*, Ordovician; *PC*, Precambrian; *Penn*, Pennsylvanian; *Perm*, *Permian*; *PL*, Pliocene; *Pt*, Pleistocene; *S*, Silurian; *ss*, sandstone; *Tr*, Triassic Rock Type.

correlation means that they have an absolute negative association, and a 1 means they have a perfect positive relationship. Table 4.6 below is the correlation coefficient for He, N_2, CO_2, and CH_4 for 115 fields or wells for which data is sufficient to analyze the relationship of the two elements and two compounds. The analysis can represent the average of several samples or one sample. Gas composition is known to vary across a field or through time. The 115 fields and wells in Table 4.5 represent a wide range of diverse geologic settings, age of reservoir, and source of gas, and the data may be biased by a greater number of samples from one area over other areas.

The literature shows He can demonstrate a weak to strong correlation with N_2 depending upon the area that is sampled and size of the dataset. Table 4.6 indicates the average, median, and standard deviation for the 115 fields and displays a wide range of He, N_2, CO_2, and CH_4 values. The correlation coefficient for He to N_2 from Table 4.6 indicates a strong correlation. He to CO_2 and CH_4 shows no correlation. Correlation between N_2, CO_2, and CH_4 indicates a weak to high negative correlation (Table 4.6).

Another way to evaluate the dataset from Table 4.5 is presented in Table 4.7 for specific areas with more than three samples. Table 4.7 indicates a strong to very strong correlation

TABLE 4.6 Correlation coefficient for all 115 fields listed in Table 4.5.

	He	**N_2**	**CO_2**	**CH_4**
Average	1.57	27.74	21.92	44.09
Median	0.75	15.5	0.80	48.86
Standard deviation	2.07	29.24	34.91	33.94
	He	**N_2**	**CO_2**	**CH_4**
He	1			
N_2	0.70	1		
CO_2	−0.29	−0.35	1	
CH_4	−0.29	−0.45	−0.65	1

TABLE 4.7 Correlation coefficient, average, median, and standard deviation (StDev) for specific subdatasets of Table 4.5.

		He	N_2	CO_2	CH_4		He	N_2	CO_2	CH_4	N_2/He
Utah from Table 46 sample Colorado Plateau	He	1.00									
	N_2	0.60	1.00			Average	0.98	30.61	28.36	32.56	34.20
	CO_2	−0.29	−0.43	1.00		Median	0.72	19.70	8.63	13.50	24.11
	CH_4	−0.23	−0.38	−0.28	1.00	StDev	1.17	29.67	35.46	31.41	30.89
New Mexico from Table 8 samples Colorado Plateau		He	N_2	CO_2	CH_4		He	N_2	CO_2	CH_4	N_2/He
	He	1.00									
	N_2	0.89	1.00			Average	5.22	61.54	3.36	27.03	10.69
	CO_2	0.01	0.11	1.00		Median	5.54	70.67	1.10	13.76	12.09
	CH_4	−0.87	−0.59	−0.20	1.00	StDev	2.35	31.02	6.75	30.43	4.32
Arizona from Table 12 samples Colorado Plateau		He	N_2	CO_2	CH_4		He	N_2	CO_2	CH_4	N_2/He
	He	1.00									
	N_2	0.84	1.00			Average	4.29	47.07	22.63	25.03	9.67
	CO_2	−0.60	−0.63	1.00		Median	4.70	45.05	7.45	9.80	9.58
	CH_4	−0.37	−0.51	−0.02	1.00	StDev	3.14	36.92	36.74	33.10	6.27
LAAHR from Table 10 samples		He	N_2	CO_2	CH_4		He	N_2	CO_2	CH_4	N_2/He
	He	1.00									
	N_2	0.26	1.00			Average	1.33	21.86	19.68	54.79	20.82
	CO_2	0.21	−0.62	1.00		Median	0.97	16.55	0.66	63.80	17.40
	CH_4	−0.51	0.15	−0.86	1.00	StDev	0.81	17.78	40.55	33.61	17.34

between He and N_2 for the samples from Utah, Arizona, and New Mexico. The correlation coefficient between He and N_2 for the Las Animas Arch Helium Region (LAAHR) indicates a weak correlation. This can be the product of a small dataset or bad data. Also of note is the N_2:He ratio for Arizona and New Mexico is much lower than in Utah and LAAHR areas.

As discussed in Chapters 1 and 3, Ontario was the area of initial He production in North America. Samples collected by Lollar et al. (1994) were used to evaluate He and the other gases from three reservoirs as indicated in Table 4.8. The Trenton and Black River (Ordovician) are major producing zones in the Michigan Basin but tend not to be hydrocarbon gas producing and are formed by low temperature hydrothermal fluids that are expelled from either the basin center or basement rocks during local reactivation of basement faults. To use Pearson's correlation coefficient, samples from the same field were averaged and treated as one sample point in order not to bias the data. The Trenton reservoir data as shown in Table 4.9 display a typically strong relationship between He and N_2 as has been shown previously with other datasets above. The Cambrian sandstones in Table 4.9 are similar in age to those found and that are productive at Swift River Field in Saskatchewan. Table 4.9 indicates that the correlation coefficient for the Cambrian sandstone between He and N_2 is essentially negative. In addition, the N_2/He ratio is very low and similar to that of Arizona and New Mexico data from Table 4.3. It can be suggested that the high He areas may be derived from a source that is releasing significant He relative to N_2 or a separate source entirely from N_2. The Black River Formation shows a similar negative relation between He and N_2 as the Cambrian sandstones. There is insufficient data to reach a conclusion as to why.

TABLE 4.8 Helium, N_2, CO_2, and CH_4 data from Lollar et al. (1994) for select fields in southwestern Ontario.

Age of reservoir	Well	Field	He	N_2	CO_2	CH_4	N_2/He
Ordovician – Trenton	1	Gosfield South	0.04	2.14	0.04	77.91	53.50
Ordovician – Trenton	2	Colchester South	0.15	8.21	0.04	60.95	54.73
Ordovician – Trenton	3	Malden 3-41-IV	0.11	14.27	0.18	57.09	129.73
Ordovician – Trenton	4	Dover	0.09	6.02	0.09	82.90	66.89
Ordovician – Trenton	5	Dover	0.09	2.41	0.20	85.63	26.78
Ordovician – Trenton	6	Dover	0.13	2.93	0.22	85.62	22.54
Ordovician – Trenton	7	Hillman	0.03	1.69	0.05	65.17	56.33
Ordovician – Trenton	8	Hillman	0.10	3.54	0.03	86.43	35.40
Ordovician – Trenton	9	Hillman	0.08	2.81	0.02	85.47	35.13
Ordovician – Trenton	10	Hillman	0.03	1.41	0.05	72.21	47.00
Ordovician – Trenton	11	Hillman	0.09	3.22	0.31	84.70	35.78
Ordovician – Trenton	12	Hillman	0.05	2.69	0.31	80.85	53.80
Ordovician – Trenton	13	Goldsmith	0.01	0.99	0.11	70.23	99.00
Ordovician – Trenton	14	Goldsmith	0.02	0.55	0.05	66.05	27.50
Ordovician – Trenton	15	Wheatley	0.04	1.82	0.24	70.94	45.50

(Continued)

TABLE 4.8 Helium, N_2, CO_2, and CH_4 data from Lollar et al. (1994) for select fields in southwestern Ontario.—cont'd

Age of reservoir	Well	Field	He	N_2	CO_2	CH_4	N_2/He
Ordovician – Trenton	16	Wheatley	0.08	3.38	0.02	92.42	42.25
Ordovician – Trenton	17	Wheatley	0.13	3.71	0.02	88.25	28.54
Ordovician – Trenton	18	Wheatley	0.09	4.16	0.03	86.95	46.22
Ordovician – Trenton	19	Wheatley	0.03	1.80	0.10	73.16	60.00
Ordovician – Trenton	20	Wheatley	0.08	2.31	0.05	84.93	28.88
Ordovician – Trenton	21	Wheatley	0.06	3.20	0.02	89.54	53.33
Ordovician – Black River	22	Fergus	0.38	4.79	0.03	86.02	12.61
Ordovician – Black River	23	Acton	0.39	7.96	0.03	82.71	20.41
Ordovician – Black River	24	Sombra	0.13	6.21	0.03	80.77	47.77
Ordovician – Black River	25	Hornby	0.39	7.12	0.02	84.20	18.26
Ordovician – Black River	26	Dover	0.11	2.71	0.18	85.62	24.64
Ordovician – Black River	27	Dover	0.13	2.87	0.25	87.61	22.08
Ordovician – Black River	28	Egremont	0.56	7.59	0.08	83.46	13.55
Ordovician – Black River	29	Willey	1.26	2.55	0.04	86.57	2.02
Cambrian	30	Innerkip	0.74	5.10	0.02	84.05	6.89
Cambrian	31	Innerkip	0.63	6.07	0.01	85.21	9.63
Cambrian	32	Innerkip	0.75	6.16	0.03	84.36	8.21
Cambrian	33	Innerkip	0.51	5.50	0.11	86.06	10.78
Cambrian	34	Clearville	2.70	3.83	0.22	80.56	1.42
Cambrian	35	Clearville	2.31	4.46	0.38	79.81	1.93
Cambrian	36	Gobles	0.73	6.06	1.00	82.82	8.30
Cambrian	37	Raleigh	1.36	5.21	0.41	80.92	3.83
Cambrian	38	Raleigh	1.07	15.76	0.08	72.50	14.73
Cambrian	39	Arthur	0.45	7.55	0.03	83.09	16.78
Cambrian	40	North Gosfield	0.55	8.78	0.06	82.57	15.96

TABLE 4.9 Correlation coefficient, average, median and standard deviation (StDev) for the Trenton, Black River, and Cambrian reservoir rocks in southwestern Ontario in Table 4.8.

	He	N_2	CO_2	CH_4		He	N_2	CO_2	CH_4	N_2/He
He	1									
N_2	0.61	1			Average	0.07	3.49	0.10	78.45	49.94
CO_2	0.01	0.07	1		Median	0.08	2.81	0.05	82.90	46.22
CH_4	0.25	−0.34	−0.08	1	StDev	0.04	3.00	0.10	10.23	25.19

	He	N_2	CO_2	CH_4		He	N_2	CO_2	CH_4	N_2/He
He	1									
N_2	−0.15	1			Average	0.42	5.29	0.09	84.42	21.25
CO_2	−0.44	−0.64	1		Median	0.39	6.21	0.04	84.20	20.41
CH_4	0.26	−0.78	0.67	1	StDev	0.41	2.47	0.09	2.36	13.87

	He	N_2	CO_2	CH_4		He	N_2	CO_2	CH_4	N_2/He
He	1									
N_2	−0.34	1			Average	1.106	6.938	0.233	81.79	9.16
CO_2	0.18	−0.27	1		Median	0.74	6.065	0.095	82.69	8.97
CH_4	−0.40	−0.69	−0.04	1	StDev	0.79	3.41	0.31	3.83	5.58

There is a very large amount of natural gas analysis data in the USA and Canada. Unfortunately, a significant percentage of that data is not usable because of missing components or poor sample collecting techniques.

The reasons are:

(1) Even though He data are available, and generally CH_4 is reported, N_2 and CO_2 are not reported;
(2) The collection point of several datasets is poorly known or unknown;
(3) Some datasets do not report the age and/or name of the formation tested, aka Alberta Geological Survey (2020);
(4) Based on data presented below, large databases may actually impede detail analysis because they contain a significant number of noncorrectable variables.

Appendix 4.1 lists the Pearson correlation coefficient; average, median, and standard deviation for datasets from Kansas, Ontario, Alberta, Saskatchewan, LAAHR area data from the COGCC (2020), Arkansas based on several databases. The datasets in Kansas for discussion purposes here are further divided into populations of geologic basins, specific pay zones and fields. In addition, two of the datasets are compared due to different dataset sizes and sources from the literature. The datasets from Alberta and Saskatchewan are divided into two subsets: wellhead and DST results. For the Alberta data, only the data that had all four elements and compounds were used in Saskatchewan and only the data having He, N_2, and CO_2 were used.

Figs. 4.1 and 4.2 are the result of analysis from Appendix 4.1. Fig. 4.1 indicates a correlation coefficient for He to N_2 that generally indicates a moderate to strong positive relationship but there are exceptions here too. The He correlation to CO_2 indicates a nonrelationship overall, but there are exceptions. The relationship between CH_4 and He indicates a negative association, which is assumed as the more He present the less hydrocarbons.

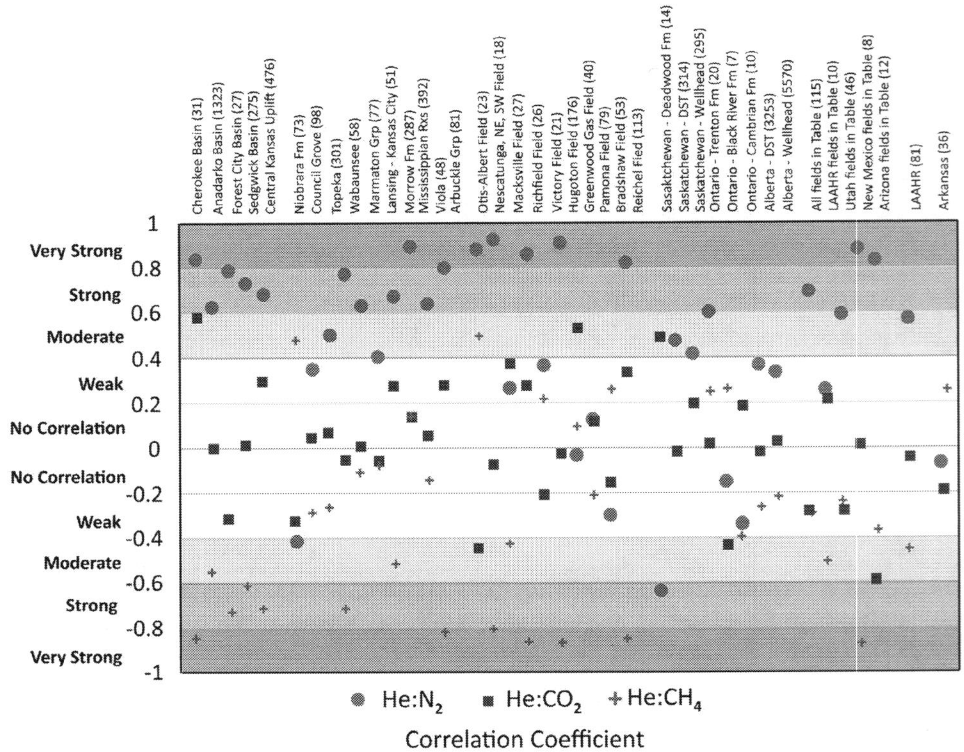

FIGURE 4.1 The Pearson's correlation coefficient for He in relation to N_2, CO_2, and CH_4 for the datasets presented in Appendix 4.1.

Fig. 4.2 is the relation between N_2, CO_2, and CH_4. Because all three are chemically related in terms of petroleum systems, clays, and associated formation water, there are generally more definitive relationships. N_2 to CO_2 relationship indicates a nonrelationship with occasional exceptions. This reflects that they are not interacting chemically. N_2 to CH_4 relationship is as expected. With increase in N_2 in the gas composition, CH_4 content decreases. Methane to CO_2 relationship indicates a similar relationship as N_2 to CO_2 suggesting that sources and chemistry are not necessarily interrelated.

The Kansas dataset is unique in that it provides the ability to evaluate different geologic variables. Comparison of five of the basins or regions, the Central Kansas Uplift and the Cherokee, Anadarko, Forest City, Sedgwick Basins, in Kansas, indicates different characteristics. The dataset size varies. All five areas indicate a strong correlation between He and N_2. The Cherokee Basin shows a moderate positive correlation between He and CO_2, which may reflect the area has considerable amount of gas produced from coal sources. The other four areas generally have a negative to positive weak to no correlation between He and CO_2. In all five regions He has a moderate to strong negative correlation with CH_4. In the five regions N_2 to CO_2 has no to a positive or negative weak correlation and N_2 to CH_4 exhibits a very strong negative correlation. Carbon dioxide's relationship to CH_4 in the five regions indicates a moderate negative relationship with CH_4 in some areas and in the other areas no to a weak correlation.

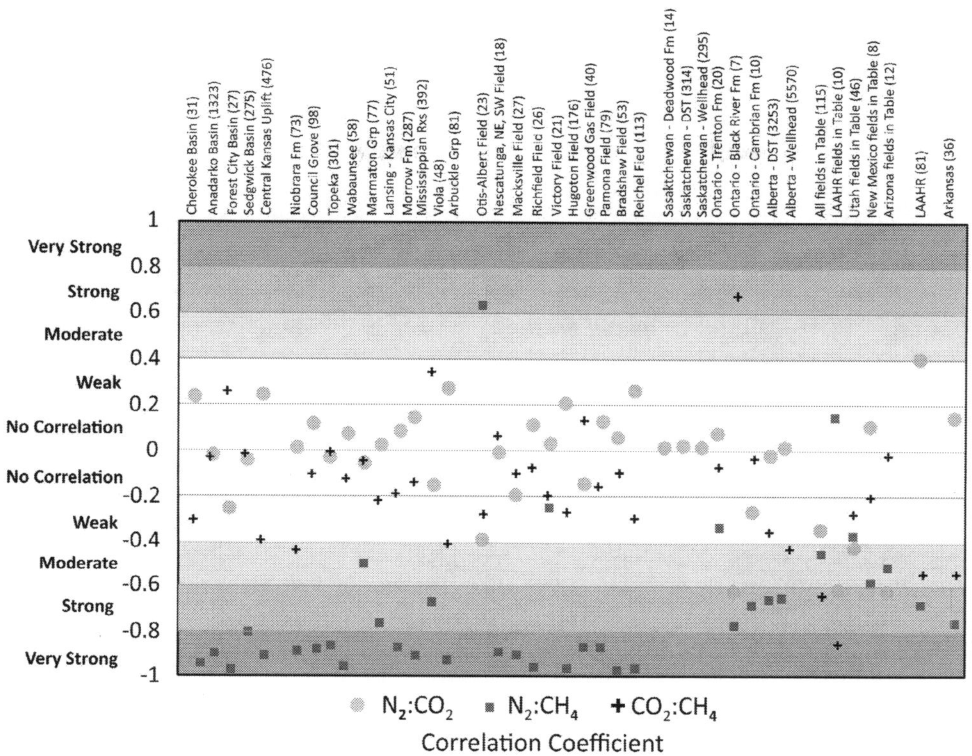

FIGURE 4.2 The Pearson's correlation coefficient for N_2 related to CO_2 and CH_4 and CO_2 to CH_4 for the datasets presented in Appendix 4.1.

When the data is separated into individual reservoirs there are significant differences. Carbonaceous mudstones or "shale" such as the Niobrara (Appendix 4.1) has a negative relationship between He and N_2. According to the literature, "shale" reservoirs typically contain minimal He, if any. The shale plays derive their oil and gas and produce directly from the source rock. The introduction of He into these systems is typically not likely, as the trap is extremely large in areal extent and would require significant generation of the element from crustal sources. The Cambrian sandstone data were not summarized as most of the data were from the Otis-Albert and Reichel fields, which are discussed below in relation to existing fields.

The correlation coefficient for the remaining pay zones indicates a general ever-increasing correlation between He and N_2 with depth. The number of samples for each zone varies, which affects the correlation coefficient and interpretation of that specific zone. The larger the number of samples in a dataset for a specific reservoir, there tends to be a more reliable interpretation. The Council Grove, Topeka and Wabaunsee reservoirs which typically are productive from the Hugoton Field and the data indicate an increasing correlation between He and N_2 with depth. The ratio of N_2 to He is not similar to areas of high He content such as Otis-Albert Field, Arizona, and New Mexico. The poor correlation with the Lansing-Kansas City probably represents a thick section, with multiple reservoirs present that can be sampled, resulting in little detail definition of individual pay zones. The Morrow sandstone (Pennsylvanian) analysis indicates a typical positive strong correlation for He to N_2, but the underlying Mississippian shows an even stronger correlation between He and N_2. The dataset from the Morrow and Mississippian strata indicates typical correlations between CH_4 and CO_2 seen elsewhere. The Mississippian zone is not generally He productive as compared to the Morrow and Reagan sandstones (Cambrian).

The analysis of the correlation coefficients of individual fields in Figs. 4.1 and 4.2 and Appendix 4.1. The zones the fields are producing from are:

(1) Otis-Albert Field (Central Kansas Uplift) – Reagan Sandstone (Cambrian), Arbuckle Group (Upper Cambrian-Lower Ordovician), Simpson Sandstone (Middle Ordovician), Pennsylvanian Conglomerate, Lansing-Kansas City carbonates (Pennsylvanian), Topeka Group (Pennsylvanian) and Chase Group (Permian). The evaluation here is strictly Reagan Sandstone.
(2) Reichel Field (Central Kansas Uplift) – Reagan Sandstone (Cambrian), Arbuckle Group (Upper Cambrian-Lower Ordovician), Simpson Sandstone (Middle Ordovician), Pennsylvanian Conglomerate, Lansing–Kansas City carbonates (Pennsylvanian), Topeka Group (Pennsylvanian) and Chase Group (Permian). The evaluation here is strictly Reagan Sandstone.
(3) Nescatunga, NE, SW Field (Anadarko Basin) – Marmaton, Cherokee (Pennsylvanian), Mississippian, and Viola (Ordovician)
(4) Macksville Field (Central Kansas Uplift) – Marmaton (Pennsylvanian), Mississippian and Viola (Ordovician)
(5) Richfield Field (Anadarko Basin) – Morrow sandstone (Pennsylvanian)
(6) Victory Field (Anadarko Basin) – Lansing – Kansas City, Marmaton, Cherokee, Atokan, Morrow (Pennsylvanian), Mississippian
(7) Hugoton Field (PHF) – Chase Grp reservoir (Permian)
(8) Greenwood Field (PHF) – Topeka age Virgil Limestone
(9) Pamona Field (PHF) – Council Grove reservoir (Permian)
(10) Bradshaw Field (PHF) – Chase Grp reservoir (Permian)

Correlation coefficient for He:N_2 indicated in Appendix 4.1, Figs. 4.1 and 4.2 for the Otis-Albert and Reichel Fields on the Central Kansas Uplift which primarily are produced from the Cambrian Reagan sandstone have a strong positive relationship. The data used are almost exclusively from the Reagan sandstone whereas the third field, Macksville from the same area, has a weak positive or no correlation between He and N_2. However, the data from this the Macksville Field covers several different pay zones, which may be the cause of the low correlation. The Hugoton Field, Anadarko Basin, has a strong positive correlation coefficient for He:N_2, however, internally the Greenwood, Pamona, and Bradshaw subfields do not show a similar correlation. All of the samples from Pamona were from the Council Grove zone of Permian age, the Greenwood Gas Field reservoir is Pennsylvanian Topeka in age and the Bradshaw Field produces from the Chase Group of Permian age. The correlation between He and the other variables also shows a wide variance in the fields. The correlation between N_2, CO_2, and CH_4 exclusive of He shows the typical relationship seen throughout the datasets. The two other fields, Richfield and Victory, indicate one has the strong correlation for He to N_2, whereas the other does not. The Richfield Field dataset was almost exclusively collected from the Morrow sandstone (Pennsylvanian) except for three samples from the Pennsylvanian age Topeka zone and one sample from the Kansas City Group. The Victory Field had several samples from the Mississippian and the Pennsylvanian age Morrow, Cherokee and Lansing-Kansas City zones. The non-correlation of He and N_2 within the subfields may indicate local small scale variations and insufficient sampling.

The N_2/He ratio for these 10 fields is consistent when compared to other areas in that the higher the He percentage the lower the ratio. Otis-Albert, Reichel, and Macksville have higher He content than the Nescatunga, Ne and SW, Richfield, Victory, Hugoton, Greenwood, Pamona, and Bradshaw and thus higher N_2/He. This fact suggests that the source of He in the low N_2/He is either rich in He or the He is acting independently from N_2.

The data from the Alberta and Saskatchewan geological surveys (2016 and 2020, respectively) are shown in Appendix 4.1 for Pearson's correlation coefficients for the DST and wellhead data. The Alberta database has 22,169 samples and the Saskatchewan dataset has 1874 samples. Unfortunately the Alberta dataset does not provide information on what zone the gas sample was collected from which can hinder analysis and interpretation. The datasets from sources were filtered into two categories, wellhead and DST-derived samples. The wellhead analysis samples consist of 5570 samples from Alberta and 295 samples from Saskatchewan. In the case of the Alberta data, there appears to be a weak to moderate positive relationship between He and N_2. The wellhead samples from Saskatchewan indicate a positive-moderate relationship between He and CO_2 and no relationship between N_2 and CO_2. The Alberta wellhead samples indicate no relationship between He:N_2 and N_2:CO_2, and a negative relationship for He, N_2, and CO_2 to CH_4 which is consistent with almost all of the other datasets. The dataset from Saskatchewan did not report any CH_4 analysis. The analysis of gas collected from DSTs (3253 samples from Alberta and 314 samples from Saskatchewan) indicates a weak positive relationship between He and N_2. The DST samples from Saskatchewan indicate a positive moderate relationship between He and CO_2 and no relationship between N_2 and CO_2. The Alberta wellhead samples indicate no relationship between He:N_2 and N_2:CO_2, and a negative relationship for He, N_2, and CO_2 to CH_4 which is consistent with almost all of the other datasets. The dataset for the Cambrian sandstones in Saskatchewan was separated out. They comprise only 14 samples, and the correlation results indicate no correlation between He and N_2, a moderate relationship between He and CO_2, and no relationship between CO_2 and N_2.

The LAAHR have two datasets in Appendix 4.1, one derived from the COGCC (2020) resulting in 81 samples and the other from Table 4.1 with 10 samples and from a different literature source. The LAAHR with 81 samples has a correlation coefficient of 0.58, which is a moderate relationship for He to N_2 whereas the dataset derived from Table 4.1 has a weak positive 0.26 correlation. The LAAHR with 81 samples, correlations for N_2, CO_2, and CH_4 are more consistent with the other datasets in Appendix 4.1 whereas the smaller dataset for LAAHR in Table 4.1 is not. This highlights the uncertainty of interpretation when using small datasets.

The Arkansas dataset was included in the analysis as it represents an area that does not have anomalous He. Fig. 4.2 indicates that He relationships with N_2, CO_2, and CH_4 have no correlation. The dataset also indicates similar correlations of N_2 to CO_2 and CH_4, and CO_2 and CH_4 as seen in other datasets, which suggests that this dataset elemental and compound relationships is probably more typical of areas with no significant He.

In summary the relationship between He and N_2 based on chemistry should be nonexistent. Based on the real-world data, there seems to be a relationship in many cases. That relationship is most likely because He and N_2 have a similar source, i.e., crustal, igneous and metamorphic rocks. Although there could be two different sources but unless isotopes are used, the different sources may not be definable. Central Kansas Uplift, Black Mesa, Holbrook, Four Corners Platform, LAAHR, parts of the Uncompahgre Uplift, Southwestern Ontario on the flank of the Michigan Basin, the Bowdoin-Sweetgrass Arch lying both in Alberta and Saskatchewan seem to be areas where the He is derived from crustal rocks. In these same areas the best He production is from the lowest trap that can accumulate He, such as the Cambrian sandstones. In reservoirs that are He productive with low percentages but large natural gas volumes, the source of He may be from multiple sources and the result of migration from multiple areas. In summary, the use of the He to N_2 relationship and ratio is probably not a definitive exploration tool but an overall first pass tool to understand the He source and system in an area.

4.5 Migration models

There are several models for petroleum migration that can be applied to He, CO_2, and N_2 migration. However, some of the models used for He, CO_2, and N_2 are not conducive for petroleum migration, as the processes destroy hydrocarbons. An example of such a process would be expulsion of He during volcanic activity. Halford (2018) developed a "Helium System" model for the Four Corners area of the Colorado Plateau that can be used in other areas (Fig. 4.3). Halford (2018) modified the typical petroleum system model based on 279 samples involving noble gases and $\delta\ ^{15}N(N_2)$ isotopes that indicated the dominant source of the He-rich gas found in the Four Corners Paleozoic reservoirs that originate from the shallow Precambrian granitic crustal rocks. Helium is derived from the radioactive decay of U and Th within the granitic bodies. Halford (2018) looked at carbonaceous mudstones as sedimentary source of helium for the Four Corners area but could not find evidence to support their contribution. Halford's model is applicable if large amounts of He accumulate during decay of uranium in granitic rocks and were subject to long durations of physical and thermal stability within the granitic trap. Granitic rocks are relatively impermeable to migrating He and so the gas will remain trapped until released by changes in the seal. Release of the dissolved He from the source rocks would be due to activity caused by deep faulting, igneous intrusions, or other mechanisms, which increase the permeability of the granitic host body (Halford, 2018). The He is subject to migration in association with carrier gases or fluid, N_2 or CO_2, moving updip and vertically (Fig. 4.3) through the overlying sedimentary strata (Halford, 2018). Noble gas geochemistry of δC (CO_2) isotopes for this model strongly indicates that distance is a controlling variable related to the migration of He via CO_2 (Halford, 2018). Danabalan (2017) suggests that there is likely an ideal distance from the accumulation or "sweet spot" to the source of the CO_2 to where economic He accumulations are discovered. Carbon dioxide reacts with the adjacent host rocks as it migrates throughout the subsurface from the mantle or crust and disappears (Halford, 2018). Secondary migration begins for He in carrier beds that intersect the migration pathways from the mantle or crust and move

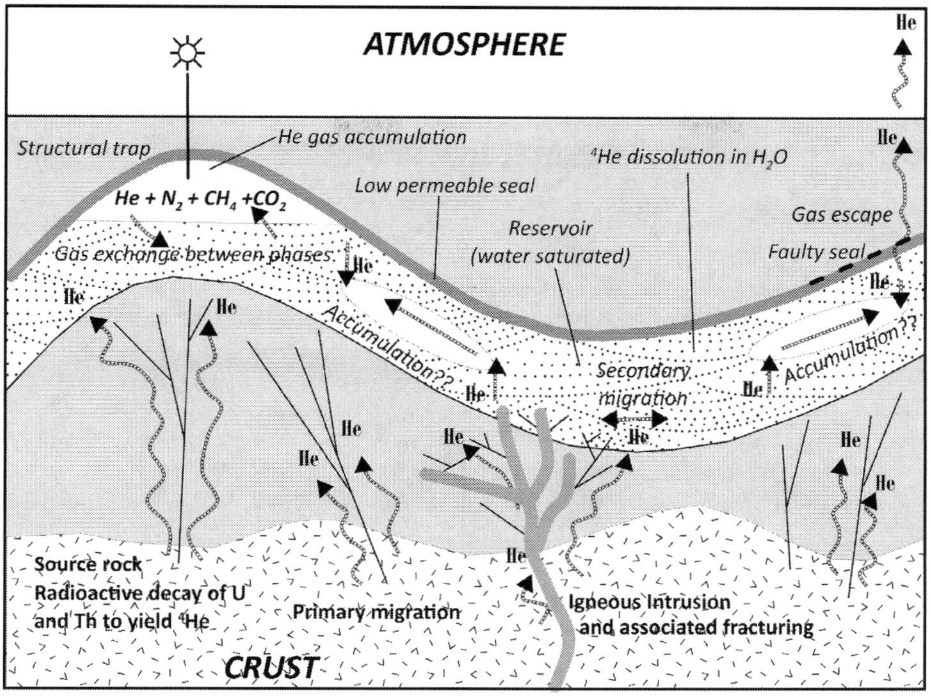

FIGURE 4.3 Proposed model for He accumulation in the Four Corners area modified from Halford, 2018.

laterally via groundwater. Groundwater is according to Halford (2018) a critical factor in the concentration of He deposits. After dissolution into the pore water, which is subject to pressure, temperature, salinity controls, He migrates with the solution through the subsurface. The model does not take into account large volumes of He moving independently of the migrating fluids. Helium in solution is very small, if nonexistent, and higher amounts of He will be accumulating independently of the host brine or gas.

Halford (2018) stated that migration is observed through the degree of gas water fractionation reactions observed in the noble gases which are geochemically inert. Fractionation is observed on a significant scale. Migration is interpreted to be the result of physical processes, which can adequately account for the degree of fractionation observed (Halford, 2018). If a migrating gas phase encounters a He rich brine or fluid it can effectively exsolve it from the solution back into the gas phase (Halford, 2018). Exsolution is plausible given the proper pressure, temperature and salinity controls in the absence of a migrating carrier gas. Exsolution is another critical point where the accumulation and concentration of He occurs as it migrates updip or vertically to accumulate in reservoirs. He and petroleum accumulations are similar in reservoir characteristics, but the best He accumulations, found in the Colorado Plateau and Central Kansas, for example, are either devoid of or have a minimal presence of liquid hydrocarbons (Halford, 2018). This model is not applicable to supergiant natural gas fields with low percentages of He. The negative relationship between He and liquid hydrocarbons is probably because He is pushed out of the trap during migration. This model is not generally applicable to low He concentrations in supergiant gas fields such as Pars North and South, Big Piney-La Barge and Texas Panhandle—Hugoton fields.

The traps in the Four Corners area, as indicated by subsurface geologic mapping Halford (2018), are both structural and stratigraphic ranging from 3- and 4-way domal closures supplemented by extensive faulting to tight stratigraphic pinch outs. Faults act as both migration pathways and seals for He accumulations. Halford (2018) proposed that organically plugged shale is the most effective seal for the accumulation of He. These types of shales are typically classified as carbonaceous mudstones and in themselves are petroleum producing zones known as "shale" plays. Along with organically plugged shales, evaporites and tight limestones are proposed as the most likely effective seals. If a seal is ineffective or breached, a strong indication is the seepage of large amounts of He measured on the surface such as hydrothermal vents or highly anomalous He in surface geochemical surveys.

Craddock et al. (2017) concluded that the high concentrations He, CO_2, N_2, and Ar in the Colorado Plateau (Fig. 4.4) are all sourced from the crustal area based on N_2/Ar, He/Ar, and $^{40}Ar/^{36}Ar$ ratios. CO_2 was sourced both from magmatic and non-magmatic sources such as oxidation of hydrocarbon gas or alteration of carbonates (Fig. 4.4). The Colorado Plateau is basically a raft of Paleozoic and late Precambrian crust floating on the lower crust and upper mantle. Breeches in the crustal raft occur through geologic time that allows venting of He, N_2, and CO_2 from the mantle and crust, or released due to chemical alteration as hot fluids which then migrate upward. According to Craddock et al. (2017), He, CO_2, N_2, and Ar are found in specific areas around the plateau and congregate generally in groups.

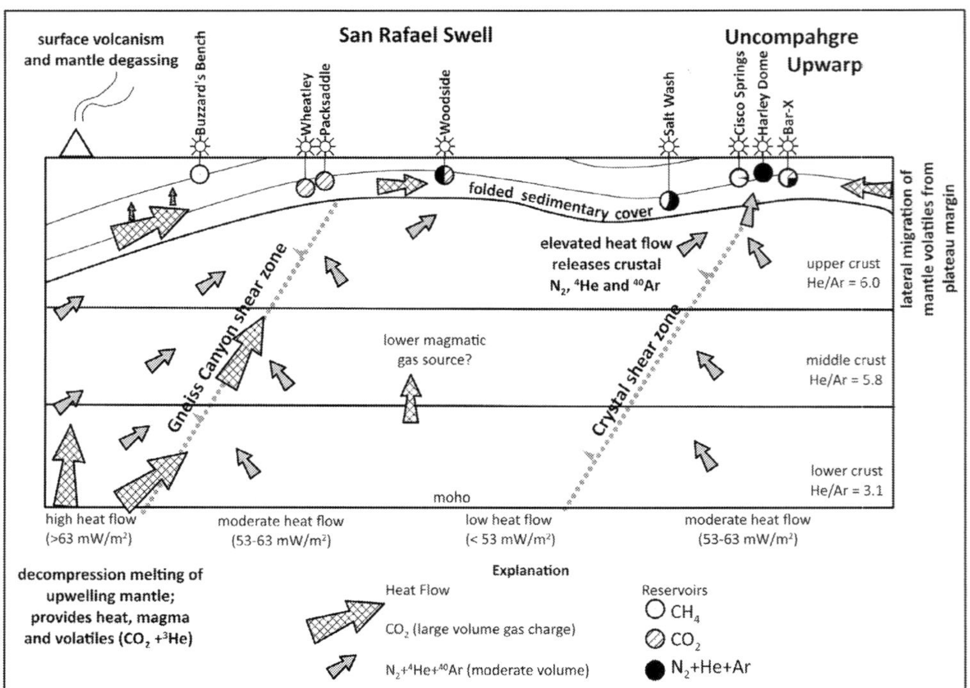

FIGURE 4.4 Proposed model for mantle and crustal gases migrating and accumulating in Post Precambrian strata within the Colorado Plateau modified from Craddock et al., 2017.

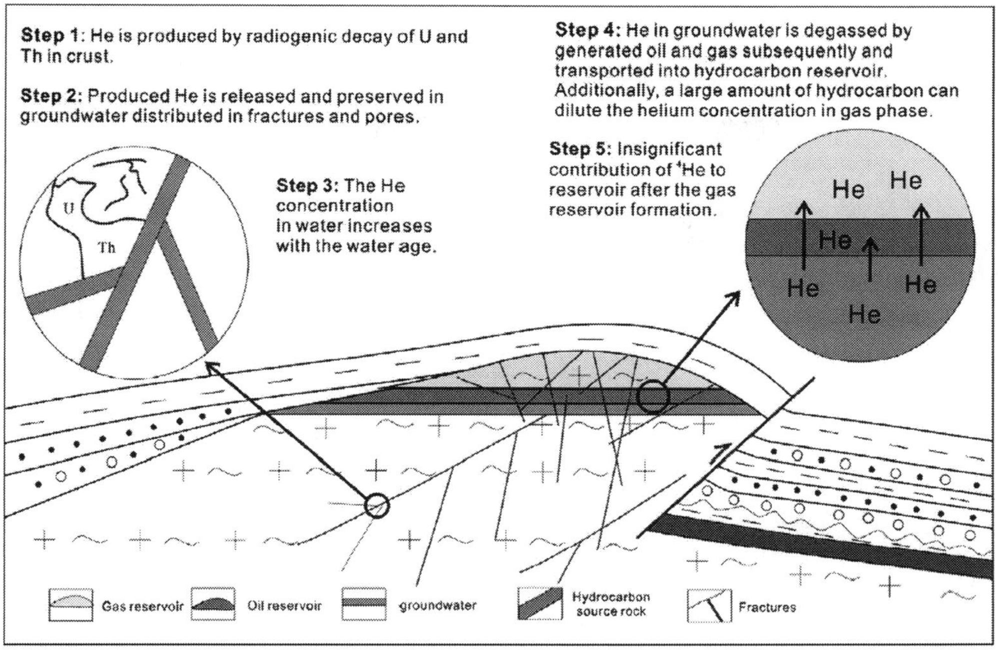

FIGURE 4.5 Proposed model for He accumulating process using noble gases (Zhang et al., 2019a).

Zhang et al. (2019a), proposed a model for development of He accumulation in a geothermal reservoir. Atmospheric ^{20}Ne and ^{36}Ar dissolved into surface or meteoric water that is recharging an aquifer (Fig. 4.5). As the recharged water moves basin ward, it picks up in solution crustal-derived He. Zhang et al. (2019a) presented noble gas (He, Ne, Ar, Kr, and Xe) isotope and gas compositional data from 10 producing wells located in the Mabei, Dongping, and Niudong gas fields in the North Qaidam Basin, China, to investigate the mechanism that causes He and hydrocarbon charging processes in a water-oil-gas system (Zhang et al., 2019a). ^{3}He/^{4}He ratios show that there is little contribution from the mantle source, indicating that the majority of noble gases in the gas reservoirs are derived from air-saturated water and the crustal strata. The air-saturated water-derived noble gases, ^{20}Ne, ^{36}Ar, ^{84}Kr, and ^{130}Xe that are present are accounted for by meteoric water recharge (Zhang et al., 2019a). The data from Mabei and Niudong gas fields were interpreted as an open-system fractionation between oil and water followed by gas and water fractionation. The Dongping data, on the other hand, can be explained by equilibration in an open oil/water system followed by fractionation in a closed gas/water system (Zhang et al., 2019a).

Zhang et al. (2019a) found a linear relationship between ^{4}He and ^{20}Ne that indicates that ^{4}He is dissolved into the groundwater and mixed with ^{20}Ne before coming into contact with the oil or gas phase. The estimated initial ^{4}He concentrations Mabei and Dongping calculated in the groundwater are less variable for than measured ^{4}He concentrations in the gas phase. Zhang et al. (2019a) suggested that CH_4 has a significant dilution effect on He and other noble gases partitioned into the gas phase. Zhang et al. (2019a) also summarized that the groundwater residence times for the He is calculated at 0.31–2.78 million years based on the external ^{4}He flux derived from the entire crust. Zhang et al. (2019a) identified the groundwater in the Niudong Gas Field as younger than that in Mabei and Dongping gas fields and this was supported by the lower He percentages in that field. Zhang et al. (2019a) concluded that little if any He was added to any of the reservoirs after the petroleum entered the reservoir. The lack of He being found generally with any oil reservoirs, except for Dineh-Bi-Keyah Field in the Four Corners area, Arizona, USA, for example, may suggest that as heavy hydrocarbons enter a reservoir and increase the reservoir pressure, they push the He and probably N_2 out of the reservoir and into the seal. The small size of the He and N_2 molecules in relation to heavier and larger hydrocarbon molecules means He and N_2 cannot move anywhere but in the direction of least pressure.

Zhang et al. (2019a) theorized that the He accumulation process can be summarized into five steps (Fig. 4.5).

(1) Helium is generated by radiogenic decay of U and Th in the crust.
(2) The majority of He is released and dissolved into groundwater present in rock fractures.
(3) Groundwater with a long residence time can accumulate high amounts of He.
(4) Helium in the groundwater partitions into the oil and gas and is subsequently transported into the hydrocarbon reservoir, where hydrocarbons have a dilution effect on the helium concentration in the gas phase.
(5) There is insignificant additional contribution of He to the reservoir after the formation of the gas reservoir. One is more likely to succeed in He exploration at gas fields associated with old groundwater systems and moderate hydrocarbon reserve.

Zhang et al. (2019b) presented a second model for a different area in North Qinling Belt, China, that was related to a geothermal reservoir (Fig. 4.6). Zhang et al. (2019b) indicated ^{20}Ne, ^{36}Ar, other air-saturated water-derived noble gases, and N_2 (2.9%–30.3% of total N_2)

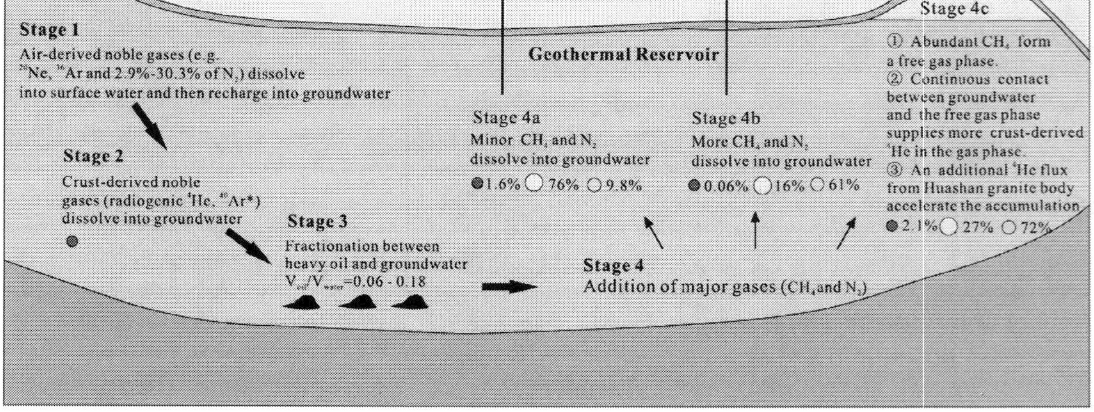

FIGURE 4.6 Proposed model for He accumulating process using noble gases (Zhang et al., 2019b).

from air dissolved into surface water at recharge conditions of 10°C, 0M NaCl, and an average altitude of 1000 m (North Qinling Belt) were incorporated. Crustal radiogenic noble gases ^{4}He and ^{40}Ar were transferred to the groundwater on a time scale between 0.3 million years and 1.98 million years. This assumes the releasing and transfer efficiency from source rock to groundwater of ^{4}He is 100%, therefore the time scales for ^{4}He accumulation could be longer if efficiency is lower. On the other hand, ^{4}He accumulation ages could be shorter if there were episodic release of ^{4}He accumulated in ancient rocks, which would produce a high ^{4}He flux. Groundwater underwent heavy oil-groundwater Rayleigh fractionation with $Volume_{oil}/Volume_{water}$ being 0.06–0.18. This process caused an increase of $^{20}Ne/^{36}Ar$ ratio (0.195–0.382) in groundwater to values higher than that in air-saturated water (0.143).

Both Zhang et al. (2019a,b) models rely on He migrating from both the atmosphere with meteoric water recharge and granitic intrusions. The age of accumulation for He in both studies that Zhang et al. (2019a,b) presented are recent. However, one of the conclusions again was that without older water the anomalous accumulation of He was unlikely. In neither study was the He indicated to be economic and is currently not being recovered.

Haines and Allen (2019) evaluated the Amadeus and Murraba Basins for hydrocarbons and He. Fig. 4.7 illustrates their model for hydrocarbon and He migration and accumulation. The He migrates up along basement faults and in this model accumulates in the Heavitree Formation which is of Precambrian age and is overlain by dense shale and thick evaporite section, forming a perfect seal. This model is consistent with other models presented here and in the literature for He accumulations.

Brown (2010, 2014 and 2017) proposes a multisource He model starting with a typical He model of generation from a radioactive source in the mantle or the crust (Fig. 4.8). Brown (2010, 2014 and 2017) indicated migration with the petroleum gases change in composition

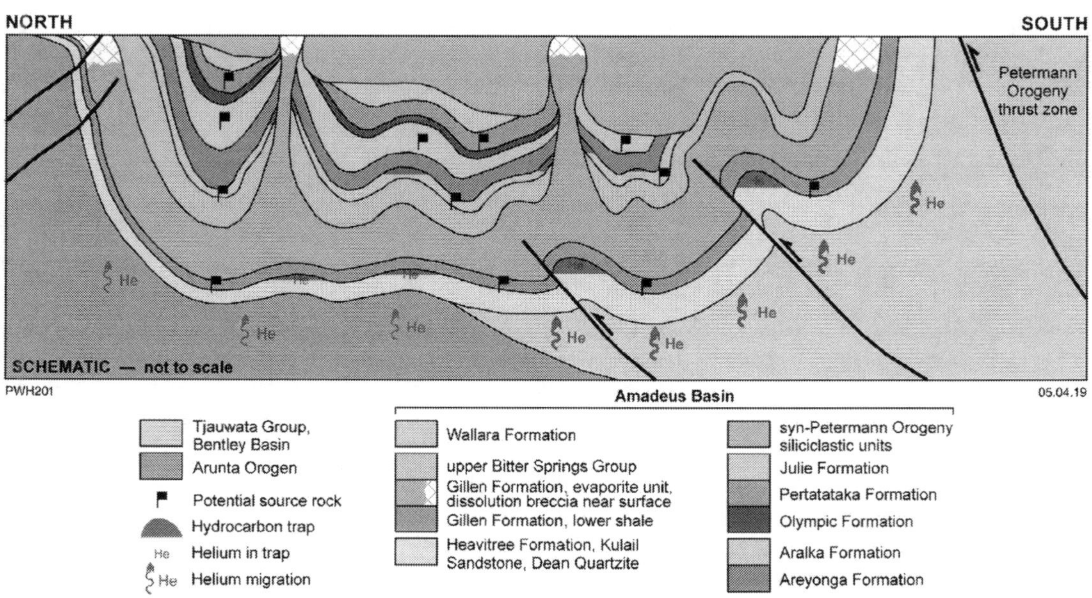

FIGURE 4.7 Proposed model for He migration and accumulation processes in Amadeus and Murraba Basins in western Australia (Haines and Allen, 2019).

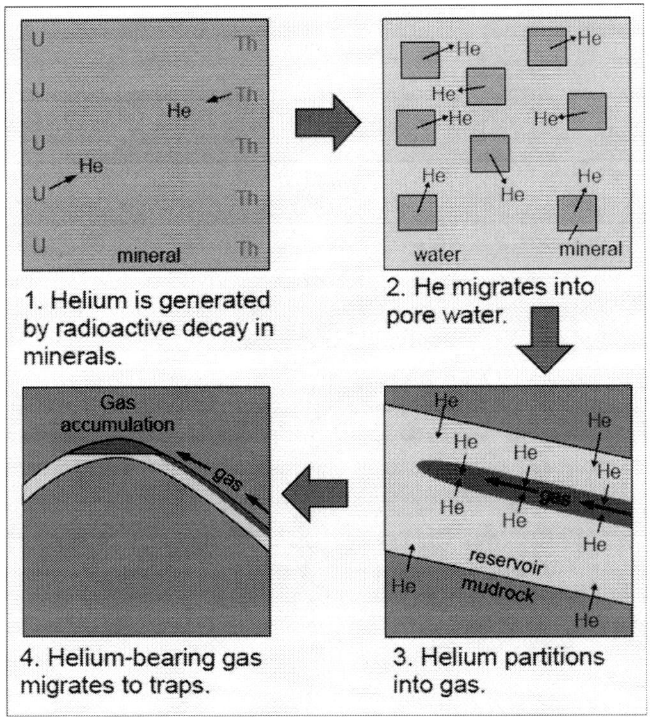

FIGURE 4.8 Proposed model for He migration process (Brown, 2010, *reprinted by permission of the AAPG whose permission is required for future use*).

during migration as they pass through different host rocks, increasing or decreasing the percentages of CO_2 and N_2 present. This hypothesis is supported by plotting He/N_2 percent ratios (Fig. 4.9), which indicates the Hugoton and Texas Panhandle parts of the Texas Panhandle-Hugoton Field (PHF) have different gas compositions. There is a saddle between the fields allowing separation and nonmixing of gases (see Chapter 3).

The He is also partly derived from migration of He in basinal fluids expelled from the deep Anadarko Basin to the southeast as well as from the Wichita Uplift (see Chapter 3). The amount of He produced from the PHF exceeds the amount of He estimated to have been migrated from the Anadarko Basin alone. This revelation led the identification for a third source of He, CO_2, and N_2 to be proposed by Brown (2010)— the PHF accumulated some of these gases through hydrodynamic flow (Fig. 4.10). This is basically similar to part of Zhang et al. (2019a,b) model in that noble gases migrate in with meteoric recharge. The hydrodynamic flow has He migrating from the Sierra Uplift, Palo Duro, and Dalhart Basins (Fig. 4.11). The hydrodynamic flow model contributes a significant amount of He to the PHF. However, the very light, unreactive and mobile nature of He, wanting to leave the planet, makes it difficult to accept this third concept easily without further study. Pushing He downdip seems to be a difficult proposition, especially over several hundreds of miles or kilometers. Hydrodynamic models do work in certain cases for petroleum and natural gas and may work for He. The absence of He in Pre-Permian reservoirs such as the Pennsylvanian Morrowan sandstone underlying the PHF which are He rich in the adjacent Keyes Dome and LAAHR is difficult to explain and hydrodynamic model maybe the answer.

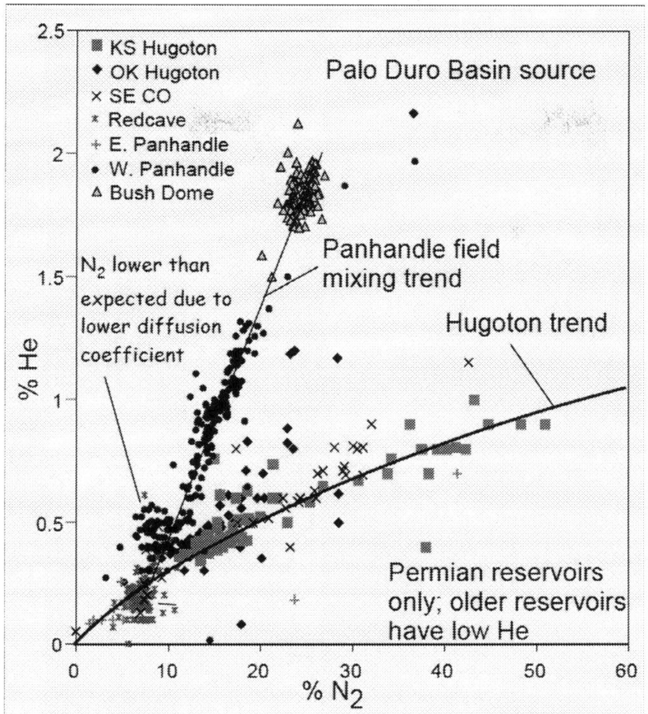

FIGURE 4.9 Helium-nitrogen concentration trends in the Panhandle-Hugoton field and adjacent areas (Brown, 2010, *reprinted by permission of the AAPG whose permission is required for future use*).

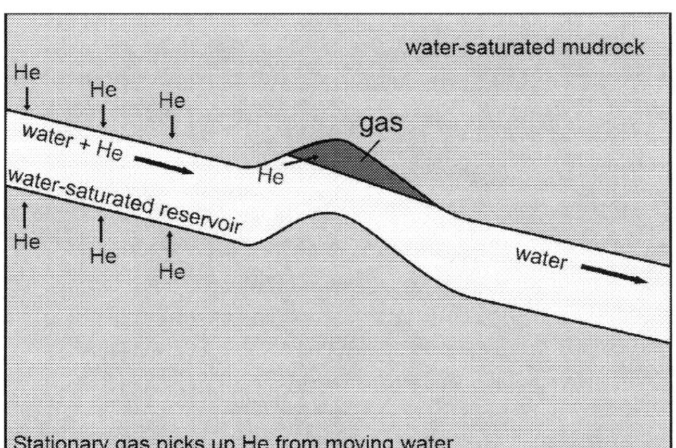

FIGURE 4.10 Proposed hydrodynamic model from Brown (2010) (*reprinted by permission of the AAPG whose permission is required for future use*).

Danabalan (2017) proposed a model for He deposits in the East Africa Rift System (EARS). The model relies on a magma chamber to be in close proximity to a series of sandstone and carbonate depositional centers with effective seals. The EARS lake deposits with associated sandstone and carbonate reservoirs have multiple fault systems and migration pathways

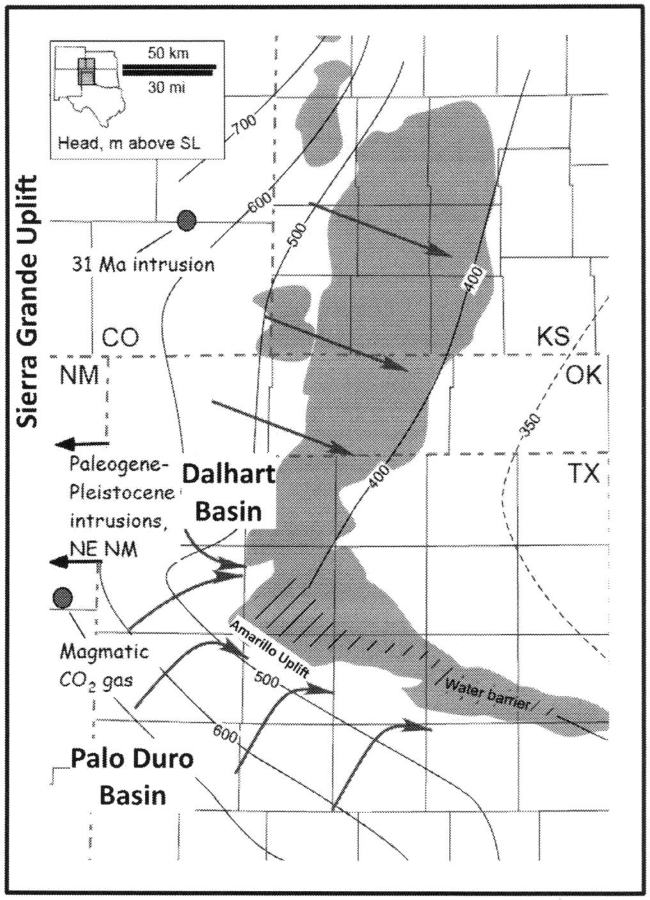

FIGURE 4.11 Potentiometric surface in Wolfcampian and Virgilian strata. *Modified from Brown, A.A., 2010. Formation of High Helium Gases a Guide for Explorationist, AAPG, New Orleans Convention Poster, Search and Discovery Article 80115, with field pressure data. Potential sources of 3He are also shown.*

that can set up perfect trapping environments for hydrocarbons, CH_4, He, CO_2 and N_2 gases (Fig. 4.12). However, the model is missing one criterion found in all other models—"old pore water"—which is typically of Paleozoic age. An example of younger grounder containing an He accumulation is Harley Dome in Grand County, UT, USA, which produces He from Jurassic reservoirs. Also the EARS is in the very front area of the magma where CO_2 and He gas are collecting in parts of the volcanic plumbing system that will terminate in a well-sealed reservoir and trap. A risk with all potential He reservoirs and seals is that they may only be effective briefly in terms of geologic time due to tectonic activity before they are destroyed or breached. As He formation, expulsion, migration and entrapment takes so long, a geologically stable area is essential to most He accumulations.

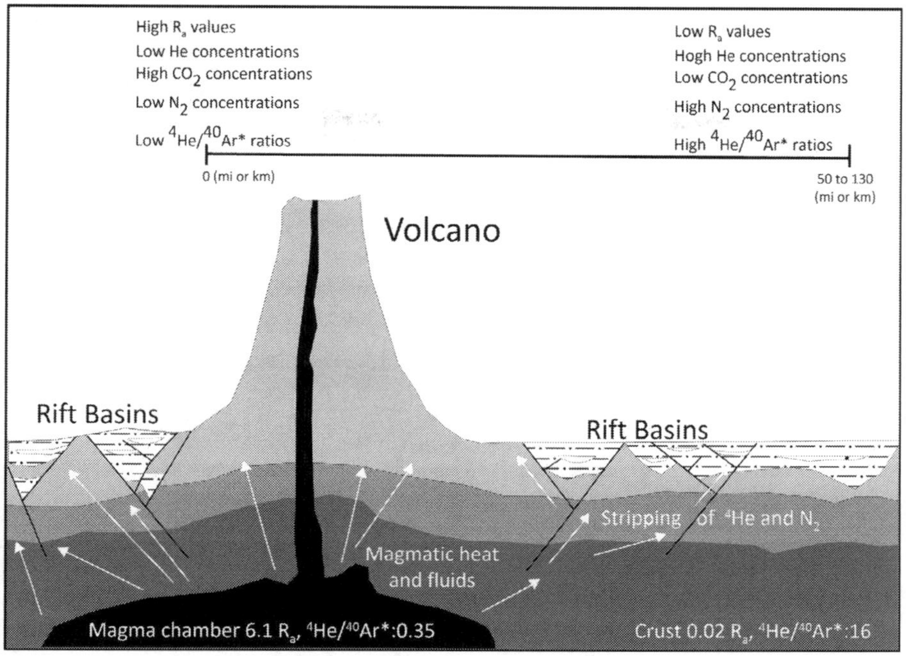

FIGURE 4.12 Proposed model for He migration process in relation to volcanic and rifting activity modified from Danabalan, 2017.

Most of the He generation and accumulation models can be summarized as follows:

(1) Sources of He are generally from crustal or magmatic sources. Atmospheric sources of He as a large source of the element and being trapped at depth probably can be ruled out due of the nature of the He molecule having difficulty migrating downward;

(2) Most He deposits of economic importance are related to old, undisturbed pore water meaning the accumulations are typically found in Paleozoic rocks;

(3) Unlike He, CO_2 and N_2 can be sourced from magmatic or nonmagmatic sources, such as through chemical alteration;

(4) Significant He, N_2, and CO_2 accumulations typically are found in natural gas deposits: (a) concentrating effect of hydrocarbons on He; (b) similar migration and trapping environments;

(5) The greater the amounts of heavy hydrocarbons found in the reservoir the lower the probability of He, N_2, and CO_2 accumulations;

(6) He, N_2, and CO_2 accumulations are related to basement faulting and fracture systems;

(7) The seals typically are thick and composed of either salts, gypsum and anhydrites, very low porosity shale (most commonly organic rich shale), or a combination of all three.

4.6 Igneous intrusions, aeromagnetics, and gravity

Helium is derived from radioactive decay of U and Th in crustal and mantle rocks and migrates upward from the source rock along fractures and faults and/or laterally along sedimentary carrier beds. The source rocks for He, CO_2, and N_2 unlike for oil and gas sources are not generally available for visual and chemical analysis. The He is trapped, relatively, temporarily within a reservoir capped by a dense seal typically composed of salts, anhydrites, clay, or organic rich shale of a significant thickness to delay leakage.

Carbon dioxide-rich reservoirs that are generally greater than 10% volume gas composition are more likely related to igneous or volcanic activity. Large volumes of CO_2 migrate ahead of the magma front along fractures and fault systems that can eventually be trapped in a reservoir such as a volcanic laccolith or dyke (igneous), or sandstone or carbonate (sedimentary) type reservoir. Some magma bodies are not rich in CO_2 so the material composition of the magma is important for it to generate the gas. Lower volumes of CO_2 rich reservoirs are probably related to release of the compound by chemical changes or alteration within the associated or nearby brines and minerology of the reservoir or adjacent sediments. There is always the potential of a combination of chemical and igneous-derived CO_2 in a reservoir. See Chapter 2 for further discussion.

Nitrogen-rich reservoirs have historically been related to venting from the crust. Brown (2017) related the high N_2 percentage found in natural gas such as the PHF to chemical changes that occur during or after fluid migration.

Mantle and crustal source rocks for He and CO_2 can be evaluated by using seismic, magnetic, gravity, and geophysical methods. Magnetic geophysical methods measure the Earth's magnetic field in search of variations that can lead to petroleum or ore deposits or construction impediments in the Earth's surface. A compass is a simple magnetic field detector but is limited to generally pointing north. The magnetometer measures the field variation in gauss, tesla, and nanotesla (also called gamma) units. The detail measurement of the Earth's magnetic field at a particular point on the Earth's surface is done by the use of a magnetometer. The magnetometer was first developed in 1833 by Carl Fredrich Gauss. Aeromagnetics was originally developed to find submarines during World War II, as they represent a large variation in the Earth's magnetic field at a particular point and moment. The use of aeromagnetics in petroleum exploration began about 65–70 years ago.

The theory of magnetics and gravity interpretation and analysis in relation to petroleum exploration is out of the scope of this book. What will be discussed here are four examples that demonstrate the use of aeromagnetics in relation to producing fields, subsurface geology and helium exploration (Tedesco, 2017).

4.6.1 The Black Mesa basin, Defiant Uplift, and Four Corners Platform

Aeromagnetics methods can be useful in areas where there has been a significant amount of structural and igneous activity in the basement affecting the overlying sedimentary strata. Because of its shallow depth, the Black Mesa, Defiant Uplift, and Four Corners Platform (BDF) is an optimal area for using aeromagnetics to define basement structure and architecture and relating that to existing production, shallow sedimentary stratigraphy and structural

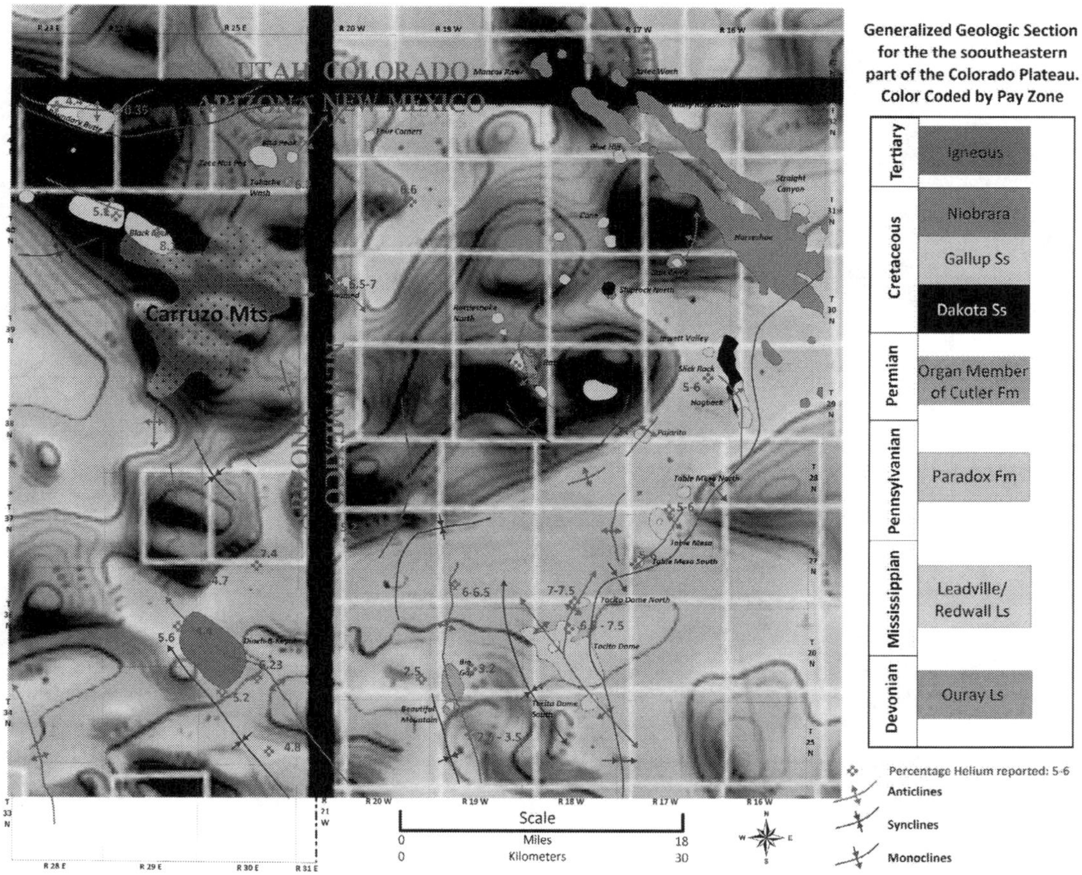

FIGURE 4.13 Total magnetic intensity for the Black Mesa, Defiant Uplift and Four Corners Platform Area, Southern Colorado Plateau from public data as processed, interpretated and reprinted with permission by Earthfield Technology. Aeromagnetics printed with permission of Earthfield Technologies.

configuration. BDF was discussed in Chapter 3 and contains numerous He fields that have produced since the 1940s in New Mexico (Broadhead and Gillard, 2004) and 1960s in Arizona (Rauzi, 2003). Fig. 4.13 is the total aeromagnetic intensity of the BDF area from public data, processed and interpreted by Earthfield Technology. The aeromagnetic data indicates several areas of high intensity and several associated hydrocarbons and He fields. There is a very linear nature to many of the anomalies with strong northwest-southeast and northeast-southwest trends (Fig. 4.13). The Horseshoe field in the upper right which produces from shallow Cretaceous reservoirs lies almost parallel to two strong, in tandem, northwest-southeastern aeromagnetic anomalies. The Carruzo Mountains on the western side of the map is closely associated with a strong northwestern-southeastern trending and expansive anomaly (Fig. 4.13). The East Boundary Butte, Dry Mesa, Rattlesnake Bita Peak, and Black Rock fields lie on intense aeromagnetic anomalies. The Dineh-B-Keyah and Hogback-Slick Rock fields, which are major He producing reservoirs, lie on a northwest-southeastern trending low that transects a strong northeast-southwest tightening of gradient (Fig. 4.13). The Tocito Dome, Beautiful Mountain, and Big Gap fields lie on an east-northeast-west-southwestern trending anomalous low with similar northwest-southeast structural cross-trends.

One of the keys to relating hydrocarbon, He, and CO_2 accumulations to basement structure is processing aeromagnetic data to extract other parameters such as identifying deep basement faults and areas of potential intrusions. Fig. 4.14 is the interpretation of the basement faults and areas that indicate intrusions in the BDF area. Interpretation of intrusions and basement faults are based on discontinuities that are defined by rocks with differing properties and/or the presence or absence of magnetic minerals in different rock units. If there is basement faulting or intrusions with similar characteristic to the adjacent or host rock, they may not be identified by magnetic means. The intrusions are indicated in gray and are large and numerous across most of the BDF area but tend to disappear to the southeast corner of the map which is the San Juan Basin (Fig. 4.14). The Carruzo Mountains lie directly over a major intrusion suggesting good correlation with this volcanic area. Many

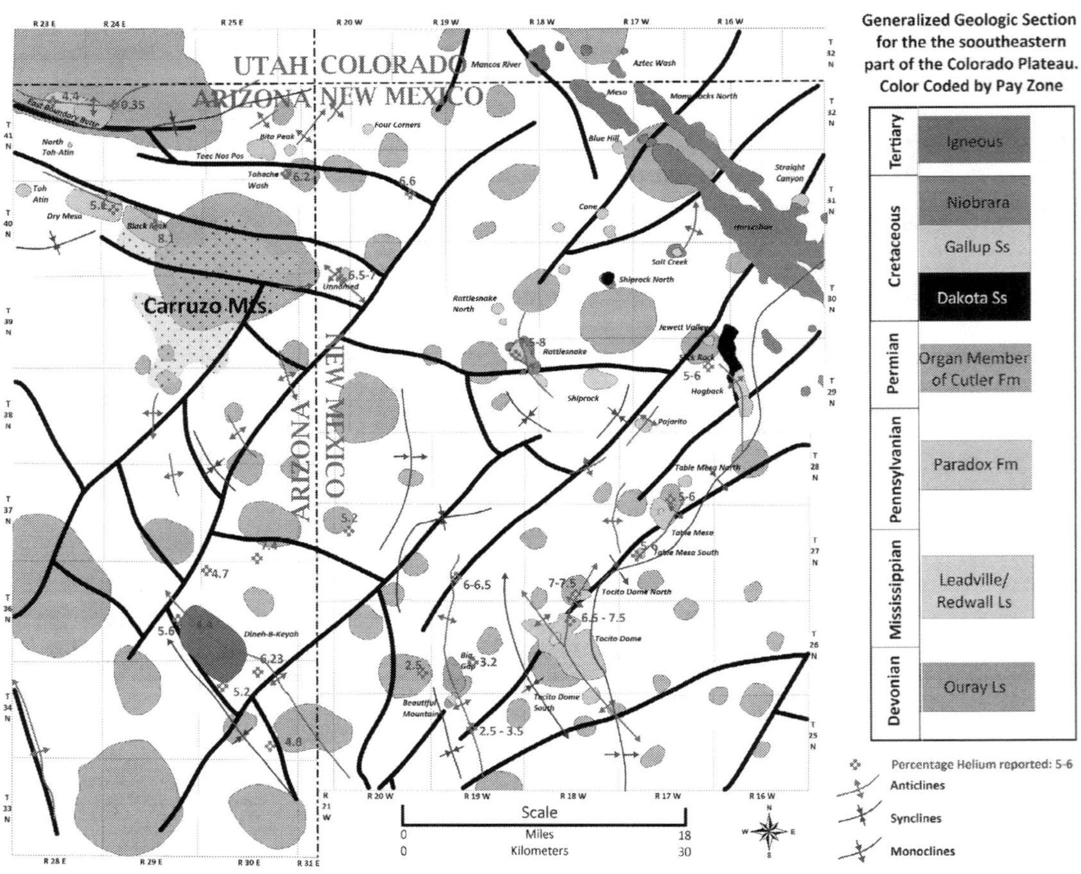

FIGURE 4.14 Deep basement faults and potential intrusions for the Black Mesa, Defiant Uplift and Four Corners Platform Area, Southern Colorado Plateau from public data as processed, interpretated and from aeromagnetic analyzed by Earthfield Technology.

of the existing He producing fields lie along basement faults and have associated intrusions (Fig. 4.14). The association of intrusion and basement faulting would allow migration of low to hot temperature hydrothermal fluids that could bring significant amounts of He, CO_2, and N_2 and then eventually hydrocarbons into the reservoirs in the Paleozoic strata.

The Horseshoe Field only produces hydrocarbon from Cretaceous reservoirs and trends perpendicular to the interpreted basement faulting (Fig. 4.14). While this may be perceived as having no relation, it can be suggested that the basement faulting could have provided petroleum migration pathways if they extended into the Cretaceous strata.

The aeromagnetic survey interpreted in Figs. 4.13 and 4.14 is public data and the data were acquired along widely spaced flight lines. As will be seen below, in more detail, a survey allows greater detail and correlation with subsurface geology.

4.6.2 Cisco Dome—Cisco Springs—Harley Dome

The Cisco Dome—Cisco Springs and Harley Dome area, Grand County, Utah, Colorado Plateau, has been discussed in Chapter 3 (Fig. 3.43). The Cisco Dome field lies on the west side of Cisco Springs in the center and Harley Dome on the east side of the gravity and aeromagnetic surveys that area indicated in Figs. 4.15 and 4.16. The data were acquired by Fugro

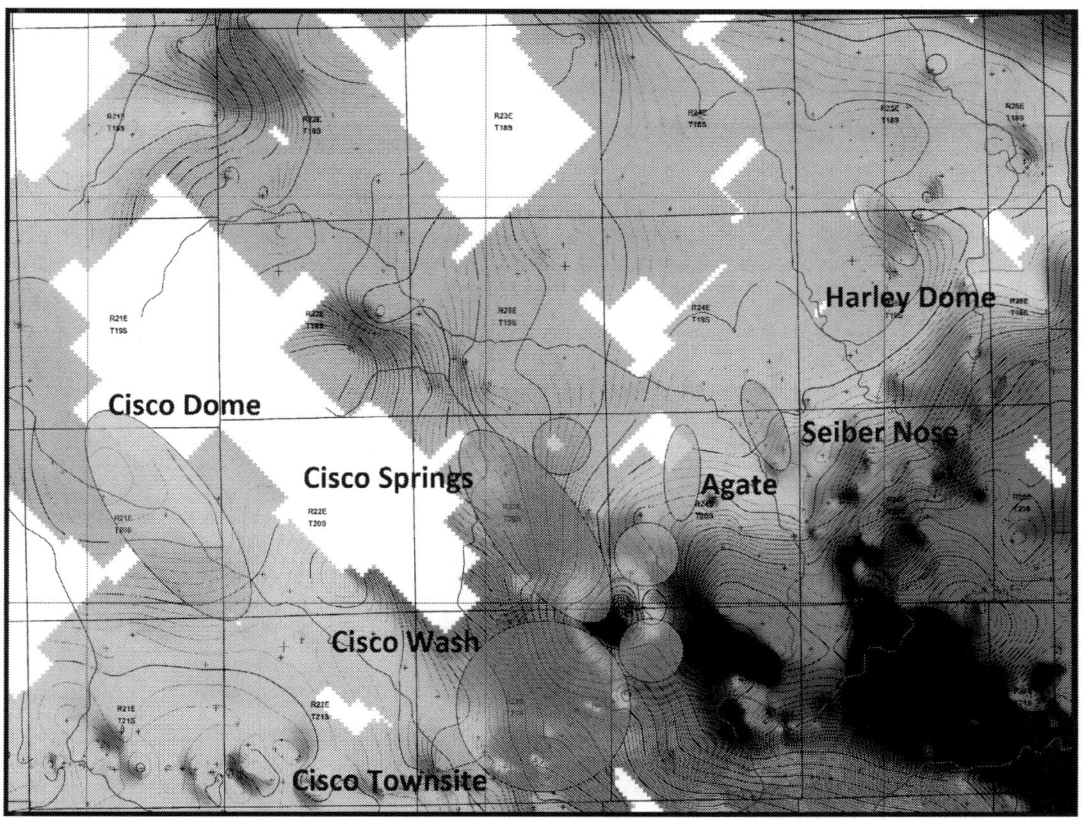

FIGURE 4.15 Gravity map indicating areas of public data and provided by Atoka Geochemical Services Corp., Centennial, Colorado and interpreted by Earthfield Technology, Houston, Texas.

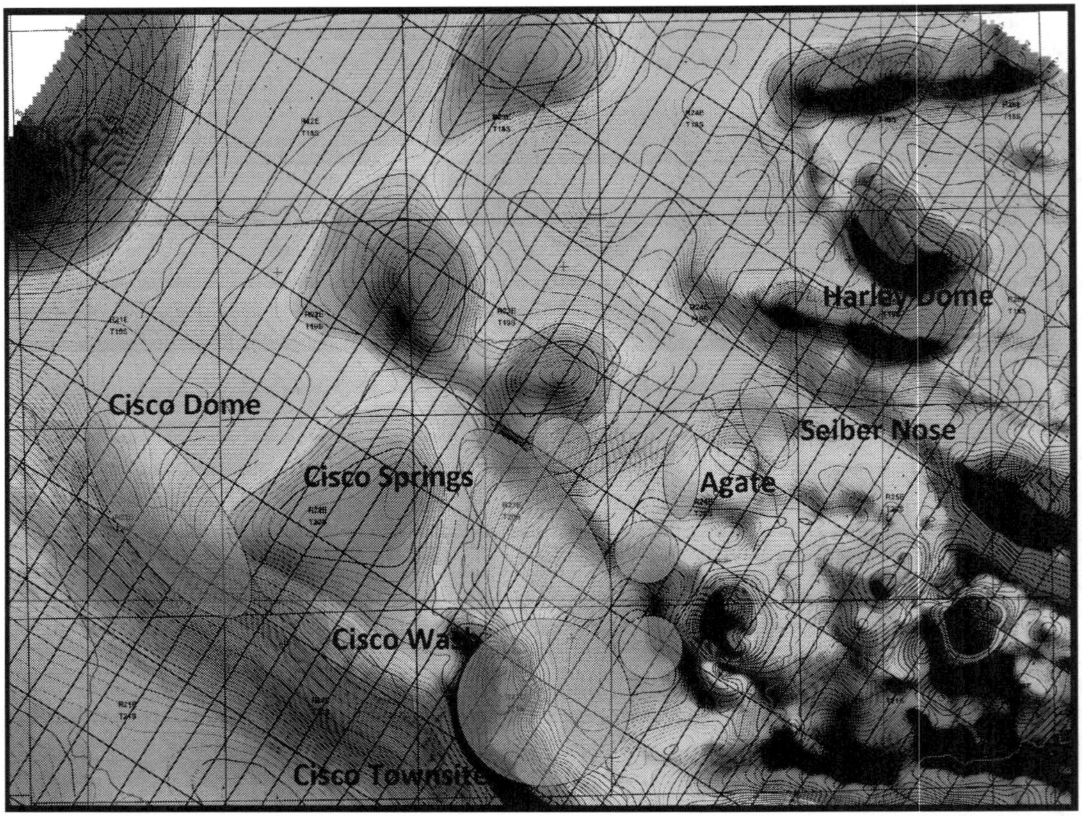

FIGURE 4.16 Total magnetic survey with flight lines over the Cisco Dome, Cisco Spring, and Harley Dome in Grand County, Utah. *Published with Atoka Geochemical Services Corp., Centennial, Colorado, and interpreted by Earthfield Technology, Houston, Texas).*

Airborne Surveys, Ottawa, Canada, for Atoka Geochemical Services Corporation and was part of an exploration program to find additional oil and gas resources. The survey acquired 2043 miles (3405 km) that has 32 degree orientation that has traverse lines spaced 4124 feet (1609 m) and cross-lines of 15,929 feet (3405 m) with a nominal terrain clearance of 120 m. A Scintrex SC-3 magnetometer was used.

4.6.2.1 *Gravity map*

The tilt derivative of the Bouguer Gravity is defined in terms of the ratio of the first vertical derivative of the potential field to the horizontal gradient of the field. This measure has the property of being positive over a source and negative elsewhere. This in essence accentuates more subtle anomalies, and positive tilt derivative anomalies are centered directly over source bodies. This technique can be applied to magnetic data as well. The public gravity data presented here for the Cisco Dome—Cisco Springs—Harley Dome area was part of the aeromagnetic data acquired in 2007 (Fig. 4.15). The existing fields are indicated as elongated ellipses.

Total magnetic intensity survey with flight lines over the Cisco Dome, Cisco Springs, and Harley Dome fields in Grand County, Utah (Fig. 4.16). The field outlines are overlain on Fig. 4.16 that are represented by elongated ellipses. In general, the fields correspond to a tightening of contours such as the Cisco Dome which is a structural closure and the Cisco Springs which lies on a northwest plunging monocline with localized structures along its trace (see Chapter 3). The Harley Dome lies on a positive anomaly that is probably related to intrusive activity.

Reduction to Pole (RTP) is a standard part of the magnetic data processing method, especially for large-scale mapping (Fig. 4.17). The RTP operation can transform a magnetic anomaly caused by an arbitrary source into the anomaly that the same source would produce if it is located at the pole and magnetized by induction only. The results are similar to the fields in relation to a tightening of contours as seen at Cisco Dome and Cisco Springs and a potential intrusion as at Harley Dome.

Depth to basement with associated deep faults is indicated in Fig. 4.18. Cisco Dome is associated with two different fault trends with a basement closure. The Cisco Springs Field that lies on a northwest trending monocline is associated primarily with a northeast trending fault

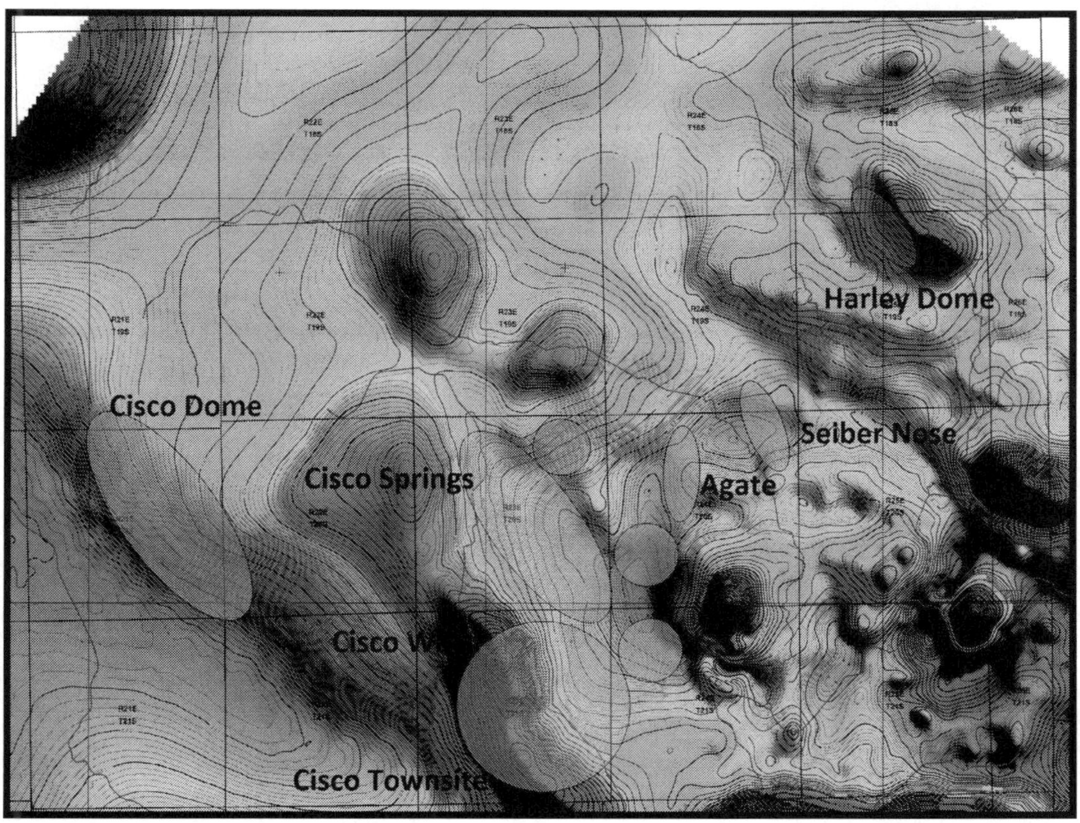

FIGURE 4.17 Reduction to Pole (RFT) published with permission from Atoka Geochemical Services Corp., Centennial, Colorado, and interpreted by Earthfield Technology, Houston, Texas.

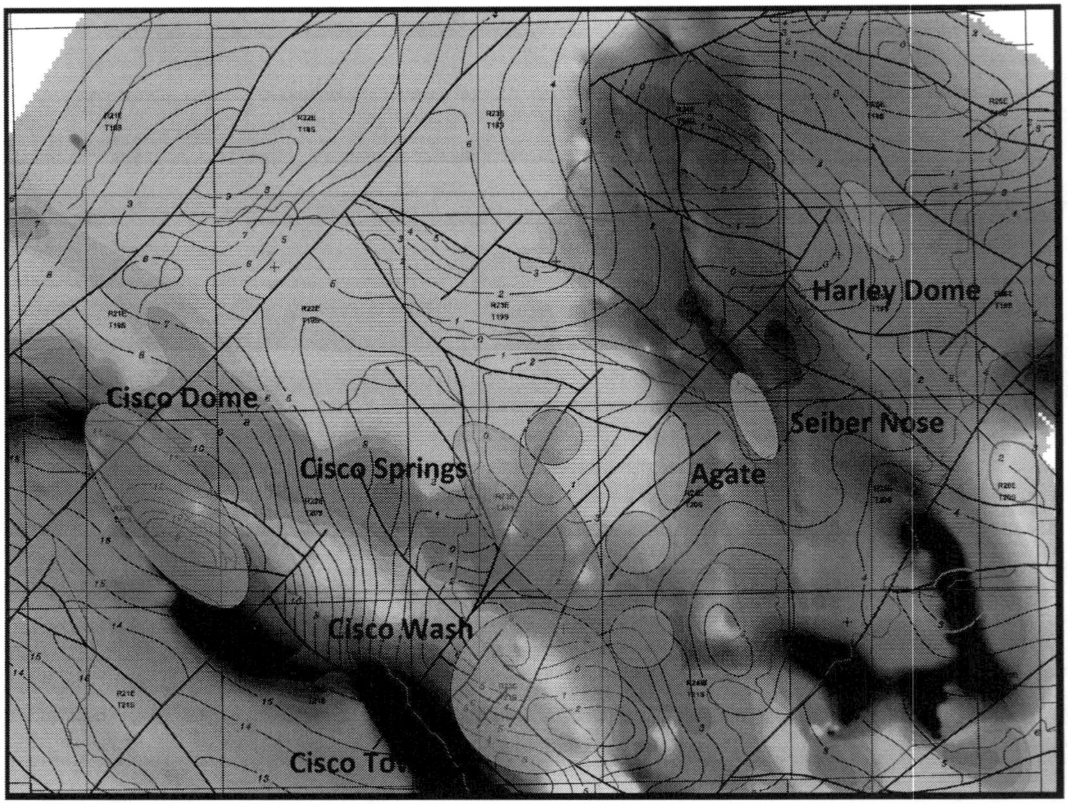

FIGURE 4.18 Depth to basement of the Cisco Dome—Cisco Springs—Harley Dome from the aermagnetic data. *Published with permission by Atoka Geochemical Services Corp., Centennial, Colorado, and interpreted by Earthfield Technology, Houston, Texas.*

system and a secondary trend to the northwest. The Cisco Wash Field has multiple short segmented faults that reflect the erratic nature of the isolated petroleum productive reservoirs present. The Harley Dome lies in the center of a northwest trending basement wrench system and is not directly related to any faulting.

RTP High pass filter residuals with basement faults are the result of high pass frequency filtering the RTP data with the filter cutoffs of 15 km and 35 km RTP data wavelengths (Fig. 4.19). For 15 km high pass grid, all RTP anomalies with wavelengths greater than 15 km are removed. For the 35 km grid, all wavelengths longer than 35 km are removed. These maps remove the effects of very deep/long wavelength-related sources. The 15 km high pass retains shallower sourced anomalies than the 35 km high pass.

RTP analytic signal is a gradient calculation that transforms the RTP data to have zero negative components. The analytic signal identifies the edges of magnetic and gravity anomalies which are often fault-related, and displays these gradients as highs.

All of the existing fields in the Cisco area seem to be associated with basement faults (Fig. 4.20). Possible intrusive anomalies, basement faults, and He data are indicated in

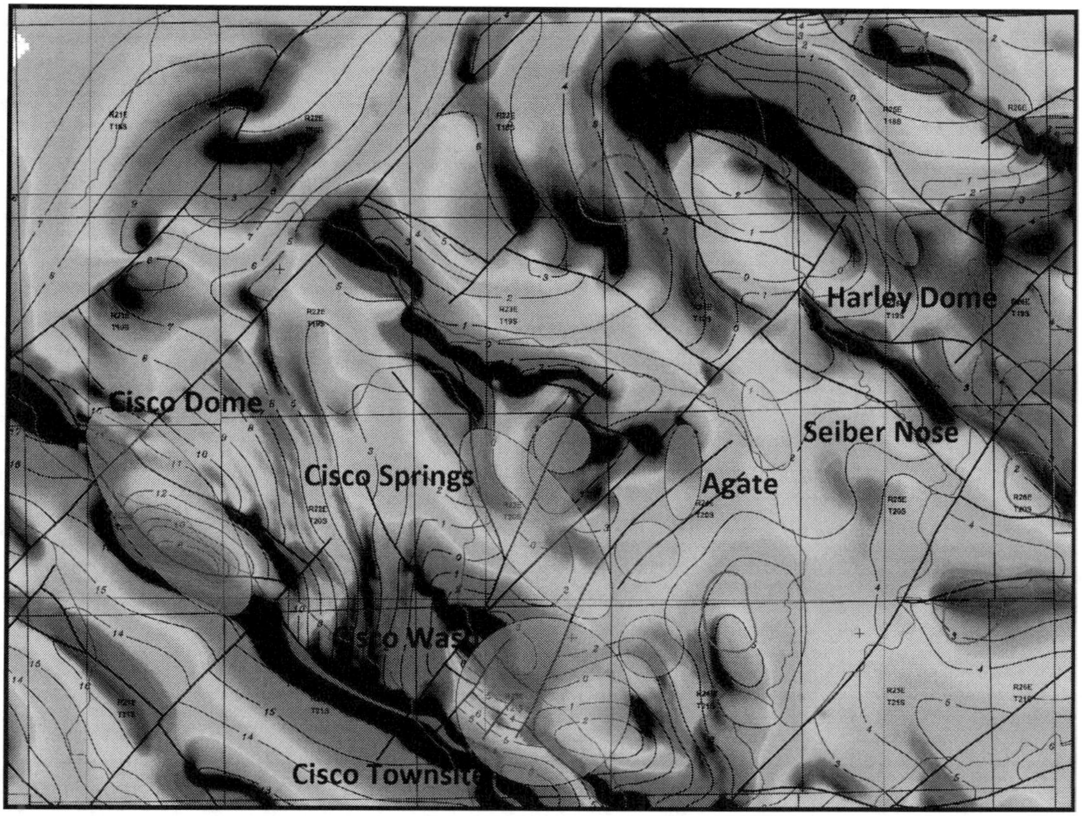

FIGURE 4.19 RTP high pass with basement faults. *Published with permission by Atoka Geochemical Services Corp., Centennial, Colorado, and interpreted by Earthfield Technology, Houston, Texas.*

Fig. 4.20 which is derived from the aeromagnetic analysis. The high He (red and orange color) at Harley Dome is associated with at least two intrusions. Lesser He anomalies are seen in the Cisco Springs area and are not closely associated with possible intrusions. The Cisco Dome, which is relatively devoid of He, and its natural gas production is not associated with any intrusions. There are several basement faults with associated intrusions related to areas that have not been drilled.

4.6.3 Sheep Mountain-Oakdale and Model Dome fields

The aeromagnetic surveys over the Sheep Mountain-Oakdale Fields were processed and interpreted from public data in an attempt to identify basement faults and depth to basement and interpret intrusions indicated in Fig. 4.21. The Sheep MountainOakdale Fields are located on the northern rim of the Raton Basin with a series of north-south intrusions to the east (see Chapter 3). In Fig. 4.21A the Sheep Mountain-Oakdale Fields lie on the east side of a northwest-southeast basement fault that is intersected with two northeast-southwest trending faults at the north and southern ends. The magnetic basement is very shallow and

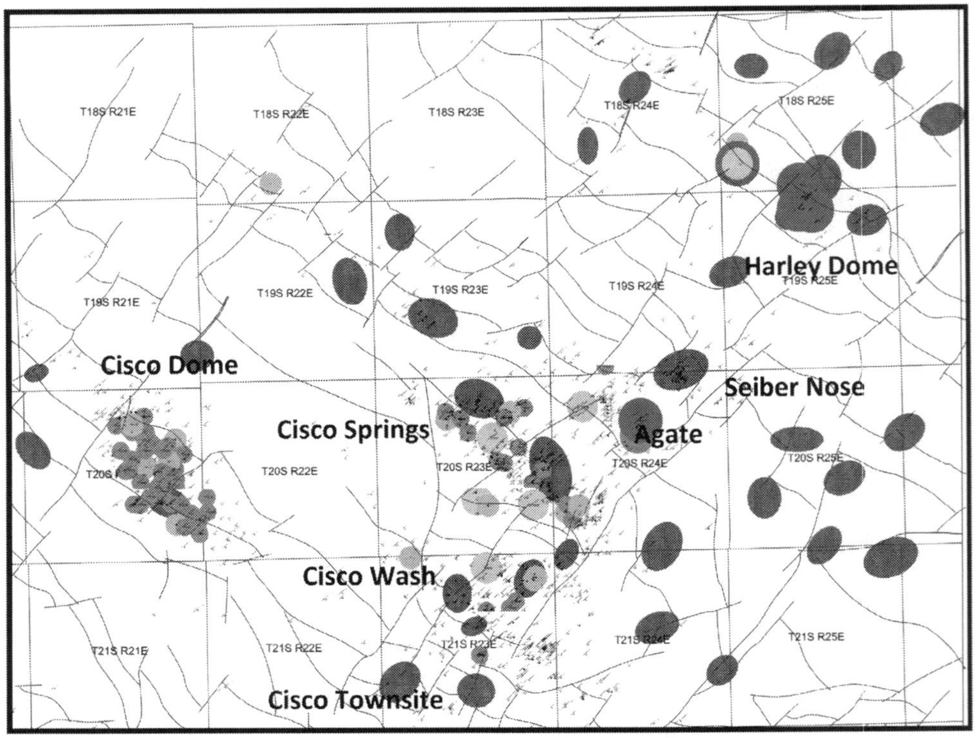

FIGURE 4.20 Possible intrusions (olive drab) and basement faults (*blue lines*) based on interpretation of aeromagnetic data with existing fields and He data indicated (gray—lowest values to red—highest values). *Published with permission by Atoka Geochemical Services Corp., Centennial, Colorado, and interpreted by Earthfield Technology, Houston, Texas.*

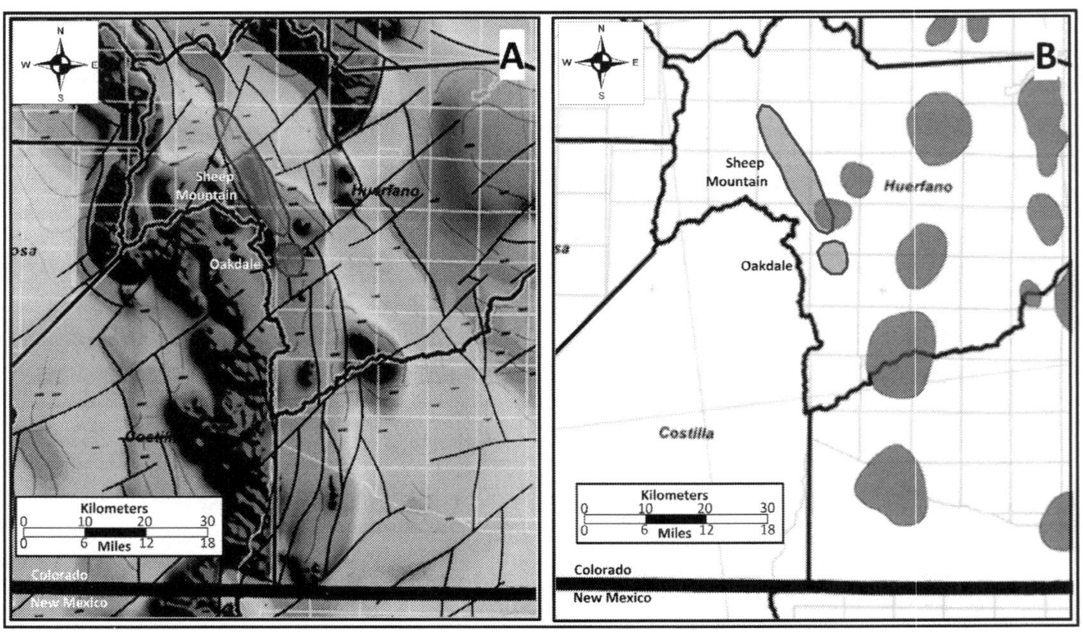

FIGURE 4.21 Depth to basement with basement faults (A) and interpretated intrusions (B) for the Sheep Mountain and Oakdale CO2 fields as derived from regional aeromagnetics (Earthfield Technologies). Color scheme for A is red is shallow basement and blue is basement at its greatest depth. In B the intrusions are red and were derived from the aeromagnetic data is publicly available. Printed with permission of Earthfield Technologies.

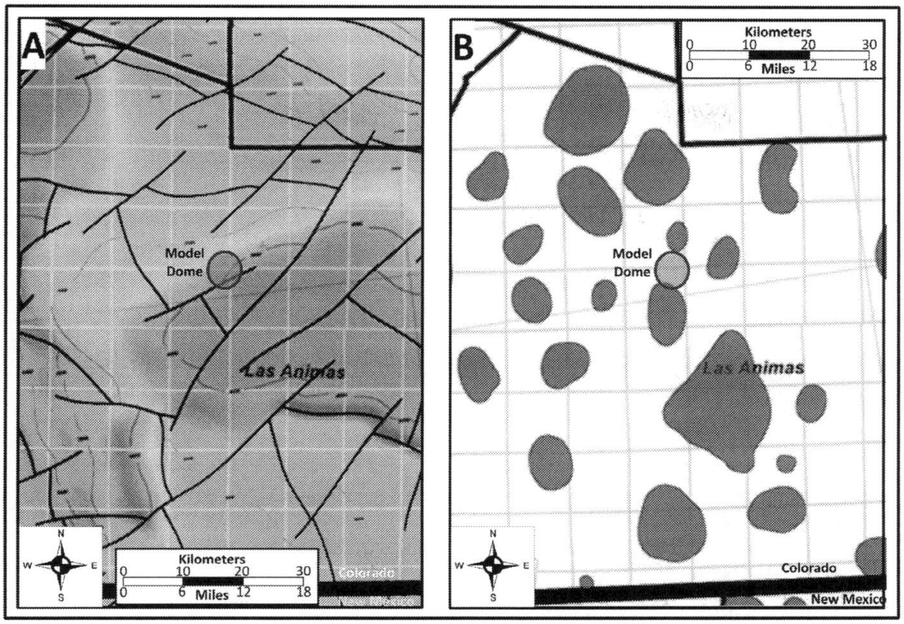

FIGURE 4.22 Depth to basement with basement faults (A) and interpretated intrusions (B) for the Model Dome He Field as derived from regional aeromagnetics (Earthfield Technologies). Color scheme for A is red is shallow basement and blue is basement at its greatest depth. In B the intrusions are red and were derived from the aeromagnetic data that is publicly available. Printed with permission of Earthfield Technologies.

coincides with the thrusting of the Sangre de Cristo Mountains as described in Chapter 3 that lies to the west of the fields. Fig. 4.21B indicates the location of the intrusions, which tend to lie at some distance to the east of the Sheep Mountain-Oakdale Fields. Two small intrusions lie either under the lower part of the Sheep Mountain-Oakdale fields or just to the east (Fig. 4.21B).

The Las Animas County, Colorado area, (see Chapter 3) has a single He producing field called the Model Dome. Fig. 4.22 displays the aeromagnetic data as processed to indicate basement faulting (A) and intrusions (B). The field is located along a northeast-southwest trending fault system (Fig. 4.22A) and between two intrusions (Fig. 4.22B). The area is similar aeromagnetically to the BDF area presented in Figs. 4.13 and 4.14 in terms of extensive basement faulting and basement intrusions.

A concern in using interpreted aeromagnetic data of basement structures and intrusions as an exploration tool is there needs to be some way to physically sample these bodies for their U and Th content. Intrusions can have highly variable chemical composition, and without a methodology to determine that composition their use as an exploration tool is limited. Basement faults are determined by rapid changes in aeromagnetic gradient and commonly represent not only discontinuities but changes in chemical composition of the basement rocks. Aeromagnetics does provide a good first pass tool to delineate areas of potential focus.

4.7 Surface geochemical methods

Petroleum exploration began by using macroseeps to find hydrocarbon accumulations (Davidson, 1963; Sittig, 1980; Hunt, 1995; Tedesco, 1994, 2021). Most fields discovered early in the development of the petroleum industry were found by surface mapping and or visible oil and gas seeps. Macroseeps at the surface are generally represented as asphalt and tar residues, such as the La Brea tar pits in Los Angeles, California, Iran, the Baku seeps of Azerbaijan in the former Soviet Union, and the seeps of Venezuela. In the pre-Christian era around Hit, Iran, there was a thriving bitumen industry (Tedesco, 1994). In southwestern Trinidad, a large seep called Pitch Lake, which is 2000 feet (606 m) across and 150 feet (45 m) deep, has yielded over 25 million tons of asphalt (Davis, 1967; Tedesco, 1994). Macroseeps have led to the discovery of numerous oil fields. This form of exploration was predominantly used prior to 1900 although oil and gas macroseeps are still major criteria for focusing on exploration of target areas such as the Basin and Range Province, Nevada (Tedesco, 1994). The seeps of Madagascar have caused international oil company activity in the past decades, and the California offshore seeps have long pointed to prolific oil production (Tedesco, 1994).

A summary of what happens in the soil substrate when petroleum and other associated gas migrate through as macroseep follows:

1. Evaporation of volatile hydrocarbons. In the first 2 weeks, a seep loses all petroleum with a boiling point up to 250°C (C, to C). Over later weeks or months, additional heavier hydrocarbons are degraded and evaporated.
2. Microbial degradation of petroleum through oxidation.
3. Leaching of water-soluble elements and compounds of S, O, and N_2 by oxidation.
4. Polymerization of petroleum molecules by eliminating H_2O, CO_2, and H. This causes the formation of even larger hydrocarbon molecules.
5. Auto-oxidation causing the formation of O-rich polymers. Petroleum components absorb sunlight and O when exposed at the surface for long periods of time.
6. Formation of rigid gel structures (gelation), which may develop for some seeps.

In the 1940s and 1950s technology and increased interest in surface geochemical methods developed to detect microseeps or leakage of hydrocarbons from depth that could not be seen with the naked eye. The presence of both macro- or microseeps at the surface is based on the concept of vertical migration. Vertical migration assumes that all oil and gas accumulations generally release micro amounts of hydrocarbons, sometimes large volumes (macroseeps), by migration through the overlying rock layers to the surface (Tedesco, 1994). Seals overlying the reservoirs are assumed to always be imperfect and to have multiple microfractures, discontinuities, and disconformities plus variations in lithology that allow leakage. The cross-sectional shape of the leaking pattern has conventionally been labeled as "chimney effect." Most chimneys are nearly vertical, although hydrodynamic flow strongly suggests that groundwater movement and the dip of the strata should deflect the ascending hydrocarbons. Evidence of that upward movement of gases is so rapid and so great that the vertical movement vector grossly dominates hydrocarbon transport (Tedesco, 1994, 2021). Fig. 4.23 is a classical representation of this concept. Many authors have noted the lack of offset between the surface geochemical at the surface and the petroleum accumulation at depth.

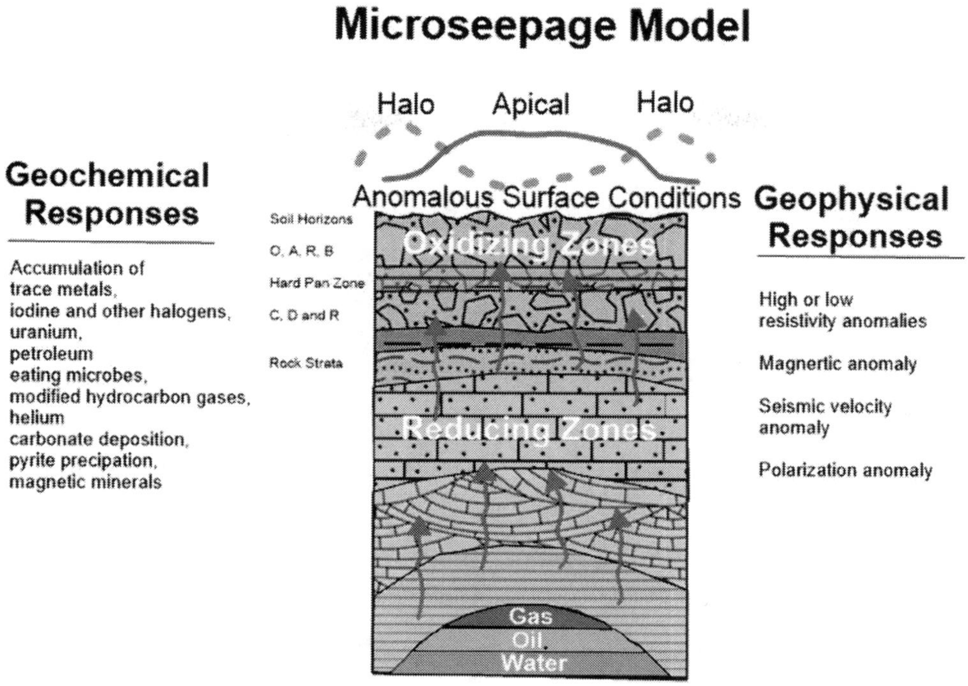

FIGURE 4.23 Microseepage model (Tedesco, 2021).

The mechanism of upward migration is through colloidal or buoyancy processes of hydrocarbons as well as other gases through the micropores, fractures, and unconformities as a result of decreasing pressure upward (Tedesco, 1994). This process deposits hydrocarbons, He and other gases in the soil and creates anomalous conditions for these elements and compounds as well as forming carbonate, oxide and uranium compounds associated with either redox or oxidation reactions. Methods have been developed to detect these migrating hydrocarbons, other gases and resulting byproducts in the soil environment (Fig. 4.23).

Microseeps are only detectable using laboratory or mobile detectors that report in the part per million or part per billion range. The sensitivity of the detectors is critical to the ability to determine the presence or absence of microseepage. The lighter hydrocarbons at the micro-level can be difficult as they are mobile, subject to barometric changes, and subject to rapid variations in concentrations at sample sites. Helium is similar to lighter-end hydrocarbons in that it is typically volatile and can be elusive depending upon barometric pressure, soil conditions, and weather.

Many surface phenomena are empirically associated with known deep hydrocarbon reservoirs. Any concept or theory for vertical migration of hydrocarbons must address the following observed conditions (not necessarily in order of importance):

1. Increases in specific type of petroleum-consuming bacteria in association with the geochemical anomaly overlying the accumulation
2. Changes in pH and Eh in the soil substrate

3. Increases in trace and major metals in various strata above the petroleum deposits
4. Increases in halogens, specifically iodine, in the soil substrate and in brines adjacent to the reservoir
5. Increases or decreases in U minerals and their daughter products in the soil substrate
6. Increases in magnetic and nonmagnetic iron minerals both in and above a petroleum reservoir
7. Increases or decreases in He and radon in the soil substrate
8. Increases in hydrocarbon gases from immediately above the reservoir to the surface
9. Cementation of clastic sediments concentrated in zones between the reservoir and the surface
10. Occurrence of high-velocity zones in overlying rock strata
11. Concentrations of iron and manganese clay minerals due to complexing by organic acids in the overlying strata
12. Increases in carbonate minerals in overlying strata
13. Interaction between calcite- and kerogen-derived carbonate causing depletion of C13 in the soil substrate
14. Decreases in gamma-ray response at the surface corresponding to the field
15. Changes in vegetation at the surface
16. Disruption of groundwater sulfate equilibriums by hydrocarbon-consuming and reducing bacteria

The use of surface geochemical methods to find hydrocarbons has direct application to finding He accumulations, as they tend to be associated with natural gas. The exploration for CO_2 and N_2 while they should from a theoretical standpoint be detectable using surface geochemical methods, their presence in the soil and atmosphere as part of biological processes make them less useful. Hanson et al. (2014) utilized a CO_2 survey to define a blind geothermal system in Taupo Volcanic Zone, New Zealand, and concluded they could define it but that the survey could not stand alone in defining the system. There are difficulties in determining anomalous or background conditions because of biological activity generating CO_2 and the atmosphere is made up of approximately 80% N_2. Helium has historically been used for detecting faults as they tend to leak the element readily. Examples of He surface geochemical surveys in relation to existing He, oil and gas fields are very sparse in the literature. The primary way to detecting migrating He accumulations at depth is to analyze for He in the soil substrate that has migrated from depth. Detecting He accumulations can be augmented by using other surface geochemical methods used to find petroleum. Some of these methods are directly related to hydrocarbons and are less expensive. Unfortunately, they do not necessarily imply the presence of a He accumulation.

4.7.1 Examples of surface geochemical methods

There are very few surveys in the literature that demonstrate the use of He for finding He deposits. In some cases, He surveys have been used to find fault systems and oil and gas fields (Tedesco, 1994) and degassing related to earthquakes or tectonic activity (Guerra and Lombardi, 2001; Deshpande and Gupta, 2013). Helium surveys have been used extensively to find U deposits (Reimer et al., 1979). Dyck (1976) presented data that concluded the He would be useful in the discovery of radioactive ore deposits, uranium deposits, oil and gas fields, and fracture zones related to mineral occurrences under the right conditions. Butt and Gole (1985) concluded that He surveys could not be used, as they were not dependable to find blind mineral deposits as they failed to define three existing deposits they had surveyed. It was determined that the He was easily flushed from the soil section. Butt and Gore (1986) found a similar concern in using He surveys from groundwater even though they were able to define existing deposits but found numerous false anomalies that seem to be related to other sources. A He survey that was collected on an evenly spaced grid over the WaKeeney area, Trego County, Kansas, even after correcting for soil moisture, soil type, and temporal variations showed a poor correlation to the existing fields (Cunningham, 1987). The fields in the survey that Cunningham (1987) collected were all oil productive and did not contain He as an economic commodity.

A soil gas survey, utilizing both hydrocarbons and noble gas analysis, across the Teapot Dome in Wyoming utilizing $^{3}He/^{4}He$ isotope analysis identified leakage from structure of up to 37.7 ppm relative to atmospheric background of 5.24 ppm suggesting a deep source for the element (Macintosh and Ballentine, 2012). The results of the study recommended that groundwater analysis is more reliable than soil gas to determine what is going on in the subsurface (Macintosh and Ballentine, 2012). While this may be true, soil gas methods are more useful as ground water samples are typically not available in undrilled areas or even within existing fields. In addition, the depth of the soil zone (A,B,C, or D) of where air flushing takes place, due to barometric changes in pressure, on a daily basis has to be considered. If sampling is within a flushed zone, the result of sampling, there will be a lack of anything to analyze except for the atmosphere.

4.7.2 Bush Dome

The US Bureau of Mines carried out a soil gas and water well survey for He across the Bush Dome (Holland and Emmerson, 1979; Tedesco, 1994) that is located in the Cliffside facility, 12 miles (18 km) north of Amarillo, Texas. The dome is part of the Texas Panhandle-Hugoton Field (discussed in Chapter 3). The Bush Dome was discovered in 1924 and as of 1978 had stored 39 BCF of 70% rich He. Currently the facility is closed. The He was stored in the Brown Dolomite of Permian age at 3200 to 3500 feet (970–1060 m) with 400 feet (130 m) of anhydrite caprock. Fig. 4.24 is the result of the surveys analyzing for dissolved He analysis from the water wells and the soil gas surveys defining He leakage from Bush Dome (Holland and Emmerson, 1979; Tedesco, 1994). The two surveys show anomalous conditions in and around the Bush Dome (Holland and Emmerson, 1979). The He soil gas survey indicates linear features generally on the flank of the Bush Dome that may represent faulting. The water well survey indicates anomalous conditions across the Bush Dome as well as off of it. However, it also has to be remembered that the Bush Dome is part of a larger regional field and that the two surveys may represent leakage not specifically related to the structure. It is unclear what time of year the samples were collected, what the reservoir pressure was at the time, or any other factors that maybe significant as defined by Tedesco (1994) in relation to interpretation surface geochemical surveys. Was the Bush Dome at maximum storage and under maximum pressure conditions or was minimum storage capability and thus minimum

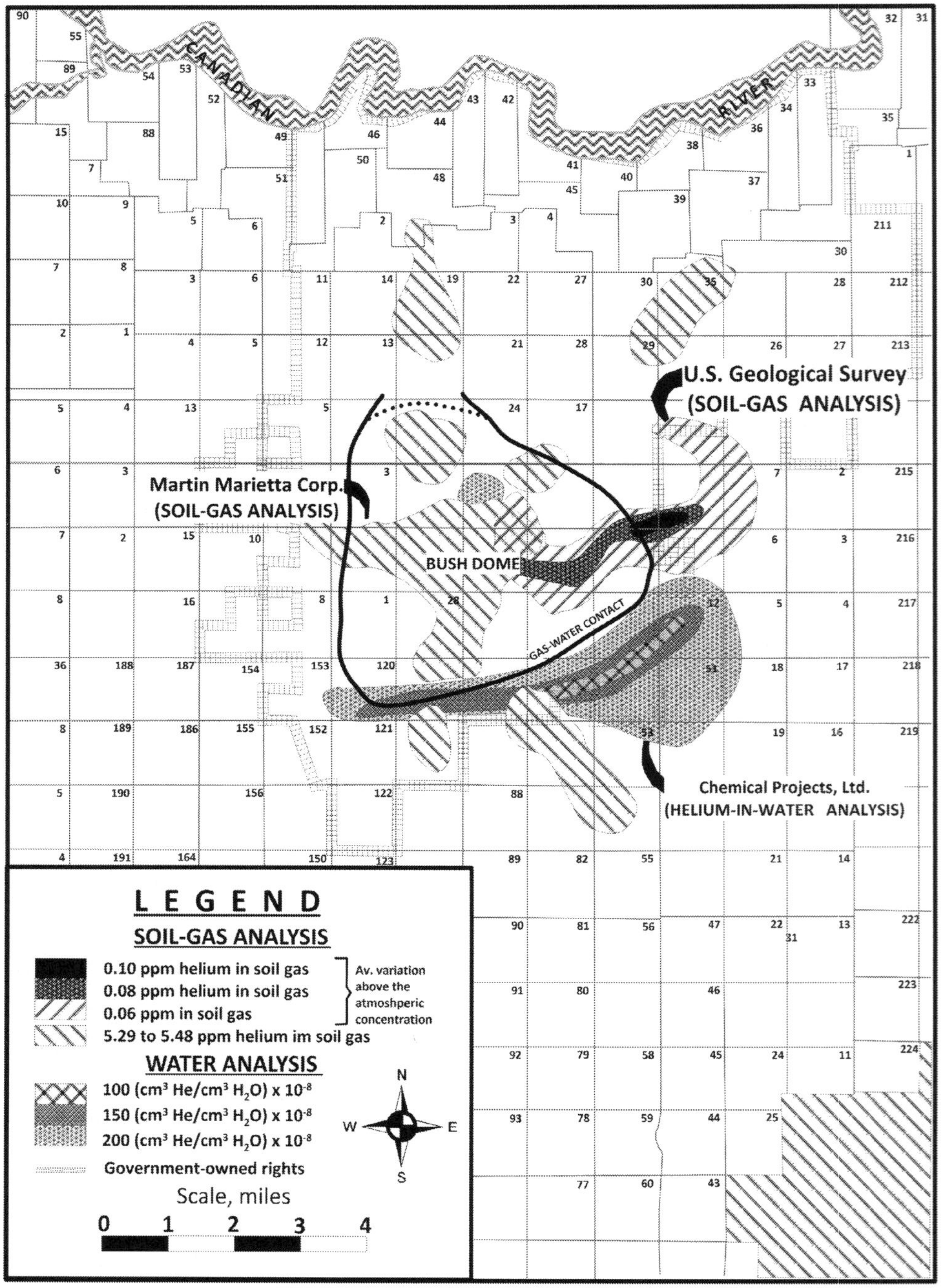

FIGURE 4.24 Soil gas and water analysis over the Bush Dome, part of the Cliffside Facility, Texas Panhandle-Hugoton Field from Holland and Emmerson, 1979 and modified for this publication.

pressure? Tedesco (1994, 2021) demonstrated that changes in injection and withdrawal from underground storage through time are reflected in changes the amount of soil gas occurring in the soil section. Anomalous hydrocarbons in the soil section and thus He above the storage reservoir are at their lowest when pressure in the reservoir is at a minimum.

4.7.3 Harley Dome

Harley Dome (discussed in Chapter 3) and is located in Grand County, Utah, has been the subject of at least one He surveys. The field produces He from the Jurassic Entrada sandstone reservoir at 965 feet (290 m). The survey presented in Fig. 4.25 was done by Roberts (1981) as part of a larger project of 15 He surveys organized to define existing oil and gas fields and

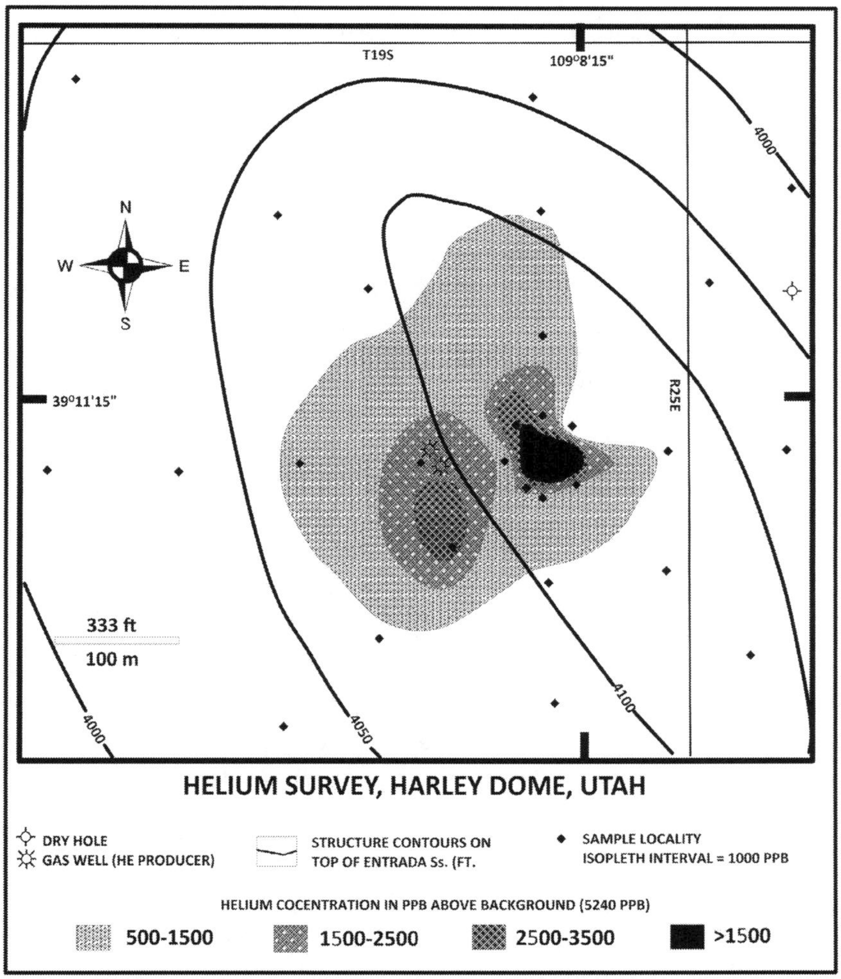

FIGURE 4.25 Helium survey over the Harley Dome Field by Roberts (1981), republished in Tedesco (1994) and modified by the author.

potentially identify undrilled productive locations. Those surveys by Roberts (1981) were across 11 fields and 4 prospects. According to Roberts, six fields were clearly defined by surface geochemical methods as existing petroleum fields, two were not seen as at all, and three showed scattered but definable anomalies. Two of the prospects did not have He or hydrocarbon soil gas anomalies associated with them and drilling resulted in dry holes. Two anomalies at the time of Robert's (1981) publication remained undrilled. The Harvey Dome survey as seen in Fig. 4.25 by Roberts (1981) indicates an anomalous pattern over the existing field.

4.7.4 Cisco Springs Field

The Cisco Springs Field is located in Township 21 South, Ranges 23 and 24 East, Grand County, Utah, and about 12 miles (20 km) to the west of Harley Dome Field (discussed in Chapter 3). The Cisco Springs Field has predominantly produced oil and gas from the Brushy Basin, Saltwash, Cedar Mountain, and Dakota sandstones. Helium percentages in the natural gas from the present gas wells in the field vary from 0.1% to 0.7% He, depending on what part of the area and what specific reservoir the gas was sampled from. Almost all of the He data presently available are from the Dakota sandstone with no analysis available from the Entrada sandstone in the field. The Entrada sandstone, the main He producer at Harley Dome Field (discussed in Chapter 3) just to the east, has had a few tests in the Cisco Springs Field area, none in areas of closure based on a recent 2D seismic survey and only limited oil or gas production from existing wells. The Entrada sandstone is a sheet-like aeolian clastic reservoir that requires structural closure or permeability/porosity changes to trap hydrocarbons and He.

A soil gas and He surface geochemical survey was conducted across the field prior to shooting the 2D seismic survey Township 20 South, Range 23 East, Grand County. Helium and soil gas samples were collected at 3 feet (1 m) using the free-air method as described by Tedesco (1994) with special containers to retain the He for a few weeks. The climate in the area is desert, soil profile is limited as in many areas the Cretaceous Mancos is the bedrock and maybe at the surface. A number of samples actually extracted soil gas for analysis from the Mancos Shale.

There were a number of potential target structural closures, as defined by the subsurface geology and 2D seismic survey, within the Cisco Springs Field for potential He production from the Entrada or deeper reservoirs. One of these areas is in sections 9 and 10, Township 21 South, Range 23 East, that is defined by structural closure and shallower gas production from the Dakota sandstone. The gas production in sections 9 and 10 is presented in Fig. 4.26 and indicates some excellent producing wells at less than 2000 feet (630 m) from reservoirs above the Entrada sandstone. Structure is indicated on the Dakota Sandstone with a 20-foot (6 m) contour interval. The closure in sections 9 and 10 (Fig. 4.26) has not been tested to the Entrada and as such a He and hydrocarbon survey was acquired and is presented as a class post or bubble map in Figs. 4.27–4.29. Fig. 4.27 indicates a classed post for He for the results of the survey in sections 9 and 10. The anomalous conditions as seen in Fig. 4.25 across the Harley Dome are not very high above atmospheric levels, Fig. 4.28 indicates the CH_4 results of the survey and indicates numerous anomalies in sections 9 and 10 and to the east where several gas wells already exist. Because of the nature of the climate

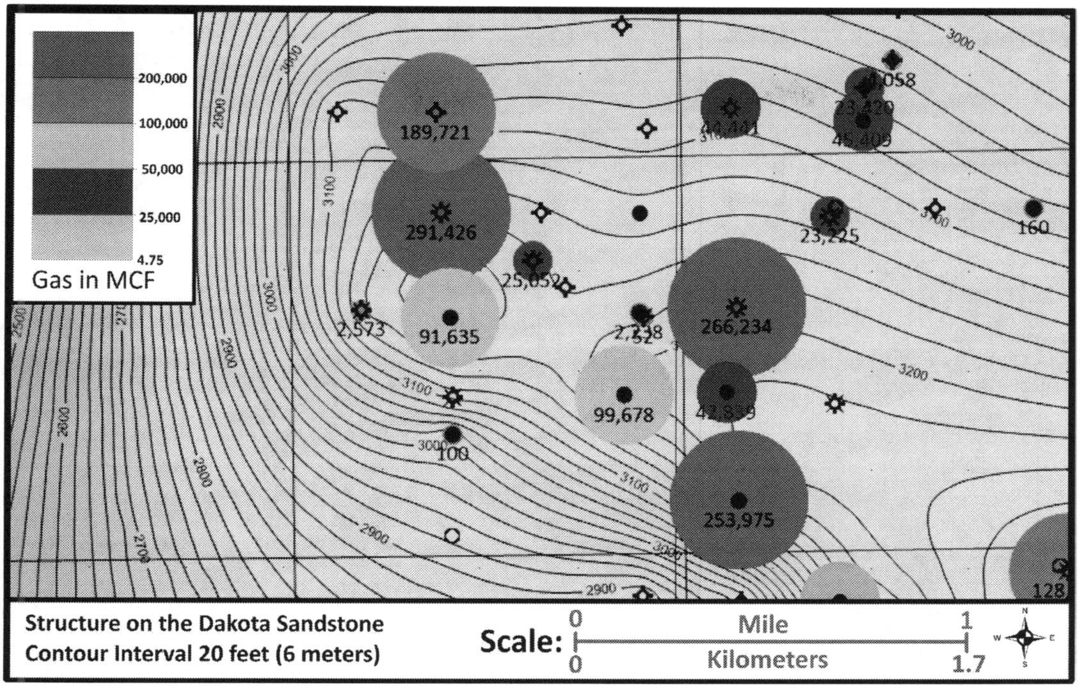

FIGURE 4.26 Gas production classed post or bubble map for Section 9 Township 20 South, Range 23 East, Grand County, Utah.

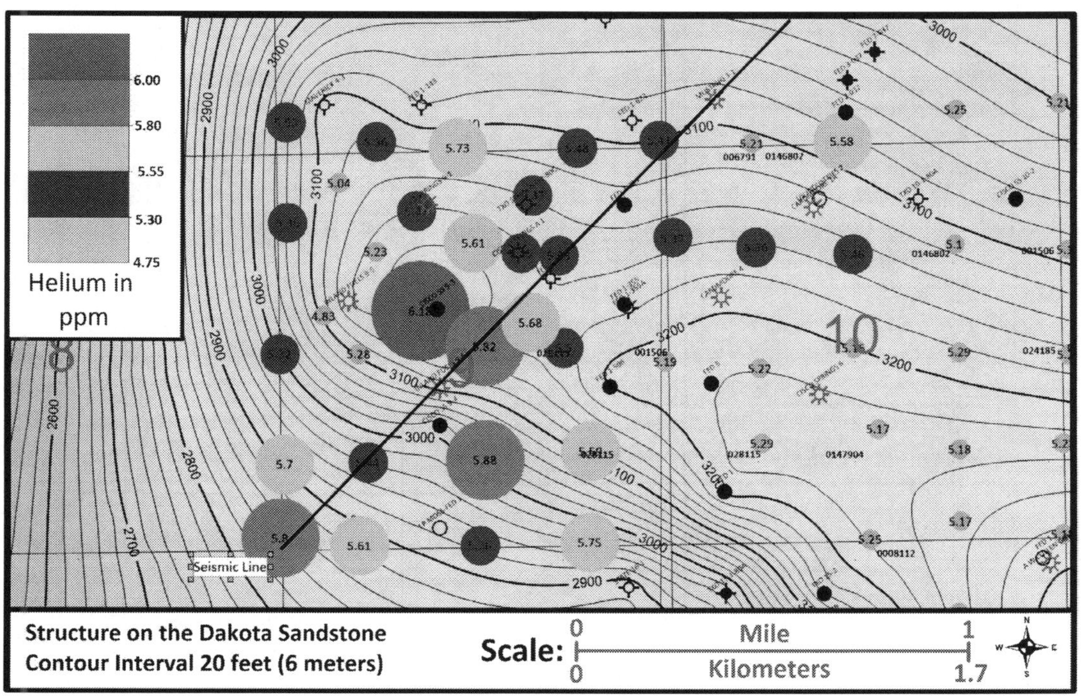

FIGURE 4.27 Helium survey classed post or bubble map for Section 9 Township 20 South, Range 23 East, Grand County, Utah. *Printed with permission from Running Foxes Petroleum.*

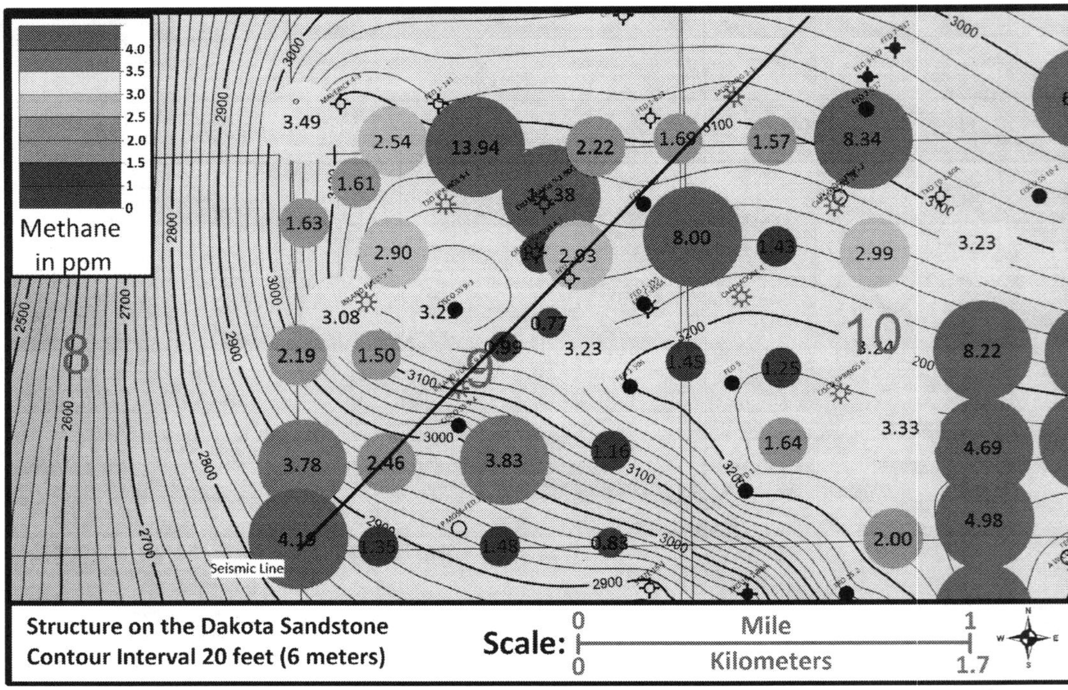

FIGURE 4.28 Methane classed post or bubble map for Section 9 Township 20 South, Range 23 East, Grand County, Utah. *Printed by permission from Running Foxes Petroleum.*

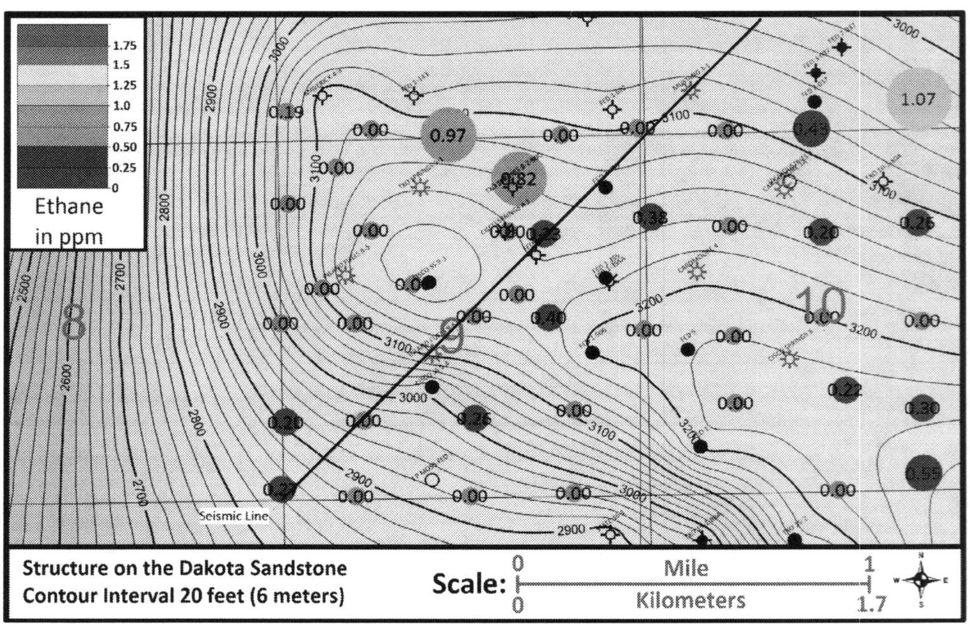

FIGURE 4.29 Ethane classed post or bubble map for Section 9 Township 20 South, Range 23 East, Grand County, Utah. *Printed by permission from Running Foxes Petroleum.*

and soil section in the area of the survey, the CH_4 results are not likely a product of biological activity generating CH_4. Fig. 4.29 is a classed post orbubble map of the C_2H_6 analysis from the survey and its general absence indicates that this a gas prone area. The present oil wells are subeconomic producers. The gas content from the producing wells in Sections 9 and 10 is typical 89% CH_4, the rest composed of heavier hydrocarbons, He, CO_2, and N_2.

4.8 Seismic

Seismic is a well-accepted methodology to find reservoirs and traps, define depositional settings and tectonic features in finding petroleum and thus He and CO_2. Seismic technology uses sound waves (pressure) waves to map rock layers in the subsurface. Different rock types have different velocities and densities which together give the layers impedance contrasts with each other. In turn, this contrast is manifested on the seismic profile as wave form changes. Sedimentary rocks generally form semi-parallel layers that present themselves in the form of contrasting and waves allow them to be mapped over large areas. These maps will display structures and stratigraphic changes in the subsurface and changes in wave form will indicate lateral changes in the impedance of the layers. These lateral impedance changes can be used to identify rock types and potential reservoirs. Seismic data can thus be used to not only identify closures and traps in the subsurface but also reservoirs, seals, and rock types. Seismic data are typically not processed or interpreted to look at strata below the top of the basement which is typically of Precambrian age and are igneous and metamorphic rocks. Precambrian rocks are typically igneous but in some areas such as the Mid-Continent Rift in the central part of the USA, they are clastic and metamorphic in nature. In some basins in California, USA, Cretaceous or older rocks are metamorphic rocks and termed "basement" for limits to oil and gas potential Chapter 3.

Seismic methods require contrast between different geologic units and within sedimentary basins this is generally impedance variations or contrasts mentioned previously. There are reservoirs that are almost invisible to seismic. Two rock types can exist where one has a high velocity and a low density and another has a lower velocity and a higher density. The impedence between these two layers may be so small as to make the layers invisible. The phenomenon is seen in some areas between sandstone and salt. An example of limits on seismic methods is the Morrow sandstones in the LAAHR (discussed in Chapter 3) where high porosity and permeable sandstones have little or no impedance contrast with the encasing shale. The Morrow valleys are generally easily defined but the sandstone reservoirs within in especially southeastern Colorado and western Kansas can be difficult to identify if not impossible.

Seismic provides information to define a drill site target location that other remote sensing methods cannot provide. Seismic is therefore critical to define the location of a potential He or CO_2 reservoir as implied in an area by other methods. Three-dimensional seismic provides a more definitive target definition than does 2D and 2D seismic swathe. Targets that are shallow, areas of difficult terrain or land with surface or political issues such federal land in the USA may dictate using 2D and 2D swathe over 3D seismic methods.

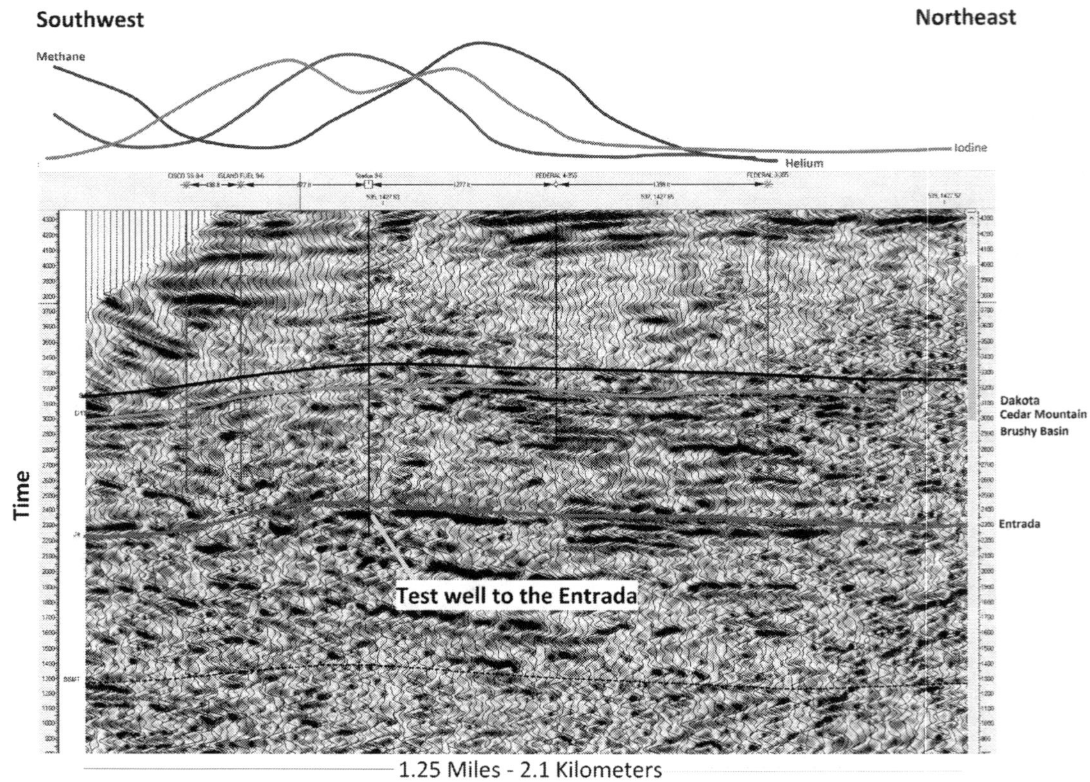

FIGURE 4.30 Seismic line through Section 9, Township 20 South, Range 23 East of the Cisco Springs Field, Grand County, Utah. The He, CH_4, and I_2 surface geochemical profiles are in abstract (no values presented) but help define the viability of the drilling the target. Seismic is measured in time. The test well will be drilled in 2022. *Printed with permission from Running Foxes Petroleum.*

An example of using seismic in conjunction with surface geochemistry is in Fig. 4.30. Approximately 100+ miles of 2D seismic data was acquired in 2021 across the Cisco Springs Field, Townships 21 and 22 South, Ranges 23 and 24 East, and Grand County, Utah (discussed in Chapter 3). A 3D seismic survey was not considered by the operator of the field because of US Bureau of Land Management requirements for wildlife and culture rules created significant increase in costs. Some clarity and detail are probably lost by not using 3D to survey over the project area but with the density of existing well bores analysis and interpretation it was concluded not to be detrimental to exploration analysis. A number of 2D seismic lines crossed over the Sections 9 and 10 mention in the surface geochemical surveys discussed in the previous section. Based on geologic data from existing wells this indicated a likely closure with potential for He and natural gas deeper than had previously drilled. The Section 9 target area described in Figs. 4.27–4.30 but only one line is presented here. While acquiring the seismic data, at every shot point a soil sample for I_2 analysis was collected. The profile in figure has the He, CH_4, and I_2 survey profiles across the target area and indicate anomalous conditions exist where the proposed drilling location is indicated. The target structure is similar in size to the Harley Dome Field for the Entrada sandstone.

Seismic typically is not used to specifically target He, and CO_2 producing reservoirs. Instead 2D and 3D seismic has been utilized to target petroleum potentially productive reservoirs. In certain circumstances this is changing as in the Central Kansas Uplift area where a few 3D seismic surveys have been shot with the intention of finding any remaining structural closures. Many of the potential targets expected from these 3D seismic surveys were typically expected to one or two wells in size. The economics of these type of surveys for small reserves is facilitated by an existing He processing infrastructure being present in the area. The results to date have been disappointing, with typical test wells finding depleted reservoirs, structural seismic busts, or lack of He or hydrocarbon in the tested structures. The Sweetwater Arch area that extends from Montana into southwester Saskatchewan and Southeastern Alberta is undergoing an intensive land play. A few small He fields exist which are being used as models in the area. In many parts of this play area there is a lack geologic or geophysical data. This play area is more akin to a typical petroleum land rush in the belief there is a resource present and the need to acquire as much land as possible. If this was a petroleum play, the land acquisition is typically followed by aeromagnetic, gravity, surface geochemical, 2D and 3D surveys. Because the targets in this play area are small to medium size basement structures or erosional remnant features with Cambrian sandstones draped over them drilling must be very focused. The Cambrian sandstones are not always present across the Precambrian structures or erosional remnants. Several Operators in the play area that have presented seismic data publicly are indicating the structures are only few hundred acres in size accommodating one to five He producers.

4.9 Models for helium exploration

There are different geologic settings where He, CO_2, and N_2 are present in significant quantities and other terranes that are lacking or barren of these two elements and/or compound. Carbon dioxide is associated with natural gas in many areas and its presence is not necessarily related to degassing of the crust, mantle or expulsion during volcanic activity. The majority of economic CO_2 fields discussed in Chapter 3 are related to significant tectonic and/or volcanic (degassing) events and the simple model for CO_2 exploration is to explore in either areas of volcanic activity such as the Salton Sea in California or Jackson Dome in Mississippi, or areas of thrusting such as Big Piney-La Barge Field. Nitrogen models are not presented individually as the element is not being pursued economically from natural gas streams.

There are several different types of models proposed for He exploration by Brown (2010, 2014), Haines and Allen (2019), Danabalan (2017), Zhang et al. (2017, 2019a,b, Craddock et al. (2017), Brown (2017), and Halford (2018). These models tend to be based on area-specific geologic and isotopic evidence. There is a strong association between He and Precambrian and Paleozoic strata when they are present. Where Paleozoic strata are absent, Mesozoic or Tertiary rocks are known to trap He, N_2, and CO_2 (aka Harley Dome, and East African Rift system). Deposits in these younger rocks have so far been exceptions proven to be small or as a teaser area with high percentage of He by volume but with no proven long-lasting reserves. The models for exploration described here are general suggestions subject to modification based on different geologic areas, or as new technologies and insights become available. To date there has been no universal model(s) proposed in the literature.

The proposed He models are as:

(1) SOAR Model: This is the simplest and most common of models and seems to be very repeatable from basin to basin. Anticlinal or domal traps, with or without some stratigraphic trap component, in relative proximity to a basement source with high U and Th, and with a very effective overlying seal. Reservoir examples include the Leadville, Madison, or Mississippian limestones; Morrow sandstones; Red River Limestones, Cherokee strata or Permian carbonates and sandstones. The presence of adjacent faulting or fracturing that connects the basement source with the potential reservoir allows migration of fluids and gases into the trap. Percentage of He in the trap typically needs to be greater than 1%, as these deposit tend to be small in areal extent and total gas volume as compared to the TPH and SGUY models described below. These are structural in nature, whether overall in form or internal configuration. The source of the He, N_2, and CO_2 can be from the mantle, crust, and/or release from minerals or chemical changes in the fluids or host rock. The source of the He and other gases is irrelevant because it generally cannot be determined until after drilling of a target and doing isotope analysis. Old (Paleozoic) pore water is an important criterion as older fluids have had more time to concentrate He. As always there are exceptions. For example, the Harley Dome produces from Jurassic reservoir, the Entrada, and has been sufficiently charged for economic production. What is consistent is the presence of closure with associated faulting and fracturing. The areas where this model is applicable includes the Sweetgrass-Bowdoin Dome in Alberta and Saskatchewan, Black Mesa and Holbrook basins in Arizona, Four Corners Platform in New Mexico, Colorado Plateau in Utah, Colorado Plateau, Las Animas Arch Region, and Central Kansas Uplift (see Chapter 3).

(2) PHF Model: This type of model is represented by the Panhandle Hugoton Field (PHF) in Texas, Oklahoma and Kansas, USA, which is a large regional reservoir that is a large-scale stratigraphic trap with multiple internal minor structural closures whereby He is in very low percentages but overall large volumes (see Chapter 3). Chapter 3 Helium, N_2, and CO_2 found in the PHF field is from a combination of mantle, crustal, and atmospheric sources modified by chemical changes in the fluids over time, through migration. The problem with this model for finding similar He accumulations is these types of fields are supergiants and are typically found early in any basin's development.

(3) SGUY Model: This model is always a structural closure across a large area with several TCF of natural gas reservoirs. South and North Pars and Hassi R'Mel supergiant gas fields are examples of this model. These fields have significantly large volumes of natural gas and a very low percentage of He within a producing structure (see Chapter 3). Future discovery of these types of fields is probably limited as most productive oil and gas basins around the world that are in mature development. However, there are underexplored areas such as the Mozambique Channel where tremendous structures are adjacent to crustal and granitic rocks. These types of deposits are economic with low He percentages but because of their large volume of natural gas.

(4) WITY Model: The model for this type of deposit is unique in that it only seems to be present in one of the world, Virginia Gas Field in South Africa. In this case, gas is associated with a major base and precious metal deposits (see Chapter 3). The world is significantly underexplored for gas deposits associated with base and precious metal deposits. The key to the WITY model is in identify shallow sandstone or carbonate reservoir containing biogenic or thermally generated methane trapped with He that has migrated from a highly fractured source mantle or crustal rock at depth in an area that is moderately tectonically active. Exploitation of this type of model is recognized where natural gas is present as a hazard to mining of industrial, precious or base metal deposits.

(5) RIFT Model: This model is specifically related to accumulations found in a rift system such as the East African volcanic and tectonic area (see Chapter 3). This model is defined by structural and stratigraphic traps created and sealed by wrench faulting and shales or salt seals. The model is only associated with active rift systems as ancient ones have so far generally been found devoid of trapped hydrocarbons, noble and acid gases.

4.10 Summary

Basin analysis for targeting He deposits are in the early stages of development. Until recently the price of He has not been significantly economic to justify specific exploration focus for the element; however, He acts independently of the petroleum systems it interacts with and as such basin analysis for He needs to modified. The sources for He are clearly different than that of hydrocarbons. Historically, the best exploration tool to find He accumulations was to analyze existing gas wells. This would be followed by producing those existing wells for He or potentially identifying nearby potential structures that maybe prospective. This is essentially the SOAR model as described above. The TPH and SGUY models presented are generally rare. Large or supergiant fields such as the TPH or SGUY models typically start out producing natural gas only. It is only when detail gas analysis is used to identify He that these types of fields or models become subject to He extraction. The remaining two models, WITY and RIFT, are in areas generally not associated with conventional petroleum resources and to date seem to be very rare type accumulations. Another simple exploration tool for identifying areas with anomalous He is avoiding areas that are not gas prone. To date, significant reserves of He generally are not found with oil except in rare instances.

Exploration methods that are useful to find He deposits, other than gas analysis from existing wells, are analysis of aeromagnetic and gravity data to help identify areas of basement discontinuities, structural or erosinal highs, faults, and intrusions. This can provide areas for mapping of subsurface data to identify or predict potential structural closures and reservoirs that could host natural gas and He. If existing seismic data are available they may assist in providing leads to potential structural closures or nosing. Once a prospect has been identified, surface geochemistry can help ascertain the presence of anomalous He and hydrocarbons seeping from the subsurface into the soil section. If there are potentially multiple subsurface targets within a prospect area a 2D or 3D seismic survey may be necessary. All these data combined will hopefully define potential drill sites.

Appendix 4.1

		Pearson's correlation coefficient				Average, median and standard deviation (StDev)					
Regions in Kansas											
Cherokee Basin 31 samples		He	N_2	CO_2	CH_4		He	N_2	CO_2	CH_4	N_2/He
	He	1.00									
	N_2	0.84	1.00			Average	0.59	13.64	1.29	79.61	49.35
	CO_2	0.58	0.24	1.00		Median	0.23	5.39	0.57	87.00	29.13
	CH_4	−0.86	−0.97	−0.31	1.00	StDev	0.84	15.23	1.77	19.46	57.79
Anadarko Basin 1323 samples		He	N_2	CO_2	CH_4		He	N_2	CO_2	CH_4	N_2/He
	He	1.00									
	N_2	0.63	1.00			Average	0.45	13.38	0.34	75.81	226.71
	CO_2	−0.01	−0.02	1.00		Median	0.36	10.49	0.13	76.06	41.32
	CH_4	−0.55	−0.90	−0.04	1.00	StDev	0.47	12.38	1.32	13.70	545.67
Forest City Basin 27 samples		He	N_2	CO_2	CH_4		He	N_2	CO_2	CH_4	N_2/He
	He	1.00									
	N_2	0.79	1.00			Average	0.39	8.72	0.47	88.98	36.08
	CO_2	−0.32	−0.25	1.00		Median	0.32	6.13	0.33	90.99	21.39
	CH_4	−0.74	−0.99	0.26	1.00	StDev	0.38	13.70	0.50	13.72	39.87
Sedgwick Basin 275 samples		He	N_2	CO_2	CH_4		He	N_2	CO_2	CH_4	N_2/He
	He	1.00									
	N_2	0.73	1.00			Average	0.33	8.36	0.16	80.79	26.31
	CO_2	0.01	−0.04	1.00		Median	0.25	6.30	0.10	82.74	23.38
	CH_4	−0.62	−0.81	−0.02	1.00	StDev	0.28	7.14	0.44	8.55	15.09
Central Kansas 476 samples		He	N_2	CO_2	CH_4		He	N_2	CO_2	CH_4	N_2/He
	He	1.00									
	N_2	0.69	1.00			Average	1.57	20.32	0.38	69.90	16.48
	CO_2	0.29	0.25	1.00		Median	1.33	17.75	0.17	72.87	11.26
	CH_4	−0.72	−0.92	−0.41	1.00	StDev	1.27	13.64	1.33	16.45	17.26
Reservoirs in Kansas											
Niobrara 73 samples		He	N_2	CO_2	CH_4		He	N_2	CO_2	CH_4	N_2/He
	He	1.00									
	N_2	−0.41	1.00			Average	0.10	6.30	1.73	91.69	92.47
	CO_2	−0.33	0.01	1.00		Median	0.10	4.53	0.67	92.69	45.27
	CH_4	0.48	−0.90	−0.44	1.00	StDev	0.02	9.08	4.47	4.60	284.54

—cont'd

	Pearson's correlation coefficient					Average, median and standard deviation (StDev)					
Council Grove 98 samples		He	N_2	CO_2	CH_4		He	N_2	CO_2	CH_4	N_2/He
	He	1.00									
	N_2	0.35	1.00			Average	0.47	16.55	0.14	72.36	36.97
	CO_2	0.04	0.12	1.00		Median	0.42	16.30	0.10	72.21	37.23
	CH_4	−0.29	−0.89	−0.11	1.00	StDev	0.19	4.88	0.16	4.76	10.78
Topeka 301 samples		He	N_2	CO_2	CH_4		He	N_2	CO_2	CH_4	N_2/He
	He	1.00									
	N_2	0.51	1.00			Average	1.00	19.32	0.16	69.71	30.34
	CO_2	0.06	−0.03	1.00		Median	0.58	19.61	0.10	68.01	31.44
	CH_4	−0.27	−0.87	−0.01	1.00	StDev	0.80	9.35	0.14	9.79	21.52
Wabaunsee 58 samples		He	N_2	CO_2	CH_4		He	N_2	CO_2	CH_4	N_2/He
	He	1.00									
	N_2	0.78	1.00			Average	0.63	20.04	0.18	71.71	32.72
	CO_2	−0.06	0.08	1.00		Median	0.53	18.73	0.10	72.87	32.72
	CH_4	−0.72	−0.96	−0.13	1.00	StDev	0.40	11.47	0.29	13.53	8.90
Marmaton 77 samples		He	N_2	CO_2	CH_4		He	N_2	CO_2	CH_4	N_2/He
	He	1.00									
	N_2	0.64	1.00			Average	0.21	7.47	0.21	76.97	43.70
	CO_2	0.00	−0.06	1.00		Median	0.10	4.98	0.10	80.88	28.07
	CH_4	−0.10	−0.52	−0.05	1.00	StDev	0.57	6.59	0.57	11.81	41.95
Lansing Kansas City 51 samples		He	N_2	CO_2	CH_4		He	N_2	CO_2	CH_4	N_2/He
	He	1.00									
	N_2	0.41	1.00			Average	0.25	21.69	0.25	68.38	25.39
	CO_2	−0.07	0.02	1.00		Median	0.20	22.41	0.20	68.06	20.80
	CH_4	−0.08	−0.77	−0.21	1.00	StDev	0.50	10.55	0.50	11.38	20.52
Morrow 287 samples		He	N_2	CO_2	CH_4		He	N_2	CO_2	CH_4	N_2/He
	He	1.00									
	N_2	0.68	1.00			Average	0.40	8.91	0.33	79.44	57.74
	CO_2	0.27	0.09	1.00		Median	0.20	6.20	0.20	81.92	25.13
	CH_4	−0.52	−0.87	−0.19	1.00	StDev	0.45	8.82	0.52	10.78	467.86

(Continued)

—cont'd

	Pearson's correlation coefficient					Average, median and standard deviation (StDev)					
Mississippian 392 samples		He	N_2	CO_2	CH_4		He	N_2	CO_2	CH_4	N_2/He
	He	1.00									
	N_2	0.90	1.00			Average	0.47	7.26	0.24	83.90	19.07
	CO_2	0.13	0.14	1.00		Median	0.22	4.37	0.10	87.13	15.65
	CH_4	−0.80	−0.92	−0.15	1.00	StDev	0.63	8.76	0.75	10.31	13.93
Viola 48 samples		He	N_2	CO_2	CH_4		He	N_2	CO_2	CH_4	N_2/He
	He	1.00									
	N_2	0.65	1.00			Average	0.22	9.69	0.22	76.24	23.86
	CO_2	0.05	−0.15	1.00		Median	0.20	7.42	0.20	76.73	20.15
	CH_4	−0.14	−0.67	0.34	1.00	StDev	0.14	6.88	0.14	10.29	22.56
Arbuckle 81 samples		He	N_2	CO_2	CH_4		He	N_2	CO_2	CH_4	N_2/He
	He	1.00									
	N_2	0.81	1.00			Average	1.25	27.08	1.25	55.89	12.44
	CO_2	0.27	0.27	1.00		Median	0.40	17.48	0.40	69.41	8.38
	CH_4	−0.83	−0.95	−0.42	1.00	StDev	2.92	23.95	2.92	31.03	14.48
Fields in Kansas											
Otis-Albert Field 23 samples Central Kansas		He	N_2	CO_2	CH_4		He	N_2	CO_2	CH_4	N_2/He
	He	1.00									
	N_2	0.89	1.00			Average	1.53	12.81	0.25	77.14	8.85
	CO_2	−0.46	−0.39	1.00		Median	1.65	1.65	0.20	77.28	8.20
	CH_4	0.50	0.64	−0.29	1.00	StDev	0.36	0.36	0.14	1.61	2.25
Reichel Field 113 samples Central Kansas		He	N_2	CO_2	CH_4		He	N_2	CO_2	CH_4	N_2/He
	He	1.00									
	N_2	0.83	1.00			Average	2.09	23.26	0.17	68.59	11.67
	CO_2	0.33	0.26	1.00		Median	1.92	1.92	0.10	71.03	9.75
	CH_4	−0.85	−0.99	−0.30	1.00	StDev	0.78	0.78	0.13	11.50	8.11
Macksville Field 27 samples Central Kansas		He	N_2	CO_2	CH_4		He	N_2	CO_2	CH_4	N_2/He
	He	1.00									
	N_2	0.27	1.00			Average	0.86	8.02	0.28	84.24	10.52
	CO_2	0.37	−0.19	1.00		Median	0.80	7.51	0.30	84.10	7.14
	CH_4	−0.43	−0.91	−0.10	1.00	StDev	0.32	3.76	0.16	3.56	7.32

—cont'd

		Pearson's correlation coefficient				Average, median and standard deviation (StDev)					
Nescatunga, NE, SW Field 18 samples Anadarko		He	N_2	CO_2	CH_4		He	N_2	CO_2	CH_4	N_2/He
	He	1.00									
	N_2	0.94	1.00			Average	0.38	9.87	0.27	79.68	29.91
	CO_2	−0.08	−0.01	1.00		Median	0.21	8.66	0.20	79.64	23.72
	CH_4	−0.80	−0.90	0.06	1.00	StDev	0.34	6.91	0.29	8.33	11.99
Richfield Field 26 samples Anadarko		He	N_2	CO_2	CH_4		He	N_2	CO_2	CH_4	N_2/He
	He	1.00									
	N_2	0.87	1.00			Average	0.41	11.41	0.31	77.27	25.01
	CO_2	0.27	0.11	1.00		Median	0.34	7.28	0.20	79.37	22.33
	CH_4	−0.87	−0.97	−0.08	1.00	StDev	0.18	9.84	0.26	12.74	14.07
Victory Field 21 samples Anadarko		He	N_2	CO_2	CH_4		He	N_2	CO_2	CH_4	N_2/He
	He	1.00									
	N_2	0.37	1.00			Average	0.38	9.81	0.17	73.59	32.42
	CO_2	−0.22	0.03	1.00		Median	0.33	9.56	0.11	74.61	23.06
	CH_4	0.22	−0.26	−0.20	1.00	StDev	0.23	5.47	0.20	9.01	21.60
Hugoton Field 176 samples Anadarko		He	N_2	CO_2	CH_4		He	N_2	CO_2	CH_4	N_2/He
	He	1.00									
	N_2	0.92	1.00			Average	0.73	19.37	0.15	69.54	38.18
	CO_2	−0.03	0.20	1.00		Median	0.72	15.40	0.02	72.54	36.91
	CH_4	−0.88	−0.98	−0.27	1.00	StDev	0.21	9.49	1.05	8.26	7.06
Greenwood Field 40 samples Anadarko		He	N_2	CO_2	CH_4		He	N_2	CO_2	CH_4	N_2/He
	He	1.00									
	N_2	−0.03	1.00			Average	0.57	20.01	0.15	65.91	36.13
	CO_2	0.53	−0.15	1.00		Median	0.55	0.55	0.10	65.41	37.68
	CH_4	0.09	−0.88	0.13	1.00	StDev	0.11	0.11	0.16	3.16	6.85
Pamona Field 79 samples Anadarko		He	N_2	CO_2	CH_4		He	N_2	CO_2	CH_4	N_2/He
	He	1.00									
	N_2	0.13	1.00			Average	0.45	17.18	0.15	70.85	39.68
	CO_2	0.11	0.13	1.00		Median	0.42	0.42	0.10	72.09	37.96
	CH_4	−0.21	−0.88	−0.16	1.00	StDev	0.12	0.12	0.17	3.73	9.65

(*Continued*)

—cont'd

		Pearson's correlation coefficient				Average, median and standard deviation (StDev)					
Bradshaw Field		He	N_2	CO_2	CH_4		He	N_2	CO_2	CH_4	N_2/He
53 samples Anadarko	He	1.00									
	N_2	−0.30	1.00			Average	0.73	35.93	0.19	53.20	96.34
	CO_2	−0.16	0.07	1.00		Median	0.70	0.70	0.10	56.81	44.05
	CH_4	0.26	−1.00	−0.10	1.00	StDev	0.20	0.20	0.26	12.15	265.32
Data from Canada											
Saskatchewan (SK), Yurkowski (2016)											
SK Deadwood		He	N_2	CO_2	CH_4		He	N_2	CO_2	CH_4	N_2/He
14 samples Williston Basin	He	1.00				Average	1.47	97.02	1.05		74.14
	N_2	−0.64	1.00			Median	1.76	96.78	1.28		54.99
	CO_2	0.49	0.01	1.00		StDev	0.50	0.66	0.54		26.39
SK - DST		He	N_2	CO_2	CH_4		He	N_2	CO_2	CH_4	N_2/He
314 samples Williston Basin	He	1.00				Average	0.18	15.75	1.22		297.18
	N_2	0.43	1.00			Median	0.05	6.57	0.20		117.50
	CO_2	0.54	0.04	1.00		St Dev	0.52	23.04	4.65		735.02
SK - Wellhead		He	N_2	CO_2	CH_4		He	N_2	CO_2	CH_4	N_2/He
295 samples Williston Basin	He	1.00				Average	0.11	9.46	1.60		202.04
	N_2	0.41	1.00			Median	0.10	5.52	0.22		60.68
	CO_2	0.00	0.62	1.00		St Dev	0.14	13.41	4.19		634.04
Ontario, Lollar et al. (1994)											
Ontario -Trenton		He	N_2	CO_2	CH_4		He	N_2	CO_2	CH_4	N_2/He
20 samples Michigan Basin	He	1.00									
	N_2	0.61	1.00			Average	0.07	3.49	0.10	78.45	49.94
	CO_2	0.01	0.07	1.00		Median	0.08	2.81	0.05	82.90	46.22
	CH_4	0.25	−0.34	−0.08	1.00	StDev	0.04	3.00	0.10	10.23	25.19
Ontario -Black River		He	N_2	CO_2	CH_4		He	N_2	CO_2	CH_4	N_2/He
7 samples Michigan	He	1.00									
	N_2	−0.15	1.00			Average	0.42	5.29	0.09	84.42	21.25
	CO_2	−0.44	−0.64	1.00		Median	0.39	6.21	0.04	84.20	20.41
	CH_4	0.26	−0.78	0.67	1.00	StDev	0.41	2.47	0.09	2.36	13.87

—cont'd

	Pearson's correlation coefficient					Average, median and standard deviation (StDev)					
Ontario -Cambrian 10 samples Michigan Basin		He	N_2	CO_2	CH_4		He	N_2	CO_2	CH_4	N_2/He
	He	1.00									
	N_2	−0.34	1.00			Average	1.11	6.94	0.23	81.79	9.16
	CO_2	0.18	−0.27	1.00		Median	0.74	6.07	0.10	82.70	8.97
	CH_4	−0.40	−0.69	−0.04	1.00	StDev	0.79	3.41	0.31	3.84	5.58
Alberta, Alberta Geological Survey (2021)											
AB - DST 3253 samples Alberta Basin		He	N_2	CO_2	CH_4		He	N_2	CO_2	CH_4	N_2/He
	He	1.00									
	N_2	0.37	1.00			Average	0.31	0.09	0.02	72.56	16.80
	CO_2	−0.03	−0.02	1.00		Median	0.18	6.78	0.30	80.00	2.97
	CH_4	−0.27	−0.67	−0.37	1.00	StDev	0.47	10.43	5.36	17.20	464.66
AB - Wellhead 5570 samples Alberta Basin		He	N_2	CO_2	CH_4		He	N_2	CO_2	CH_4	N_2/He
	He	1.00									
	N_2	0.34	1.00			Average	0.19	5.83	1.23	83.18	559.55
	CO_2	0.02	0.02	1.00		Median	0.14	4.61	0.15	90.00	343.72
	CH_4	−0.22	−0.65	−0.44	1.00	StDev	0.19	5.59	3.83	12.76	4841.34
Samples from the Colorado Oil and Gas Commission (2020) and the references given below											
All in table 115 samples various locations		He	N_2	CO_2	CH_4		He	N_2	CO_2	CH_4	N_2/He
	He	1.00									
	N_2	0.70	1.00			Average	1.57	27.74	21.92	44.09	31.89
	CO_2	−0.29	−0.35	1.00		Median	0.75	15.50	0.81	48.86	17.57
	CH_4	−0.29	−0.45	−0.65	1.00	St Dev	2.07	29.24	34.91	33.94	57.51
Utah from table 46 sample Colorado Plateau		He	N_2	CO_2	CH_4		He	N_2	CO_2	CH_4	N_2/He
	He	1.00									
	N_2	0.60	1.00			Average	0.98	30.61	28.36	32.56	34.20
	CO_2	−0.29	−0.43	1.00		Median	0.72	19.70	8.63	13.50	24.11
	CH_4	−0.23	−0.38	−0.28	1.00	Stdev	1.17	29.67	35.46	31.41	30.89
New Mexico from table 8 samples		He	N_2	CO_2	CH_4		He	N_2	CO_2	CH_4	N_2/He
	He	1.00									

(*Continued*)

—cont'd

	Pearson's correlation coefficient					Average, median and standard deviation (StDev)					
Colorado Plateau	N_2	0.89	1.00			Average	5.22	61.54	3.36	27.03	10.69
	CO_2	0.01	0.11	1.00		Median	5.54	70.67	1.10	13.76	12.09
	CH_4	−0.87	−0.59	−0.20	1.00	Stdev	2.35	31.02	6.75	30.43	4.32
Arizona from table 12 samples Colorado Plateau		He	N_2	CO_2	CH_4		He	N_2	CO_2	CH_4	N_2/He
	He	1.00									
	N_2	0.84	1.00			Average	4.29	47.07	22.63	25.03	9.67
	CO_2	−0.60	−0.63	1.00		Median	4.70	45.05	7.45	9.80	9.58
	CH_4	−0.37	−0.51	−0.02	1.00	Stdev	3.14	36.92	36.74	33.10	6.27
LAAHR from table 10 samples Las Animas		He	N_2	CO_2	CH_4		He	N_2	CO_2	CH_4	N_2/He
	He	1.00									
	N_2	0.26	1.00			Average	1.33	21.86	19.68	54.79	20.82
	CO_2	0.21	−0.62	1.00		Median	0.97	16.55	0.66	63.80	17.40
	CH_4	−0.51	0.15	−0.86	1.00	Stdev	0.81	17.78	40.55	33.61	17.34
LAAHR 81 samples Las Animas Arch		He	N_2	CO_2	CH_4		He	N_2	CO_2	CH_4	N_2/He
	He	1.00									
	N_2	0.58	1.00			Average	1.40	39.05	1.65	32.87	49.46
	CO_2	−0.05	0.40	1.00		Median	0.68	47.46	1.32	27.53	46.46
	CH_4	−0.45	−0.69	−0.56	1.00	StDev	1.65	19.32	1.05	15.65	32.78
Arkansas 36 samples		He	N_2	CO_2	CH_4		He	N_2	CO_2	CH_4	N_2/He
	He	1.00									
	N_2	−0.07	1.00			Average	0.10	4.14	0.63	91.64	74.83
	CO_2	−0.20	0.14	1.00		Median	0.10	1.25	0.40	96.80	12.36
	CH_4	0.26	−0.78	−0.56	1.00	StDev	0.08	7.50	0.73	9.83	174.08

Data used for evaluation from: Anderson and Hinson (1951), Boone (1958), Wagner (1961), Zartman and Wasserburg (1961), Munnerlyn and Miller (1963), Miller and Norrell (1964a,b), Moore et al. (1964), Miller and Norrell (1965), Moore and Shrewsbury (1966), Cardwell and Benton (1967, 1968, 1969, 1970, 1971, 1972, 1973), Magloire (1970), Moore (1973, 1975, 1976a,b, 1977, 1978a,b, 1979, 1980, 1981, 1982), Miller and Hertweck (1982, 1983), Hertweck and Fox (1984), Moore and Hamak (1985), Jenden et al. (1988), Hamak and Sigler (1989, 1990, 1991a,b, 1993), Sonnenberg and van Drehle (1990), von Drehle (1990), Hamak and Gage (1992), Hamak and Driskill (1996), Gage and Driskill (1998, 2001, 2003, 2004, 2005), Rauzi (2003), Broadhead and Gillard (2004), Driskill (2008), Brown (2005, 2017), Alberta Geological Survey (2020), Singh et al. (2020), Wiseman and Eckels (2020), COGCC (2020), Newell (2021).

References

Alberta Geological Survey, 2020. Helium Concentrations from Natural Gas Analysis Tests in Alberta (Tabular Data, Tab-Delimited Format). https://ags.aer.ca/publication/dig-2020-0033.

Allen, P.A., Allen, J.R., 2013. Basin Analysis: Principles and Application to Petroleum Play Assessment. John Wiley & Sons, p. 619.

Anderson, C.C., Hinson, H.H., 1951. Helium-bearing Natural Gases of the United States. USGS Bull., p. 486

Boone Jr., W.J., 1958. Helium-bearing natural gas of the United States analysis and analytical methods, Supplement to Bulletin 576. USBLM Bull. Can. Pet. Geol. 576, 129.

Broadhead, R.F., Gillard, L., 2004. Helium in New Mexico Geological Distribution and Exploration Possibilities NMGS, 483. OFR, p. 62p.

Brown, A.A., 2005. Origin of High He Concentrations in Dry Gas by Water Fractionation, vol. 89. AAPG.

Brown, A.A., 2010. Formation of high helium gases a guide for explorationist. In: AAPG. New Orleans Convention Poster, Search and Discovery Article 80115.

Brown, A.A., 2014. Compositional gas migrating through water-saturated rock. In: AAPG. April 6–9, Annual Convention and Exhibition, Houston, Texas, Search and Discovery 41391, p. 5.

Brown, A.A., 2017. Possible Origins for Low Thermal Maturity, High Nitrogen Natural Gases. AAPG. Search and Discovery, p. 9. Article #42065.

Butt, C.R.M., Gole, M.J., 1985. Helium in soil and overburden gas as an exploration pathfinder – an assessment. J. Geochem. Explor. 24, 141–173.

Butt, C.R.M., Gole, M.J., 1986. Groundwater Helium surveys in mineral exploration in Australia. J. Geochem. Explor. 25, 309–344.

Cardwell, L.E., Benton, L.F., 1967. Analysis of Natural Gases, 1966. BM-IC-8356. USBM, p. 130.

Cardwell, L.E., Benton, L.F., 1968. Analysis of Natural Gases, 1967. BM-IC-8395. USBM, p. 187.

Cardwell, L.E., Benton, L.F., 1969. Analysis of Natural Gases, 1968. BM-IC-8443. USBM, p. 169.

Cardwell, L.E., Benton, L.F., 1970. Analysis of Natural Gases, 1969. BM-IC-8475. USBM, p. 134.

Cardwell, L.E., Benton, L.F., 1971. Analysis of Natural Gases, 1970. BM-IC-8475. USBM, p. 130.

Cardwell, L.E., Benton, L.F., 1972. Analysis of Natural Gases, 1971. BM-IC-8554. USBM, p. 130.

Cardwell, L.E., Benton, L.F., 1973. Analysis of Natural Gases, 1972. BM-IC-8607. USBM, p. 163.

Colorado Oil and Gas Conservation Commission (COGCC), 2020. https://cogcc.state.co.us/maps.html#/gisonline.

Craddock, W.H., Blondes, M.S., DeVera, C.A., Hunt, A.G., 2017. Mantle and crustal gases of the Colorado Plateau Geochemistry, sources, and migration pathways. Geochem. Cosmochim. Acta 213, 346–374.

Cunningham, K.I., 1987. Helium Concentration in Soil-gas Over Known Petroleum Accumulations. WaKeeney Area, Trego County, Kansas, USGS, OFR, 87–254, p. 26.

Danabalan, D., 2017. Helium: Exploration Methodology for a Strategic Resource. Ph.D., Durham University, United Kingdom, p. 296.

Davidson, M.J., July 1963. Geochemistry Can Help Find Oil If Properly Used. World Oil, pp. 94–106.

Davis, J.B., 1967. Petroleum Mocrobiology. Elsevier Publishing, Co., New York, p. 604.

Deshpande, R.D., Gupta, S.K., 2013. Groundwater helium: an indicator of active tectonic regions along Narmada River, central India. Chem. Geol. 344, 42–49.

Driskill, D.L., 2008. Analysis of Natural Gases, 2005–2007. Technical Note 427. USBLM, p. 208.

Dyck, W., 1976. The use of helium in mineral exploration. J. Geochem. Explor. 5 (1–2), 3–20.

Einsele, G., 1991. Sedimentary Basins: Evolution, Facies, and Sediments Budget. Springer-Verlag, p. 792.

Gage, B.D., Driskill, D.L., 1998. Analysis of Natural Gases, 1996–1997. Technical Note 404. USBLM, p. 80.

Gage, B.D., Driskill, D.L., 2001. Helium Resources of the United States–2001. Technical Note 408. USBLM, p. 40.

Gage, B.D., Driskill, D.L., 2003. Analysis of Natural Gas 1998–2001. Technical Note 412. USBLM, p. 179.

Gage, B.D., Driskill, D.L., 2004. Helium Resources of the United States–2003. Technical Note 415. USBLM, p. 42.

Gage, B.D., Driskill, D.L., 2005. Analysis of Natural Gases 2002–2004. Technical Note 418. USBLM, p. 252.

Guerra, M., Lombardi, S., 2001. Soil-gas Method for Tracing Neotectonic Faults in Clay Basins: The Pisticci Field (Southern Italy).

Haines, P.W., Allen, H.-J., 2019. Hydrocarbon and helium prospectivity of the Amadeus and Murraba basins in western Australia. In: KEEP, M., MOSS, S.J. (Eds.), The Sedimentary Basins of Western Australia V: Proceedings of the Petroleum Exploration Society of Australia Symposium, Perth, WA, p. 17.

Halford, H.D.T., 2018. Isotopic Analysis of Helium from Wells Located in the Four Corners Area. Ph.D. Dissertation. Southwestern US CSM, p. 376.

Hamak, J.E., Driskill, D.L., 1996. Analysis of Natural Gases - 1994–1995. USBLM, p. 76. Technical Note 399.

Hamak, J.E., Gage, B.D., 1992. Analyses of Natural Gases, 1991. IC9318, p. 97.

Hamak, J.E., Sigler, S., 1989. Analyses of Natural Gases, 1988. IC 9225. USBM, p. 66.

Hamak, J.E., Sigler, S., 1990. Analyses of Natural Gases, 1989. IC 9256. USBM, p. 60.

Hamak, J.E., Sigler, S., 1991a. Analyses of Natural Gases, 1990. IC 9290. USBM, p. 56.

Hamak, J.E., Sigler, S., 1991b. Analyses of Natural Gases, 1986–1990. IC 9301. USBM, p. 315.

Hamak, J.E., Sigler, S.M., 1993. Analyses of Natural Gases, 1992. IC 9356. USBM, p. 62.

Hanson, M.C., Oze, C., Horton, T.W., 2014. Identifying blind geothermal systems with soil CO_2 surveys. Appl. Geochem. 50, 106–114.

Hertweck Jr., F.R., Fox, D.D., 1984. Analyses of Natural Gases, 1983. IC 8993. USBM, p. 127.

Holland, P.W., Emerson, D.E., 1979. Helium in Ground Water and Soil Gas in the Vicinity of Bush Dome reservoir, Cliffside Field, Potter County, Texas, US Bureau of Mines, 22p.

Hunt, J.M., 1995. Petroleum Geochemistry and Geology, second ed. Freeman Press, p. 743.

Jenden, P.D., Newell, K.D., Kaplan, L.R., Watney, W.L., 1988. Composition and stable-isotope geochemistry of natural gases from Kansas. Chem. Geol. 71, 117–147.

Kleinspehn, K.L., Paola, C., 1988. New Perspectives in Basin Analysis. Springer-Verlag, p. 453.

Lerche, L., 1989. Basin Analysis Quantitative Methods, vol. 1. Elsevier Publishing, p. 576.

Lollar, B.S., Weise, S.M., Frape, S.K., Barker, J.F., 1994. Isotopic constraints on the migration of hydrocarbon and helium gases of southwestern Ontario. Bull. Can. Pet. Geol. 42 (3), 283–295.

Macintosh, S.J., Ballentine, C.J., 2012. Using $3H_3/4He$ isotope ratios to identify the source of deep reservoir contributions to shallow fluids and soil gas. Chem. Geol. 304–305, 142–150.

Magloire, P.R., 1970. Triassic gas field of Hassi er R' Mel, Algeria, AAPG. In: Geology of Giant Petroleum Fields, vol. 14. Memoir, pp. 489–502.

McDowell, B.P., Milkov, A.V., Anderson, D.S., 2017. The He System - a Modification of the Petroleum System for Inert Gases. AAPG. Search and Discovery, p. 42098.

Miall, A.D., 1990. Principles of Sedimentary Basin Analysis. Springer-Verlag, p. 668.

Miller, R.D., Hertweck Jr., F.R., 1982. Analyses of Natural Gases, 1981. IC 8890. USBM, p. 84.

Miller, R.D., Hertweck Jr., F.R., 1983. Analyses of Natural Gases. IC 8942. USBM, p. 100.

Miller, R.D., Norrell, G.P., 1964a. Analyses of Natural Gases of the United States, 1961. IC 8221. USBM, p. 148.

Miller, R.D., Norrell, G.P., 1964b. Analyses of Natural Gases of the United States, 1962. IC 8239. USBM, p. 120.

Miller, R.D., Norrell, G.P., 1965. Analyses of Natural Gases of the United States, 1963. IC 8241. USBM, p. 102.

Moore, B.J., 1973. Analyses of Natural Gases, 1973. IC 8658. USBM, p. 96.

Moore, B.J., 1975. Analyses of Natural Gases, 1974. IC 8684. USBM, p. 122.

Moore, B.J., 1976a. Analyses of Natural Gases, 1917–74. CP 1-76. USBM, p. 889.

Moore, B.J., 1976b. Analyses of Natural Gases, 1975. IC 8717. USBM, p. 82.

Moore, B.J., 1977. Analyses of Natural Gases, 1976. IC 8749. USBM, p. 94.

Moore, B.J., 1978a. Analyses of Natural Gases, 1977. IC 8780. USBM, p. 95.

Moore, B.J., 1978b. Analysis of Natural Gas, 1978. Info Circular 8810. USBM, p. 136.

Moore, B.J., 1979. Analyses of Natural Gases, 1978. IC 8810. USBM, p. 113.

Moore, B.J., 1980. Analyses of Natural Gases, 1979. IC 8833. USBM, p. 100.

Moore, B.J., 1981. Analyses of Natural Gases, 1980. IC 8856. USBM, p. 236.

Moore, B.J., 1982. Analysis of Natural Gas, 1917–80. Info Circular 8870. USBM, p. 1074.

Moore, Hamak, 1985. Analyses of Natural Gases, 1984. IC9046. USBM, p. 102.

Moore, B.J., Shrewsbury, R.D., 1966. Analyses of Natural Gases of the United States, 1965. IC 8316. USBM, p. 181.

Moore, B.J., Miller, R.D., Shrewsbury, R.D., 1964. Analyses of Natural Gases of the United States, 1964. IC 8302, 1966. USBM, p. 144.

Munnerlyn, R.D., Miller, R.D., 1963. Helium-bearing Natural Gases of the United States: Analyses, Second Supplement to Bulletin 486. USBM Bull., p. 617 93

Newell, K.D., 2017. Compositional and Isotopic Analysis of Natural Gas from a Shallow Water Well in Southeastern Cowley County, Kansas, KGS, OFR 2017-9. www.kgs.ku.edu/Publications/OFR/2017/OFR17_9/index.html.

Newell, K.D., 2021. Stratigraphic and Regional Trends in Chemistry and Quality of Natural Gas in Central and Western Kansas, Mid-continent, USA, KGS (in press).

Rauzi, S.L., 2003. Review of Helium Production and Potential in Arizona Arizona Geological Survey, OFR 03-05, p. 31.

Reimer, G.M., Denton, E.H., Freidman, I., Otton, J.K., 1979. Recent developments in uranium exploration using the U.S. Geological Surveys mobile helium detector. J. Geochem. Explor. 11, 1–12.

Roberts, A.A., 1981. Helium manometry for exploring for hydrocarbons: Part II. In: Unconventional Methods in Exploration for Petroleum and Natural Gas, Symposium II. Southern Methodist University Press, Dallas, TX, pp. 136–149.

Singh, A., Rokosh, C.D., Branscombe, P., Filewich, C., 2020. Helium Concentrations of Natural Gas Analysis Tests in Alberta (Tabular Data, Tab-Delimited Format). Alberta Geological Survey. https://ags.aer.ca/publication/dig-2020-0033.

Sittig, M., 1980. Geophysical and Geochemical Techniques for Exploration of Hydrocarbons and Minerals. Noyes Data Corp, Newark, NJ.

Sonnenberg, S.A., van Drehle, W.F., 1990. Morrow gas composition in southeast Colorado. In: Sonnenberg, S.A., Shannon, L.T., Rader, K., von Drehle, W.F., Martin, G.W. (Eds.), Morrow Sandstones, RMAG, pp. 233–235.

Tedesco, S.A., 1994. Surface Geochemistry in Petroleum Exploration. Chapman and Hall.

Tedesco, S.A., 2017. The use of aeromagnetics and micromagnetics to identify potential areas of hydrocarbons in the Midcontinental United States. In: Gaci, S., Hachay, O. (Eds.), Oil and Gas Exploration: Methods and Application. John Wiley & Sons, pp. 237–259.

Tedesco, S.A., 2021. Microseeps as pathfinder and regional filtering tool in petroleum exploration. In: Methods and Applications in Petroleum and Mineral Exploration and Engineering Geology. Elsevier (Chapter 2).

von Drehle, W.F., 1990. Barrel Springs field, Prowers County, Colorado. In: Sonnenberg, S.A., Shannon, L.T., Rader, K., von Drehle, W.F., Martin, G.W. (Eds.), Morrow Sandstones, RMAG, pp. 183–189.

Wagner, C.R., 1961. Keyes field. In: Oil and Gas Fields of Texas and Oklahoma Panhandles. Panhandle Texas Geological Society, pp. 239–243.

Wiseman, T., Eckels, M.T., 2020. Proven and Hypothetical Helium Resources in Utah, SITLA and Utah DNR, vol. 174. Misc Pub, p. 43.

Yurkowski, M.M., 2016. Helium in Southwestern Saskatchewan: Accumulation and Geological Setting. Saskatchewan Ministry of the Economy. Saskatchewan. Geological Survey, Open File Report 2016-1, p. 20.

Zartman, R.E., Wasserburg, G.J., 1961. Helium, argon, and carbon in some natural gases. J. Geophys. Res. 66 (1), 277–306.

Zhang, M., Guo, Z., Zhang, L., Sun, Y., Cheng, Z., 2017. Geochemical constraints on origin of hydrothermal volatiles from understanding the degassing sys in the India-Asia continental subduction zone. Chem. Geol. 469, 19–33.

Zhang, W., Li, Y., Zhao, F., Han, W., Li, Y., Wang, Y., Holland, G., Zhou, Z., 2019a. Using noble gases to trace groundwater evolution and assess helium accumulation in Weihe Basin, Central China. Geochem. Cosmochim. Acta 251, 229–246.

Zhang, W., Li, Y., Zhao, F., Han, W., Zhou, J., Holland, G., Zhou, Z., 2019b. Quantifying the helium and hydrocarbon accumulation processes using noble gases in the North Qaidam Basin, China. Chem. Geol. 525, 368–379.

CHAPTER

5

Processing procedures to extract helium, nitrogen, and carbon dioxide

5.1 Introduction

The expensive aspect of natural gas production, if required, is the extraction of He, N_2, and CO_2 from the natural gas flow. Extraction of He and N_2 from large volumes of natural gas is typically via a cryogenic plant for smaller volumes of natural gas by pressure spring adsorption process or PSA. The greater the volume of these impurities, the greater the cost. Extraction units associated with higher-grade He in small areal size deposits such as Harley Dome can vary from $750,000 to $5,500,000 (US). They are typically on skids or semi-mobile so they can be moved from one location to another. This assumes that the He and natural gas composition is relatively similar between deposits. Cryogenic extraction units associated with lower-grade large areal extent deposits such as Texas Panhandle-Hugoton can be in the range of tens of millions of dollars. These tend to be completely immobile and require significant long-term reserves.

Facilities that remove CO_2 are generally known as *amining plants and* are very common in the petroleum industry. In some cases the major constituent of the natural gas is CO_2 which is sold for industrial uses such as food processing or enhanced oil recovery. The hydrocarbon gas is therefore an impurity or byproduct. Carbon dioxide can be very corrosive to pipelines, power plants, and equipment. Small amounts of the CO_2 maybe allowable in natural gas sold through the pipeline system, but this is usually in areas where its corrosive effects are minimal on equipment because of dilution by the natural gas.

Nitrogen is similar in that it has to be removed by a cryogenic or PSA process unit to allow the sale of natural gas. Nitrogen, however, is inert, noncorrosive, and can be released into the atmosphere with minimal environmental impact.

The development of extracting He from natural gas was first attempted by the British during World War I to provide the element for the war effort (Nuttall et al., 2012; and Sears, 2015). The British military determined that several natural gas wells in western Ontario (Michigan Basin) would yield He in the amounts of 0.1%–0.33% range. The first plant

https://doi.org/10.1016/B978-0-323-90988-4.00001-3

was established by a French Company called Air Liquide in Hamilton, Ontario (Nuttall et al., 2012; and Sears, 2015). The process was a modified separation unit that utilized cold temperatures to separate He from the air. The plant's subsequent use and development were not without problems, as at the time it was difficult to remove the heavier hydrocarbons. Through trial and error by 1918, it was able to achieve 87% pure He. Then the amount of natural gas being produced from the existing fields in western Ontario began to decline. Southwestern Ontario is part of the Michigan Basin and produces gas from several Paleozoic reservoirs of Cambrian, Ordovician, and Devonian age at less than 3500 feet (1060 m). The plant was subsequently moved to Bow Island Field in Alberta (Nuttall et al., 2012; and Sears, 2015). The Bow Island Gas Field, Southern Alberta, yielded 0.29%−0.34% He from the Bow Island sandstone at 1900 feet (575 m).

For large-scale extraction, He is removed from natural gas by fractional distillation from natural gas, which can contain as much as 7% helium. Helium has the lowest boiling point of any element, extremely low temperature and high pressure are used to liquefy nearly all the other gases during the extraction process. Usually, N_2 and CH_4 with the heavier hydrocarbons are generally not present or are removed prior to entering the cryogenic unit to extract He (Nuttall et al., 2012; and Sears, 2015). Propane and C_2H_6 can be very valuable byproducts of this process are removed prior to the natural gas entering the cryogenic unit so as not to be wasted. Helium gas is purified by successive lowering temperatures to remove the N_2 and any other gases. The final stage of the purification process uses activated charcoal, generally resulting in 99.95% pure Grade-A helium. Interestingly, the principal impurity in Grade-A He is Ne, which is far more valuable than He. Diffusion of crude natural gas through special semipermeable membranes using changes in pressure (PSA) another method to recover and purify He.

Helium as part of the cryogenic process will be liquefied in the final step of extraction to allow easier transportation, move more product, and lower costs per mcf for transportation. In a liquefied form, five times the amount of He can be moved compared to as a gas.

Helium is extracted only from CH_4^+, N_2, or CO_2 rich gases because it is generally present in significant concentrations compared to its concentration in the atmosphere of 5.6 ppm. If He was extracted from the atmosphere, this would provide only 1% of the world's He needs using the world's air distillation plants (EPCM, 2021). Helium is also synthesized by bombardment of Li or B with high-velocity protons, or by bombardment of Li with deuterons, but these processes are uneconomic at present.

Helium is commercially available in either liquid or gaseous form. As a liquid, it can be supplied in small insulated containers called *dewars* which hold as much as 1000 L of helium, or in large ISO containers which have nominal capacities as large as 42 m^3 by volume or approximately 200 barrels (EPCM, 2021). In gaseous form, small quantities of helium are supplied in high-pressure cylinders holding as much as 8 m^3 or 282 standard cubic feet, while large quantities of high-pressure gas are supplied in tube trailers which have capacities of as much as 4860 m^3 or approximately 172,000 standard cubic feet (EPCM, 2021).

Presented here is a general discussion of processing methods for extracting for He, CO_2, N_2 and H_2S from natural gas. The processing details are far more complex then presented here and the reader is recommended to read and review Bahadori (2014) and Kidnay et al. (2019).

5.2 Carbon dioxide

Carbon dioxide is a major source of environmental concern but unfortunately, it is critical to modern society in many industries and products (Miller and Zawaki, 1978; Gary and Handwerk, 1984; Kohl and Nielson, 1997; Kidnay et al., 2019). Amines are used to remove CO_2 in various areas ranging from natural gas production to the food and beverage industry, and have been used for over 60 years. Biogas production requires removal of CO_2 to make it useable as a natural gas.

Amines which are derivatives of NH_4 whereby one H atom has been replaced or substituted by alkyl or aryl groups (Miller and Zawacki, 1978; Kidnay et al., 2019; Gary and Handwerk, 1984; Kohl and Nielson, 1997). Important organic amines include amino acids, biogenic amines, trimethylamine, and aniline and inorganic derivatives of ammonia are also called amines, such as monochloramine ($NClH_2$). The concentration of amine in the absorbent aqueous solution is an important parameter in the design and operation of an amine gas treating process (Kidnay et al., 2019). Treating natural gas with an amining plant to remove H_2S and CO_2 is a common process used in refineries, and is also used in petrochemical plants, natural gas processing plants, and other industries. The process is also referred to as *amine scrubbing, gas sweetening,* and/or *acid gas removal.* It refers to a series of processes that use aqueous solutions of various alkylamines or amines to remove H_2S and CO_2 from gases.

These are typical amine concentrations with approximate weight percent of pure amine in the aqueous solution (Miller and Zawaki, 1978; Gary and Handwerk, 1984; Kohl and Nielson, 1997; Kidnay et al., 2019):

- Diethanolamine (DEA) varies from 20% to 25% for removing H_2S and CO_2.
- Monoethanolamine (MEA) approximately 20% for removing H_2S and CO_2, and approximately 32% for removing only CO_2.
- Methyldiethanolamine (MDEA) approximately 30%–55% for removing H_2S and CO_2.
- Diisopropanolamine (DIPA) approximately 50% for removing H_2S and CO_2.
- Aminoethoxyethanol or diglycolamine (DGA) approximately 50% for removing H_2S and CO_2.

The most commonly used amines in industrial plants, oil refineries, and natural gas processing are the alkanolamines DEA, MEA, and MDEA. The Girbotol process (Fig. 5.1) includes an absorber unit and a regenerator unit with associated equipment. The amine solution flows down in the absorber and absorbs H_2S and CO_2 removing these elements from the gas stream and the amine solution will be in the absorbed acid gases. The resultant "rich" amine is then routed into the regenerator which is a stripper with a reboiler to produce regenerated or "lean" amine solution that is recycled for reuse in the absorber (Miller and Zawaki, 1978; Gary and Handwerk, 1984; Kohl and Nielson, 1997; Kidnay et al., 2019). The stripped overhead gas from the regenerator is concentrated with H_2S and CO_2.

Variations in stripper configurations include matrix, internal exchange, flashing feed, and multipressure with split feed. Many configurations can offer more energy efficiency for specific solvents or operating conditions (Miller and Zawaki, 1978; Gary and Handwerk, 1984; Kohl and Nielson, 1997; Kidnay et al., 2019). The type of amine will affect the circulation

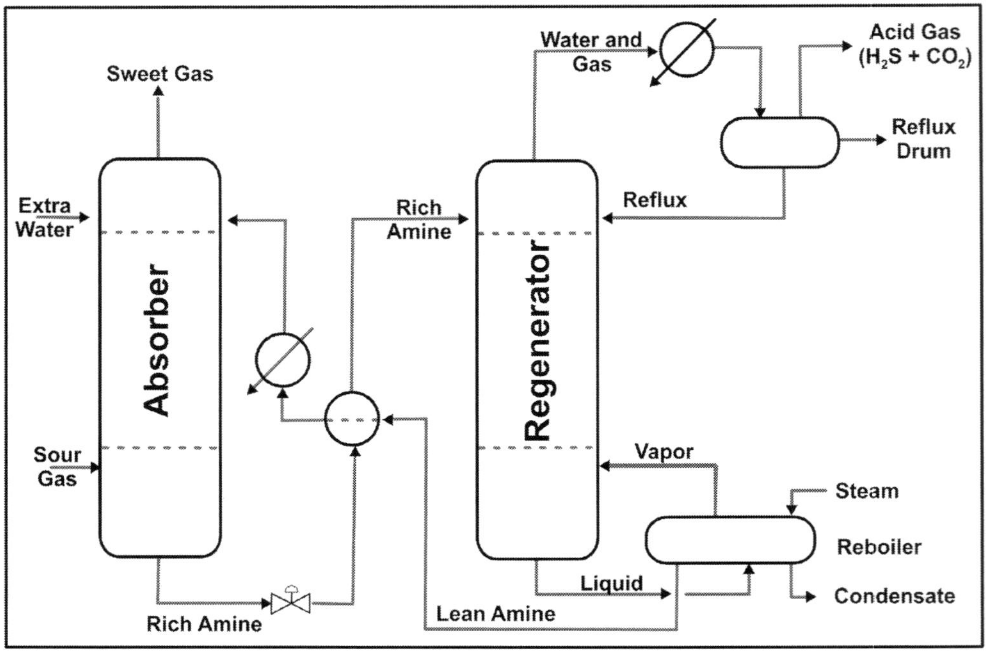

FIGURE 5.1 Typical amining gas flow diagram in natural gas process.

rate of amine solution, the regeneration energy consumption, and the ability to selectively remove either H_2S alone or CO_2 alone if desired. Vacuum operation favors solvents with low heats of absorption while operation at normal pressure favors solvents with high heats of absorption (Kidnay et al., 2019). Solvents with high heats of absorption require less energy for stripping from temperature swings at fixed capacity. A matrix stripper recovers 40% of CO_2 at a higher pressure but does not have inefficiencies associated with multipressure stripper. Energy and costs are reduced since the reboiler duty cycle is slightly less than normal pressure stripper (Miller and Zawaki, 1978; Gary and Handwerk, 1984; Kohl and Nielson, 1997; Kidnay et al., 2019). The internal exchange stripper has a smaller ratio of H_2O vapor to CO_2 in the overhead stream, and therefore, less steam is required. The multipressure configuration with split feed reduces the flow into the bottom section, which also reduces the equivalent work. Flashing feed requires less heat input because it uses the latent heat of H_2O vapor to help strip some of the CO_2 in the rich stream entering the stripper at the bottom of the column (Miller and Zawaki, 1978; Gary and Handwerk, 1984; Kohl and Nielson, 1997; Kidnay et al., 2019). The multipressure configuration is more attractive for solvents with a higher heat of absorption.

The critical part in the amining unit is treating produced natural gas versus petroleum refinery byproduct gases that contain relatively low concentrations of both H_2S and CO_2 which are acidic gases that corrode carbon steel. Carbon dioxide is the stronger acid and H_2S forms iron sulfide film on the surface of the steel that protects the steel. When treating gases with a high percentage of CO_2, corrosion inhibitors are often used and that permits the use of higher concentrations of amine in the circulating solution (Miller and Zawaki, 1978; Gary and Handwerk, 1984; Kohl and Nielson, 1997; Kidnay et al., 2019).

The economic costs of any amining plant are driven by energy costs to drive the process, minimizing the amining solutions replacement but also maximizing the amount of CO_2 or H_2S that is removed (Miller and Zawaki, 1978; Gary and Handwerk, 1984; Kohl and Nielson, 1997; Kidnay et al., 2019). In any natural production situations, where H_2S is not present, this will significantly reduce costs, and wear and tear on equipment.

5.3 Nitrogen rejection units

A nitrogen rejection unit or NRU selectively removes the element from natural gas. Pipeline specification typically require less than 5% of inserts in the saleable gas stream. There are two methods of removing N_2 from the gas stream using either or both cryogenic and pressure swing desorption. There are four basic NRU cryogenic processes and they are multistage flash separators, single column heat pumped process, double-column process, and dual column cycles (Agrawal et al., 2003). The process chosen depends on the flow rate of the feed gas, the composition of the feed and in discussion here, the He concentration in order to make its subsequent recovery economic (Kidnay et al., 2019). Multiflash separator processes typically have higher energy requirements compared to cryogenic distillation column processes but have lower capital and production costs than overhead vapor streams with low He concentrations. A single-stage flash separator process produces N_2-rich overhead vapors containing 1% to 3% He, depending on the feed gas composition (Feda, 2021). Complex, double-column NRU processes can recover 50%–70% He when operating in partial condensation mode, where N_2 and small amounts of CH_4, H, Ne, Ar, and CO_2 are also present (Agrawal et al., 2003; West, 2009). High flow rates above 15 MMCFG use the cryogenic process of distillation. Methane and N_2 have a boiling point of −161.6°C and −195.69°C (Feda, 2021). Through a series of compression and distillation columns the temperature is drastically reduced whereby the CH_4 is liquefied and removed and the N_2 is released into the atmosphere. As an alternative in the case of lesser applications is a series of heat exchangers may be used.

The distillation columns and heat exchangers operate at very low temperatures are installed inside a very large insulated vessel called the "cold box." The refrigeration cycle operates on Joule Thomson effect which is also called *throttling effect* (Fig. 5.2). During the throttling process, the gas is passed through an insulated valve or an insulated porous plug whereas the temperature of the gas changes due to change in pressure (Feda, 2021).

PSA is an adsorption-based process that has been used for various gas separation and purification purposes (Kidnay et al., 2019; Feda, 2021). Separation in these processes depends on the ability of one or more adsorbents which are porous solids to selectively adsorb one or more of the components of a gas/liquid mixture. Other adsorption-based processes are TSA (Temperature Swing Adsorption), VSA (Vacuum Swing Adsorption), VPSA (Vacuum Pressure Swing Adsorption), and CSA (Concentration Swing Adsorption). For example the PSA uses adsorbent with separation through an aperture slightly smaller than the molecule of the larger species which in this case is CH_4 that is 3.8 Å in size. Nitrogen is 3.6 Å in size

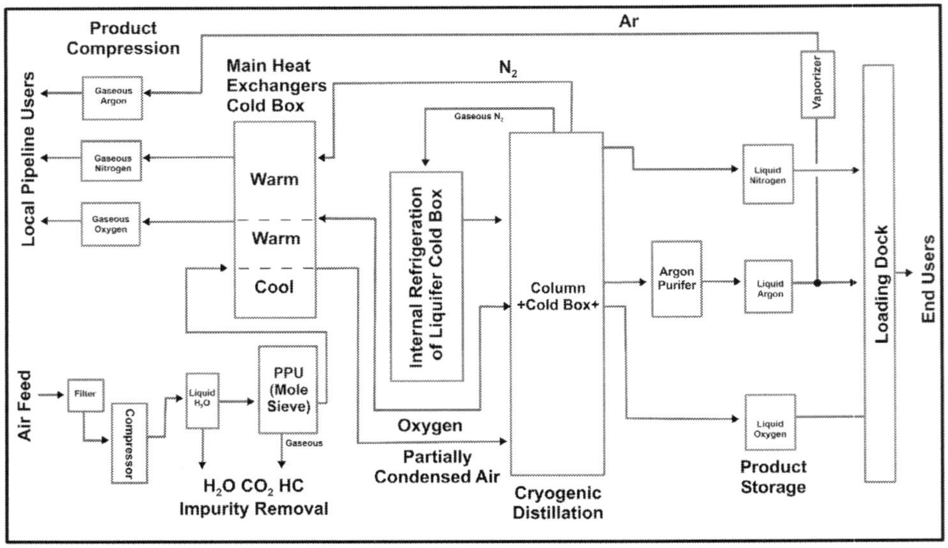

FIGURE 5.2 Cryogenic unit for removing N_2 modified from Chemical Engineering World, 2021. This is one of many configurations for removing N_2.

(Kidnay et al., 2019; Feda, 2021). The PSA is a flexible method that can be used for various flow rates. The N_2 diffuses through the adsorbent, filling adsorption sites, whereas CH_4 does not. The CH_4 stream becomes purified and the adsorbent is regenerated and leaves as a highly pure stream.

The operating conditions of PSA units are quite variable and depend on the requirements of removing other impurities such as H_2S, H_2O, CH_4^+, and CO_2 to minimize damage to the absorbents. Equipment can vary in terms for achieving CH_4 purity and recovery (Kidnay et al., 2019; Feda, 2021). All the processes are similar in alternating between two main steps: (1) adsorption step during which the separation occurs and (2) the desorption step when regeneration of the adsorbent occurs. The major difference between processes is in the method that is used for regenerating the adsorbent:

- PSA/VSA/PVSA: the adsorbent is regenerated by lowering the pressure to atmospheric for PSA or even lower for the VSA/PVSA processes (Kidnay et al., 2019; Feda, 2021);
- TSA: heating is used to regenerate the adsorbent (Kidnay et al., 2019; Feda, 2021);
- CSA (used in liquid separations): by using a desorbent liquid to purge the adsorbent which regenerates the column (Kidnay et al., 2019; Feda, 2021).

The adsorption-based process is separated into at least two gas streams of which one is enriched in the desired component(s). In the PSA process, the gas mixture is fed to one or more adsorption columns at P_{Feed} (P_{high}) which can be atmospheric or higher. During this step, one/more components are adsorbed more than the others, so the less adsorbed components leave the system at P_{Feed} (P_{high}) (minus the bed pressure drop). The saturated adsorbents are then regenerated by reducing the pressure to a pressure less than P_{feed}. As a result, the more strongly adsorbed heavier components leave the system through its feed end, at P_{low} which can be atmospheric or lower (Feda, 2021). PSA systems usually consist

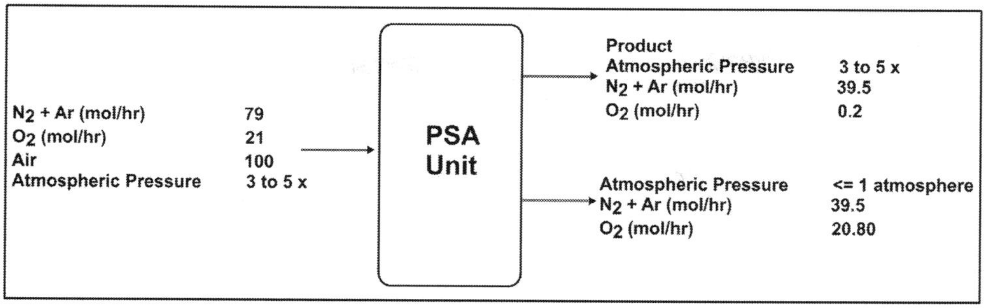

FIGURE 5.3 Nitrogen PSA generator modified from Feda, 2021.

of two or more adsorption vessels; while one vessel is going through adsorption, the other goes through regeneration (desorption). Examples of PSA gas separation processes are air-separation for N_2/O_2 generation and H purification (Feda, 2021). The type of adsorbent and the PSA process design determine performance and results.

PSA Process Performance Product purity, recovery, and the energy needed for the process are the indicators of the performance of a PSA process (Kidnay et al., 2019; Feda, 2021). The energy required for a PSA unit is the energy needed for the control system and gas compression and/or vacuum creation in cases where P_{low} is less than atmospheric. In this case, the process is usually referred to as *VSA* or *VPSA* depending on the feed pressure. However, the energy needed for the control system is negligible compared to the energy needed for compression. Fig. 5.3 is a diagram of a PSA generator. Product purity and recovery are defined by Eqs. (5.1) and (5.2) (Feda, 2021):

$$\text{Product Purity}(\%) = \frac{\text{Desired Component in the Product(usually in mols)}}{\text{Total Product(usually in mols)} \times 100} \tag{5.1}$$

$$\text{Product Recovery}(\%) = \frac{\text{Desired Component in the Product(usually in mols)}}{\text{Desired Component in the Feed(usually in mols)} \times 100} \tag{5.2}$$

For example, in a PSA process that separates nitrogen from air (shown in Fig. 5.3) the recovery and purity are as follows:

$$N_2 \text{ Purity}(\%) = \frac{39.50(\text{mol/h})}{(39.5 + 0.2)(\text{mol/h})} \times 100 = 99.5\% \tag{5.3}$$

$$N_2 \text{ Recovery}(\%) = \frac{39.50(\text{mol/h})}{79(\text{mol/h})} \times 100 = 50\% \tag{5.4}$$

PSA Process Design: the following parameters should be determined in designing a PSA process (Kidnay et al., 2019; Feda, 2021):

- Type of adsorbent(s)

- Temperature and pressure
- Number and size of vessels
- Number and type of steps
- Cycle and steps times

A variety of different types of adsorbents are available for the gas separation and purification industries such as activated C, zeolites, activated Al, Si gel, and molecular sieve carbons (Kidnay et al., 2019; Feda, 2021). Choosing the proper adsorbent is the most important factor and the center piece of designing the PSA process. The feed gas dictates the type of adsorbent to be used (Kidnay et al., 2019; Feda, 2021).

The selection of the adsorbent is a function of the following mechanisms (Feda, 2021):

- Steric effect: separation is a result of molecular sieving; only components that have molecules with specific shape and size diffuse into the sorbent;
- Kinetic: separation is a result of the difference in diffusion rates of components. This is critical for understanding proper application of any adsorbent (Kajjumba et al., 2018);
- Equilibrium: separation is a result of the difference in components' equilibrium adsorption.

To help determine the specific adsorbent to be used, the following experimental data are needed for using isotherms and breakthrough curves:

- Isotherms shows the amount of adsorbed gas per unit volume/mass of adsorbent at different pressures at a constant temperature. Isotherms are used to determine the amount of adsorbent needed for the separation, which determines the required pressure swing;
- Breakthrough curve is defined by passing a gas mixture with a known concentration through an adsorption bed and then recording the gas concentration at the bed outlet versus time;
- Breakthrough curve provides information about the kinetics and the potential working capacity of the adsorption bed.

In gas mixtures with more than two components, layered beds may be used in which layers of different adsorbents are used to adsorb different components. Temperature and pressure control the selectivity of the absorbent and PSA systems performance is better when the pressure ratio is higher, and the temperature is lower (Kidnay et al., 2019; Feda, 2021). Adsorption is an exothermic process and it is typically enhanced by lowering the temperature (Fig. 5.4).

There are different steps that may be used in a PSA process to optimize the system in terms of cost: both operating and capital expenditures (Kidnay et al., 2019; Feda, 2021). Fig. 5.4 is a schematic of each step with a brief summary. The pressure in the system remains constant in some steps while in others the pressure will change (Feda, 2021). Feed pressurization of the feed gas is used to pressurize the bed either from the lowest pressure or an intermediate pressure from previous step called *equalization* or *backfill*. Duration of each step may have a significant impact on the system performance. A combination of all or some of the steps is shown in Fig. 5.5, depending upon the required product(s), may be used in a PSA process design (Feda, 2021).

Fig. 5.6 is a flow diagram for a N_2 rejection unit (NRU) as outlined by Air Liquide (2021). Fig. 5.7 is a commercial carbon membrane sieve example modified from South-tek Systems (2021).

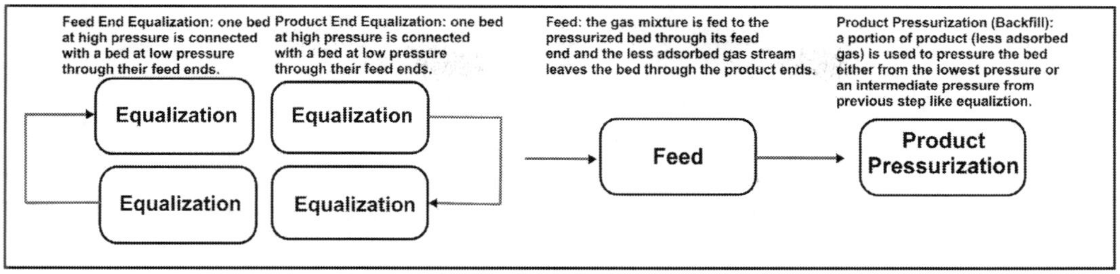

FIGURE 5.4 Effect of temperature and pressure on adsorption modified from Feda, 2021.

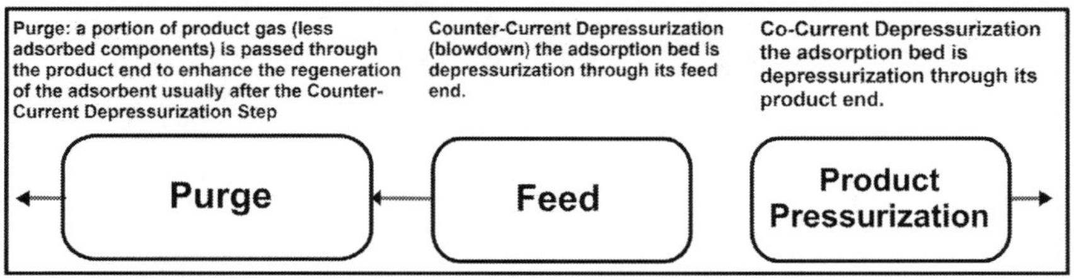

FIGURE 5.5 PSA processing steps (Feda, 2021).

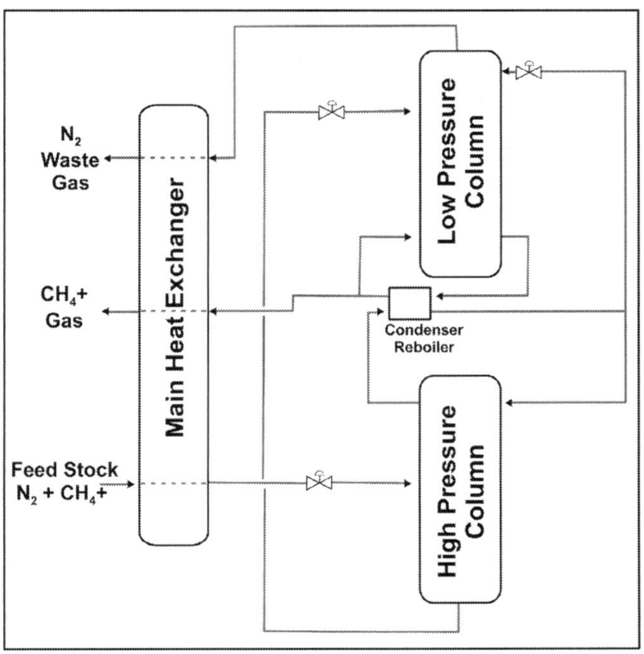

FIGURE 5.6 Nitrogen rejection unit flow diagram from Air Liquide (Nitrogen Rejection Unit | Air Liquide (engineering-airliquide.com)).

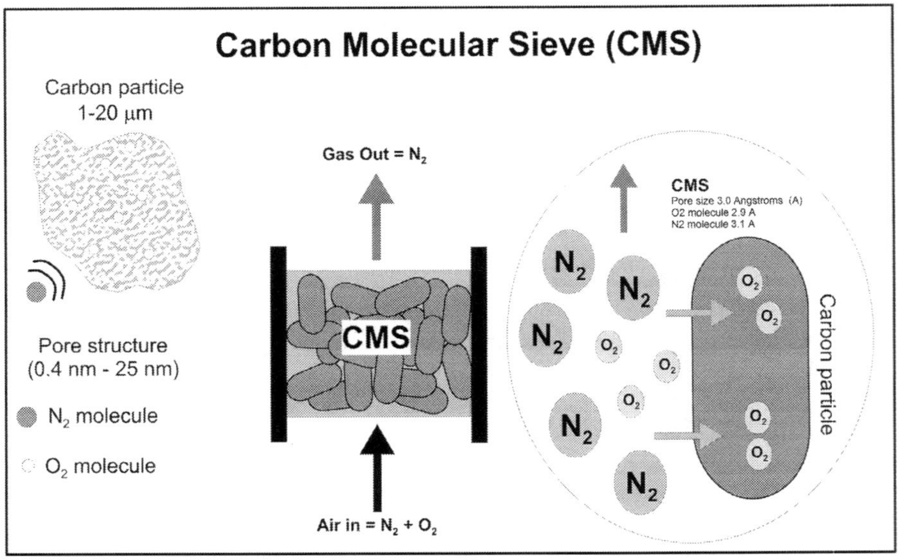

FIGURE 5.7 Is a pressure swing unit absorbents process for a N_2 generator with Pressure Swing Adsorption Filtration (PSA) (South-tek Systems, 2021).

5.4 Helium processing

Despite its status that He is the second-most abundant element in the universe, the only commercially viable source for element recovery is natural gas, CO_2 or N_2 production (Sunarso et al., 2016). Extraction of He from gas sources containing more than 0.03% by volume, He has been proven economic (Kim, 2014). The goal of separating He from natural gas is to decrease the heating value of natural gas to make it pipeline acceptable and increase the He recovery for higher economic returns (Mehrpooya and Shafaei, 2016). The global demand in 2010 for He reached 30,000 tons amounting to US$1billion (Soleimany et al., 2017), where demand in the next two decades is expected increase from 5% to 7% per annum (Sunarso et al., 2016). Table 5.1 indicates the present known plants based on public information but it is not complete. Over 60% of global helium production plants are located in the United States, followed by Algeria, Qatar, Poland, and Russia.

In He extraction from natural gas, He is first separated and then recovered from the bulk fluid. This allows further purification to be done. A flow diagram of the general process for He recovery from natural gas is indicated in Fig. 5.8 (Soleimany et al., 2017).

There are three basic gas separation processes: cryogenic processes, pressure swing adsorption (PSA), and membrane separation (Bakhsh et al., 2007; Crawford et al., 2010). A normal natural gas composition flowing through a cryogenic unit is 10%–90% CH_4 which typically includes C_2H_6 plus heavier hydrocarbons along with potentially N_2, H_2S, H_2O, CO_2 and potentially traces of heavy metals (Hosseini, 2009; Mehrpooya and Shafaei, 2016). Helium requires all other impurities and contaminants to be removed for commercial sale. Six steps are required in extracting and producing liquefied He from natural gas or N_2 gas (Soleimany et al., 2017):

TABLE 5.1 List of Global He Extraction Plants from West (2009) and updated by this author.

Owner or operator	Location	Year commissioned	Type of He product	Capacity t/d
Plants located in the USA				
Cimarex	Big Piney, WY	2011	Pure He	
Air Products Helium Inc.	Hansford County, TX		Pure He	
Air Products Helium Inc.	Liberal, KS	1966, 1991	Pure He	1.3
BP America. Inc.	Sunray, TX	1995	Crude He	
BP America. Inc.	Ulysses, KS	1998	Crude He	
Tumbleweed	Cheyenne Co., CO		Crude and pure He	
DCP Midstream, LLC	Hansford County, TX		Crude He	
DCP Midstream, LLC	Liberal, KS		Crude He	
DCP Midstream, LLC	Borger, TX		Crude He	
EnCana Oil & Gas (USA) Inc.	Moab, UT		Crude and pure He	
ExxonMobil Gas & Power Marketing Co.		1986	Crude and pure He	
IACX Energy 2	Otis, KS		Crude He	
IACX Energy 2	Harley Dome, UT	2020	Crude He	
K-L Energy Partners, LLC3	Lakin, KS	1995	Crude He	
Linde Global Helium Inc.	Otis, KS		Pure He	
Midstream Energy Services, LLC	Keyes, OK	1996	Crude and pure He	
Nacogoches Oil and Gas, L.L.C.4	Shiprock, NM		Pure He	
ONEOK Field Services Co.3	Bushton, KS		Crude He	
ONEOK Field Services Co.3	Scott City, KS		Crude He	
Pioneer Natural Resources Co.	Fain, TX			
Pioneer Natural Resources Co.	Satanta, KS		Crude He	
Praxair, Inc.			Bushton, KS	
Praxair, Inc.	Ulysses, KS	1962, 1998		
SemGas, LP6	Dodge City, KS		Crude He	
Plants located outside the USA				
Linde	Darwin, Australia	2010	Pure He	2.6
Helison	Skikda, Algeria	2006 (2012)	Pure He	10

(Continued)

TABLE 5.1 List of Global He Extraction Plants from West (2009) and updated by this author.—cont'd

Plants located in the USA				
Owner or operator	Location	Year commissioned	Type of He product	Capacity t/d
RasGas	Qatar	2005 (2013)	Pure He	8.65 (17.7)
Orenburg Gazprom	Orenburg, Russia	Before 1993	Pure He	3.0
Polish Oil & Gas Co.	Odolanbw, Poland	1977	Pure He	1.6
HELIOS	Arzew, Algeria	1995	Pure He	7.4
ONGC (PSA pilot plant)	Kuthalam, India	2008	Pure He	0.00023
People's Republic of China	Ningxia, China	2020		

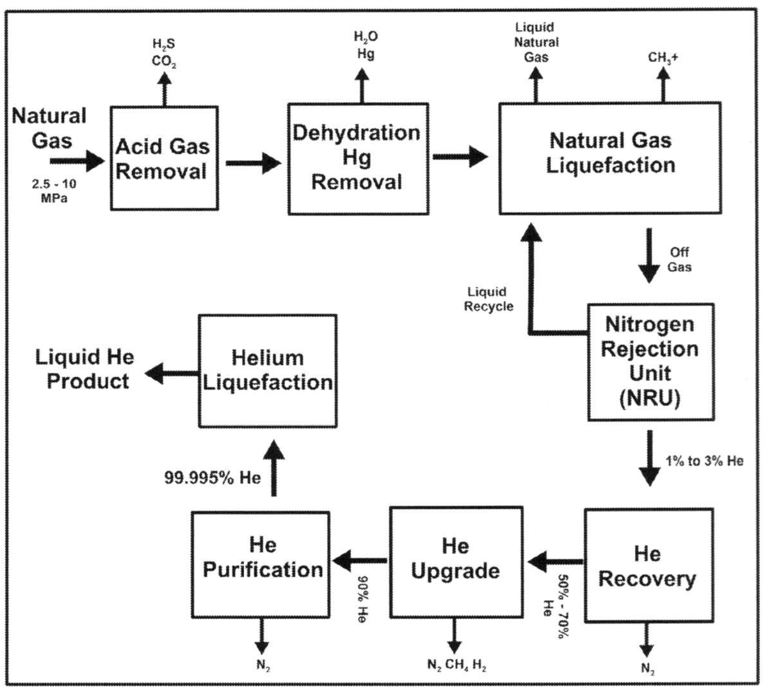

FIGURE 5.8 Schematic representation of a typical helium recovery process from natural gas (Soleimany et al., 2017).

1. Natural gas processing/pretreatment with the removal of H_2S, CO_2 (amining), H_2O (dehydration), and any heavy metals;
2. Natural gas refrigeration (removal of heavier hydrocarbons if any) and liquefaction (production of liquefied natural gas);
3. N_2 rejection and then He recovery from natural gas;
4. He upgrading;

5. He purification;
6. He liquefaction.

The pretreatment process is imperative for removing acid gases such as CO_2 and H_2S, H_2O, and heavy metals such as Hg prior to entering the refrigeration and liquefaction process. Helium's low boiling point allows it be concentrated in rejection units such as a reboiler or a NRU. This process is conventionally a cryogenic distillation or PSA process where the He recovery is integrated with the NRU (Rufford et al., 2014; EPCM, 2021).

PSA is typically used in pretreatment stages and membrane separation and is in its infancy in terms of He extraction for large cryogenic operations (Sunarso et al., 2016). The cryogenic process at temperatures below −65°C can have up to a 90% He recovery. Cryogenic processes are divided into two groups: multiflash cycles and high-pressure distillation column processes (Victory et al., 2009).

Liquid Natural Plants (LNG) use cryogenics for NRU which allows recovery of He off the feed gas. Four basic cryogenic processes used for NRUs are multistage flash separators, single column heat pumped process, double-column process, and dual column cycles (Agrawal et al., 2003; EPCM, 2021). The choice of process configuration depends on the flow rate of the feed gas, the composition of the feed and He concentration in order to make the recovery economically feasible. Multiflash separator processes have higher energy requirements compared to cryogenic distillation column processes but have lower capital costs and production overhead vapor streams with low helium concentrations. A single-stage flash separator process produces N_2-rich overhead vapors containing 1%−3% He, depending on the feed gas composition. He recovery of 50%−70% can be achieved from more complex, double-column NRU processes when operating in partial condensation mode, where N_2 and small amounts of CH_4, H, Ne, Ar, and CO_2 are also present (Agrawal et al., 2003; West, 2009; EPCM, 2021). The dissolved N_2 and He are removed in a multistage flash process from the gas feed by reducing the pressure over a series of flash vessels (Fig. 5.9). The He is vaporized with the N_2 in each flash stage, where the liquid natural gas is used to preheat the feed. This process can also be applied to the after the NRU being cooled (partially condensed) for separating the gases (West, 2009). Crude He concentrations in the final product are dependent on the feed composition, pressure drops, and temperature changes.

In the single-column heat-pumped process (Fig. 5.10), a purified natural gas feed is precooled in the main cryogenic heat exchanger (MCFHE; against the rejected N_2) and then fed to a high-pressure column (13−28 bar). The vapor ejected from the top of the column contains N_2 and He. A closed-loop with CH_4 as working fluid supplies heat to the reboiler and cooling duty to the condenser. The CH_4 is condensed and drawn from the bottom of the tower whereafter it is flashed over a control valve (reduction in pressure). The feed gas and rejected N_2 from the top of the column are used to heat the CH_4 prior to being compressed into downstream facilities. This process produces a N_2-rich stream containing approximately 1%−3% He, depending on feed gas conditions (Rufford et al., 2014).

Häussinger et al. (2005) described a simplified process flow scheme of a modern, double-column process for NRU and He recovery from natural gas (Fig. 5.11). The feed gas is cooled against cold product streams in a cryogenic heat exchanger and is fed to the bottom of the

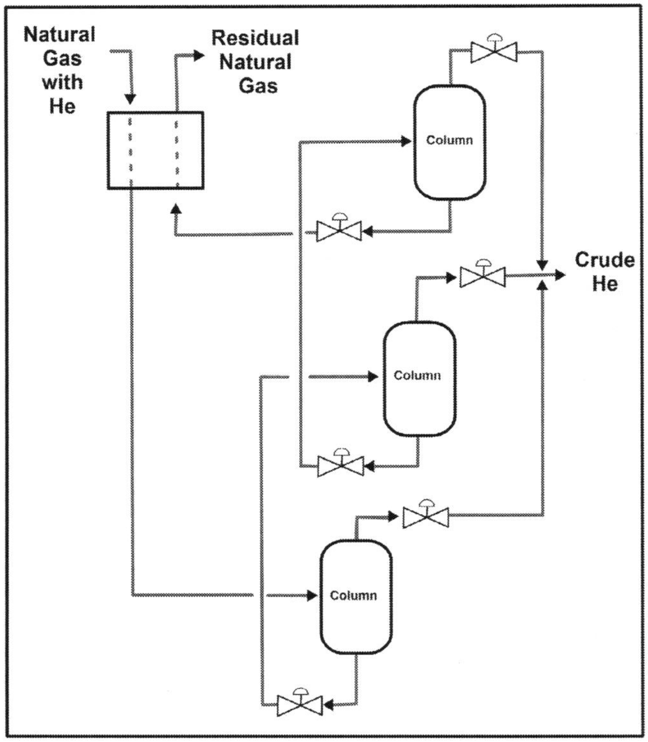

FIGURE 5.9 Schematic of a multiflash process for the recovery of He from natural gas (Rufford et al., 2014).

high-pressure column (EPCM, 2021). In this high-pressure column, the He is recovered from the feed at typical operating pressures between 10 bar and 25 bar. Reflux for the low-pressure and the high-pressure column is provided by means of the column's partial condenser (not presented). The He is contained in the noncondensed part of the high-P column (EPCM, 2021).

The final separation of N_2 from CH_4 that has accumulated in the bottom of the high-pressure column is performed in the low-pressure column. The N_2-rich overhead product is heated in the low-pressure column against the bottom of the high-pressure column. To achieve a low concentration of N_2 in the CH_4-rich residual product, a reboiler is required at the bottom of the low-pressure column (Rufford et al., 2014; EPCM, 2021). The reboiler takes the condensation from the top N_2 stream from the high-pressure column. The necessary heat transfer is possible because of the higher temperature of the N_2 stream when compared with the temperature of the CH_4 from the bottom of the low-pressure column. A CH_4-rich stream is produced at the bottom of the low-pressure column which is then elevated in pressure, vaporized, and heated against the natural gas feed (Rufford et al., 2014; EPCM, 2021).

Modern cryogenic He recovery processes are significantly more complicated than the simplified single-column and double-column systems (Rufford et al., 2014; EPCM, 2021). The dual-column cycle shares many common features with the double-column process but

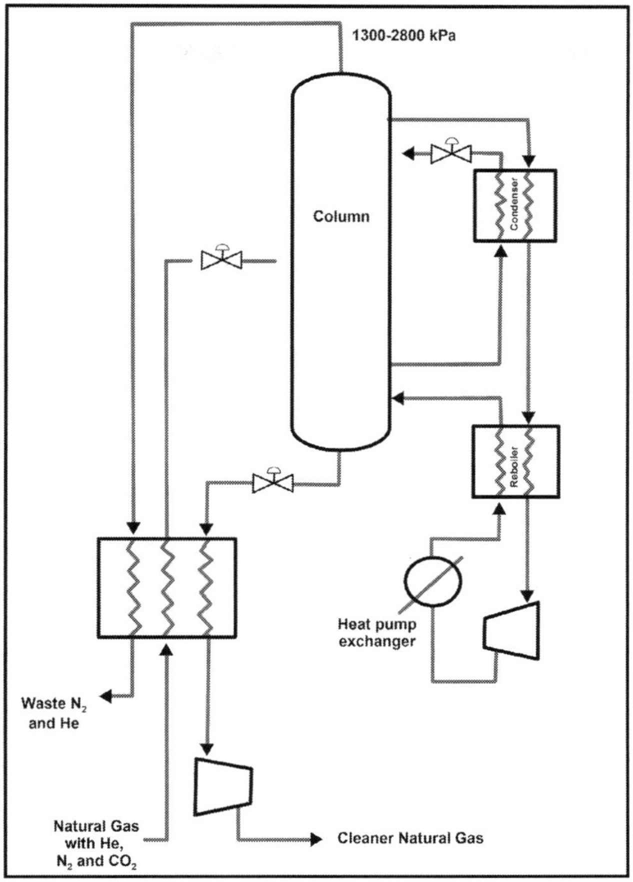

FIGURE 5.10 Single column heat-pumped process for He extraction (Agrawal et al., 2003).

has a more complex level of integration between process and refrigeration streams (Rufford et al., 2014; EPCM, 2021). Depending on the local feed gas composition, and available product markets, the complete, integrated cryogenic process may include the recovery of C_2H_6+ and N_2, in addition to the recovery of crude He (Rufford et al., 2014).

The crude He from the NRU must be upgraded to a concentration of at least 90% pure He prior to liquefaction. The impurities still contained in the crude He are N_2, CH_4, H and sometimes Ne, must be removed in several stages (EPCM, 2021). The upgrading process includes condensation of the bulk components, catalytic oxidation of any trace H, separation of H_2O, CO_2, and O_2 from the reactor by condensation of the H_2O water and then within a PSA unit, and removal of the final traces of N_2 in another PSA unit. The product from these steps can approach He purities of 99.995% (Daly, 2005).

Fig. 5.12 is a schematic representation of a typical process to purify and upgrade He (EPCM, 2021). The upgraded He is mixed with air for combustion with O_2, heated to above 300 K and compressed through a catalyst bed to oxidize traces of H or remaining CH_4^+. The product from the reactor is cooled to condense any H_2O formed and sent through a separator

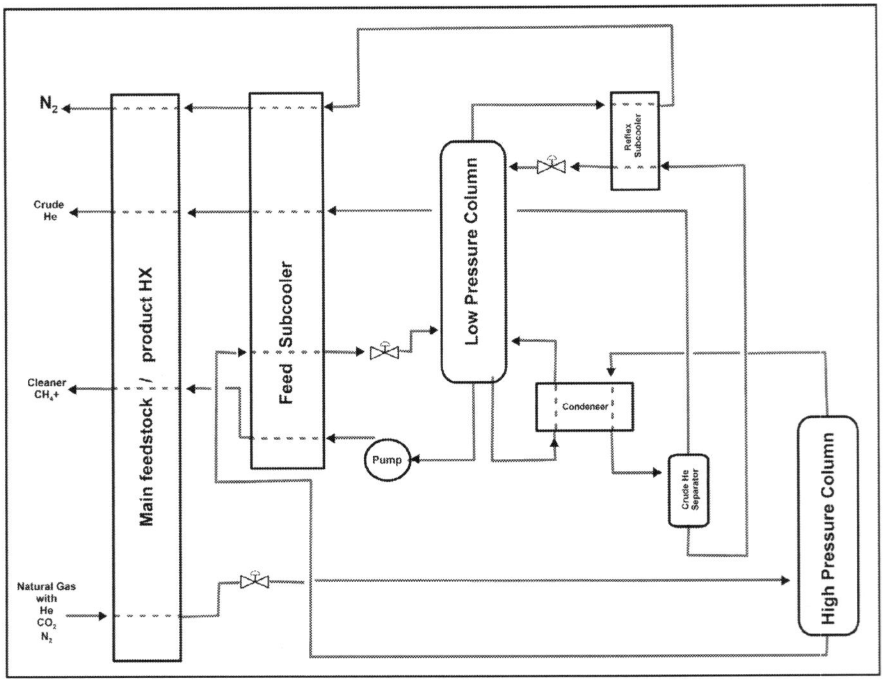

FIGURE 5.11 Double-column process for N_2 rejection and He recovery (Häussinger et al., 2005).

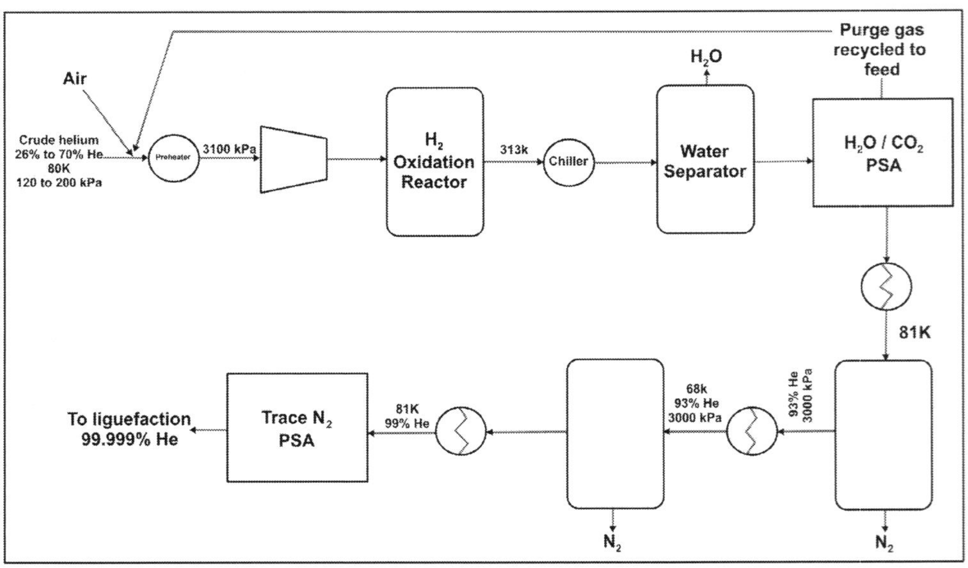

FIGURE 5.12 Schematic of the upgraded He purification process (Agrawal et al., 2003).

from where the top gases are fed to a PSA unit. Molecular sieves in the PSA unit may be used for further dehydration along with CO_2 and O_2 capture (Rufford et al., 2014).

Helium purity of 99.995% can be attained when the H-free gas needs to flow through a low-temperature PSA unit (Lindemann et al., 2010) and/or additional cryogenic condensation processes to remove N_2 to less than 10 ppm (Agrawal et al., 2003). Four-bed PSA molecular sieve units containing adsorbent such as zeolite 4A are commonly used (EPCM, 2021). The N_2 blowdown gas from the PSA purification unit is compressed, dried, and recycled to the inlet of the upgrader where it is combined with the crude He feed (Rufford et al., 2014).

Common industry processes for liquefaction of He are based on isenthalpic throttling of the purified He across a Joule-Thomson valve (EPCM, 2021). The purified He is compressed (to 20 bar) and precooled to 80 K with exhaust from the He expander or liquid N_2, then cooled to 20 K with H refrigerant or below 80 K again with exhaust from the He expander (Agrawal et al., 2003; Häussinger et al., 2005). The final liquid He is produced by the free expansion of the compressed gas.

A case study is presented by Lindemann et al. (2010) on a recently commissioned He production facility in Australia based on direct He separation from natural gas using cryogenic separation is discussed here. The He plant at the Darwin LNG production facility has a crude feed of 3% He from an NRU (downstream of main cryogenic heat exchangers). The plant is capable of producing 2.6 ton per day of liquid He (860 L per hour) with a purity of 99.999% He. This process utilizes a two-stage cryogenic flash process, H oxidation reactor, and two PSA units. The N_2-rich feed is compressed to 2 bar and upgraded by cooling to 80.5 K in the first N_2 removal stage, where part of the N_2 is condensed to give a He-enriched stream of 26% (EPCM, 2021). The gas is then warmed to ambient temperature, compressed to 31 bar, and mixed with air to feed the H oxidation reactor. The H_2O and CO_2 formed are removed in the PSA unit. After H removal, the He-enriched gas is cooled to 81 K (second N_2 condensation stage), producing a 93% He vapor stream. The upgraded He stream is then flashed to provide a product of 99% He. Traces of N_2 are removed (<5 ppm) in the final purification stage by means of a cryogenic PSA unit. The world's largest production facility, Ras Laffan Plant in Qatar, has a production capacity of 17.7 tons of He per day, and feed gas composition of 0.04% He is recovered via the same cryogenic distillation principles. Each of its seven LNG production trains (how a cryogenic unit is refered too) has its own He recovery unit installed at the cold end of the LNG process to upgrade crude He that is produced (Daly, 2005).

Adsorption-based processes are mainly used during the pretreatment process to remove H_2O, CO_2, H_2S, and other impurities from the feed natural gas in cryogenic facilities. PSA units are becoming popular and applicable for small He deposits whether N_2 or CH_4 rich gas feed as they are more economical than cryogenic units. This process is also used for removing trace impurities of N_2 and CH_4 during He purification (Tagliabue et al., 2009). Gas separation using adsorption comprises of two steps: adsorption and desorption. During adsorption, a porous solid bed selectively adsorbs the higher affinity gas to the adsorbent bed and produces a gas stream enriched with the less strongly adsorbed gas component(s). Once the solid adsorbent is saturated, it then must be regenerated. In the desorption column, the gaseous product is enriched with the strongly adsorbed component (Rufford et al., 2014).

Technologies that deploy this method are TSA, PSA, fluidized adsorbent beds and moving adsorbent beds. TSA methods regenerate the gas using external heat to increase the temperature in the desorber. Conversely during PSA methods regeneration is conducted under low temperatures but releasing pressure and purging the bed (Tagliabue et al., 2009). Fluidized and moving bed operations are less common for industrial gas separations than the cyclic-batch fixed bed operations of TSA and PSA (Seader and Henley, 2006).

The molecular sizes of He (2.60 Å) and H (2.89 Å) are similar and both having very low boiling temperatures. The PSA processes used for H purification potentially have many common features to He purification processes. Rufford et al. (2014) suggests the potential of new technologies that are under development for H purification could also be also applicable to He purification (EPCM, 2021). In general, modern PSA units for H purification use layered adsorbent beds with three to four adsorbents (Si, Al for H_2O, activated C for CO_2, and 5A zeolite for CH_4 and N_2 (Ritter and Ebner, 2007)). Similar multilayer bed designs are presently being evaluated for He recovery processes according to Bakhsh (2010).

Commercially available zeolites and narrow pore activated C with reasonable adsorption capacities for N_2 make suitable adsorbents for He purification. Table 5.2 lists commercial adsorbents used in industrial processes of H purification that potentially can be used in PSA processes to separate heavier components from He. Bakhsh (2010) discussed the use of Ca and Li-exchanged 13× adsorbent.

Rufford et al. (2014) discussed two case studies of direct He recovery from natural gas using adsorption-based processes (ECPM, 2021). The first is India's Oil and Natural Gas Corporation PSA Pilot Plant with a production capacity of 23 kg per day of He, having a feed gas

TABLE 5.2 Commercial adsorbents with potential for PSA units for He upgrading and purification (Rufford et al., 2014; EPCM, 2021). Some of the products may be no longer available or have changed names.

Adsorbent type	Trade names	Species adsorbed
Zeolites	ZSM-5	CH_4, CO_2, N_2
	HISIV 3000 (UOP)	
	Zeolite 5A (UOP, Sigma)	
	Zeolite 13× (UOP)	
Wide pore-activated carbons	BAX-1100 (Westvaco)	CH_4, CO_2, N_2
	RB3, R2030, GAC 1240 (Norit)	
	BPL 4 × 10 (Calgon)	
	Acticarb EA 1000 (Activated Carbon	
	Technologies, Australia and New Zealand)	
Narrow-pore activated carbons	Maxsorb (Kansai Coke & Chemicals Co.)	N_2 in final purification stage
Ca- and Li- exchanged 13× (Baksh, 2010, Das et al., 2008)		N_2 in final purification stage

containing 0.06% He, 88.5% CH_4, 9.86% $C_2H_6^+$, 1.18 % N_2, and 0.4% CO_2. This process has four stages: (1) pretreatment of the feed gas using PSA, (2) recovery of CH_4 gas onto an adsorption bed, (3) upgrade of He from N_2 on a zeolite 13× adsorption bed, and (4) He purification also using zeolite 13× adsorption bed. Although this plant indicates the possibility of using PSA methods throughout the process, the plant only recovers 65% He in the feed gas. Conventional cryogenic and PSA processes are still more effective recovering more than 95% He from the feed gas.

The second case study discussed is a US patent no. 5542966 where a natural feed gas containing 4% He, 26% CH_4^+, and 70% N_2 flows through a two-stage activated carbon PSA process (Rufford et al., 2014). The production capacity is 550 kg per day where recovery of 95% purity He is reported. Although the He recovery of 95% is approaching cryogenic distillation process results, the N_2 in the feed gas is high compared with conventional natural gas fields which typically are less than 10% N_2 and 70%–90% CH_4.

Gas separation performed by means of membrane methods is the permeation of gases through a homogeneous membrane (EPCM, 2021). This is a solution-diffusion phenomenon where the ability of a membrane to separate components of a gas mixture is dependent on the separation factor of gaseous components which in turn is a function of the gas solubility and diffusivity coefficients (Häussinger et al., 2005). The small molecular diameter of He compared with other natural gas components in most membranes is greater than the molecular size of the impurities allowing for separation to be possible.

Membrane processes have been developed over the the past 40 years and can be found in a wide range of patents and research articles (Stern et al., 1965; Scholes and Ghosh, 2017). Recovering He directly from natural gas, membranes have demonstrated that this separation is possible when combined as two or three stages in a series, with recycle streams (Fig. 5.13). Multistage systems exhibit large pressure drops for enrichment of the He-rich gas which permeates across various membrane units, requiring interstage compressors (ECPM, 2021). The recompression between membrane stages increases both capital and operating costs of membrane systems. However, these designs can enable natural gas containing as low as 0.3% He to be purified to a very high concentration, while utilizing existing membranes that have high separation of He and CH_4 (Scholes and Ghosh, 2017).

No data on the performance of He recovery plants from natural gas utilizing membrane technology have been published in the open literature that this author has found, though there are a range of different processes that have been patented in the literature (Rufford et al., 2014; Scholes and Ghosh, 2017).

Two- and three-stage membrane processes in series with recycles have been reported feasible to recover and purify He from the overhead gas of NRUs (Scholes and Ghosh, 2016). This is due to the high N_2 concentration and reasonable He/N_2 compression ratios above 20. Alders et al. (2017) conducted a technoeconomic comparison between various membrane-based processes for He recovery and enrichment. Two scenarios were considered: (1) He and N_2 were recovered after cryogenic distillation of the natural gas, and (2) direct separation of He from natural gas. Scenario (1) investigated the separation of N_2 and He by comparing (i) a pressure relief distillation, (ii) two-stage membrane process, and (iii) second low-temperature distillation. Scenario (2) investigated direct helium removal from natural gas by comparing (i) multistage pressure relief, (ii) two-stage membrane process, and (iii) a three-step membrane process (ECPM, 2021).

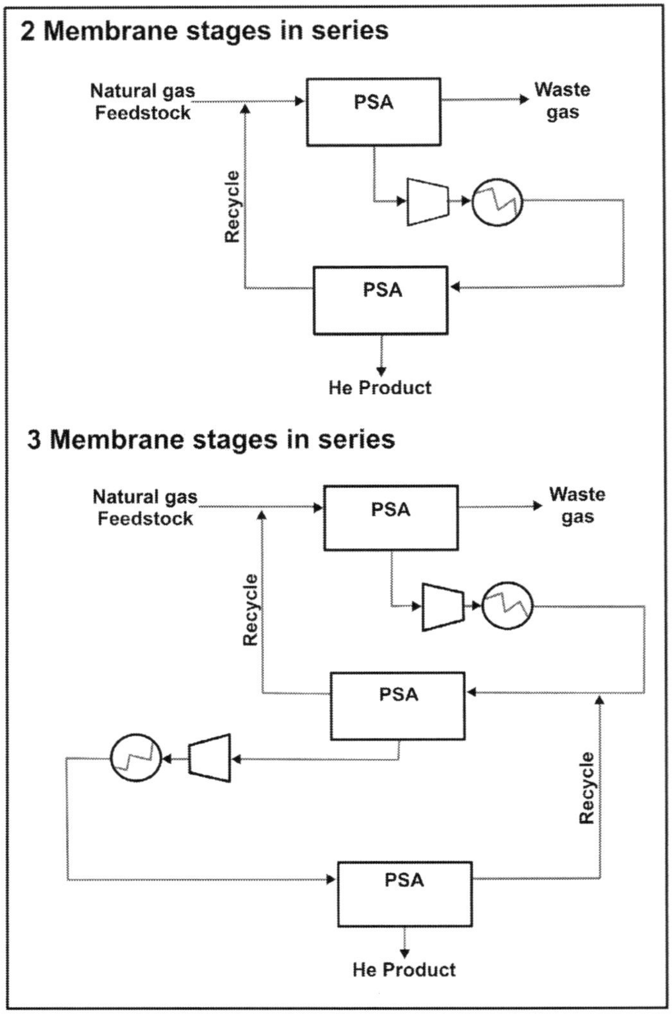

FIGURE 5.13 Two- and three-stage membrane process He extraction (Scholes and Ghosh, 2017).

The processes for Scenario (1) are indicated in Fig. 5.14 for the pressure-relief distillation after cryogenic distillation, Fig. 5.15 for the two-stage membrane process after cryogenic distillation, and Fig. 5.16 two-stage distillation system (Alders et al., 2017).

The evaluation confirmed that the hybrid process integrating the distillation and membrane technologies resulted in the lowest treatment costs and were the most favorable concerning operating costs for short amortization periods and shown the highest He recovery of 94.2%. All three processes were based on the same natural gas feed conditions of 500 kmol/h, 80% CH_4, 19% N_2, and 1% He.

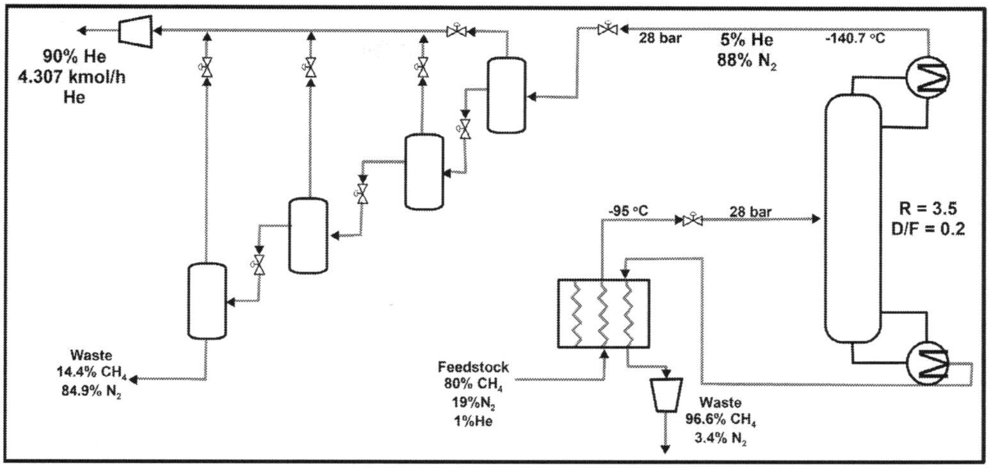

FIGURE 5.14 Combined distillation and pressure relief process for He extraction (Alders et al., 2017).

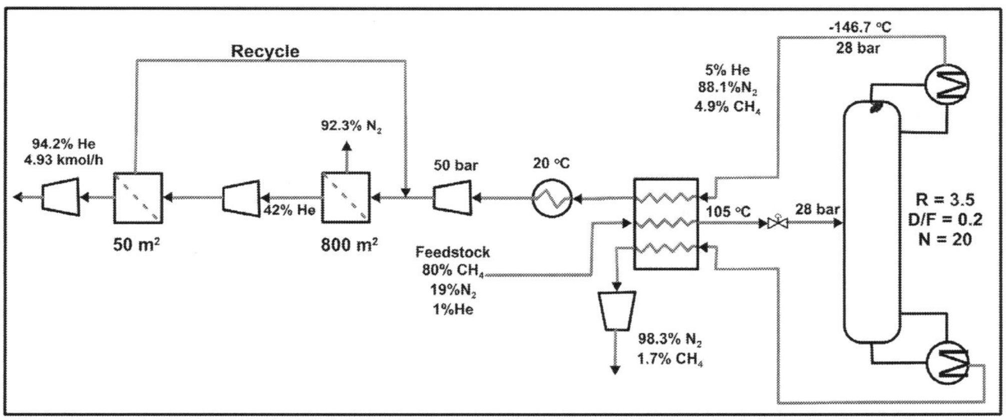

FIGURE 5.15 Hybrid process integrating distillation and membrane gas separation for He extraction (Alders et al., 2017).

Processes for Scenario (2) are indicated in Fig. 5.17 for multistage pressure relief, Fig. 5.18 for two-stage pressure relief and Fig. 5.19 for three-step membrane process (Alders et al., 2017).

The results illustrated that the three-stage membrane process required the lowest treatment and cost, and remained the most favorable option for all investigated prices while still achieving a He recovery of 90.2%. Scholes et al. (2017) state that membrane materials with even higher selectivities usually do not lead to a further reduction in treatment costs.

Extensive research has been undertaken to investigate the most favorable membrane materials for He recovery from natural gas (Barboiu et al., 2006; Bernardo et al., 2009; Li et al., 2011; Schrier, 2010). The first He extraction membranes included tubular silicates and

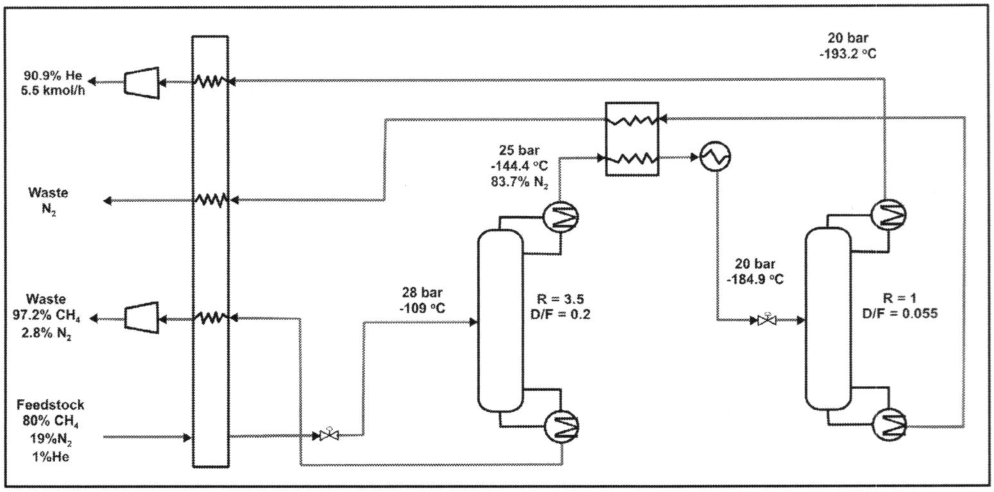

FIGURE 5.16 Two-stage low-temperature distillation system for He extraction (Alders et al., 2017).

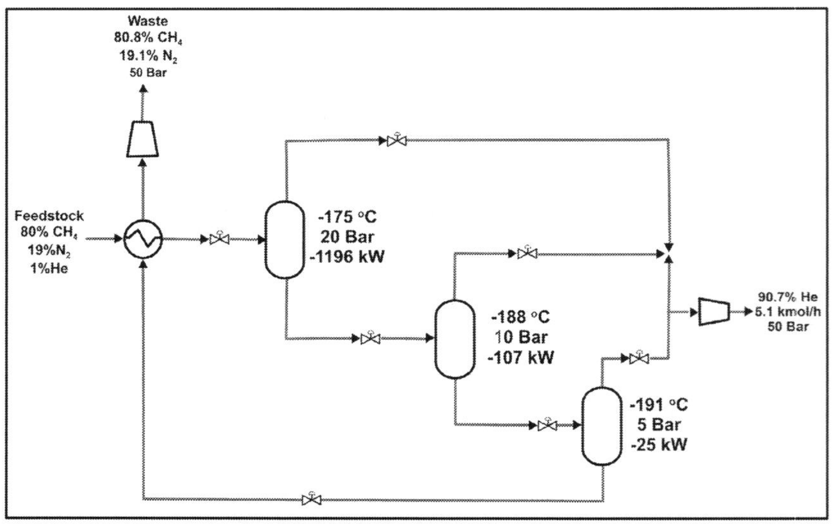

FIGURE 5.17 Multistage pressure relief system for the extraction of He (Alders et al., 2017).

quartz glass membranes and later, polymeric membranes. Recently reported membranes to include those constructed from ultramicroporous silica (Barboiu et al., 2006), molecular sieve C, porous graphene (Schrier, 2010), titanium silicates (Li et al., 2011), polyamides, and mixed matrix membranes of polyimides and zeolitic imidazolate frameworks (Bernado et al., 2009).

Sunarso et al. (2016) surveyed five different membrane materials in terms of their H and He permeation performance and the related stabilities during permeation processes.

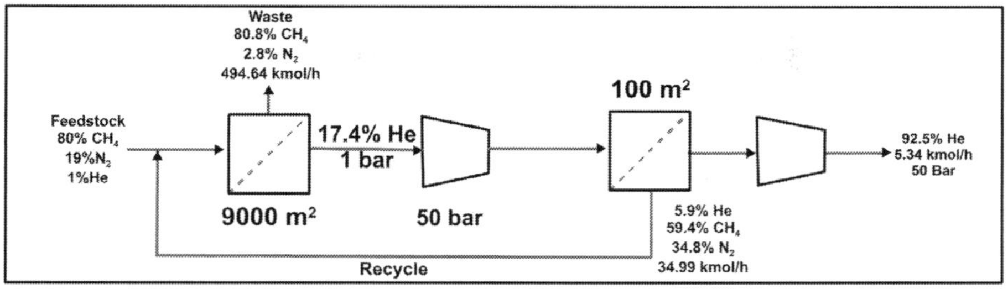

FIGURE 5.18 Two-stage membrane process for the extraction of He (Alders et al., 2017).

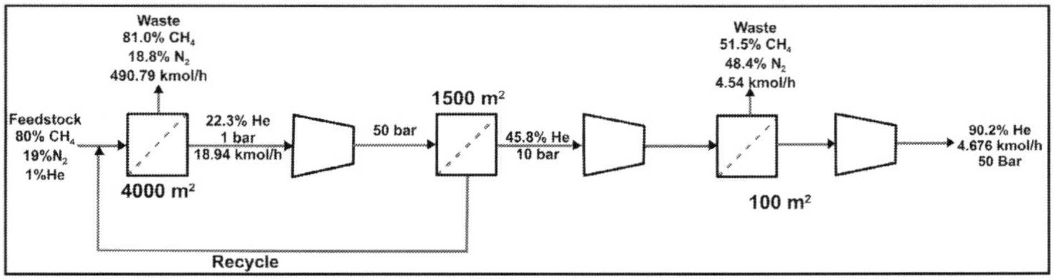

FIGURE 5.19 Three-stage membrane process for He extraction (Alders et al., 2017).

Membranes that were evaluated included: Si, polymers, zeolite, metal-organic framework, and mixed matrix membranes. The results illustrated that each membrane can be used for recovering He, but each membrane has advantages and disadvantages in terms of permeability, selectivity, stability, cost, synthesis procedures, and reproducibility (ECPM, 2021).

Ultramicroporous inorganic membranes and glassy polymeric membrane processes appear to be the most promising membranes for achieving successful He recovery (ECPM, 2021). A variety of polymeric membranes have been implemented successfully for the upgrading of crude He to 90% purity, which has energy requirements comparable with other crude He separation technologies (Scholes and Ghosh, 2017).

The future success of membrane-based He production from natural gas process technologies will depend on the development of high-performance membranes and the design of applicable membrane separation processes. Table 5.3 is a brief summary of the advantages and disadvantages for all three methods for He recovery (EPCM, 2021). The focus for He extraction must be on implementing low-cost modules with optimal designs. There are considerable opportunities to improve the efficiency of PSA processes for He recovery through using improved designs: optimized processing conditions by operation, for example, cryogenic temperatures; and improvements in adsorbent materials (Rufford et al., 2014). A challenging approach is to develop He-selective adsorbent materials which can possibly reduce the size and energy requirements of PSA beds. Although detailed technical feasibility on energy requirements and process economics of these integrated system designs still needs to be conducted, it would appear that integration is the key for economic extracting He extraction from natural, N_2 and CO_2 rich gases (EPCM, 2021).

TABLE 5.3 Advantages and limitations of He recovery technologies (EPCM, 2021).

Technology	Advantages	Limitations
Cryogenic fractionation	• Advanced technology widely used for direct recovery of He from natural gas; • High He recovery purity (>95% up to 99.999%); • Easy scale-up for increased capacities; and • Small- to microscale facilities are economic.	• High capital requirements; and • Intensive energy requirements – high operational expenditures
Adsorption-based processes	• No fluid phase changes resulting in lower energy requirements; and • Low direct He recovery from natural feed gas (<65% reported for plants in operation).	• Recommend for He purification, not for direct recovery from natural gas; • Lower He purity recovery from direct separation; • Requires high purity feed, i.e., crude He (impurities cause adsorption bed saturation leading to reduced efficiencies).
Membrane-based processes	• No fluid phase changes resulting in lower energy requirements; • Small footprint (lower impact on environments); and • Lower capital costs	• Requires "cleaned" high purity feed gas to prevent membrane fouling and damage; Membranes have not been commercialized as there is limited data available; • Requires high pressure ratios resulting in high operational costs.

5.5 Conclusions

The key in helping to increase supply of He is developing new processes that are energy efficient, optimally designed, cost effective, and able to handle very low percentage of He in the gas stream economically. As extraction costs for He per MCF rise, this will encourage research and development of new processes to reduce costs. Adsorption-based processes, specifically PSA and excluding cryogenics are only recommended for He recovery from natural, N_2 and CO_2 rich gases in small to medium accumulations as cryogenic methods are to costly for these types of deposits. Cryogenic methods which are very efficient are more applicable to signficant volumes of natural, N_2 and CO rich gases with low He percentage such as the North and South Pars Field in Qatar as described in Chapter 3. At present PSA membrane-based technologies are not yet capable of competing efficiently with cryogenic separation in relation to large-scale deposits. PSA have a greater economic viability than cryogenic processes where the He volume maybe high but small in overall reserves.

References

Agrawal, R., Herron, D.M., Rowles, H.C., Kinard, G.E., 2003. Cryogenic Technology, Kirk-Othmer Encyclopedia of Chemical Technology, vol. 8. John Wiley and Sons, pp. 40–65.

Air Liquide, 2021. Nitrogen Rejection Unit | Air Liquide. engineering-airliquide.com.

Alders, M., Winterhalder, D., Wessling, M., 2017. Helium recovery using membrane processes. Sep. Purif. Technol. 189, 433–440.

Bahadori, A., 2014. Natural Gas Processing. Elsevier, p. 896.

Bakhsh, M.S.A., Jaynes, S.E., Neu, B.T., Smolarek, J., Emley, T., 2007. Helium Recovery, US Patent No. 20040237789.

Baksh, M.S.A., 2010. Methods and Systems for Helium Recovery, US Patent No. 2010/0251892 A1.

Barboiu, C., Mourgues, A., Sala, B., Julbe, A., Sanchez, J., de Perthuis, S., Hittner, D., 2006. Ultra-microporous silica membranes for He purification. Desalination 200 (1–3), 89–91.

Bernardo, P., Drioli, E., Golemme, G., 2009. Membrane gas separation: a review/state of the art. Ind. Eng. Chem. Res. 48 (10), 4638–4663.

Chemical Engineering World, 2021. Cryogenic Distillation of Air. Chemical Engineering World.

Crawford, D., Coyle, D.A., Anantharaman, B., 2010. Method for Nitrogen Rejection and or Helium Recovery in a Liquefaction Plant.

Daly, J.W., 2005. Helium Recovery from LNG, International Petroleum Technology Conference, American Association of Petroleum Geologists (AAPG); the European Association of Geoscientists and Engineers (EAGE); the Society of Exploration Geophysicists (SEG); and the Society of Petroleum Engineers (SPE), Doha, Qatar, pp. 1–4.

Das, N.K., Chaudhuri, H., Bhandari, R.K., Ghose, D., Sen, P., Binha, B., 2008. Purification of helium from natural gas by pressure swing adsorption. Curr. Sci. 95 (12), 1684–1687.

EPCM, 2021. https://epcmholdings.com/process-technologies-for-helium-recovery-from-natural-gas-a-review/.

Feda, 2021. Pressure Swing Adsorption: Gas Separation and Purification. In: https://fedanitrogen.com/wp-content/uploads/2018/09/Pressure-Swing-Adsorption-A-Gas-Separation-Purification-Process.pdf.

Gary, J.H., Handwerk, G.E., 1984. Petroleum Refining Technology and Economics. CRC Press, p. 488.

Häussinger, P., Glatthaar, R., Rhode, W., Kick, H., Benkmann, C., Weber, J., Wunschel, H.J., Stenke, V., Leicht, E., Stenger, H., 2005. Noble gases. In: Walter, G. (Ed.), Ullmann's Encyclopedia of Industrial Chemistry. Wiley-VCH Weinheim, Baden-Wärttemberg, Germany.

Hosseini, S.S., 2009. Membranes and Materials for Separation and Purification Hydrogen and Natural Gas, Department of Chemical Engineering and Biomolecular Engineering, National University of Singapore.

Kajjumba, S., Emik, A., Ongen, H., Ozcan, K., Aydin, S., 2018. Modelling of Adsorption Kinetic Processes – Errors, Theory and Application. https://doi.org/10.5772/intechopen.80495 (Modelling of Adsorption Kinetic Processes—Errors, Theory and Application | IntechOpen).

Kidnay, A.J., Parrish, W.R., McCartney, D.G., 2019. Fundamentals of Natural Gas Processing. CRC Press, p. 498.

Kim, D., 2014. Helium Extraction Form LNG End Flash. MS Thesis. Department of Energy and Process Engineering, Norwegian University of Science and Technology.

Kohl, A., Nielson, R., 1997. Gas Purification. Gulf Publishing, p. 1408.

Li, X., Zhou, C., Lin, Z., Rocha, J., Lito, P.F., Santiago, A.S., Silva, C.M., 2011. Titanosilicate AM-3 membrane: a new potential candidate for H_2 separation. Microporous Mesoporous Mater. 137 (1–3), 43–48.

Lindemann, U., Boeck, S., Blum, L., Kurtcuoglu, K., 2010. Turnkey helium purification and liquefaction plant for Darwin, Australia. AIP Conf. Proc. 1218, 271.

Mehrpooya, M., Shafaei, A., 2016. Advanced exergy analysis of novel flash-based helium recovery from natural gas processes. Energy 114, 64–83.

Miller, L.N., Zawacki, T.S., 1978. Process for Acid Gas Removal from Gaseous Mixtures. Institute of Gas Technology.

Nuttall, W.J., Clarke, R.C., Glowacki, D.G., 2012. The Future of Helium as a Natural Resource. Routledge Publishing, p. 376p.

Ritter, J.A., Ebner, A.D., 2007. State-of-the-art adsorption and membrane separation processes for hydrogen production in the chemical and petrochemical industries. Separ. Sci. Technol. 42 (6), 1123–1193.

Rufford, T.E., Chan, I.K., Huang, S.H., May, E.F., 2014. A review of conventional and emerging process technologies for the recovery of helium from natural gas. Adsorpt. Sci. Technol. 32, 49–72.

Scholes, C.A., Ghosh, U.K., 2016. Helium separation through polymeric membranes: selectivity targets. J. Memb. Sci. 520, 221–230.

Scholes, C.A., Ghosh, U.K., 2017. Review of membranes for helium separation and purification. Membranes 7 (9). https://doi.org/10.3390/membranes7010009.

Scholes, C.A., Ghosh UK, U.K., Ho, M.T., 2017. Ind. Eng. Chem. Res. 56, 335–383.

Schrier, J., 2010. Helium separation using porous graphene membranes. J. Phys. Chem. Lett. 1 (15), 2284–2287.

Seader, J.D., Henley, E.J., 2006. Separation Process Principles. John Wiley and Sons, Hoboken, NJ.

Sears, W.M., 2015. Helium the Disappearing Element. Springer International Publishing 138p.

Soleimany, A., Hosseini, S.S., Gallucci, F., 2017. Recent Progress in Developments of Membrane Materials and Modification Techniques for High-Performance Helium Separation and Recovery: A Review, Chemical Engineering and Processing: Process Intensification. https://doi.org/10.1016/j.cep.2017.06.001. June 2017, URL:

South-tek Systems, 2021. Carbon Molecular Sieve (CMS). https://www.southteksystems.com/psa-technology/.

Stern, S.A., Sinclair, P.J., Hareis, N.P.V., Mohr, P.H., 1965. Ind. Eng. Chem. 57, 49.

Sunarso, J., Hashim, S.S., Lin, Y.S., Liu, S.M., 2016. Membranes for helium recovery: an overview on the context, materials and future directions. Separ. Purif. Technol. 176, 335–383.

Tagliabue, M., Farrusseng, D., Valencia, S., Aguado, S., Ravon, U., Rizzo, C., Corma, A., Mirodatos, C., 2009. Natural gas treating by selective adsorption: material science and chemical engineering interplay. Chem. Eng. J. 155 (3), 553.

Victory, D., Miles, M.W., Oelfke, R.H., 2009. Helium Recovery from Natural Gas Integrated with NGL Recovery. Google Patents.

West, J.E., 2009. Helium Extraction and Production Techniques. URL: (Accessed: January 27, 2011).

CHAPTER

6

Reserves and future trends in helium

The most exciting phrase to hear in science, the one that heralds new discoveries, is not 'Eureka! [I've found it!]', but 'That's funny...' ***Isaac Asimov.***

6.1 Introduction

Approximately 17.6% of the natural gas wells in the USA contain greater than 0.3% mole percent He which is considered economic at present prices (Broadhead and Gillard, 2004). The percentage of He in natural gas that is economic may change downward with increase in prices and advances in extraction technology. ^{3}Helium is sourced from the USA and Russian military and is 1000 times more expensive than ^{4}He. Broadhead and Gillard (2004) estimate of wells in the US having economic He is probably on the high side as there are many areas where there is no analysis available. Most gas sampling has occurred with a bias in areas where He is potentially and economic commodity. Even though 17.6% of the wells sampled have what would be considered economic extractable He, most are uneconomic because they as not associated with significant reserves of the host gas.

In general terms there are two approaches to He resource exploration: (a) Very specific exploration for small to medium high-grade deposits, and (b) more general exploration for lower grade deposits in supergiant natural gas fields. The majority of high-grade deposits tend to be concentrated in a small area and generally have a much higher chance of failure due to limited reservoir. The low-grade deposits produce from supergiant fields tend to have a lower rate of failure due to the significant natural gas reserves and flow rates present. This relationship is certainly true for base, precious metals, and petroleum exploration in the historical past as well. Presently petroleum and mining industry tend to focus on low-grade deposits such as shale (petroleum) and porphyry (copper and gold). These types of deposits have overall minimized exploration and production risk and allow economies of scale. These low-grade deposits have low rate of return but have large reserves and longevity in terms of production than the smaller highly productive accumulations. This situation seems to be true for He deposits as well. The five main producing deposits or facilities in the world are related to relatively low-grade He accumulations. Natural gas prices can have a major impact to extraction even when He prices remain high.

This chapter discusses the economics of the He industry and looks at previous production as a guide to future trends. The main caveat in any discussion of He production and supply is

Geology and Production of Helium and Associated Gases
https://doi.org/10.1016/B978-0-323-90988-4.00005-0

the limited information available from many areas of the world. The lack of available data compared to petroleum information and analysis especially from several countries is a major problem in projecting economics of any particular deposit. The USA, Australia, and Canada tend to generally have databases that provide useful data on He that result in a reasonable conclusion for viability of any project in those areas. As new data become available, they may change analysis considerably and the economics of one or more projects.

6.2 Reservoir characteristics for an economic project

Helium is a very small molecule, and therefore low permeability and porosity in the reservoir are not necessarily a limiting factor for economic flow rates of the element. The small size of the He molecule allows significant flow through reservoir rock or many other lithologies with minimal impediment and is only restrained by the flow characteristics of the host gas. Helium production is rarely associated with oil production but is in certain areas associated with natural gas production. Almost all He production tends to be tied to reservoirs with good porosity and permeability. Fracture stimulation if used seems to be limited rather than common place such as the large fracture stimulations that are typically utilized in shale plays.

The characteristics of a reservoir to consider for petroleum as well as He and CO_2 are (Ahmed, 2006; Lake, 2007):

1. Height (thickness) of reservoir
2. Area of reservoir
3. Porosity
4. Water saturation
5. Permeability
6. Percentage of He
7. Age of pore water

The smaller the areal extent of the reservoir, the greater the amount of He that needs to be present, aka Harley Dome, Grand County, Utah, USA, for the accumulation to be economic. The larger the areal extent of the reservoir, the lower the percentage of He that is needed to be economic such as Otis-Albert and Reichel fields in Kansas, USA; Salt Wash Field, Grand County, Utah, Battle Creek, Saskatchewan Canada and Bledsoe Ranch, Colorado, USA.

The age of the pore water is critical for an He accumulation, as previously stated, old pore water is advantageous as the concentration of He takes many millions of years. The older the pore water, the more time there is to accumulate He and therefore the more likely it will develop into an economic deposit. This assumes all of the other characteristics of area, thickness, water saturation (low), porosity, and permeability are favorable as well. There are exceptions such as Harley Dome and Grand County, where the pore water is younger and the accumulation is probably the result of remigration from an older trap or local release from basement rocks.

The type of trap for He accumulations is predominantly structural or a structural-stratigraphic trap. Because of the small size of the He molecule, an excellent seal, such as salt, anhydrite or thick, organic or clay rich shale is very important and a common attribute of most He accumulations. There is no particular depositional setting for He accumulations. Sabkha porosity zones, oolitic zones, shoreline sandstones, fluvial channel systems, reef build-up, fractured granite, karst, etc., all can host economic He accumulations. A typical example of a structural trap is the Big Piney-La Barge (BPLB), North and South Pars Field (NSP) in Qatar and Iran, the Dineh-Ki-Bey Field, Harley Dome, Battle Creek in Saskatchewan, Canada and Ryersee Field in Kansas, USA. A combination structure-stratigraphic type would be Sorrento-Mt. Pearl in Colorado, Panhandle-Hugoton Field (PHF) in Texas, Oklahoma-Kansas, Salt Wash in Utah, and Otis-Albert in Kansas. In these cases, the structure was formed sometime after deposition and the porosity zones are present but not uniform. The Salt Wash Field is an example from Grand County, Utah, that produces from porosity zones that have limited extent and are draped across a structural nose or closure.

6.3 Play characteristics

Play characteristics for a He accumulation usually include the following: (1) Reservoir has reasonable porosity and permeability; (2) Excellent seal—commonly salt, anhydrite or thick shale with high clay content; (3) Old (Paleozoic) rock with old pore fluids—indicating that the He has several millions of years necessary for the concentration of the element; (4) Reservoirs tend to be shallow and generally pressure; (5) The source for He is in the crust and mantle so it is out of sight with high radioactivity increases the likelihood of He production from radioactive minerals. In looking at Canada and the USA, the setting for most He fields is on the edges of basins overlying areas of shallow Precambrian cratonic rocks. These areas may be either in tectonically active or tectonically dormant regions. The larger fields such as the BPLB and PHF fields are unique super-sized He-producing oddities rather than the norm of accumulations to be found elsewhere in the two countries.

For a new area, the major play characteristic would be similar areas such as the Las Animas Arch Helium Region (LAAHR), Sweetgrass Arch–Bow Island Arch–Bowdoin Dome area, the flanks of the Black Mesa and Holbrook basin in association with the Defiant Uplift, Central Kansas Uplift, and the Uncompaghre Uplift. The difficulty working in the western part of the USA is that it is predominantly owned by Federal and State governments. Some states are pro-industry like Utah and others are anti-industry like Colorado. The Federal lands are undergoing turmoil because the present administration intend to stop leasing (which has been delayed) and significantly change regulations and royalty structure.

As previously stated, a criteria for most if not all He deposits is that they lie above the relatively shallow Precambrian craton. Fig. 6.1 indicates where Precambrian rocks are present and exposed. This bodes well for the East Africa Rift System (EARS) that is presently testing for He resources. What it also says is that many areas of the world can be excluded from exploration. These areas include the majority of western Europe, Gulf Coast, Mexico, Eastern Australia Japan, Philippines, Indonesia, northern Middle East, western most South and North America, Alaska, Scandinavia, central Russia, Laos, Vietnam, Thailand, Myanmar, the Himalayas, northwest Siberia, Central America, northeastern Canada, and Iceland. The areas that are shaded for Ar Precambrian rocks in Fig. 6.1 are exposed for the most part and are devoid

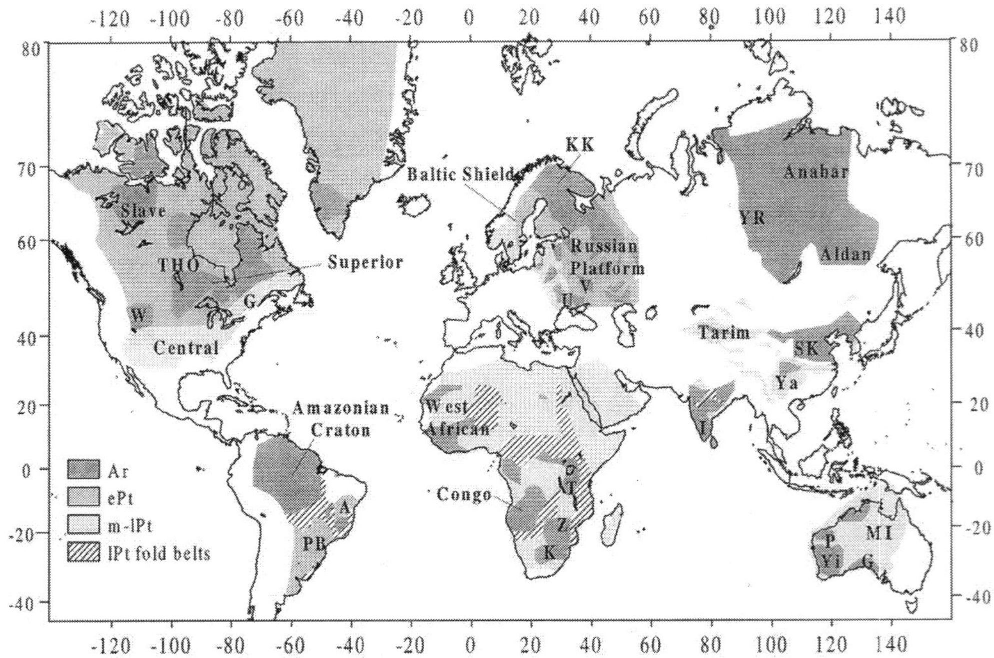

FIGURE 6.1 Location of Precambrian cratons around the world modified from Artemievai and Moody, 2001.

of potential reservoir and traps for He. What Fig. 6.1 does show is that there are numerous areas that have not been explored. Areas like Greenland and Antarctica are probably not accessible under the present political climate let alone the presence of several miles of ice.

6.4 Example of field economics

The Arrowhead Field discussed in Chapter 3 located in Cheyenne and Kiowa counties, Colorado, USA, located in the LAAHR is an example of the typical He field, partially rich in hydrocarbons and nitrogen that is likely to be found in the United States and Canada in the future. The field has produced over 3.5 BCFG with minor oil. The potential of the field is 4.5 BCF with a reported He content of 2.554%. The field theoretically (these numbers are general and for discussion purposes only) could produce 114,930 MCF He, which is valued at \$300 an MCF for a gross total revenue of \$34,479,000 (US). The natural gas has an approximate value as of February 2021 at \$2.50 an MCF of \$54,701,000 (US) after N_2 removal. The present cost of a portable He extraction unit varies from \$750,000 (US) to \$3,000,000 (US). The plant will remove CO_2 and N_2 prior to extracting the He. The value of He is significantly greater than the value of the CH_4.

In addition, based on present industry estimates, the cost of processing and extracting He from a gas stream on average will cost \$150 to \$200 (US) an MCF. The lease operating expenses should be minimal compared to the cost of processing. The transportation costs to move the He to an end user or another plant for further high-grading are not included here. The trailers that transport He are reported to cost $1M (US) each.

The USA, States of Arizona and Utah, Alberta and Saskatchewan provincial governments take a percentage of royalties from the sale of He. All other states do not take a royalty from He revenue. Private mineral owners only receive a royalty from the sale of He if it is specified in the lease. There is a monetary negative to the Operator when the government or private mineral owners benefit from He production.

6.5 World helium economy

Analysis of He economics and trends are difficult to assess because of the lack of public information and transparency. Oil and gas production data are generally available in many western countries, less so in communist, oligarchy, autocratic, socialist, or authoritarian type governments. For He, the data is even more scarce in almost all countries even with very public, accessible and complete oil and gas data. Usually reporting of He production or gas analysis data by companies is based on the need to raise monies for investment in their projects (Edison Investment Research, 2020). The He industry is essentially an oligopoly with a small group of companies and nations controlling supply and distribution. These various distribution companies that purchase and sell He do have a greater insight in terms of shortages and surplus over the short and long term. There are no independent authorities to regulate or oversee He production and sales. The United States Geological Survey (USGS), Energy Information Administration, Bureau of Land Management (BLM), Alberta and Saskatchewan provincial governments are the closest to a public regulatory source for He data that generally reports production and sales for their respective areas. Major corporations such as ExxonMobil operator of BPLB facility, or Linde/Air Products, one of the largest purchasers and distributors of He typically do not report specific sales of the element as it is not significant to separate out for financial reporting or required to be publicly disclosed. National gas producing organizations owned by countries such as Qatari, Algeria, Russia, Poland and China countries rarely release He production and sales data (Edison Investment Research, 2020). As a result, there is no definitive or verifiable picture of supply or demand for He but rather a significant amount of speculation, anecdotal data, and inference (Edison Investment Research, 2020; Stratistics MRC, 2020). This results in estimation of the production capabilities of a field or processing facility will vary between analysists. Outages of He production capacity or expansion are rarely reported. The small reservoirs of He being targeted by companies in the USA and Canada will not likely offset the decline in PHF (Edison Investment Research, 2020; Stratistics MRC, 2020). If the small reservoirs begin producing as a group, they may impact future supply, but that is probably years away. Mega projects in Qatar, Tanzania, and Amur in Russia may offset temporary shortfalls in the market but based on perceived or unknown uncertainties associated with all three, there could still be potential shortages of He in the future. An interesting note on the demand side is the blockade of Qatar in 2012 and 2017 saw no increase in prices. Iran is the great unknown as it reports it has the largest He reserves in the world, potentially true, with its ownership in the supergiant NSP with Qatar. But with the present sanctions, it may not be able to capitalize on the He asset for some time.

The five present major sources of helium are the BPLB—Exxon-Mobil, PHF—BLM facility, NSP-Qatar, HRM-Algeria, and Amur in Russia. The USA has controlled the price and supply of He for several decades and thus created a monopoly market for the element. With the recent surge in industrial demand, upstream exploration, and development, the He market has become a semi-free market to free market in He according to Edison Investment Research (2020) in certain areas. The reality is the He market is controlled by essentially four countries, the majority of production from five fields or facilities and distribution from less than six companies or countries this makes it less of a free market. Many countries are attempting to develop internal sources as a matter of national security. Based on present knowledge of He geology, the majority of these countries will likely fail in achieving a sufficient reserves to be able to secure for 100% of their needs.

The production and storage from 1960 to 2016 is indicated in Fig. 6.2 by Hamak (2020). The BLM since 2018 has seen a notable tightening in the He market, driven by lower global output and potential supply deferments (Edison Investment Research, 2020) (Fig. 6.2). The decline starting in 2012 in Fig. 6.2 represent depletion of the government stocks at the BLM facility. According to the USGS (2020), domestic USA production and use of Grade-A He (99.997% or greater) extracted during 2020 by private industry was about $322 million in value. Presently, 14 plants (one in Arizona, two in Colorado, five in Kansas, one in Oklahoma, four in Texas, and one in Utah) extracted He from natural gas and produced crude He that ranged from 50% to 99% helium (USGS, 2021). One plant in Colorado and one in Wyoming extract He from natural gas and produced Grade-A He. Three plants in Kansas and one in Oklahoma

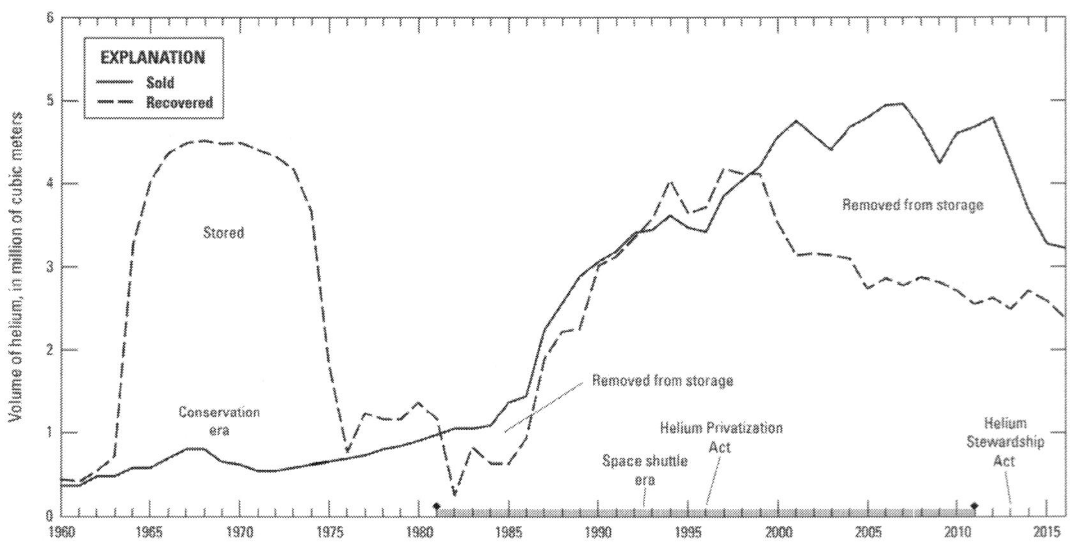

FIGURE 6.2 The production and storage for He recovery in the USA by the US government from 1960 to 2016 (Hamak 2020b).

TABLE 6.1 Estimate of US reserves by the USGS (2020 and 2021).

	Proven		Potential			Total
Region	**(MCM)**	**(BCF)**	**(MCM)**	**(BCF)**	**(MCM)**	**(BCF)**
Mid-Continent						
The Helium Fields[a]	1018	36.7	11.1	0.4	1029	37.1
BLM (Res and Stored)[b]	406	14.6	–	–	406	14.6
Subtotal	1424	51.3	11.1	0.4	1435	51.7
Rocky mountain						
Wyoming	1689	60.9	1.4	49.6	1691	110.5
Four Corners	50	1.8	0.1	4.0	50	5.8
Arizona/New Mexico	11	0.4	0.0	0.2	11	0.6
Subtotal	1750	63.1	1.5	53.8	1752	116.9
Total regions	3174	114.4	12.6	54.2	3186	168.6

[a]*While the Mid-Continent has proven reserves of 65.7 BCF (1.77 MCM) of helium, the He Fields that serve the BLM crude pipeline have 1.0 BCM of He crude located in the PHF field.*

[b]*Approximately 10.9 BCF (301.4 mcm) of Conservation Crude, 1.3 BCF (36.9 MCM) of privately owned crude, and 2.4 BCF (67.5 MCM) of native gas contained He at 1.8%. Estimated US He Reserves and resources (BCF/MCM).*

accepted crude He from other producers and the BLM pipeline and purify it to Grade-A helium. In 2020, estimated domestic apparent consumption of Grade-A helium was 1.4 BCF (40 MCM). Helium extraction and sales are summarized in Table 6.1. Note the decline in worldwide sales of He in 2020 due to the COVID-19 pandemic worldwide.

Poland (Fig. 6.3) and Russia (Fig. 6.4) demonstrate the changing sources of He around the world. Poland is experiencing a slow, steady decline as their present fields are depleting (Fig. 6.3). Russia with the establishment of the Amur Facility will experience a surge in He extraction and potentially to dominate the market in certain areas of the world. In Fig. 6.5 is Edison Investment Research (2020) decline curve for the Hugoton portion of the PHF

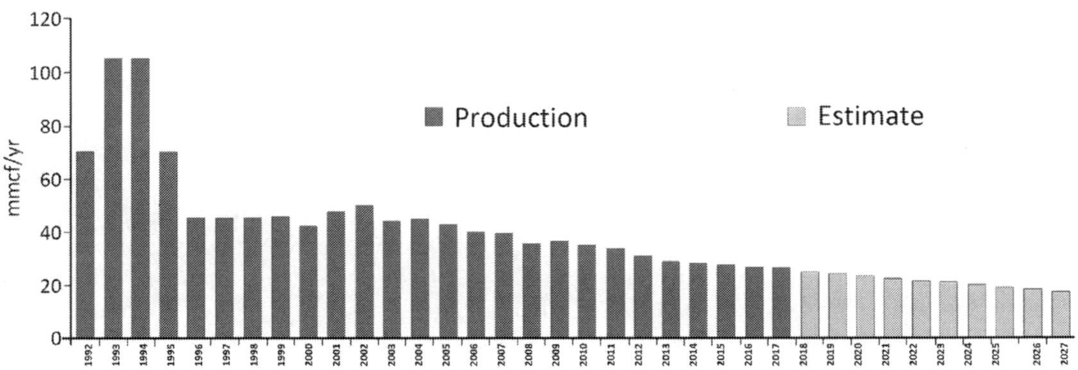

FIGURE 6.3 Polish production and projected decline curve modified from Edison Investment Research, 2020.

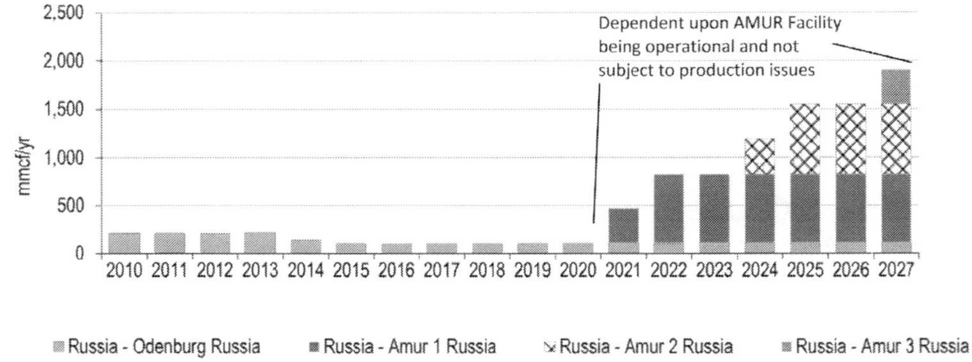

FIGURE 6.4 Russian decline curve augmented by projected increase in production based on the operational start of the Amur Facility modified from Edison Investment Research, 2020.

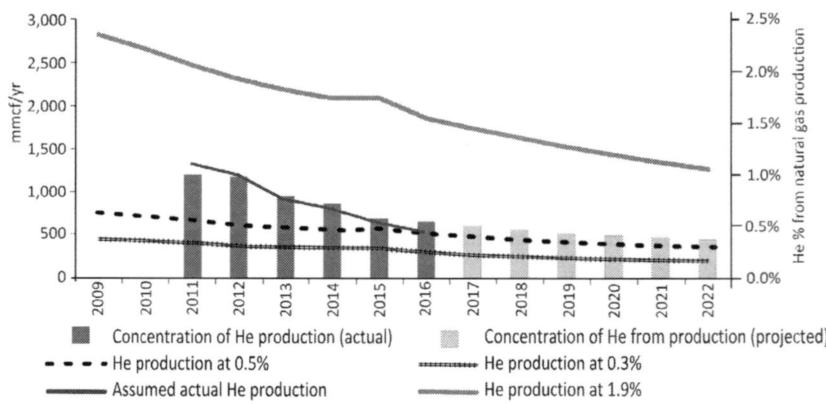

FIGURE 6.5 Decline curve and projected decline curve for the Hugoton portion of the PHF modified from Edison Investment Research, 2020.

and projected decline curve beyond 2019. The field is in terminal decline and there is no solution that can revitalize it. One can expect over the next decade several of the He processing facilities related to PHF will be shut-down, scrapped, or moved to other areas. Edison Investment Research (2020) projected production from several different sources in Fig. 6.6 for worldwide production.

Edison Investment Research (2020), used BLM and USGS data to graphically express worldwide reserves of He (Fig. 6.7A and B). Helium estimates are changing through time and with variations in sources. Much of the information used in constructing these types of graphs rely on public statements of sovereign government energy companies that can be self-serving and misleading. The China estimate is unverifiable as there is very little public data to support the projected reserves. South Africa data is based on the Virginia Field which at present is a one-off resource in that country. Data for Iran is missing. Argentina, while the government is trying to find He resources, so far does not seem to be geologically endowed with resources of the element.

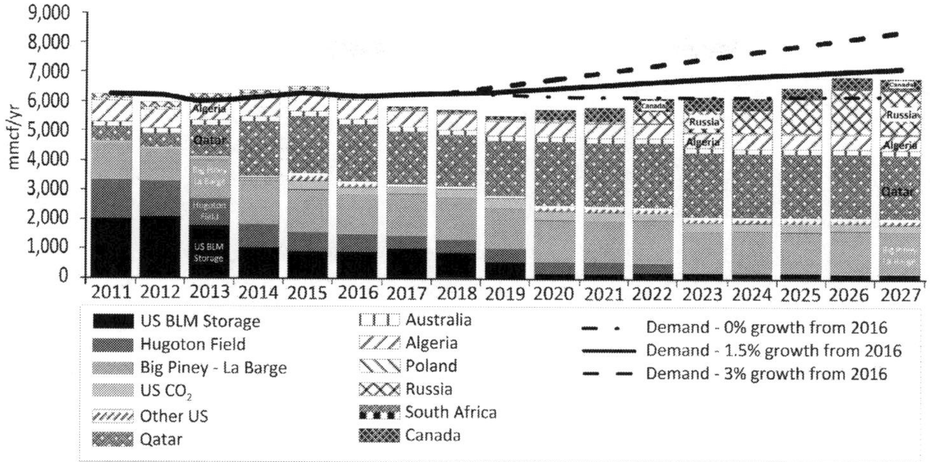

FIGURE 6.6 Production curve and projected production curve for various producing sources of He through 2027 modified from Edison Investment Research, 2020.

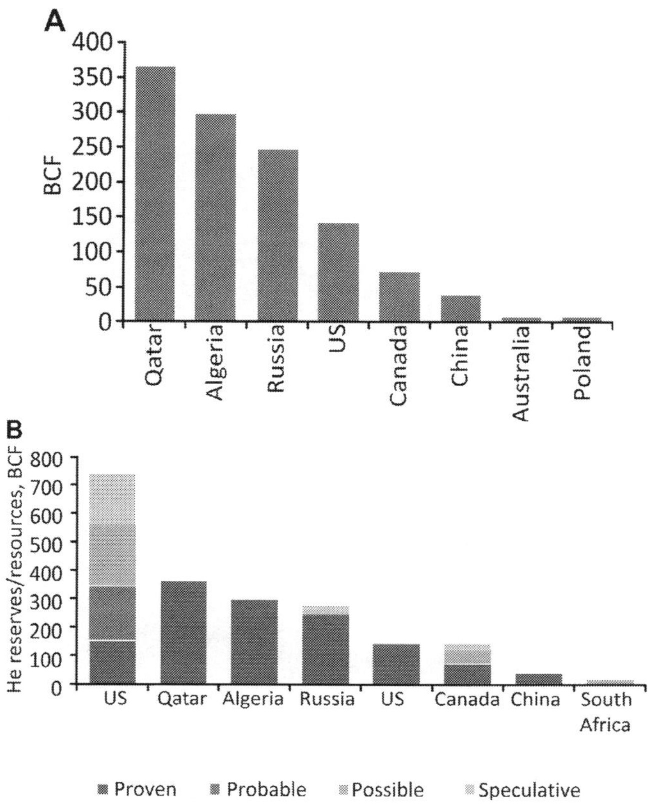

FIGURE 6.7 (A) Reserves of He globally, including from reserves natural gas fields, estimates made by the BLM in 2013, modified from Edison Investment Research, 2020. (B) USGS summary of global He (intermedia findings). USGS Mineral Commodity Summaries, January, 2016, modified from Edison Investment Research, 2020.

The USGS has been collecting information from various operators over nearly 100 years from US companies and governmental agencies such as the USGS and, the BLM, etc. Table 6.1 is an estimate of those reserves by the USGS (2020 and 2021). The BPLB in Wyoming presently has the largest amount of He reserves. Table 6.2 is the USGS (2021) estimate of world He reserves with the addition of Iran. Note there has been a decline due to the pandemic.

The USGS every year compiles the production and consumption of He in the USA (Table 6.3). Imports are declining as the USA maintains its position as the world source of He. Withdrawals from storage are not declining due to price or lack of sales but due to depletion and the PHF being near the end of its life.

The global He market by end user is divided into energy and power, defense, healthcare, electronics and semiconductors, aerospace and aircraft, and metal fabrication (Stratistics MRC, 2020). The global He market can also be divided into six markets (Fig. 6.8). On Fig. 6.8 is the projected Compounded Annual Growth Rate (CAGR) from 2019 to 2027 by Stratistics MRC (2020). Table 6.4 indicated the sales of He in 2019, expected sales in 2027, and (CAGR) Market share increases with Compounded Annual Growth Rate (CAGR) from 2019 to 2027 according to Stratistics MRC (2020).

Helium is extensively utilized in the manufacture of semiconductors, owing to its ability to offer high thermal conductivity, which helps control the temperature of silicon (Si) and germanium (Ge) during growth, and low temperature and inert gas features that are not available from over elements and compounds. This use is expected to increase tremendously in the coming decades. The Asian region manufactures the majority of the worlds electronics

TABLE 6.2 Is the USGS (2021) estimate of world He reserves with the addition of Iran. Note there has been a decline due to the pandemic.

	Mine production		
	2019	**2020[e]**	**Reserves[a]**
United States (extracted from natural gas)	68	61[e]	3900
United States (from Cliffside Field)	21	13	[b]
Algeria	14	14	1800
Australia	4	4	NA
Canada	<1	<1	NA
China	NA	NA	NA
Poland	1	1	23
Qatar	45	45	Large
Iran	NA	NA	Large
Russia	5	5	1700
World total (rounded)	160	140	NA

[a]*See Appendix C for resource and reserve definitions and information concerning data sources in USGE (2021).*
[b]*Included in United States (extracted from natural gas) reserves.*
e, estimate; *NA*, not available.

TABLE 6.3 Salient statistics for He extraction and consumption in the United States for 2020 (USGS, 2021).

	2016	2017	2018	2019	2020e
Helium extracted from natural gas[a]	66	63	64	68	61
Withdrawn from storage[b]	23	28	26	21	13
Grade-A helium sales	89	91	90	89	75
Imports for consumption	23	19	8	7	5
Exports	62	74	84	106	100
Consumption, apparent[c]	50	36	40	40	40
Net import reliance as a percentage of apparent consumption	E	E	E	E	E

[a]*Both Grade-A and crude He.*
[b]*Extracted from natural gas in prior years.*
[c]*Grade-A He. Defined as Grade-A helium sales + imports − exports. However, substantial increases in exports reported in 2018, 2019, and 2020 suggest that domestic consumption declined, although no significant decline in the US. He consumption is thought to have taken place. For that reason, apparent consumption for 2018–20 was estimated to have remained at about 40 MCM.*
E, estimate.

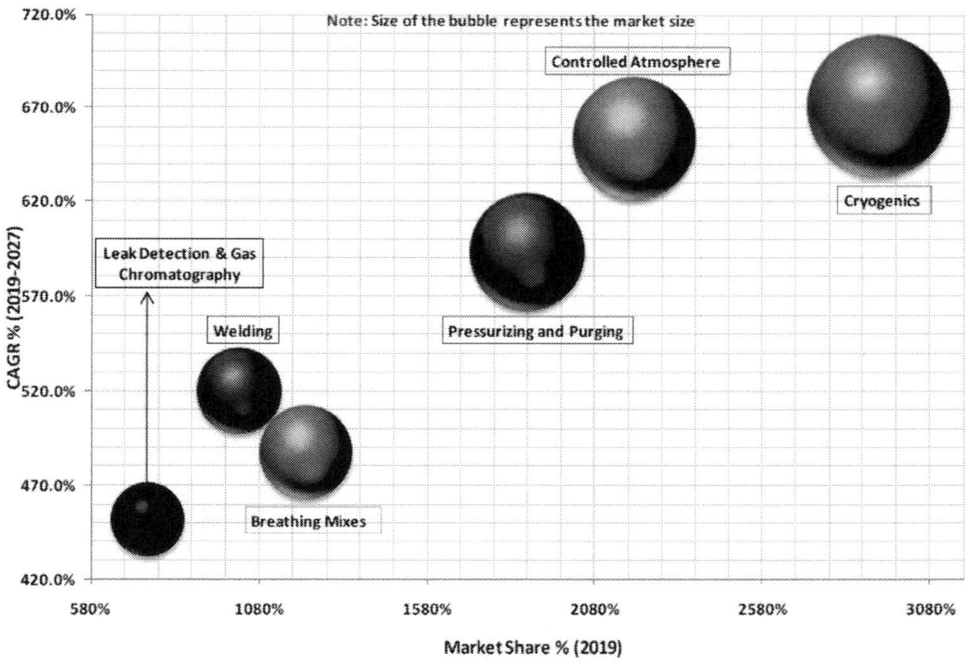

FIGURE 6.8 The He market share type of application for 2019 versus expected growth to 2027 according to Stratistics MRC (2020).

TABLE 6.4 Global helium market analysis, by end user, 2019 versus 2027, US$ in millions, Stratistics MRC (2020).

Industry	Revenues in 2019 (US) in millions	Projected Revenues in 2027 (US) in millions	Compounded Annual Growth Rate (CAGR)
Energy and power	\$302.08	\$489.60	6.2%
Defense	\$440.32	\$701.76	6.0%
Healthcare	\$529.92	\$840.48	5.9%
Electronics and semiconductors	\$250.88	\$408.00	6.3%
Aerospace and aircraft	\$670.72	\$1054.64	5.8%
Metal fabrication	\$366.08	\$587.52	6.1%

which has a critical need for He in the construction of flat-panel displays, as a coolant in manufacturing operations of Si and Ge wafers and devices for semiconductors, and optical-fibers.

Helium price increases are also associated with strict safety regulations in the manufacturing and transportation of He gas and have been a major restraint on the market. These substances are toxic and hazardous when exposed to extreme temperatures and pressures (Stratistics MRC, 2020). Companies are having to invest in equipment, technologies, and processes to limit the toxic levels and chemical concentrations. Moreover, storage of He requires complex tanks with double-walled, insulated pressure containers, multilayer vacuum insulation, and external radiation shield chilled with liquid N_2 to regulate the storage quantity. These regulations increased the operating costs limiting the scope for investment relating to the launch of new products and expansion into new markets, thereby affecting market growth. For example, The US government regulatory agencies OSHA and EPA specify mandates for carriage and storage of cryogenic liquids in all storage forms. Compliance with these standards has been the increase handling and transportation costs for companies (Stratistics MRC, 2020).

Figs. 6.8 and 6.9 graphically represent the growth expected in each industry or use from 2019 to 2020 as projected by Stratistics MRC (2020). The Stratistics MRC (2020) report came out prior to the pandemic and that abrupt change in consumption is not taken into account. Fig. 6.10 (Stratistics MRC, 2020) is the projected growth from 2019 to 2027 by country exclusive of the USA. In all cases as the world continues to increase its use of hi-tech electronic equipment, He will increase in demand.

Table 6.5 is a breakdown of the He market by application. The market is projected to have steady growth over the next seven years according to Stratistics MRC (2020).

Helium is the predominate gas used in cryogenic applications and has exceptional cooling abilities proven useful for the development of various cryogenic devices. The development of cryogenic turbines and other devices utilizing titanium (Ti) and stainless-steel parts in a He atmosphere or liquid is critical for the aerospace and aircraft industries in vacuum brazing

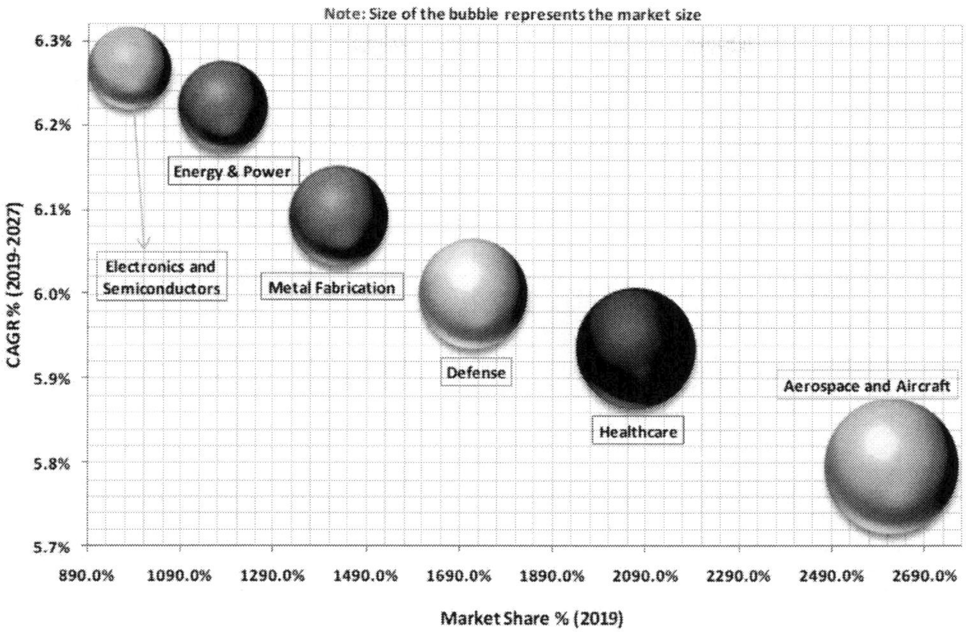

FIGURE 6.9 The He market share for industry type for 2019 versus expected growth to 2027 according to Stratistics MRC (2020).

and the process for nickel (Ni) coating. Helium is used in electron beam welding in, cryogenic compressors and pumps, and aerodynamic gas bearings. The He market can be divided into gas and liquid with projected consumption to the year 2027 (Stratistics, 2020) (Table 6.6).

The gas segment of He accounted for $1830.40 M in 2019 and is expected to reach $3055.92 million by 2027 growing at a CAGR of 6.6% from 2019 to 2027 (Stratistics, 2020) (Table 6.7). Helium gas dominated the market and is projected to reach the highest CAGR from 2020 to 2027. Helium gas consumption is the most widespread market as it is used in welding by either gas tungsten arc welding or gas metal arc welding, pressuring and purging, fiber optics, semiconductors, leak detection, and as a heat transfer medium. Helium gas is consumed in aerospace, automotive, electronics, healthcare, welding, party balloons, and other industries.

The He liquid market generated $729.60 million in revenues in 2019 and is expected to reach $1024.08 million by 2027 growing at a CAGR of 4.3% from 2019 to 2027 (Table 6.8) (Stratistics MRC, 2020). Liquid He shows superfluidity that is used in a variety of applications such as to keep the magnets in particle accelerators cooled, MRI applications, cryogenic refrigerant, and nuclear magnetic resonance (NMR) Spectroscopy also utilizes the superconducting properties of liquid He. Liquid He is used in the healthcare, scientific research aerospace, and electronics industry. There is no substitute for He in applications where temperatures below −261.1°C (−429°F) are required. The cooling power of liquid He has meant that it is a vital and irreplaceable element in the manufacture of many critical modern electronic components.

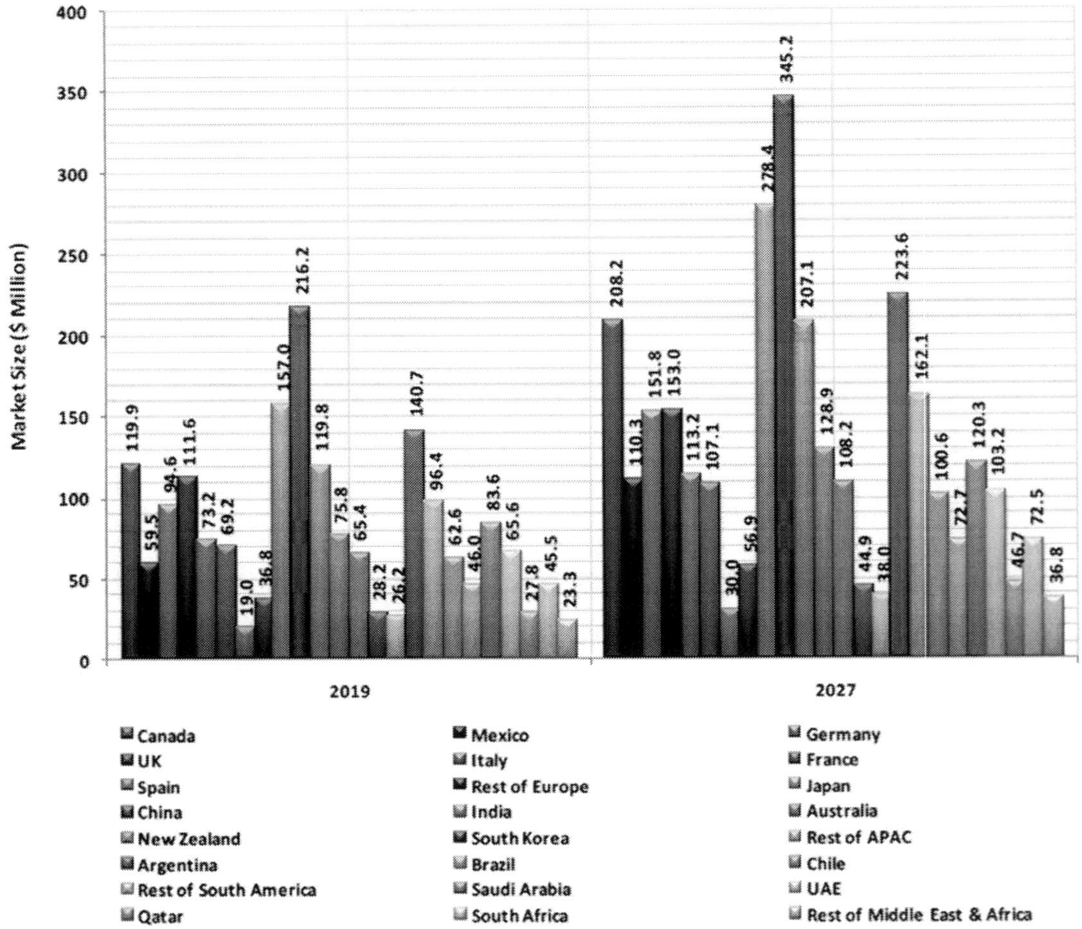

FIGURE 6.10 The He market share for various countries excluding the USA for 2019 versus expected growth to 2027 according to Stratistics MRC (2020).

TABLE 6.5 Global helium market by application, market forecast, 2018–27. US$ in millions, Stratistics MRC (2020).

Application	2018	2019	2020E	2024E	2027E	CAGR% (2019–27)
Cryogenics	680.3	750.1	824.6	1124.5	1260.7	6.7
Controlled	52.6	563.2	615.0	832.5	934.3	6.5
Atmosphere						
Pressurizing and purging	444.1	481.3	515.7	682.8	763.0	5.9
Breathing mixes	297.6	312.3	328.2	412.6	457.0	4.9
Leak detection and gas	181.9	192.0	198.6	248.3	273.4	4.5
Chromatography						
Welding	245.7	261.1	275.8	350.5	391.7	5.2
Total	2362.3	2560.0	2757.7	3651.1	4080.0	6.0

E, estimate.

TABLE 6.6 Global He by market phase, market forecast, 2018–27, in US$ in millions, (Stratistics MRC, 2020).

Phase	2018	2019	2020E	2024E	2027E	CAGR% (2019–27)
Gas	1670.1	1830.4	1991.1	2687.2	3055.9	6.6
Liquid	692.2	729.6	766.6	963.9	1024.1	4.3
Total	2362.3	2560.0	2757.7	3651.1	4080.0	6.0

E, estimate.

TABLE 6.7 Global helium market by gas, market forecast, 2018–27, US$ in millions, Stratistics Analysis (2020).

Region	2018	2019	2020E	2024E	2027E	CAGR% (2019–27)
North America	571.2	626.0	679.0	913.7	1032.9	6.5
Europe	267.2	289.2	310.6	411.1	458.4	5.9
Asia Pacific	444.3	492.4	543.6	744.4	861.8	7.2
South America	225.5	247.1	268.8	365.5	418.7	6.8
Middle East and Africa	162.0	175.7	189.2	252.6	284.2	6.2
Total	1670.1	1830.4	1991.1	2687.2	3055.9	6.6

E, estimate.

TABLE 6.8 Global He market for liquid, market forecast, 2018–27, US in millions, Stratistics Analysis (2020).

Region	2018	2019	2020E	2024E	2027E	CAGR% (2019–27)
North America	236.7	249.5	261.4	327.7	346.1	4.2
Europe	110.7	115.3	119.6	147.5	153.6	3.7
Asia Pacific	184.1	196.3	209.3	267.0	288.8	4.9
South America	93.4	98.5	103.5	131.1	140.3	4.5
Middle East & Africa	67.1	70.0	72.8	90.6	95.2	3.9
Total	692.2	729.6	766.6	963.9	1024.1	4.3

E, estimate.

Forecasting of He production and consumption remain difficult, but because of continued and ever-increasing growth in various He industries, the market will most certainly tighten. Large-scale supply additions from mega projects in Qatar and Russia, which are planned from 2020 to 2022 and beyond are probably going to be partially delayed by the pandemic and also due in construction difficulties. A projected annual demand growth of 1.5% will put pressure on the supply/demand balance in the longer term. This should support prices, motivating further development and He exploration. This result is being seen presently in

Canada, Tanzania, and the USA. Iran has a significant supply but is constrained by sanctions and lack of the necessary technical capabilities. Supply constrained, short-term deficits are likely as He is a vital and irreplaceable element in many critical modern components industries. With the shutdown of the BLM–PHF facility, the almost 100 years of US strategic reserves controlling the market is nearly at an end, and replacement of this supply is not simple. The majority of companies in the USA and Canada are generally small and targeting small-scale He assets that do not greatly impact the overall market. Many of these companies have announced ambitious plans, and the biggest stumbling block is putting in an extraction facility which can take up to two years to be financially burdensome. As discussed above, He data in terms of supply, demand, and prices are not widely disclosed and this creates tremendous market uncertainty for predicting future supply and demand. Analysis of project developments by Edison Investment Research (2020) and Stratistics MRC (2020) suggests a lack of new supply to satisfy increasing (price inelastic) demand. The current He price ranges generally within \$200 to \$350 an MCF. Helium lacks any effective substitute in the main industries, so prices should increase over time. In the years 2011 to 2013, there was an He shortage that caused price spikes resulting in exploration for, and development of, new supplies, mostly associated with liquid natural gas processing facilities expansion. If various facilities expand, various new fields come on line and existing sources such as Iranian's part of NSP come on line, then this can cause prices to return to normal or lower levels. Present supply constraints should continue to support existing prices and encourage the exploration and development of new reserves.

Projects can be delayed indefinitely due to lack of financing. Part of the issue can be the market perception in terms of He supply and demand dynamics (Edison Investment Research, 2020). In addition, many projects are not proven economic until actual production. There is anecdotal data for a number of projects that began with projected economic production and reserves actually saw actual He percentage declines gradually or significantly sooner than expected (Edison Investment Research, 2020).

A factor that is generally not discussed by many public companies is the cost and source of electricity to power the He extraction facilities. The He extraction process requires that N_2, CO_2, and any hydrocarbons be removed. Helium accumulations that contain only N_2 have an advantage of costing less in processing as they do not require addition, an expensive steps, to remove CO_2 and generally CH_4. This is an advantage not only from lower costs for extraction but the environmentally friendly effects of venting only N_2 gas. The downside to these types of He accumulations is the increased cost in finding power to run the processing plants. Power required can be in the form of propane, diesel or electrical sources. Helium accumulations that have CH_4 have an advantage in that the natural gas can be used to generate electricity at very low cost.

Helium supply is structurally fragile as an unexpected down time for any of the big five fields or facilities can have a major effect on the worldwide price of He. With increasing prices many fields or processing facilities will continue at present output or increase production of He despite declining production percentages and recovery factor. In many countries, other stumbling blocks are government regulations and processes that can delay projects for years. Leasing of land from governments is also a drawn-out process which can be derailed by environmental, nonindustry, or governmental groups. On the converse, states like Wyoming, Utah, Montana, Colorado, Kansas, most of the eastern half of the USA, southwestern western Ontario, Montana, parts of Alberta and Saskatchewan have private lands where the mineral rights including He are privately owned. This allows leasing of said lands in a short period of time with little public knowledge and thus less competitor interference. The provinces of Alberta, Saskatchewan, the states of Utah and Arizona and the USA government typically take a royalty on the resources produced including He. The majority of states in the USA has no royalty specific to He and thus do not impede revenue to the Operator of a field. Most other countries get their royalty portion of any He sales.

Shale plays, while prolific in the USA and Canada, do not contain significant He. This result is not surprising as the hydrocarbon source and trapping mechanism is not conducive to He concentration and entrapment (Edison Investment Research, 2020; Stratistics MRC, 2020). Helium associated with CO_2 production is typically less reliable as that compound relies on oil prices for economic viability. When oil prices are low, secondary recovery using CO_2 declines. A number of CO_2 accumulations are also not located near any oil productive areas, such as Bravo Dome in New Mexico, USA. These projects suffer from the lack of available pipelines to transport the product, and thus limit their economic viability.

In most fields, large quantities of natural gas and/or CO_2 have to be extracted for commercial volumes of He to be separated, and as a result the commerciality of a He project generally does not rely on He volume itself (Edison Investment Research, 2020). In recent years, natural gas prices have declined or remained low worldwide, particularly in the USA, causing natural gas projects to become less attractive. The current surplus of LNG supplies (and low prices) has led to postponement of large LNG projects prior to 2020 from which He could have been extracted. Low oil prices prior to 2021 caused CO_2 flooding (from which helium can be extracted) to less or uneconomic, reducing the demand for CO_2. Shales reservoirs to date do not seem to trap He, which means that the supply of the element has not benefited from the massive shale boom in the USA. A resource shale play economics tend to inhibit producing natural gas from conventional reservoirs but if He is present, these reservoirs maybe economic.

Future He reserves and production are likely be subject to steady, price inelastic demand growth. Lack of investment resulted in a supply shortage of He in 2011–2013. The resulting price rise caused many industries to replace (or recycle) He, where possible. Because of He's unique properties, it is irreplaceable in many applications. The demand destruction seen in the industry in 2011–2013 may have left demand more inelastic to price movements (Edison Investment Research, 2020). Despite the current looser supply/demand dynamics, pricing has not retreated and BLM auction prices rose around 137% in August 2018 (vs. previous year). More clearly it is a fact that He is being required more and more in new Hi-Tech, aerospace, military, and medical technologies. Helium provides an efficiency in use unseen with other noble gases or other elements or compounds. The He industry is likely to continue to grow, as many fields likely to be found in the USA and Canada are small and only they will have an impact on supply only as an aggregate. Fields discovered in the past two years and the next five years will only make significant impact after 2027.

6.6 Exploration outlook for various areas for helium

6.6.1 Super-size fields

Super-size fields such as the PHF (Fig. 3.1), BPLB (Fig. 3.9), NSP (Fig. 3.97), and Hassi R'Mel (HRM) (Fig. 3.102) supply the majority of the world's He. Amur in Russia (Fig. 3.103) is a processing facility and cannot be considered in field analysis even if it draws

He from several large size He fields. HRM and PHF are in significant decline. Expansion at NSP and Amur can offset some of the loss, but it is uncertain whether their production significantly offset the decline in other fields or, the expected continued growth in consumption of the element. The main issues for Amur and NSP are operational issues that seem to be a constant problem. Also natural gas production, including He sales, from Russia is typically a political tool in relation to the European Union.

Super-sized fields that remain undiscovered are probably limited to some very remote areas where petroleum and gas is not presently not being explored for and/or undeveloped. Examples include the Nevada Basin and Range, failed rift systems such as the Mid-Continent of the Central USA, Greenland, parts of onshore Africa, the Mozambique Channel, Antarctic, and parts of the Artic.

6.6.2 Central Kansas Uplift

The Central Kansas Uplift area (Fig. 3.14) in the Mid-Continent of the USA is one of the oldest areas of He production. The area is in close proximity to Dexter, Kansas, where the discovery of subsurface He was first made. The area has had numerous discoveries since the 1920s, several smaller discoveries in later years, and is still currently He productive. The area continues to be explored and exploited for new reserves, but based on the historical and present oil and gas drilling density of wells it is unlikely that any medium to large size fields will be found. However, where there is a Precambrian high that has not been evaluated, such as along the Nemaha Ridge and Central Kansas Uplift, there may be opportunities to find smaller gas fields with potentially economic He.

6.6.3 Arizona—New Mexico—Southern Colorado Plateau

Parts of the southern Colorado Plateau in Arizona and New Mexico (Figs. 3.27 and 3.28) have been referred by some clever marketers as being the "Saudi Arabia of Helium." The area has had a hand full of fields that have produced significant amount of He since the 1940s in New Mexico and the 1950s in Arizona (Fig. 3.27) (Rauzi, 2003; Broadhead and Gillard, 2004). The area has multiple wells that have yielded significant He gas percentages. However, presently, only one field in Arizona, Dineh-bi-Keyah, at this time is producing He (Arizona Oil and Gas Conservation Commission, 2021). Many of the oil and gas accumulations in the Southern Colorado Plateau have high percentage He but have not had significant gas or He production. The area has had significant drilling over the years and still has no large deposits. Most of the productive deposits were found early on in exploration in the area for oil. The Rattlesnake Field in New Mexico had a lot of promise but water encroachment early on in its life terminated production. Some of the recently discovered reservoirs in Arizona will produce freshwater along with the He and is a problem as that resource is far more critical in the area. The majority of land is held by various Indian tribes, the State of Arizona, and New Mexico and the BLM which can inhibit development due to regulation and other nontechnical factors. State lands can also be problematic depending upon the politics of a specific state. Some state governments in the USA are trending toward shutting down oil and gas drilling and production that makes the exploration of He difficult as it is commonly associated with natural gas. The argument of whether natural gas is a green resource or related to nasty hydrocarbons is out of the purview of this book as it is a political or philosophical question. Based on science, natural gas is a hydrocarbon. The American Indian tribes can be excellent partners in finding hydrocarbons and He, such as the southern Utes, or the Navajo who each maintain separate oil companies that are run by professionals independent of the tribe and are not subject to internal tribal politics. Whereas other tribes have internal politics and agreements that can be terminated or revised when the previous council members are voted out and a new administration is put in place. This leads to significant uncertainty for petroleum or He companies.

The Holbrook Basin has had two recent tests that have been touted as having significant flow rates by a company called Desert Mountain Energy (2021):

The State 10-1 well located in Section 10 Township 15 North Range 23 East, Navajo County, Arizona, with an estimated flow rate of 24,214 MCFGPD (Desert Mountain Energy, 2021). There is no indication of whether it was a ten-minute test or 48 h. The longer the test, the more valid the flow rate. The permitted depth was 2790 ft (845 m) but no actual depth from where the gas flow was from was reported by the Operator.

The average gas analysis according to the operator:

He: 7.1321%
N_2: 77.0837%
CO_2: 4.0183%
CH_4+: 2.6512%

The State 16-1 well located in Section 15, Township 15 North, Range 23 east, Navajo County, Arizona with an estimated flow rate of 1251.2 MCFGPD (Desert Mountain Energy, 2021). There is no indication of the duration of the test.

The average gas analysis according to the operator:

He: 4.0904%
N_2: 90.2742%
CO_2: 0.0063%
CH_4+: 3.5535%

Note the different amounts of He, N_2, CO_2, and CH_4 between the two wells that are approximately one mi (1.6 km) apart with the same pay zone. This difference suggests that there is compartmentalization in the reservoir or one well is structurally lower than the other. There has also been drilling permits approved for reopening the Pinta Dome, Navajo Springs, and Navajo Springs Field, which all have high He percentage tests and previous production.

6.6.4 Utah-Southwestern Colorado—Northern Colorado Plateau

The He potential in the Northern Colorado Plateau is considered more promising than the southern portion (Fig. 3.27). The Utah part of the Plateau has productive He assets at Harley Dome (Fig. 3.43) and has had other fields that have yielded high He tests but were not recovered (Salt Wash Field, Grand County, Fig. 3.47). The majority of He discoveries have been related to significant CO_2-producing fields or accumulations such as Gordon Creek, Farnham Dome, Escalante, Doe Canyon (Colorado), and McElmo Dome (Colorado). The predominate area for natural gas with He has been in Grand County, Utah. The Uinta and Piceance Basin probably have potential in Paleozoic rocks but these are at depths of 20,000 to 32,000 ft (6060 to 9,700 m) (Geldon, 2003). The area is undergoing a number of He tests at present and hopefully will find new reserves in the area.

The major stumbling blocks for exploration in areas where Paleozoic rocks are reachable are the BLM, the State of Colorado, and various tribal governments. The State of Utah encourages exploration and development of oil, gas, and He, but their lands are limited in many areas. There are also several national parks whose presence deters exploration and extraction except by an Act of Congress. Environmental groups recently have been attempting to inhibit all development (Maffly, 2021).

6.6.5 Las Animas Arch Helium Region

The Las Animas Arch Helium Region (LAAHR) is almost exclusively in southeastern Colorado with a small area in the panhandle of Oklahoma and extreme western Kansas, USA (Fig. 3.51). The LAAHR area has a long history of He production that has been overshadowed by the PHF to the east and southeast. Some of the same pipelines and processing facilities that service PHF also service the LAAHR. The LAAHR is a mature area for He. The land ownership is typically private, State of Colorado or part of the Union Pacific Land Grant (now owned by Occidental Petroleum). Historically these lands have been available for leasing and drilling with generally minimal interference from governmental agencies. However, the State of Colorado has changed governmental focus and significantly has become anti-oil and gas and thus intentionally or inadvertently anti-He. The State of Colorado, based on the present administration, has announced plans to phase out oil and gas over time and promote and convert the state to renewable energy sources. The cause of anti-oil and gas attitudes in the state developed because of shale drilling in the suburbs north of Denver, Colorado. In this area conflicts between the industry, homeowners, and home developers developed a negative image of petroleum development. Permitting of oil and gas wells now takes six months or more in the State of Colorado versus five days in the adjacent State of Kansas. In addition, the fees for drilling a well in the State of Colorado are in the tens of thousands of dollars compared to minimal monies in the State of Kansas. There are attempts by the industry to segregate various petroleum-producing areas in the State of Colorado in order to have different permitting process, generally a shorter period for areas such as the LAAHR, classified depending upon population density. The LAAHR is in an area that is rural, lightly populated, resource based, agriculture and petroleum, and pro-hydrocarbon industry.

The prospect potential of He in the LAAHR area is still considered good, but the region is in the mature stage of exploration and exploitation. The area has had a number of new fields found over the past decade and a number of large 3D surveys acquired that may be able to aid in exploration. The primary reservoir that contains He is the Pennsylvanian Morrow sandstone deposited in a fluvial system extending from LAAHR area to the Anadarko Basin to the east. The sandstones reservoirs can be erratic in areal extent, pay thickness, porosity, permeability, hydrocarbon percentage, and He content as discussed in Chapter 3. Some of the defined channel trends in the LAAHR are barren of hydrocarbons and He whereas others are gas prone but lack any significant He.

The underlying Mississippian reservoirs in the LAAHR have to date proven to be lacking in He accumulations. The seal above these carbonate reservoirs, which are typically oil prolific, is not effective for trapping He. The shale that encases the Morrow reservoir overlain by the thick rock section of low permeable shales and most of the carbonates of the Atokan and Cherokee formations provide significantly better sealing mechanisms for He. Shallower He targets have to date not been viable in the LAAHR area. Deeper reservoirs such as in the Cambrian Reagan sandstone are potential hydrocarbon and He traps, but so far, drilling has not found any indications of gas or actually any viable seal or trapping mechanisms. The Reagan sandstone in the Central Kansas Uplift at the Reichel and Otis-Albert fields (see Chapter 3) is a major He-producing reservoir and the reservoirs are equivalent to the Reagan sandstone in LAAHR.

6.6.6 Sweetwater Arch-Bow Island Arch—Bowdoin Dome-Alberta, Saskatchewan, and Northern Montana

The Sweetwater Arch-Bow Island Arch—Bowdoin Dome-Alberta, Saskatchewan, and North Central Montana area was discussed in Chapter 3, Figs. 3.79 and 3.88. The area is undergoing a resurgence in 2021 because of the He data published by the Alberta Geological Survey and presented in Fig. 3.96 as well as data from the Saskatchewan Geological Survey in Section 3.15, Chapter 3. Helium has historically been produced from the Bow Island Field in Alberta, Swift Current area in Saskatchewan, and recently the Battle Creek Field in Saskatchewan. Recently, high He prices have caused a land rush on the Canadian side of the border (French, 2020; Jaremko, 2020; and Gallant, 2021) which is spilling over into North-Central Montana. The Rudyard Field in Hill County, Montana has significant He, but at present has not been successfully developed. There are supposed to be several wells in southeast Alberta and southwestern Saskatchewan that have tested excellent He results, several 3D seismic surveys, and additional test wells planned. Presently implied field size seems to be 1 to 5 wells or 80 to 1500 acres. The majority of the exploration efforts in the play area are targeting Cambrian sandstones, but Ordovician and Devonian rocks are also indicating significant He percent in the natural gas. The play could be similar, but with lower He percentage, to the Central Kansas Uplift and Arizona—New Mexico area of the Colorado Plateau.

6.6.7 Rift systems

Rift systems found around the world encompass a variety of ages from the Precambrian to the present day. As discussed in Chapter 3, the East Africa Rift System (EARS) is presently of particular focus for He exploration and exploitation. There are numerous other present-day rift systems that are active tectonically and like EARS have ongoing migration and venting of mantle and crustal gases. In some of these rift systems, there is the potential that these migrating gases may be trapped in isolated reservoirs such as fluvial and terrestrial sand, and carbonates. There are other rift systems that do not have lake deposit and are purely or almost all volcanic in nature and are unlikely to accumulate any gases. As discussed in Chapter 3, present-day EARS play area has a number of risks but a number of positive indications for He accumulations.

Older rift systems that have greater potential for He accumulations based on the age of pore water would be the Precambrian Mid-Continent Rift system (MCR) (Fig. 6.11) which has Precambrian oil being expelled in the White Pine copper mine at its northern end in

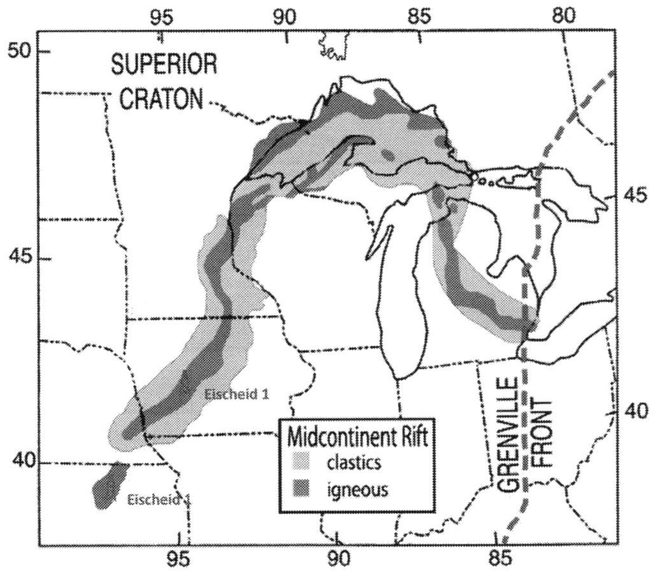

FIGURE 6.11 The location of the Mid-Continent Rift System in the Central USA (Stein et al., 2011).

upper peninsula of the State of Michigan (Palacas et al., 1990). This potential petroleum play has been tested with less than a dozen wells in Kansas, Oklahoma, Minnesota, Michigan, and Iowa. At one time, MCR was considered a potentially major oil and gas province based on production from other rifts and Precambrian rocks worldwide. There are coarse to fine clastic, low porosity (1% to 8%) reservoirs in the Precambrian sediments that were drilled in the Amoco Eischeid 1 well in Carroll County, Iowa (Anderson, 1990) and the Texaco Poersch 1 well in Washington County, Kansas (Ludvigson et al., 1990; Palacas et al., 1990; Berendsen et al., 1988) (Fig. 6.11). The rift has been determined to be barren as a potential hydrocarbon province because the Nonesuch shale in the two deep wells, the Eisched 1 and Palacas 1, are overmature and probably expelled hydrocarbon prior to the formation of any traps (Palacas et al., 1990). However, the generation of He in the crustal rocks is always ongoing and may have accumulated in traps that were formed sometime after hydrocarbon generation.

Other rift systems include the New Madrid Fault system (Mississippi Valley Graben)–Rough Creek Graben, Rome Trough and associated Wabash Valley Fault system located in the states of Tennessee, Arkansas, Illinois, Kentucky, West Virginia, and Indiana and the East Continent Rift Basin located in western Ohio and Eastern Indiana that are located to the west of the Grenville Front whose expression is the Cincinnati Arch all have potential He. Numerous structures exist in these rift and fault systems. The question is, were traps created sufficiently early to allow accumulation of He and other gases?

6.6.8 Model Dome, Las Animas County, Colorado

The Model Dome, Las Animas County, Colorado was discussed in Chapter 3, Section 3.7, and Fig. 3.76 is a historical He-producing field that has not until recently been adequately evaluated. A company has leased over 100,000 acres in the area and has permitted test wells. The Model Dome is located on the flank of the Raton Basin along the Apishapa Uplift and produces from Glorieta sandstone (Permian age). The Glorieta sandstone generally lies close to basement and the thin nature of sedimentary section across the Apishapa Uplift and flank of the Raton Basin can preclude significant reserves. The thin sedimentary section makes acquiring seismic surveys difficult as the cost will be greater than if the targets were deeper. This is an area where a well-focused exploration drilling program may be more productive than seismic-driven exploration as shallow depths make costs of drilling shallow depths and collecting of actual geologic data to put together a depositional and tectonic framework. Initial target focus could be defined by detailed aeromagnetic surveys and looking for basement faults and where intersection of basement faults occurs. These intersections may cause local structures that may be charged with He. This would be followed by stratigraphic or water wells to acquire data. From this test, wells that would produce He could be drilled more effectively.

6.6.9 Other USA and Canadian areas of potential He production

What other areas have potential for He? Probably quite a few actually. The data base that the USGS, United States Bureau of Mines (USBM), and BLM have assembled and published over the years has been focused on some very specific areas. There are.representative samples of many basins but not with the intensity that is in the PHF and other parts of Texas, southeastern Colorado, central Kansas Uplift, Arizona, northeastern New Mexico, and Utah. Based on data from the USGS, USBM, and BLM, there are many areas that are under sampled such as the Cincinnati Arch in southwestern Ontario-Ohio-Kentucky and Tennessee, Kankakee Arch in Indiana, the Rough Creek Fault System—Rome Trough in Kentucky and southern Illinois, the Nemaha Uplift in Kansas and Oklahoma, the Paleozoic section in the Wattenberg area part of the Denver Basin, the Utah part of the Colorado Plateau, central Montana, to name a few. Deeper areas for testing the Paleozoic rocks would be Rocky Mountain structures like the Pinedale Anticline in Wyoming.

6.6.10 International areas

There are a number of areas internationally outside Canada that have potential. Australia is a promising place for finding He because of the indications of a good seal, He in significant quantities in a number of tests. Poland is in decline and does not look as if significant additional He targets still exist. Russia, based on limited data can probably deliver significant new fields and reserves. Africa with the testing of the EARS and the Rufiji basin in Tanzania also has the promise of significant He reserves. The negative is that no He resources so far have been proven to exist in active rift systems. Unknown potential is fields like Iehi in Papua New Guinea that can extract He present in a low percentage from fields that produce world class natural gas reserves. India was not mentioned in Chapter 3 as there is no geologic or He production data to evaluate. India does have presently a processing facility that extracts He as a waste product from natural gas processing.

6.6.11 Carbon dioxide fields

What are the criteria for a successful He project? It always revolves around the funds available to develop and then produce the accumulation, extract He and get it to market. Carbon dioxide has different economic parameters. Carbon dioxide produced from the earth is typically used in secondary oil recovery programs. When oil prices decline the use of CO_2 declines as well causing the shutting in of wells and fields producing the gas. Oil prices are cyclic and to develop new CO_2 requires long term planning and commitment. Carbon dioxide requires highly specialized tubulars, facilities, and processes for production, extraction, and transportation. Carbon dioxide is generally transported via pipeline and putting in new or adding pipelines to existing ones can be very expensive. There are a number of shut-in CO_2 fields or natural gas fields with CO_2 that may be activated based on the present rise in crude prices.

6.7 Climate change

Climate change is a major concern in society today and has a major impact on the extraction of He and CO_2. The earth geologically is in an interglacial period, warming of the planet would be expected prior to cooling and going back to another glacial age in a few thousand years. There have been seven previous glacial cycles in the Pleistocene and Holocene with the most recent one ending 11,700 years ago. The retreat and warming of the earth at the end of the last cycle led to the rise of civilization. It can be said, with some certainty, that man has caused an increase in CO_2 and other gases that may be helping to accelerate damage to the earth's biosphere. However, the earth has had much greater amounts of CO_2 in its atmosphere historically and life has thrived. What the media and everyone else forgets is a single volcanic eruption can eject into the atmosphere in a short period of time more CO_2 than humanity can produce in a year. Amazingly the release of additional CO_2 through the industrial age has been tempered by a dramatic increase in plant growth. Some of this growth can be attributed to moving away from wood as a fuel to coal, oil, and gas as energy sources. However, plastic pollution, a product of the petroleum industry, is also major problem in the seas and on land. Mankind has to address emission issues but can do it in a way where the use of hydrocarbons, nuclear, coal, and hydro continue and are augmented by renewables. Eventually energy development hopefully will move toward fusion.

The climate discussion can be one of viscous reactions on both sides of the issue. But that is not the focus here. Despite media statements and reports from political organizations such as the United Nations, the data do not overwhelming imply the end of the world. The reason that climate change is mentioned here is that He and CO_2 are either byproducts or the main products extracted from natural gas that directly help industrial society, human health, and advances in human society. One of many petroleum's uses is in the manufacturing of poly pipe that transports natural gas or CO_2 that helps in He and CO_2 extraction. By ending the petroleum industry, which is extremely unlikely regardless of the rhetoric, considering the limitation of renewables, this would eliminate some large-scale sources of He and CO_2. Recycling of He and CO_2 is being done and can be expanded, but its contribution to the overall supply of both products is limited at present. Helium is critical to modern medicine, fusion reactors, welding, computer technology and of course party balloons for voters (such as soccer moms), and children's birthdays. Carbon dioxide has a variety of industrial applications, dry ice for shipping critical vaccines. Being able to continue extracting He and CO_2 to maintain society is a critical consideration for policy makers.

6.8 Military

One of the industries that is a major consumer of He that we typically hear little about is the military. The US military is the most high-tech armed forces in the world. It is followed closely by many NATO countries, China, Russia, and Japan. Defense use falls into two types, indirect and direct. Indirect usage by the military is often direct usage by other groups, third party contractors such as Lockheed Martin, Boeing, etc. One example is in weather monitoring, which is necessary for a number of military operations even when it is not the specific mission of a military group. This is one of the most important direct uses of helium by the US Military and represents about a fifth of yearly consumption in the USA. Both NASA and US Defense agencies need He to operate heat-guided missiles and rockets (NASA, 2006; Norton, 2019; Stratistics, 2020). It remains the only gas that can purge and pressurize the propulsion systems and tanks of rockets that rely on liquid H and O as fuel. The use of an incompatible gas could cause engines to become nonfunctional.

The use of drones and cyber-technology leads to the use of He by utilizing silicon and germanium chips grown in He atmospheres (Stratistics, 2020). Helium atmospheres are being used to envelop high-tech GPS systems, cryogenics, nuclear reactors, etc., which leads to many more countries intentionally or inadvertently using He on an ever-increasing worldwide scale for their military (Keith, 2013).

The events of 9—11 caused US Defense organizations to increase surveillance efforts. Surveillance systems based on operating balloons and dirigibles require ready access to helium (JLENS, 2009). The US military at one time sent many He-filled spy blimps to Afghanistan. That the industry suffered an He shortage according to Defense Logistics Agency, the Pentagon office responsible for keeping vital supplies flowing to the warzone. A blimp has the ability to stay in the air for days at a time and can carry more spy gear than any drone, aerostats, and airships were at one time, surveillance tools of choice in the Afghanistan War (JLENS, 2009).

Ground troops rely on He to cool some types of thermographic cameras and other equipment (Keith, 2013; Stratistics, 2020). Helium is used to stabilized ammunition and keep track of chemical leaks (Hurst, 2017). The Army in particular considers a supply of He critical to its cryogenic research (Keith, 2013; Stratistics, 2020). Some of its many military detectors for surveillance and equipment monitoring require liquid He. Navy submarine detectors use He to eradicate noisy signals in their detection equipment (Carter, 2021). The use of He in all militaries and not just the US military will continue to grow with the expanding use of cyberwarfare, electronic surveillance, use of rockets, and increasing sophisticated hi-tech GPS and weaponry.

6.9 Summary

In summary to find He is to embark upon a conquest of frustration and intellectual exercise like no other. The nature of He is never easy to completely understand as it is an element that is never a solid and does not play well with others. Helium does not follow the typical source, expulsion to migration to trap process seen in petroleum. Helium likes to hang around in old waters, cannot be biodegraded, never oxidized, does not change political views, never takes sides, does not change viscosity or gravity, leaves the trap first, and is rarely found with oil. Helium exploration, development, and production unlike oil, may take up a decade to come on line. Many of the fields that will be found will probably be small but very lucrative for the Operator of said project. Exceptions could be the EARS area but, that may be at least a decade away from significant development. AMUR, Qatar, and Iran are plagued by political, sanctions, operational issues, and distribution problems. As demand continues to grow, many areas will expand in exploration and development.

References

Ahmed, T., 2006. Reservoir Engineering Handbook. Elsevier, p. 1360.

Anderson, R.R., 1990. The Amoco M.G. Eischeid #1 Deep Petroleum Test. Iowa, IGS: Carroll County.

Arizona Oil, Gas Conversation Commission, 2021. Current Operators | Arizona Oil and Gas Conservation Commission. az.gov.

Artemieva, I.M., Mooney, W.D., 2001. Thermal thickness and evolution of Precambrian lithosphere: a global study. Journal of Geophysical Research 106 (B8), 387–414.

Berendsen, P., Borcherding, R.M., Doveton, J., Hutchison, D.R., Lee, M., Mikereit, B., Agena, W., Spencer, C., 1988. Crustal structure and deep rift basin of the Midcontinent Rift System – results from GLIMPCE deep seismic reflection profile. Geology 16, 81–85.

Broadhead, R.F., Gillard, L., 2004. Helium in New Mexico geological distribution and exploration possibilities. NMGS, OFR 483, 62p.

Carter, T., 2021. Helium-3 Is Scarce, This Navy Invention Stretches it for Radiation Detectors. Techlink. https://techlinkcenter.org/news/helium-3-is-scarce-this-navy-invention-stretches-it-for-radiation-detectors.

Desert Mountain Energy, 2021. http://desertmountainenergy.com/wp-content/uploads/2021/06/DME-Investor-Presentation-06282021-FINAL-.pdf.

Edison Investment Research, 2020. Helium – Macro View, 21. www.edisongroup.com/wp-content/uploads/2019/02/HeliumMacroUpdate2019.pdf.

French, J., 2020. Alberta's New Helium Royalties Could See Rise of Extraction Industry, CBC. May 17th. https://www.cbc.ca/news/canada/edmonton/alberta-s-new-helium-royalties-could-see-rise-of-extraction-industry-1.5572451.

Gallant, C., 2021. Helium Industry Begins to Take off. Medicine Hat News. August 10th. https://medicinehatnews.com/news/local-news/2021/06/22/helium-industry-begins-to-take-off/.

Geldon, A.L., 2003. Geology of Paleozoic rocks in the upper Colorado River Basin. In: Arizona, Colorado, New Mexico, Utah and Wyoming, Excluding the San Juan Basin, USGS, Professional Paper (1114A), p. 112.

Hamak, J.E., 2020. 2016 Minerals Yearbook Helium (Advance Release), 9. USGS.

Hamak, J.E., 2020. Helium, USGS, Minerals Yearbook. https://prd-wret.s3-us-west-2.amazonaws.com/assets/palladium/production/atoms/files/myb1-2016-heliu.pdf.

Hurst, A., 2017. Light Gas Supports Heavy Mission. AMRDEC Public Affairs. October 17th. https://www.army.mil/article/195211/light_gas_supports_heavy_mission.

Jaremko, G., 2020. Albert's Helium Industry Ballooning as Prices Soar Far above Natural Gas, Natural Gas Intel, May 20th. https://www.naturalgasintel.com/albertas-helium-industry-ballooning-as-prices-soar-far-above-natural-gas/.

JLENS, 2009. Big Tactical Aerostat Flies for First Time, AMRDEC Public Affairs, April 2nd. https://www.army.mil/article/19141/big_tactical_aerostat_flies_for_first_time.

Keith, R., 2013. Helium critical supply to Army research, AMRDEC Public Affairs. August, 16th. https://www.army.mil/article/108491/Helium_supply_critical_to_Army_research/.

Lake, L.W., 2007. Petroleum Engineering Handbook, vol. V. Society of Petroleum Engineers, 643. 1.

Ludvigson, G.A., McKay, R.M., Anderson, R.R., 1990. Petrology of Keweenawan sedimentary rocks in the M.G. Eischeid #1 drillhole. In: Wilson, L.J., The Amoco, M.G. (Eds.), Eisheid #1, Deep Petroleum Test Carroll County, Iowa, Special Report Series, No. 2, pp. 77–112.

Maffly, B., 2021. Judge Rejects Last-Ditch Move to Block Helium Drilling in Utah Wilderness, Salt Lake Tribune, January 13th. https://www.sltrib.com/news/environment/2021/01/13/judge-rejects-last-ditch/.

NASA, 2006. Space Shuttle Use of Propellants and Fluids. NASA. https://www.nasa.gov/centers/kennedy/pdf/167433main_Propellants08.pdf.

Norton, K., 2019. NASA Awards Two Contracts for Supply of Gaseous, Liquid Helium. NASA Kennedy. September 27th. https://www.nasa.gov/press-release/nasa-awards-two-contracts-for-supply-of-gaseous-liquid-helium.

Palacas, J.G., J.W. Smoker, T.A. Davis, M.J. Pawlewicz, and R.R. Anderson, Petroleum source rock assessment of Middle Proterzoic (Keweenawan) sedimentary rocks, Eischeid #1 well, Carroll county, Iowa, in eds. L.J. Wilson, The Amoco M.G. Eisheid #1, Deep Petroleum Test Carroll County, Iowa, Special Report Series, No. 2, p. 119–134.

Rauzi, S.L., 2003. Review of Helium Production and Potential in Arizona, Arizona Geological Survey, OFR 03-05, p. 30.

Stein, S., van der Lee, S., Jurdy, D., Stein, C., Wiens, D., Wyesession, M., Revenaugh, J., Frederiksen, A., Darbyshire, F., Bollmann, T., Lodewyk, J., Wolin, E., Merino, M., Tekverk, K., 2011. Learning from failure: the SPREE Mid-Continent Rift experiment. GSA Today (Geol. Soc. Am.) 21 (Issue 9), 4p.

Stratistics Market Research Consulting (MRC), 2020. Helium – Global Market Outlook (2019-2027), Research and Markets, Dublin, Ireland, p. 173. https://www.strategymrc.com.

USGS, 2020. Helium – Fact Sheet, 2. USGS msc2020. https://www.usgs.gov/centers/nmic/helium-statistics-and-information.

USGS, 2021. Helium – Fact Sheet, 2. USGS msc2021. https://www.usgs.gov/centers/nmic/helium-statistics-and-information.

Index

Note: 'Page numbers followed by *f* indicate figures and *t* indicate tables.'

W

X

Z

Made in United States
Troutdale, OR
10/09/2025

40067911R00219